Food Safety and Nutrition

Food Safety and Nutrition

Edited by **Dorothy Green**

SYRAWOOD
PUBLISHING HOUSE
New York

Published by Syrawood Publishing House,
750 Third Avenue, 9th Floor,
New York, NY 10017, USA
www.syrawoodpublishinghouse.com

Food Safety and Nutrition
Edited by Dorothy Green

International Standard Book Number: 978-1-68286-254-4 (Hardback)

Printed in China.

Contents

Preface

Consumption of unsafe and low quality nutritional food products is a major threat to public health and welfare. It is thus crucial to understand the value of food safety practices and impact of dietary habits and food borne diseases on global population. This book addresses major food safety and nutrition problems, provides various strategies and methods to maintain optimal nutrition and notable levels of food safety. It covers various approaches in food safety, nutrition, dietetics and management. It collates knowledge on emerging aspects, practices and applications of food science. Students and researchers will find this text beneficial to develop various food safety measures and technologies.

Significant researches are present in this book. Intensive efforts have been employed by authors to make this book an outstanding discourse. This book contains the enlightening chapters which have been written on the basis of significant researches done by the experts.

Finally, I would also like to thank all the members involved in this book for being a team and meeting all the deadlines for the submission of their respective works. I would also like to thank my friends and family for being supportive in my efforts.

Editor

Detection of some chemical hazards in milk and some dairy products

Karima M. F. Ahmed, Ragaa S. Hafez, Morgan, S. D. and Abeer A. Awad

Milk Hygiene and Control, Department of Food Hygiene and control, Faculty of Veterinary Medicine, Cairo University, Egypt.

Chemical contaminants in milk comprise of hazardous chemicals that may be introduced during milking, dairy processing or packaging. They possess some hazards to humans who consume milk and other dairy products. Total of one hundred and fifty (150) samples were collected; thirty each of UHT milk, yogurt, soft, hard and processed cheese. The samples were analyzed to investigate the presence of some chemical hazards. Chemical analysis indicated that tetracyclines were present in samples of UHT milk with variable percentages, while aflatoxin M1 was detected in all of the examined UHT milk samples and cheese. Sorbic acid, benzoic acid, sulphur dioxide and nitrite were detected in some samples of cheese. The potential to cause toxicological harm to the consumers is common for all of the detected chemical contaminants.

Key words: Chemical hazards milk, dairy products, antibiotic, mycotoxin, preservatives.

INTRODUCTION

The presence of chemical contaminants in milk are very harmful for the consumers and it can be a matter of public health concern because milk and dairy products are widely consumed by humans throughout the world (Jahed, 2007). They are usually classified as naturally occurring chemicals and added chemicals.

Naturally occurring chemicals include toxins that are produced by some microorganisms (e.g. mycotoxins) which can enter through animal feed and deposited some residues in milk and dairy products.Aflatoxins are groups of toxic fungal metabolites produced by certain molds of the genus *Aspergillus* growing in number of raw-food commodities. Aflatoxin M1 (AFM1) may be found in the milk of animals that are fed with aflatoxin B1 (AFB1) containing feed (Wood, 1991; Cathey et al., 1994). These toxins are very stable and may pass through quite severe processes. For this reason they may constitute a problem in processed foods (Lawley et al., 2008).

International Agency for Research on Cancer (IARC) of WHO classified AFB1 as primary and AFM1 as secondary groups of carcinogenic compounds (Cathey et al., 1994; Dragacci et al., 1995). Chronic toxicity is probably more important from a food safety point of view; ingestion of low levels over a long period has been implicated in primary liver cancer, chronic hepatitis, jaundice, cirrhosis and impaired nutrient conversion (Lawley et al., 2008).

The added chemicals include chemicals that are intentionally added such as permitted food additives and non-intentionally added chemicals to a food such as cleaning and sanitizing chemicals. Food and Agricultural Organization (FAO, 1995) describes preservative as any substance which is added to food and enables the physical properties and chemical composition of food to remain unaffected by microbial or other spoilage, so that the milk retains its original wholesomeness and nutritional value. The most likely preservatives to be found in milk are formaldehyde, hydrogen peroxide and neutralizers such as sodium bicarbonate.

Chemical additives such as sodium carbonate and sodium bicarbonate added to milk as neutralizers to preserve it for longer time and to prevent curdling. The continuous use of such milk may cause health hazards to the society (Iswariah and Guruswami, 1973; Satoskar et al., 1999).

Benzoic acid, sorbic acid and their salts are commonly used as chemical preservatives in food products including cheese to prevent alteration and degradation by microorganisms during storage. However, excessive addition of these preservatives may be harmful to consumers, because of the tendency to induce allergic contact dermatitis and convulsion. Therefore, the development of convenient and inexpensive analysis methods of these preservatives is of great importance for food safety (Fang et al., 2008).

The widespread use of antimicrobials in dairy cattle management may result in the presence of their residues; such residues in animal products may be toxic to humans, or may cause serious reactions in sensitive individuals (Lawley et al., 2008) and presents technological difficulties in the milk processing industry (Heeschen and Blüthgen 1991), so many countries have enacted regulations that limit the level of chemical residues in milk and dairy products; milk chemical safety is important for public health (Tenant, 1997). Tetracyclines display a wide spectrum of antimicrobial action: it has a stronger action against gram-positive bacteria and weak action against gram-negative ones. They possess exercise against mycoplasmas, chlamydiae, rickettsias, spirochetes, actinomyces, and some protozoa (Sundin, 2003).

MATERIALS AND METHODS

Collection of samples

UHT milk, yogurt, hard, soft and processed cheese samples were randomly collected from dairy shops and supermarkets in Giza governorate. Collected samples were transferred to the laboratory in an insulated ice box with a minimum of delay to be immediately examined.

Chemical examination

Detection of Inhibitory substances in milk

Using general test (Wynther, 1927) 10 ml of milk sample were transferred aseptically into sterilized cotton pluged test tube. Ten ml of milk free from preservatives "control" were transferred to another test tube under the same condition. Two milliliters of litmus solution (10%) "as indicator" were added to both test tubes. Two test tubes were boiled in a water bath for 20 min to kill all lactic acid producing microorganisms then cooled at room temperature. 1 ml of diluted broth culture of lactic acid producing microorganisms (diluted sour milk) was added to both test tubes, then incubated at suitable temperature and kept under observation till coagulation of milk and color reduction of the litmus solution occurs. Result was recorded as positive if the difference in time was 2 h or more between the control and tested sample.

The positive samples were subjected to the following tests:

Detection of chemical preservatives

Detection of formalin: Hehner's test (Ling, 1963)

Two milliliters of milk sample were mixed with 2 ml of distilled water in a test tube, and then sulphuric acid (90 % containing a trace of ferric chloride) was poured down the side of the tube. Development of violet ring at the junction between the two layers indicates the presence of formalin.

Detection of Hydrogen peroxide (Pien, 1953)

In a clean test tube, 2 ml of milk sample were added and 2 ml of HCL (1%) were added, thoroughly mixed, then 2 ml of potassium iodide (10%) were added. The tube was immersed in hot water (80-90°C) for 1 min after which the tube was quickly cooled in running water. 2 ml of starch solution (1%) were added as an indicator which develops blue color in the presence of hydrogen peroxide in the milk sample.

Detection of Alkalinizer (sodium bicarbonate): (Foley et al., 1974)

One milliliter of milk sample was pipetted into a test tube; the concentrated hydrochloric acid (0.5 ml) was added into the sample, the positive samples showed effervescence due to formation of carbon dioxide gas, when this gas was directed into lime water, it makes it turbid.

Quantitative analysis of tetracycline residues in milk

Analysis of tetracycline residues in milk was carried out using ELISA test kit (competitive enzyme immunoassay for the quantitative analysis of tetracycline in milk) (MaxSignal).

Preparation of samples

One and half milliliter of the cold milk samples (100C) were centrifuged at 10000 rpm for 10 min and the upper fat layer was discarded. 50 µl of milk sample was taken then 9.95 ml of 10 mM PBS buffer (pH 7.4) was added. The tubes were vortexed for 30 s, and then 75 µl of the sample were used for the assay.

Result measurement by photometer was at wavelength of 450 nm. Standard curve was constructed by plotting the mean relative absorbance (%) obtained from each reference standard against its concentration in ng./ml.

$$\text{Relative absorbance (\%)} = \frac{\text{Absorbance standard (or sample)}}{\text{Absorbance zero standard}} \times 100$$

The mean relative absorbance values were used for each sample to determine the corresponding concentration of the tested drug in ng/ml from the standard curve.

Detection and determination of preservatives in cheese

Detection of benzoic acid

The test was done using a qualitative test (ferric chloride test) as described by AOAC (2000).

Determination of sorbic acid

Determination of sorbic acid was done using oxidation method as described by AOAC (2000).

Preparation of standard curve

Sorbic acid standard solution was pipetted (5, 10 and 15 ml) into separate 500 ml volumetric flask. Then each was diluted to 500 ml and mixed. Two millimeters of each solution (2 ml of H_2O for blank) were pipetted into separate 15 ml test tubes, then 1.0 ml of 0.15 H_2SO_4 and 1.0 ml of $K_2Cr_2O_7$ solution was added to each tube and heated in boiling water bath exactly for 5 min after that the tubes were immersed in ice bath and 2 ml of thiobarbituric acid solution was added. Then the tubes were replaced in boiling water bath and heated for additional 10 min, cooled, then the absorbance (A) of each solution was determined at 532 nm against blank, using matched 1 cm cells. The absorbance was plotted against µg sorbic acid.

Preparation of test portion

Test sample was cut into strips and passed through food chopper for grinding. Then 300-600 test sample were blended (< 150°C) in high speed blender till homogenous mixture was obtained. Then 1.5 - 2 g of prepared test portion were weighed into distill tube containing SiC chips then 10 ml of (1 M) H_2SO_4 and 10 g. $MgSO_4$. $7H_2O$ were added. Steam distillation was done. 100-125 ml. were collected in 250 ml volumetric flask within 45 min. The prepared sample was treated as in the case of standard curve preparation and then the absorbance was plotted against µg sorbic acid/ml.

Determination of nitrite using sulfanilic acid method (Edward, 1981)

Preparation of standard curve

Into 50 ml volumetric flasks, 0 (blank), 1.0 ml, 2.0 ml, and so on up to 10 ml. of the dilute standard nitrite solution was transferred then water was added to each flask up to 40 ml and mixed. Two milliliters of the solution (3) of Griess reagent were added to each flask and made to volume (50 ml) with distilled water hich was mixed thoroughly and allowed to stand at room temperature for 1 h.

Suitable portion of each solution was transferred to separate cuvet (Blank + 10 samples); the transmittance of the blank (without nitrite) was adjusted to 100% transmittance (zero optical density) at 520 nm, then the optical density of each standard solution was measured. The optical density readings were plotted on coordinate paper against the nitrite concentration.

Procedure

Five grams of prepared sample were weighed in volumetric flask (225 – 250 ml) and mixed with 100 ml of water, and then the flask was heated in water bath (80°C). At the end of heating, 5 ml of saturated mercuric chloride solution were pipetted into each sample and cooled, then made to volume of 225 or 250 ml with distilled water. The solution was filtered through 12.5 cm filter paper. Ten milliliters of filtered solution were transferred to 50 ml volumetric flask. Water was added up to 40 ml and mixed with 2 ml of Griess reagent and made to volume (225 or 250 ml) with water.

The color was allowed to be developed within 40 min to 1 h at room temperature. A suitable portion of the solution was transferred to the cuvet, and the optical density was determined at wave length of 520 nm and adjusted to zero optical density with blank which contained 2 ml of reagent in 50 ml water. The sodium nitrite content was obtained in ppm through the prepared plot.

Detection of Sulphur Dioxide in different types of cheese (AOAC, 2000)

The test portion was placed in distilling flask, and then diluted to 400 ml with H_2O, and then 90 ml of HCl was added to separator which was forced into flask with gentle pressure. The flask was heated until condensation was showed in the first U-tube then the separator was removed and the heat was turned off.

One drop of methyl red was added, and titrated with NaOH (0.1 M) till clear yellow color was obtained (0.1 M NaOH =3.203 mg SO_2.). Similarly, the second U-tube was titrated. After titration, gravimetric determination was made by adding four drops of 1 M HCl and excess of filtered 10% $BaCl_2$ solution, then stood overnight. The precipitate was washed by decantation 3 times with hot H_2O through weighed gooch, then washed with 20 ml alcohol and 20 ml ether, and then dried at 105 - 110°C (mg. $BaSO_4$ × 274.46/g. test portion = SO_2 µg. /g).

The blank was determined, both by titration and gravimetrically, and the results were corrected accordingly.

Quantitative analysis of aflatoxin M1 in milk and cheese

Using ELIZA Kit (RIDASCREEN® Aflatoxin M1 30/15) which is a competitive enzyme immunoassay for the quantitative analysis of aflatoxin M1 in milk and cheese (Riedel de-Haen, 1997).

RESULTS AND DISCUSSION

It is evident from the obtained results that the chemical preservatives are the most prevalent type of inhibitory substances. Some unscrupulous producers tend to add

Table 1. Incidence of inhibitory substances in examined samples of UHT milk

Test	No. of examined samples	Positive samples		Negative samples	
		No.	%	No.	%
Wynter blyth test	30	11	36.67	19	63.33
Hydrogen Peroxide residues test	30	6	20.00	24	80.00
Formalin residues test	30	0	0.00	30	100
Alkalinizer (Na_2CO_3)	30	2	6.67	28	93.33
Tetracycline residues	30	9	30.00	21	70.00

Table 2. Statistical analytical results of tetracycline residues in the examined samples of UHT milk.

Tetracycline residues concentration (ppb.)	No. of examined samples	Positive samples		Tetracycline residues		
		No.	%	Min.	Max.	Mean ± S.E.M.
	30	9	30	10.00	78.71	23.62 ±7.01

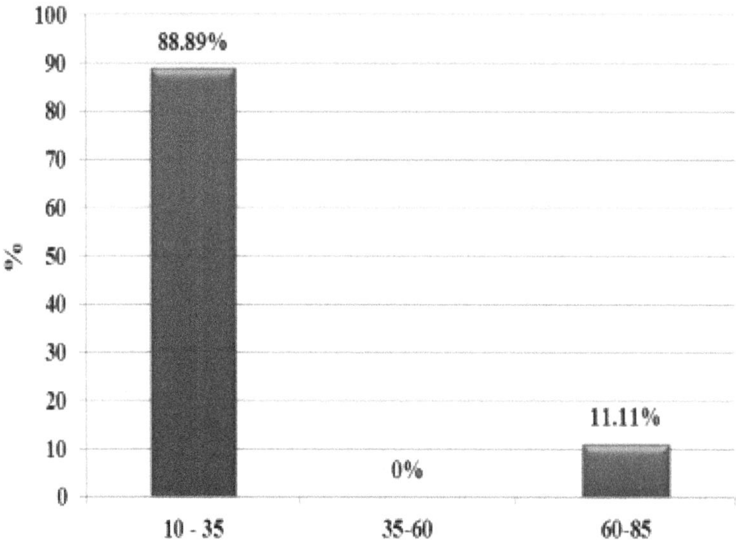

Figure 1. Frequency distribution of the examined samples of UHT milk based on their tetracycline residues concentration (ppb).

preservatives to milk in order to mask the neglected sanitary measures and to improve its keeping quality. Chemical preservation works either as direct microbial poisons or as acid neutralizer to prevent the microbial growth (Australian Academy of Science, 2004).

The results recorded in Table 1 show that 11 (36.67%) out of 30 examined UHT milk samples contained inhibitory substances; 6 samples (20%) contained hydrogen peroxide, 2 samples (6.67%) contained sodium bicarbonate (alkalinizer), while formalin was not detected in any of the examined samples. Tetracycline residues were detected in 9 samples (30%).

It is known that long continuous intake of sodium carbonate or bicarbonate and formalin might be harmful, as it causes damage to the gastrointestinal tract, primarily the stomach and lower oesophagus (Resmini et al., 1988). In addition, it may have a carcinogenic effect on human (Haddad and Winchester, 1990) and severe acidosis which result from rapid conversion of formaldehyde to formic acid (Mohanan et al., 2002).

Table 2 reveals that the mean value of tetracycline residues in the examined samples of UHT milk was 23.62 ± 7.01 with a minimum value of 10.00 and a maximum value of 78.71 ppb. The highest frequency distribution of tetracycline residues in the examined samples of UHT milk 8 (88.89 %) lies in the range of 10 to 35 ppb (Figure 1).

Table 3 represents the incidence of inhibitory

Table 3. Incidence of inhibitory substances in the examined samples of cheese.

Type of samples	No. of samples	Benzoic acid		Sorbic Acid		Nitrite		Sulphur Dioxide	
		No. of positive samples	%	No. of positive samples	%	No. of positive samples	%	No. of positive samples	%
Hard cheese	30	15	50.00	6	20.00	20	66.67	0	0.00
Soft cheese	30	5	16.67	13	43.33	17	56.67	3	10.00
Processed cheese	30	0	0.00	16	53.33	29	96.67	0	0.00

Table 4. Statistical analytical results of sorbic acid and nitrite in examined samples of cheese.

Types of samples	Sorbic acid (ppm.)			Nitrite (ppm.)		
	Min.	Max.	Mean ± S.E.M.	Min.	Max.	Mean ± S.E.M.
Hard cheese	9.00	13.30	2.12± 0.8	1.40	8.00	1.59 ± 0.31
Soft cheese	1.00	10.3	1.64 ± 0.5	0.40	3.20	0.67± 0.14
Processed cheese	2.00	14.00	1.71 ± 0.5	0.60	9.00	1.33 ± 0.027

Table 5. Statistical analytical results of aflatoxin M1 concentration (ppt.) in the examined samples.

Types of samples	No. of examined samples	Positive samples		Aflatoxin M1 conc. (ppt.)		
		No.	%	Min.	Max.	Mean±S.E.M.
Hard cheese	15	14	93.33	3.33	80.00	22.93 ± 6.31
Soft cheese	15	14	93.33	14.00	80.00	62.26 ± 7.03
UHT milk	15	15	100	7.5	80.00	53.30 ± 7.12

substances in cheese samples as benzoic acid was detected in 50% of hard cheese and 16.67% of soft cheese samples but it was not detected in the examined processed cheese samples. Sorbic acid was detected in 53.33% of processed, 43.33% of soft and 20% of hard cheese samples. The nitrite was detected in all examined cheese samples (100%) with highest incidence (96.67%) in the processed cheese samples, while sulphur dioxide was detected in examined soft cheese samples with incidence of 10%, it was not detected in the rest of the examined cheese samples.

Sorbic acid was detected in the examined hard, soft and processed cheese samples with mean values of 2.12± 0.8, 1.64 ± 0.5 and 1.71 ± 0.5, respectively; while nitrite content was detected with mean values of 1.59 ± 0.31, 0.67 ± 0.14 and 1.33 ± 0.027 ppm., respectively (Table 4).

Excessive addition of these preservatives may be harmful to consumers, because of the tendency to induce allergic contact dermatitis and convulsion (Fang et al., 2008). Sorbic acid is permitted in cheese, at levels below 3000 mg/kg. At these concentrations, sorbic acid acts as an efficient fungicide (Skovgaard, 1992) while Glass et al. (1998) mentioned that addition of sorbate with concentration of 0.1% can inhibit the growth of *Staphylococcus aureus* in processed cheese slices.

Nitrate is used in Europe to prevent late blowing of certain cheeses due to *Clostridium butyricum* and *C. tyrobutyricum* (Skovgaard, 1992; Glaesser, 1989).

The toxic effect of SO_2 in human is variable; some persons may tolerate up to 50 mg/kg body weight, while others feel headache, nausea and diarrhea at this concentration (Schroeter, 1966).

According to the obtained results in Table 5, it was cleared that aflatoxin M1was detected in all examined samples (100%) of UHT milk, (93.33%) of hard and soft cheese samples, with mean values of 53.30 ± 7.12, 22.93 ± 6.31and 62.26 ± 7.03, respectively.

Results represented in Figure 2 show the highest frequency distribution of AFM1concentration in the examined soft cheese and UHT milk samples (60 and 33.33%, respectively) lies in the range 40-80 while it lies in the range 0-10 in 40 % of examined hard cheese samples.

The higher concentration of aflatoxin M1 in cheese may be attributed to the association of aflatoxin M1 with casein fractions during manufacture (Egmond, 1989) while lower levels of aflatoxin M1 in some milk and milk

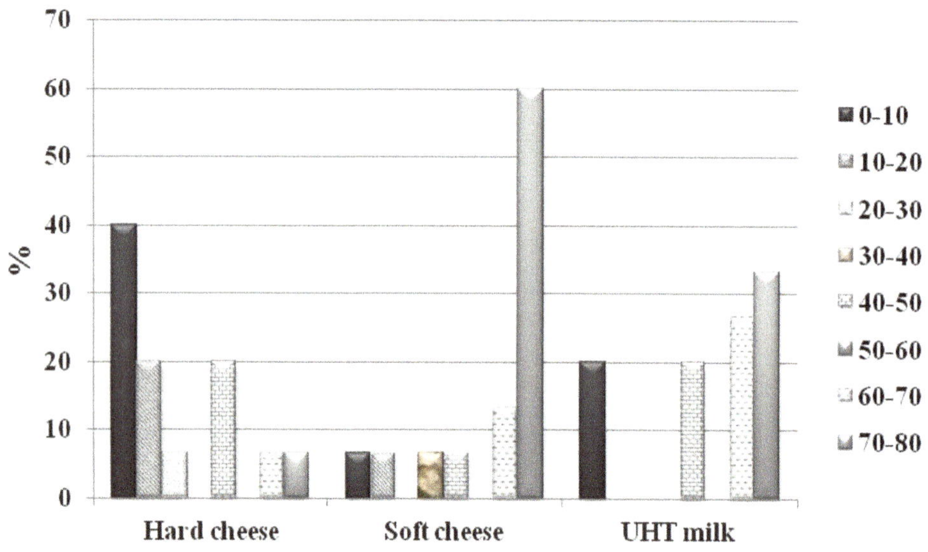

Figure 2. Frequency distribution of the examined samples based on their aflatoxin M1 concentration (ppt).

products may be attributed to greater amount of grass consumed by dairy cattle and lower amount of concentrates (Brown, 1982; Egmond, 1991; Dragacci and Fremy, 1996).

Conclusion

The results obtained from this work allow us to concluded that the majority of yogurt, hard, soft and processed cheese exposed for sale in Giza governorate are considered being hazardous to the consumers.

Tetracyclines were present in samples of UHT milk with variable percentages, while aflatoxin M1 was detected in all examined samples of UHT milk and cheese, sorbic acid, benzoic acid, sulphur dioxide and nitrite were detected in some samples of cheese. Therefore, in order to produce safe milk products, contamination should be avoided in order to reduce economic losses and expenses. To ensure safety in dairy industries the following suggestions should always be observed:

1) Strict hygienic measures should be adopted in dairy farms to ensure production of high quality milk.
2) Good sanitary conditions should be applied during production and processing of milk. Educational programs should be imposed to producers, processors and
and handlers to improve the quality of the product and to ensure the maximum safety to the consumers.
3) Application of effective technological measures (pasteurization, sterilization, acidification in technological processes to prolong the product shelf life and decreases or eliminations of pathogens in milk and milk products.
4) Application of adequate control measures through periodical examination of market milk and milk products by specialists to ensure maximum safety to the consumers.
5) Good manufacturing practices should be maintained throughout and HACCP should be applied to ensure safety of the finished products.

Conflict of interests

The authors did not declare any conflict of interest.

REFERENCES

Association of Official Analytical Chemists, AOAC (2000). Official methods of Analysis of AOAC international, 17 th ed. Published by AOAC international.
Australian Academy of Science (2004). Chemical preservation. Cited in: http://www.yahoo.com/ search. Publications / abstracts / september 2004.html.
Brown CA (1982). Aflatoxin M1 in milk. Food Technol. 34:228-231.
Cathey CG, Huang ZG, Sarr AB, Clement BA, Philips TD (1994). Development and evaluation of a mini-column assay for the detection of aflatoxin M1 in milk. J. Dairy Sci. 77:1223-1231.
Dragacci S, Fremy JM (1996). Application of immunoaffinity column clean up to aflatoxin M1 determination and survey in cheese. J. Food Protect. 12:59-65.
Dragacci S, Gleizes E, Fremy JM, Candlish AAG (1995). Use of immunoaffinity chromatography as a purification step for the determination of aflatoxin M1 in cheeses. Food Addit. Contam. 12(1):59-65.
Edward SK (1981). Handbook of meat analysis, 2 nd. ed. Avery publishing group inc., Wayne, New Jersy.
Egmond VHP (1991). Mycotoxins. In: Monograph on residues and contaminants in milk and milk products, International Dairy Federation, p. 131.

Egmond VHP (1989). Current situation on regulations for mycotoxins. Overview of tolerances and status of standard methods of sampling and analysis. Food Addit. Contam. 6(2):139-88.

Fang H, You-Zhao H, Lian L, Guo-Ni F, Hai-Yang X, Wu-Er G (2008). Determination of benzoic acid and sorbic acid in food products using electro-kinetic flow analysis–ion pair solid phase extraction–capillary zone electrophoresis. Anal. Chim. Acta 6(18):79-85.

Foley J, Buchley J, Murphy MF (1974). Commercial testing and product control in the dairy industry. University College, Cork.

Food and Agricultural Organization, FAO (1957). Report on the meeting of experts on the use of H2O2 and other preservatives in milk. FAO 157/11/8655.

Glaesser H (1989). Use of nitrate in cheese production. Dairy Ind. Int. 54:19-233.

Glass KA Kaufman KM, Johnson EA (1998). Survival of bacterial pathogens in pasteurized process cheese slices stored at 30ºC. J. Food Prot. 61(3):290-294.

Haddad LM, Winchester JF (1990). Clinical management of poisoning and drug overdose. W. B. Saunders Company, Philadelphia, PA, 19106.

Heeschen WH, Blüthgen A (1991). Veterinary drugs and pharmacologically active compounds. Monograph on Residues and Contaminants in Milk and Milk Products. IDF, Brussel.

Iswariah V, Guruswami MN (1973). Pharmacology and pharmacotherapeutics (P Varadacharya and Co, India), pp. 416, 427, 491.

Jahed KR Gh (2007). Chemical contaminants in milk and public health concerns: A review. Int. J. Dairy Sci. 2(2):104-115.

Lawley R, Curtis L, Davis J (2008). The Food Safety Hazard Guidebook ,Food Safety Info, London, Ukm.

Ling ER (1963). A text book of dairy chemistry. 3rd ed. Chapman and Hall LTD, London.

Mohanan S, Thomas PG, Lilly I, Laila M, Dominiand BI (2002). A new ultrasonic method to detect chemical additivesin branded milk. Indian Acad. Sci. 59(3):525-529.

Pien J, Desirant J, lafontaine D (1953). Detection of hydrogen peroxide in milk. Ann. Falsif, Fraudes, Paris, Dairy Sci. Abst. 46:539/540, 416-426.

Resmini P, Saracchi S, Bernardi G, Panari G (1988). Determination of spinacine (6-carboxy 1,2,3,4-tetrahydroimidazopyridine) in cheese to estimate the formaldehyde used in cheese making. Scienza-e-Tecnia-Lattiero-Casearia, 39(1):7-19.

Riedel de-Haen AG (1997). Enzymeimmunoassay zur quantitative bestimung von aflatoxin M1, ELISA system, diagnostic catalog, p. 2.

Satoskar RS, Bhanderkar SD, Ainapure SS (1999). Pharmacology, 16th Ed. Popular Prakashan, Mumbai, p. 591.

Schroeter LC (1966). Sulfur dioxide. Applications in foods, Bevearges and pharmaceuticals- pergamon, Oxford.

Skovgaard N (1992). Microbiological aspects and technological need: technological needs for nitrates and nitrites, Food Addit. Contam. 9(5):391-397.

Sundin GW (2003). Tetracycline. Encyclopedia of Agrochemicals, John Wiley & Sons, New Jersey, pp.1521-1522.

Tenant DR (1997). Food Chemical Risk Analysis. Blakie Academic and Professional. Chapman and Hall. UK. 1 st. Ed., pp. 381-398.

Wood GE (1991). Aflatoxin M1. In: Mycotoxins and Phytotoxins (chapter 6). Sharma, R.P. and Salunkhe, D.K. (Eds). London, CRC Inc., pp. 145-163.

Wynther BA, Meredith WB (1927). Foods: Their composition and analysis. 7th Ed., Charles, Griffin and Company, Ltd., London.

2

Antioxidant-mediated protective effect of hawthorn (*Crataegus mexicana*) peel extract in erythrocytes against oxidative damage

Banderas-Tarabay José Antonio[1,2], Cervantes-Rodríguez Margarita[2], Grada-Sánchez Miriam[2], Espíndola-Lozano Marlen[2,5], Cuevas-Romero Estela[3], Navarro-Ocaña Arturo[4] and Méndez-Iturbide Daniel[2]

[1]Doctorado en Ciencias Biológicas, Universidad Autónoma de Tlaxcala. Tlaxcala, México.
[2]Laboratorio de Química de Alimentos, Facultad de Nutrición y Facultad de Ciencias de la Salud, Universidad Autónoma de Tlaxcala, Tlaxcala, México.
[3]Centro Tlaxcala de Biología de la Conducta, Universidad Autónoma de Tlaxcala. Tlaxcala, México.
[4]Facultad de Química, Universidad Nacional Autónoma de México, CU, México DF.
[5]Maestría en Ciencias Biológicas, Universidad Autónoma de Tlaxcala. Tlaxcala, México.

Hawthorn (*Crataegus Mexicana*) is a traditional fruit in Mexican gastronomy and is used to treat many ailments. Previous studies have shown that acetone extracts derived from hawthorn peel (HPE) possess a strong antioxidant activity in chemical and biological model systems *in vitro*, attributable to their polyphenolic content. The main objective of this study was to investigate the ability of HPE to protect erythrocytes against oxidative damage, *in vitro*. The protection rendered by the HPE in erythrocytes was studied in terms of protection to oxidative damage by the inhibition of thiobarbituric acid reactive substances (TBARS) assays, morphological changes by light microscopy, and electrophoretic analysis of banding pattern of the red blood cells (RBCs). $FeSO_4$ was chosen to induce lipid peroxidation in human RBCs membranes and cytoskeleton proteins. The total polyphenol content in HPE was found to be 0.68 mg/g (SD 0.001) as the equivalent of gallic acid per gram. Trapping of DPPH was calculated by IC_{50} in 15.26 mg/L (SD 0.20). Better inhibition of TBARS formation by HPE was 16.78 mg/L (SD 0.33). HPE retards the morphological alteration in the erythrocytes-eryptosis. The electrophoretic pattern showed that some protein bands were not altered during a long period of incubation in HPE. Furthermore, it was found that HPE offers significant protection to human membrane erythrocyte up to for 28 days from the oxidative damage. In conclusion, our results indicate that HPE is capable of protecting erythrocytes against oxidative damage and morphologic changes by acting as a strong antioxidant.

Key words: Antioxidants, *Crataegus mexicana*, free radicals, Hawthorn, red blood cells, thiobarbituric acid reactive substances (TBARS).

INTRODUCTION

The pathophysiology of many blood diseases is associated with an increase of free radicals derived from reactive oxygen species (ROS) in the blood cells. Oxidative stress in red blood cells (RBCs) induces damage by injuring the protein cytoskeleton (Berlett and Stadtman, 1997; Dean et al., 1997), and the phospholipid cell membrane (Kowalczyk et al., 2012) causing directly and indirectly morphologic and microrheologic changes (Hebbel et al., 1990), thus premature eryptosis (Nagababu et al, 2008; Kempe et al., 2006; Lang et al., 2006). These changes and the reduced life span of the cells are involved in the pathogenesis of a number of blood diseases, including different types of anemia, altered vascular disorders, coagulopathies or clog veins alterations, also it is important in the life span of RBCs' bags (Baek et al., 2012). ROS are also involved in a number of other chronic human diseases (Kowalczyk et al., 2012).

In the bi-lipid cell membrane of the erythrocytes oxidation occurs by ROS, and malondialdehyde (MDA) formation can be measured and determined due the altered redox status of the cell (Singh and Rajini, 2008). The higher production of ROS is capable of causing oxidative damage in proteins by oxidation of amino acid residue side chain, formation of protein-protein cross-linkages, and oxidation of the protein backbone resulting in protein fragmentation (Berlett and Stadtman, 1997). Because of RBCs do not have nucleus or mitochondria organelles, their defense and reparation mechanisms against ROS became vulnerable (Pandey and Rizvi, 2011). However, RBCs have an effective anti-oxidant system that includes high levels of the enzyme catalase (CAT) (Cañas et al., 1988; Aebi et al., 1968), lower levels of superoxide dismutase (SOD) (Lynch et al., 1976), and the reduced glutathione peroxidase (GSH) (Van Asbeck et al., 1985), that make this cells mobile free radical scavengers.

A large number of fruits containing high amounts of nutrients are capable of inhibiting the oxidative stress and ROS formation; this is possible mainly by the action of antioxidant molecules, such as vitamin C, polyphenols, flavonoids and other antioxidant enzyme systems (Edwards et al., 2012; Cai et al., 2004). The clinical benefits after the intake of fruits have shown that they can protect and prevent the complications of chronic degenerative diseases, such as atherosclerosis, chronic inflammation, diabetes mellitus, cataract, coronary diseases and some types of cancer (Edwards et al., 2012; Willis and Wians, 2003; Middleton et al, 2000). Antioxidant activity of ethanolic plant extracts on cells has been demonstrated (Salawu et al., 2011; Nugraheni et al., 2011), including the inhibitory effects in RBCs hemolysis by the action of antioxidant extracts of cocoa plant (Zhu et al., 2005, 2002), potato peel (Singh and Rajini, 2008), and flaxseed oil (Liu et al., 2012). The activity has been related to flavonoids (Lam et al., 2007; Asgary et al., 2005; Sorata et al., 1984) to protect the erythrocytes membrane against deformation.

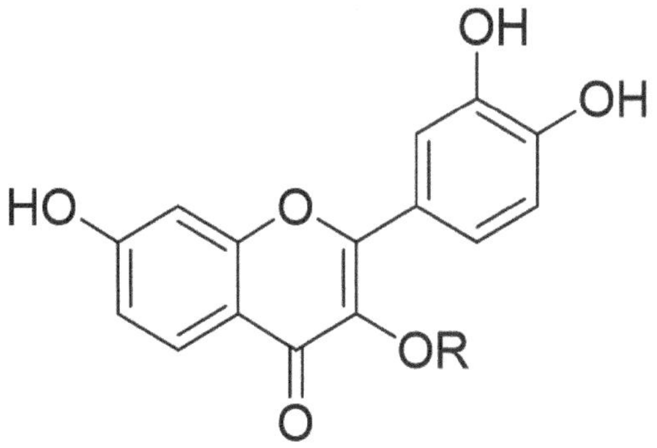

(R)

R = H

R = Rham-Rham-Glu-

R = Rham-Glu-

R = Glu-

R = Rham-

Figure 1. Structure of the quercetin glucosides in hawthorn.

The hawthorn *Crataegus mexicana*, is a traditional fruit which is a medicinal herb and in the Mexican gastronomy since ancient times. The wild type also named "tejocote" is widespread in almost all around Mexico. This variety of hawthorn has been the subject of little clinical studies for its medicinal properties and benefits (Arrieta et al., 2010; Andrade-Cetto and Heinrich, 2005) and only one study has been conducted to determine its phytochemical composition in their flowers (García-Mateos et al., 2012) showing the presence of polyphenols such as quercetin 3-*O*-glucoside, quercetin 3-*O*-rhamnoside, quercetin 3-*O*-rhamnosyl-(1→2)-[rhamnosyl-(1→6)]-glucoside, and quercetin 3-*O*-rhamnosyl-(1→6)-glucoside (rutin) (Figure 1).

In a previous study, we have demonstrated the antioxidant capacity of the hawthorn through lipid peroxidation inhibition in *in vitro* rat brain cells (Méndez-Iturbide et al., 2013). In this context to gain insight into the potential interaction of erythrocytes with the antioxidants found in Mexican fruits the present work aimed to assess the antioxidant activity of the acetone extract derived from the hawthorn peel, to inhibit lipid peroxidation and morphology changes in the red blood cell membrane during the life span of the erythrocyte *in vitro* experiments. The purpose of the study was to test the hypothesis that lipid peroxidation on the erythrocyt cell membrane is inhibited by HPE from *C. mexicana*.

MATERIALS AND METHODS

For TBARS analysis, the following reagents were purchased from Sigma-Aldrich Co (St Louis, MO, USA): Polyphenol (quercetin), ferrous sulphate ($FeSO_4$), NaCl, NaOH, KCl, $NaHPO_4$, KH_2PO_4, trichloroacetic acid, ethylenediaminetetraacetic acid (EDTA), and 2-thiobarbituric acid (TBA).

For Lowry protein concentration, the following reagents were acquired: Bovine Serum Albumin (BSA) [A-7906], Folin–Ciocalteu reagent, phosphate buffer saline (PBS) (9.5 mM, pH 7.4), Na_2CO_3, 2%; NaOH, 0.1 N; tetrahydrate potassium sodium tartrate 2% [$KNaC_4H_4O_6 \cdot 4H_2O$] and $CuSO_4$, 1% (Sigma-Aldrich Co. St Louis, MO, USA).

For scavenging effect on free radicals, 2,2-diphenyl-1-picrylhydrazyl (DPPH) was purchased from Sigma-Aldrich Co. (St Louis, MO, USA).

To measure total phenolic compounds, Folin-Ciocalteu reagent, 0.3% HCl, sodium carbonate (2% Na_2CO_3), and gallic acid, were purchased from Sigma-Aldrich Co. (St Louis, MO, USA).

Cell staining was achieved by preparing 10% neutral buffered formalin, and Haematoxylin-Eosin (HE) stain solutions (Sigma-Aldrich Co. St. Louis, MO, USA), ethanol, xylene, and mounting resin were purchased from J.T. Baker Chemical Co. (Phillipsburg, NJ, USA).

For SDS-PAGE, the following reagents were purchased: Sodium chloride, isobutyl alcohol, methanol, and acetic acid were acquired from J.T. Baker Chemical Company (Phillipsburg, NJ, USA). SDS, Tris, glycine, coomassie brilliant blue R-250, silver stain kit, TEMED, ammonium persulfate, N,N-1-methyl-bis-acrylamide, acrylamide and low molecular weight markers were from Bio-Rad (Richmond, CA, USA).

All other reagents including 2*d*-deionized water (18.3 MΩ, System Milli Q-II. Millipore, USA) were high quality analytical grade and pyrogenous free.

Source of samples

Four kilograms of hawthorn fruits were gathered in the State of Tlaxcala, México, in the month of December 2013. All samples had the same morphological characteristics in size, color, and damage free of the surface. Only skin samples of hawthorn were studied. The fruits were hand washed and the skin was manually peeled off, frozen and lyophilized. A total of 36.0 g of peel was obtained and then mashed (mesh size 20). Samples were maintained frozen at -20°C until use.

Preparation of hawthorn fruits peels extracts

Thirty grams of the resulting powder were homogenized at room temperature. Then, the homogenate was sequentially extracted for 72 h using 300 mL of the following solvents: *n*-hexane, dichloromethane, ethyl acetate, acetone, ethanol and finally methanol to obtain from the same homogenate one extract per solvent. All the extracts were decanted and evaporated until dry and kept inside a closed amber flask at 5°C until use. Only the acetone HPE was used for this study.

Scavenging effect on DPPH

The scavenging effect of antioxidants in HPE on free radicals was determined by using DPPH (Kim et al., 2005). Briefly, 100 µL of HPE were prepared at a concentration of 20 mg/L in DMSO. Then, 4000, 400, and 40 mg/L final dilutions in cool ethanol were prepared. DPPH was dissolved in ethanol (1 mM). A total of 50 µL of HPE and 150 µL of DPPH (final concentration of 1000, 100, and 10 mg/L of HPE) were mixed and incubated for 30 min at 37°C in an orbital shaker. The absorbance at 517 nm was recorded using an Elisa microplate reader (Bio-Tek EL800, USA). The percent-reduced activity of DPPH was calculated and expressed as the IC_{50} (Khattak et al., 2008) using six different concentrations of HPE (5.6 - 100 mg/L). Quercetin was used as standard.

Determination of total phenolic compounds

The total phenol content of the acetone HPE was determined using the Folin-Ciocalteu reagent method (Folin and Ciocalteu, 1927) with minor modifications as described previously (Méndez-Iturbide et al., 2013). Briefly, 3 mg of HPE were added to 40 mL of methanol/water (50:50, v/v). The mixture was left to stand for 1 h at 25°C, and the supernatant was then collected and stored (aqueous methanol extract). The residual sample was added to 40 mL of acetone/water (70/30, v/v) and treated as described before. The aqueous methanol and acetone extracts were mixed, evaporated and re-dissolved in 10 mL of ethanol (aqueous ethanol extract). Finally, the residual sample was dissolved in 10 mL of H_2O (aqueous extract). One mL of each aqueous and ethanol extract was brought to a volume of 2.5 mL using 0.3% HCl. Subsequently, 50 µL of either this sample or the standard quercetin was added to 1 mL of 2% Na_2CO_3, followed by the addition of 50 µL of Folin-Ciocalteu reagent diluted in water (1:1). After 30 min incubation at 25 °C, the absorbance at 750 nm was recorded using a spectrophotometer (Genesys 10 UV, Thermo Electron Co. USA). The total phenol content was expressed as milligrams of gallic acid equivalent per gram of HPE (GAE/g) using a standard curve (concentrations 0.2 - 1.0 mg/mL) for a freshly prepared solution of gallic acid.

Subjects selection and clinical examination

This study included eight healthy males. Their average age was 32 years old (SD 11.1). Exclusion criteria included that none of them had any chronic disease such as diabetes mellitus, cardiovascular disease, anemia; they were not under drug treatment or having history of alcoholism and/or smoking, and neither excessive exercise, which induce oxidative stress. The study was explained to all subjects, and their consent signature was taken on a voluntary bases. Also, this study was approved by the Institutional Ethical Committee for Human Research, Faculty of Health Sciences and Nutrition in the State University of Tlaxcala. Body mass index (BMI) was obtained by accurately measuring height and weight of each individual (BMI = weight [kg] / height [m^2]).

Blood sample collection and preparation

Seven mL of fasting venous blood sample was collected at 10:00 am from each subject under aseptic conditions, in sterile EDTA-vacutainer tubes (Becton Dickinson Co. NJ, USA). RBCs and serum were separated immediately by centrifugation at 3000 *g* for 10 min at 25 °C. Hematocrit and total hemoglobin were determined in each individual blood sample. After centrifugation, RBCs were washed and centrifuged 3 times in PBS [0.1 M (0.2 g KCl, 0.2 g KH_2PO_4, 8 g NaCl, and 2.16 g $NaHPO_4 \cdot 7H_2O$)/L), pH 7.4 (with 10% NaOH] and white cells were pipetted out and discharged. RBCs samples were stored in PBS as the initial hematocrit conditions (1:1) and used for all experiments. All samples were prepared and analyzed the same day of collection, unless otherwise specified.

Ghost cells (GC) preparation

Four mL of separate RBCs in PBS were used to prepare hemoglobin-free erythrocyte membranes according to a hypotonic-hypertonic lysis procedure (Matteucci et al., 1995; Krishnaveni et al., 2013; Jangde, 2011).

Lipid peroxidation induced by $FeSO_4$ in RBCs and GC (TBARS)

Antioxidant activity from the HPE was estimated by using the spectrophotometric method described by Ohkawa et al. (1979). Briefly, extract samples at 10, 30, 50, 100, and 1000 mg/L concentration were prepared in order to inhibit lipid peroxidation induced by Fe^{2+} in RBCs and GC.

Ghost cells and RBCs were washed and homogenized separately in 10 mL of PBS (pH 7.4) at 5°C (1:10 w/v) (Dominguez et al., 2005). The homogenate was centrifuged for 10 min at 3000 g, and the supernatant was adjusted at 2.66 mg of protein/mL in PBS to be used for total protein determination according to the method of Lowry et al. (1951). BSA was used to elaborate the standard curve. Then 375 µL (1 mg protein content) of RBCs or GCs were previously incubated in an orbital shaker with 50 µL of EDTA 10 µM, and 50 µL of each one of the different extract concentrations of HPE at 37 °C for 30 min for a working dilution concentrations of 3-50 mg/L. Lipid peroxidation was initiated by the addition of 50 µL of $FeSO_4$ (final concentration = 25 µM) (Ng et al., 2000). After 1 h, samples were cooled on ice, and 0.5 mL of TBA (1% 2-thiobarbituric acid in 0.05 N NaOH and 30% trichloroacetic acid, 1:1) was added. Samples were centrifuged 10 min at 3000 g. Lipid peroxidation and their final products in RBCs and ghost cells, were assayed by malondialdehyde (MDA) determination using TBARS assay method (Kibanova et al., 2009). Samples were heated at 94 °C for 30 min in a boiling water bath. After cooling on ice, 200 µL of each sample and a blank sample, were placed in an Elisa microplate reader and the absorbance was recorded at 540 nm. TBARS results were expressed as the equivalent of MDA, compared with 1,1,3,3-tetramethoxipropane as the standard. The percentage of inhibition is defined as the decrease of TBARS formation, due to the extract quelation (Esterbauer and Cheeseman, 1990). HPE was evaluated to determine the minimal concentration (µg/mL) to inhibit lipid peroxidation up to 50% (IC_{50} µg/mL).

Lipid peroxidation induced by $FeSO_4$ in RBCs to morphologic analysis

HPE was diluted at 5, 10, 100, and 1000 mg/L in PBS with the pro-oxidant $FeSO_4$ (100 µM). Then, 25 µL (~ 125 x 10^6 cells) of RBCs were added in each dilution and incubated at 25 °C during 5, 30, 120, 180 min, 1, 2, 15, 22, 24 and 28 days. The samples without $FeSO_4$ or HPE were prepared as control blanks. Experiments were carried out in triplicate in RBCs.

Cytological analysis

RBCs samples (2 □L = ~10 x 10^6 cells) were taken from different incubation times, and they were extended on a glass slide and immediately fixed with 10% formaldehyde buffer. Cells were stained with HE and a glass cover was mounted for microscope analysis at a magnification of 40x, and 100x using an oil immersion objective. An Olympus CH30RF100 microscope (Olympus Optical Co. LTD, Japan) with a digital camera connected at one PC and an image analysis software system (Motic Images Plus 2.0, Motic China Group Co. LTD) was used to quantify static and kinetic RCBs and the morphology changes.

Histometrical measurements

Histometrical measurements were performed at a magnification of 40x and 100x. The examiner was blinded to which group each specimen belonged to. RCBs were counted in three standard areas of each slide (0.1 mm x 0.1 mm each). In each area three different squares of 70 µm were selected to cell counting. Intensity of colour stain (Anisochromy), size (Anisocytic), shape (Poikilocytic), inclusions and other morphological changes in RBCs were determined as described by Mackenzie (2000) and Brecher and Bessis (1972) as follow:

HE Stain:	Normochromic
	Hypochromic
	Hyperchromic
Size:	Normocytic (7 – 8 µm)
	Microcytic (less than 7 µm)
	Macrocytic (more than 8 µm)
Shape:	Acanthocyte (crenated)
	Echinocyte
Inclusions:	Heinz bodies
Others:	Hemolysis
	Schistocyte

Dynamic flow RBCs assay

The dynamic flow of RBCs was assayed using a flow channel (20 µm wide, 10 µm high, 5 mm long) cut on a glass microscope slide made in the laboratory and covered with a glass-cover slide to permit direct visualization during the entire flow study (Hillery et al., 1996). The experiments were performed at 25°C. A clean glass slide was put with a 5° inclination on the microscope slide to permit continuous and free flow of RBCs and to avoid shear strength. Two microliters of the above mentioned 2 h-incubated RBCs in PBS or HPE, were dropped on one extreme of the chamber and let to flow freely for 120 min. The number of adherent cells, crenation (acanthocytes and echinocytes), and deformability against time, were counted by direct microscopic visualization and a video image system (20 frames per second) providing multidirectional pictures of the erythrocytes. In order to avoid glass near deformation of RBCs or echinocytes formation (Brecher and Bessis, 1972; Bronkhorst et al., 1995), only cells flowing in the middle of the channel were analyzed.

Polyacrylamide gel electrophoresis

Protein determination and standardization of RBCs concentration was determined as described above by the method of Lowry. After washing with PBS, 25 µL of each incubated sample of RBCs were reconstituted under non-denaturing conditions in 20 µL of sample solubilizing buffer (0.125 M Tris-HCl, 4% SDS, 20% glycerol, pH 6.8, 0.002% bromophenol blue) and sonicated for 3 min prior to running the gels. 1 µL of each individual RBC sample (considering that in 1 µL of blood exist ˜5 x 10^6 red blood cells) were subjected to SDS-PAGE electrophoresis at room temperature in a discontinuous gel system according to the method described by Laemmli (1970), using a 3% stacking gel and a 10% separating gel. Molecular-weight standards were used in all gels. Electrophoresis was carried out at 100V constant current until the dye front of the gel reached the bottom of the chamber. A minigel apparatus (Mighty Small II vertical slab unit; Hoefer Scientific Ins., San Francisco CA., USA) was used for this purpose.

Gel staining

Gels were fixed and stained successively with 0.1% Coomassie brilliant blue R250 in methanol-water-acetic acid (1:8:1), and silver stain. To enhance and obtain a good pattern of staining bands, gels were initially stained with the solution of Coomassie blue for 1 h and distained gradually with several changes of distain solution (methanol-water-acetic acid, 1:8:1) for approximately 60 min until the bands showed the appropriate stain.

Identification of membrane proteins

Proteins were identified according to their relative mobility in the gel and stain patterns following the criteria described by Ballas (1977), Mackenzie (2000), and Singh and Rajini (2008). Membrane and cytoskeletal molecules were scored according to the size and stain intensity of the bands as: absent (-), present (+), low intensity and size (-+), and high intensity and size (++). However, because of the subjectivity inherent in these parameters, protein bands were also scored only as present or absent.

Statistical analysis

TBARS analysis

Data in the text, tables and figures were expressed as means with standard error of mean (SEM). One-way analysis of variance (ANOVA) was used to compare the antioxidant capacity of HPE in RBCs, GC, and quercetin/GCs.

RBCs counting

All data to count the RBCs were presented as mean and the standard error of the mean (SEM). The significance between RBCs control group and RBCs experimental group with HPE, was determined using two-way ANOVA repeated measures analysis. HPE was considered as the first variable, and the time as the second variable. A subsequent Bonferroni correction test was performed to determine difference between groups and baseline. A $P<0{\cdot}05$ was considered to be statistically significant. In all cases, each experiment was run in triplicate within two control samples. The GraphPad Prism version program (*v*.5.01 for Windows) was used for this purpose.

RESULTS

Total phenolic compounds, free radical capture of DPPH and IC_{50}

A total of 2.2 g of HPE was obtained from 30 g of hawthorn dried skin. The total polyphenol content in HPE was found to be 0.68 (SD 0.001) mg of gallic acid/g. The amount of DPPH reduced was quantified measuring the decrease in absorbance of DPPH by the HPE. The IC_{50} of the HPE to scavenging free radicals was 15.26 ± 0.20 mg/L as compared with the IC_{50} of the pure antioxidant quercetin [5.37 (SD 0.14) mg/L]. It was shown that acetone extract is three times less active than quercetin in scavenging DPPH radical.

MDA determination in RBCs and ghost cells by TBARS

Red blood cells have both saturated and unsaturated lipids and fatty acids on its membrane, which are susceptible to be peroxidized by the ROS. Malondialdehyde (MDA) is one of the final products of cell membrane lipoperoxidation. In order to determine the antioxidant capacity of hawthorn fruit peel extracts, $FeSO_4$ was used to induce the lipoperoxidation in RBCs and erythrocyte ghost. The ferrous ion (Fe^{2+}) is capable of forming lipid peroxides that induce the formation of highly reactive substances as hydroxyl radicals, perferryl and ferryl species.

To evaluate the effectiveness of hawthorn extracts to protect the RBCs and ghost cells from the lipoperoxidation, different concentrations of extracts were used as described in material and methods. The acetone HPE had a better performance to inhibit lipid peroxidation in RBCs (46.84%) and in ghost cells (47.91%) in a concentration of 50 mg/L, as compared with GC/quercetin (41.58%) (Figure 2).

RBCs quantitation and morphological changes

The body mass index and RBCs samples from 8 subjects were analyzed to determine their good health status to standardize experimental conditions. Means of BMI was 21.4 kg/m^2 (SD 5.2), hematocrit 0.49% (SD 0.05), and hemoglobin 12.5 mg/dL (SD 1.8). All experiments were run immediately after each blood collection.

RBCs were incubated in HPE with the prooxidant $FeSO_4$ at different concentrations and analyzed for morphological changes, eryptosis and hemolysis. The extent of hemolysis of RBCs in the different HPE concentrations, showed a time dependent concentration in the erythrocyte damages during the 28 days of incubation. RBCs with higher HPE concentration were less damaged as compared with a lower concentration of the extract as visually demonstrated by the presence of less hemoglobin accumulation in the incubation media (no quantified). This suggests a low-level of hemolysis and damage on the membranes of the erythrocytes, keeping the colour, and less cell adhesion in the erythrocyte-clot formation (Figure 3). These observations suggest an antioxidant effect of HPE in RBCs, and are similar to those found by Baek et al. (2012) that described the attenuation of cell lesion using haptoglobin therapy in guinea pigs red blood cells.

RBCs exposed to the different concentrations of HPE retained in greater or lesser extent their normal morphology and phenotypic characteristics during all the time of the experimental conditions. A concentration of 100 mg/L showed a better and repeatability pattern under the microscopic analysis. The hemolysis process was faster in untreated RBCs (Figure 4A to F) and started at day 7 of incubation, as compared with the RBCs treated with HPE (Figure 4G to L), where the hemolysis process started later. All morphological cell changes are described independently and also the sum of these changes.

Normocytic and normochromic erythrocytes

Microscopic counting of RBCs was similar in both cases, showing normocytic cells at the beginning of the

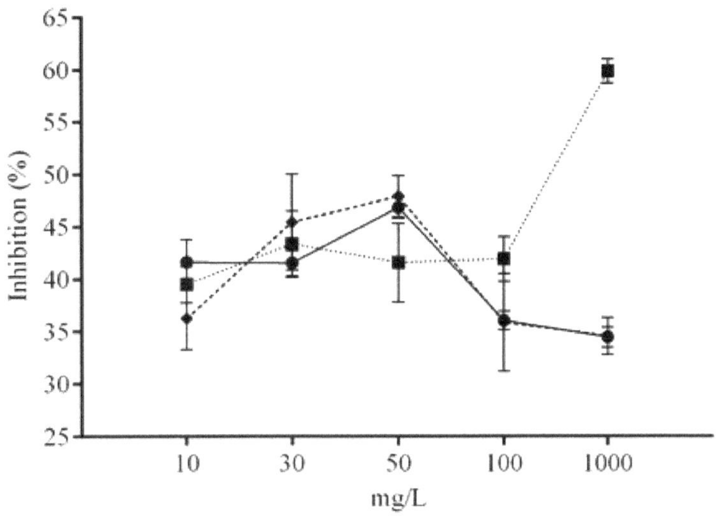

	10 mg/L		30 mg/L		50 mg/L		100 mg/L		1000 mg/L	
	Inhibition %	SME	Inhibition %	SME	Inhibition %	SME	Inhibition %	SME	Inhibition %	SME
Total RBCs –●–	41.59	2.15	41.52	1.22	46.84	1.01	36.03	0.91	34.45	0.95
Ghost cells –♦–	36.25	2.94	44.45	4.59	47.91	1.99	35.86	4.66	34.56	1.75
Quercetin –■–	39.50	1.74	43.35	3.20	41.58	3.75	41.91	2.13	59.84	1.16

Figure 2. Increasing concentrations of HPE from *Crataegus mexicana* on the production of TBARS in RBCs, and ghost cells. Data is the percent of inhibition and the SME from nine different experiments; each one by triplicate. Statistical differences among groups were at $P<0{\cdot}0001$.

experiment incubated a non-incubated with the HPE. The counting of cells was easier in treated RBCs than in untreated RBCs, because the cytological smear in untreated RBCs showed large amounts of aggregated cells that was made difficult to separate each cell from the other.

Smear from treated RBCs with HPE was homogeneous and the cells showed a better separation among them; this could be due to the antioxidant activity since the first moment that the cells were incubated in the extract. Treated RBCs were longer hydrated, that suggests no loss of water inside cells by keeping the cell membrane intact for longer periods of time. Evident morphologic changes and hemolysis of RBCs began at 180 min either treated and untreated erythrocytes with HPE, and hemolysis started at 48 h in untreated RBCs (Figure 5a). By contrast, the beginning of hemolysis in RBCs treated with HPE started later than in untreated cells, maintaining the morphology for 28 days (Figure 4K to L). Total RBCs quantitation showed the effect of HPE ($P<0{\cdot}01$), during time of incubation in min and days ($P<0{\cdot}001$). A comparable dose-response relationship was observed in RBCs at different time points of incubation.

Anisocytic RBCs

An increased number of microcytic erythrocytes was observed during the first 120 min in both incubated PBS/$FeSO_4$ cells and in PBS/$FeSO_4$/HPE cells. Following this microcytic formation, a decrease in these cell number was observed until they became lysed. However, the cells treated with the HPE remains up to for 28 days. This suggests that the crenated cell formation and the eryptosis process is retarded with the HPE extract (Figure 5b). No effect in the formation of microcytic cells was observed by the treatment ($P<0{\cdot}05$), however the time of incubation ($P<0{\cdot}0001$) and interaction ($P<0{\cdot}005$) induce microcytic formation.

Crenated, acanthocytes

The oxidation of the main structural proteins spectrin and actin that form the submembrane cytoskeletal meshwork of RBCs, induced morphological and viscoelastic changes in the cell membrane that were observed as blebs formation on the surface of the cells (Figure 4B to C, and 5c). RBCs with the HPE showed an early inhibition in the formation of these continuous bebbling oxidative phenomenon, thus also inhibited the eryptosis and hemolysis process in the life span of the erythrocyte (Figure 4I to L, and 5c) as compared with the RBCs from the control group. These data suggest that the acetone extract is acting to reduce membrane cytoskeletal deformation caused by the $FeSO_4$ oxidative effect on the lipoperoxidation of phospholipid on the membrane bilayer, and besides inhibit the oxidation of amino acid

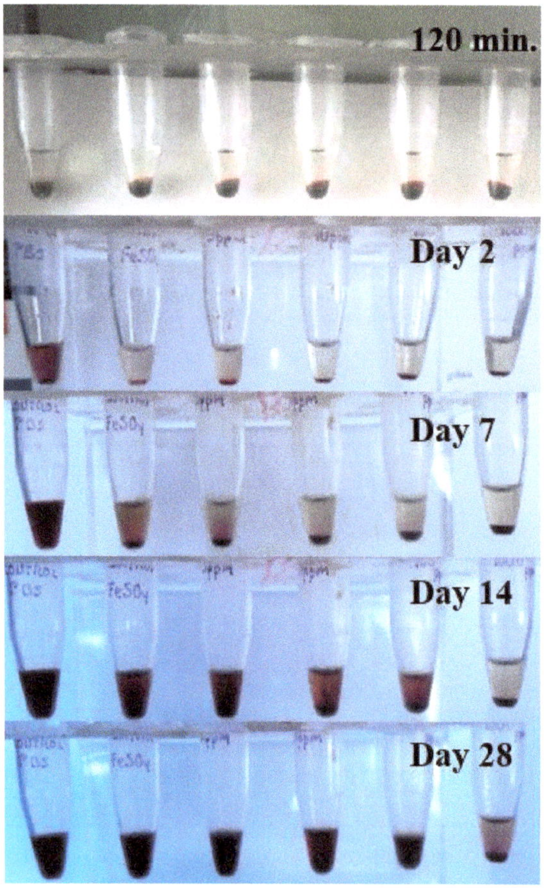

Figure 3. Effect by increasing concentrations of HPE from *Crataegus mexicana* in the lesion of RBCs at different times of incubation (left to right): RBCs/PBS, RBCs/PBS/$FeSO_4$, RBCs/PBS//$FeSO_4$/ with 5, 10, 100, and 1000 mg/L HPE respectively. Samples showed less hemoglobin accumulation in the incubation media at ongoing time of incubation and increase HPE concentration.

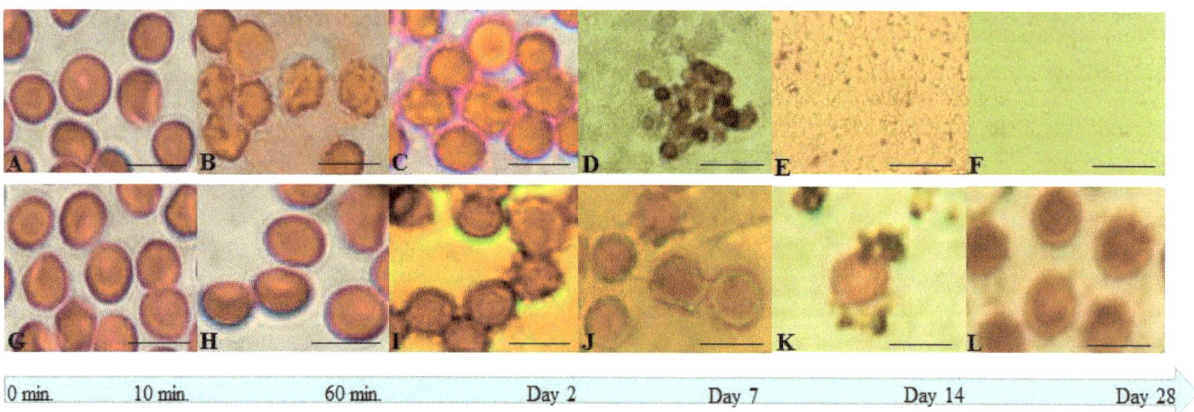

Figure 4. Effects of HPE on the contour and morphology of erythrocytes. RBCs incubated only with PBS (upper) and RBCs incubated in 100 mg/L of HPE (lower) at different time, both containing $FeSO_4$ 100 μM. Cell morphology was examined by light microscopy. Panel A-F RBCs incubated in PBS/$FeSO_4$; Panel G-L RBCs incubated in PBS/$FeSO_4$/100 mg/L of HPE. Incubation time is indicated. Further experimental details are given in material and methods. HE 100x, using different microscopy filters, Bar = 10 μm.

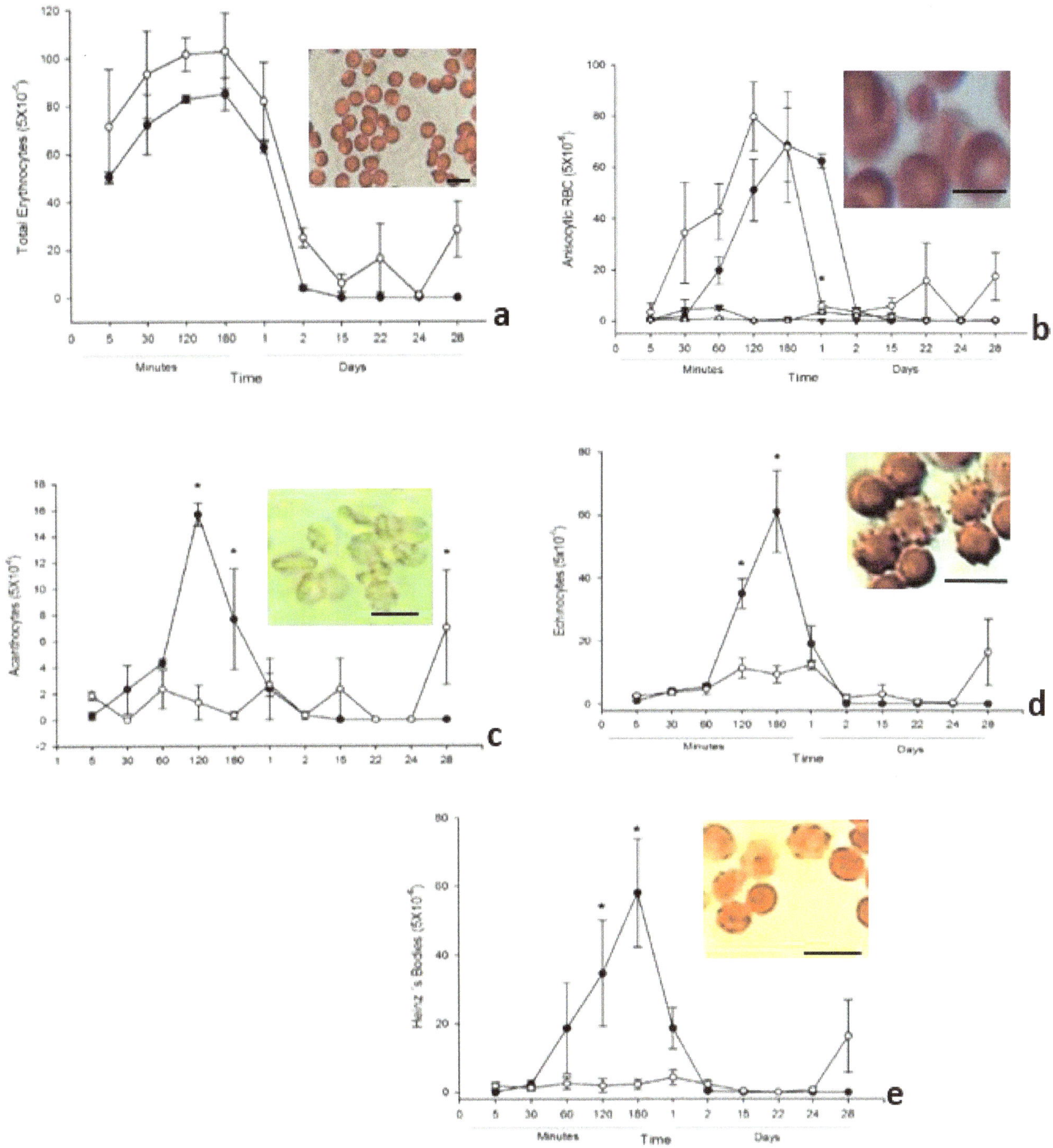

Figure 5 (**a**) Comparative quantitation between total RBCs in PBS/$FeSO_4$ (•), and RBC in PBS/$FeSO_4$/HPE (o), and time. Insert: Light microscopy of normocytic and normochromic RBCs at 5 min. HE 40x, bar = 10 µm. (**b**) Interaction between microcytic RBCs in PBS/$FeSO_4$ (•), and microcytic RBCs in PBS/$FeSO_4$/HPE (o), and time. Also macrocytic RBCs in PBS/$FeSO_4$ (▼), and macrocytic RBCs in PBS/$FeSO_4$/HPE (⌂). Insert: Light microscopy of a microcyte at 180 min. HE 100x, bar = 10 µm. (**c**) Interaction between RBCs in PBS/$FeSO_4$ (•), and RBC in PBS/$FeSO_4$/HPE (o), and time. Comparing the presence and absence of acanthocytes. Insert: Light microscopy of acanthocytes. Bebbling formation is observed at 120 min. *In vitro* 100x, bar = 10 µm. (**d**) Interaction between RBC in PBS/$FeSO_4$ (•), and RBC in PBS/$FeSO_4$/HPE (o), and time. Comparing the presence and absence of echinocytes. Insert: Light microscopy of echinocytes. Spiny form structures are observed on the surface of the cell membrane at 180 min. HE 100x, bar = 10 µm. (**e**) Interaction between RBC in PBS/FeSO4 (o), and RBC in PBS/FeSO4/HPE (•), and time. Comparing the presence and absence of Heinz bodies at different time points. Insert: Light microscopy of RBCs. Heinz bodies are observed in the cells at 180 min. HE 40x, bar = 10 µm. values are Means and SEM of three different experiments; each one by triplicate. Statistical differences was with Bonferroni correction test*

residue side chains, formation of protein-protein cross-link, and oxidation of the cytoskeletal backbone of proteins in the erythrocyte. These morphological changes observed during the first 24 h reached the maximun peak at 120 min of the untreated red blood cells. A less formation of crenated cells is related to the HPE (P<0•01), time (P<0•0001), and interaction factors (P<0•0001). These morphological changes have been described as Type I and II (Brecher and Bessis, 1972) in the eryptosis process of the RBCs. As we already mentioned, HPE-treated RBCs did not increase these changes in a time dependent manner.

Crenated, echinocytes

As the time increased the morphology changes in RBCs became more evident. The oxidative process and the crenation in RBCs induced the formation of cytoplasmic prolongations due the membrane deformation by cytoskeletal protein and lipid oxidation. These changes became non-reversible and were determinant in the eryptosis process of RBCs (Figure 4C, and 5d). The HPE from *C. mexicana* inhibited echinocytes formation following the time point in the acanthocytes formation, retarding cell deformation and the hemolysis process (Figure 4I – K, and 5d). A time dependent action in echinocyte formation (P<0•0001) and its interaction with the HPE (P<0•01) was observed. This morphology change in the cell membrane of the erythrocyte has been denominated as Type III (Brecher and Bessis, 1972) with many different forms in membrane deformation, and is considered as a non-reversible morphological alteration during the life span of the erythrocyte and therefore in the eryptosis process. The biological process in acanthocytes formation by radical-mediated damage in proteins and membrane, is similar in echynocites when they became $FeSO_4$ oxidized.

Heinz bodies

The oxidation of hemoglobin produced in their primary structure disulphide cross-links between adjacent globin chains which become distorted and eventually lead to visible precipitates in the inner membrane of RBCs. HPE was able to inhibit the formation of these bodies from the beginning of *in vitro* experiments up to the day 24. RBCs without the acetone extract initiated the formation of Heinz bodies at min 30, increasing the formation in the next min and days. The increase in hemoglobin precipitation was associated by the distortion of the cell at the final phase of acanthocytes formation during the eryptosis process. This data suggest that HPE inhibited the oxidative process of hemoglobin for a longer period of time (P<0•0001), this interaction permits the extension of the life span of the cell (P<0•0001). Afterwards, the hemolysis process continued and began the formation of schistocytes. This changes last longer than the control group in which the Heinz body formation and hemolysis took less time (Figure 4B, 4I, and 5e).

Schistocyte

The schistocytes formation was observed on the first day of incubation, this is due to the deformation, bebbling formation and separation of liposomes from the surface of the cell membrane to be completely degraded (Figure 4D to E). However, schistocytes were observed in the RBCs incubated in HPE up to for 28 days, probably because the cells keeps their membrane for longer time as is deduced observing all the graphics in the different experiments (Figure 4K).

Dynamic flow of RBCs assay

To critically evaluate the potential contribution of HPE to inhibit any cellular morphologic change by protein and lipid membrane oxidation, RBCs flow rate, adhesion, and morphologic changes were quantified from video recording of nine individual cells in the flow channel over a period of 180 min. RBCs incubated in HPE demonstrated a longer period of time of flowing through the flow channel up to for 180 min as compared with RBCs in PBS whose cells were dehydrated in less than 6 min. Morphologic changes such as acanthocytes and echinocytes formation were observed, and an increase of adhesion among cells was demonstrated in RBCs incubated only in PBS during the first 60 min as compared with RBCs incubated with the HPE (Figure 6A to J) without changes in the morphology up to 120 min. It is particularly notable in these experiments and suggest that RBCs maintain their morphologic and the micro-rheologic characteristics in the micromolar concentrations of HPE. This could be because there is not lipid or protein oxidation in the cell membrane, thus no dehydration and crenation, and therefore inhibition of cell deformity and viscosity and better flow of RBCs. However, life span of RBCs with or without the extract is going to continue its eryptosis process, thus the cell will lose their biological properties and finally will be cell lysed.

Polyacrylamide gel electrophoresis

The banding pattern of some cytoskeletal proteins of RBCs were stained and identified in SDS-PAGE. Band staining intensity was observed during almost all time points of incubation of RBCs with the different concentrations of the hawthorn extract.

The qualitative analysis showed that some bands keep the stain intensity in relation to a major concentration of the HPE. In these experiment seven main bands were observed at day 1 of the experiments (Figure 7). The colour intensity and size of some bands were maintained over a longer period of time with increasing the extract concentration and suggest an antioxidant protective effect of the HPE on the cytoskeletal proteins of the RBCs. This data is related to what we observed in the

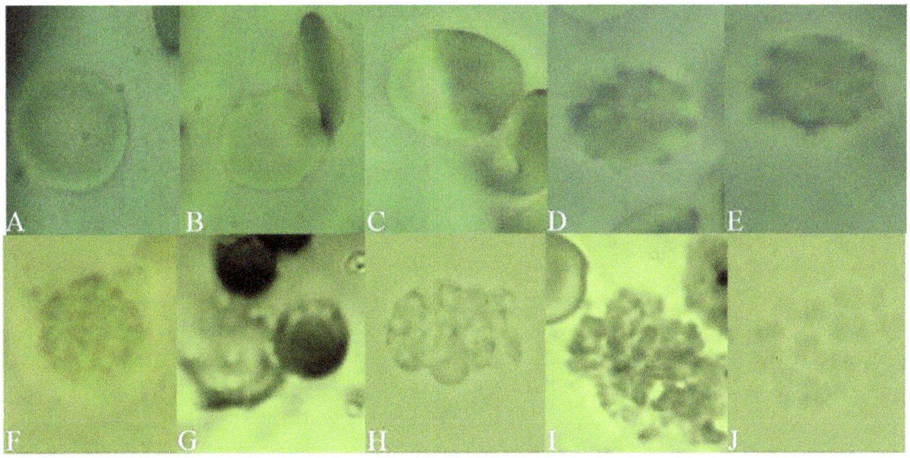

Figure 6. Morphologic changes of a representative RBC *in vitro* after exposure in HPE at 100 mg/L in a time lapse of 180 min in the flow channel. Eryptosis was retarded as compared with control RBCs. (A) normocyte; (B) beginning of Phase I; (C) beginning of crenation in phase II; (D) achantocyte and bebbling formation in phase II; (E) echinocyte in the beginning of phase III; (F, G, and H) beginning of vesicles formation and cell membrane disruption; (I) schistocytes formation, and (J) schistocytes and cell lysis.

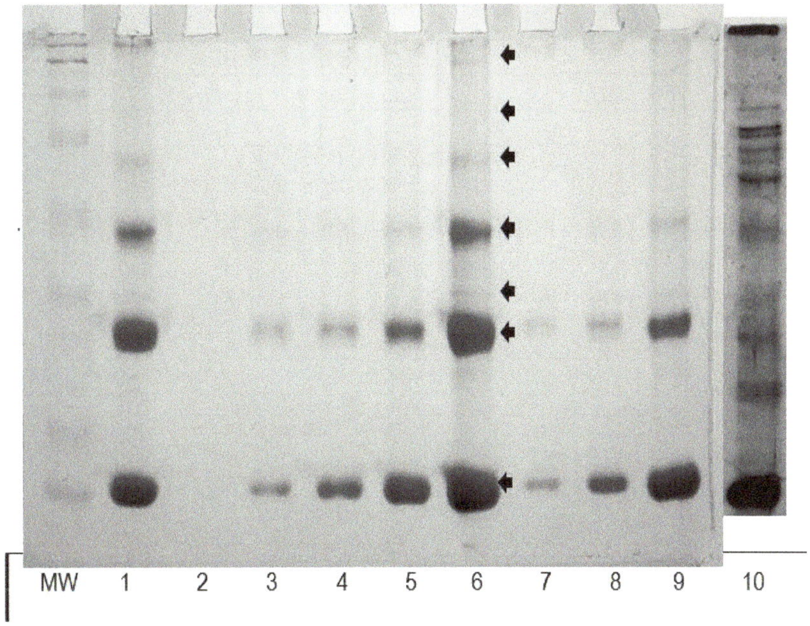

Figure 7. Determination of membrane proteins in RBCs in a representative SDS-PAGE stained with coomassie brilliant blue R250. Differences and variability in the molecular pattern (arrows) between different concentrations of HPE at day 7 (channels 1- control RBC, channel 2- PBS/$FeSO_4$, channel 3- RBC/HPE 10 mg/L, channel 4- RBC/HPE 30 mg/L, channel 5- RBC/HPE 50 mg/L, channel 6- RBC/HPE 100 mg/L), and day 14 (channel 7- RBC/HPE 10 mg/L, channel 8- RBC/HPE 30 mg/L, channel 9- (RBC/HSE 50 mg/L), channel 10- RBCs at day 1. One µL per channel according to material and methods. MW= molecular weight standards.

course of incubation, where the higher is the concentration, the lesser is the lesion in RBCs. This supports the fact that the cytoskeleton is protected against the oxidative process induced by the $FeSO_4$, therefore the cells will keep their morphology and rheology characteristics over a longer period of time.

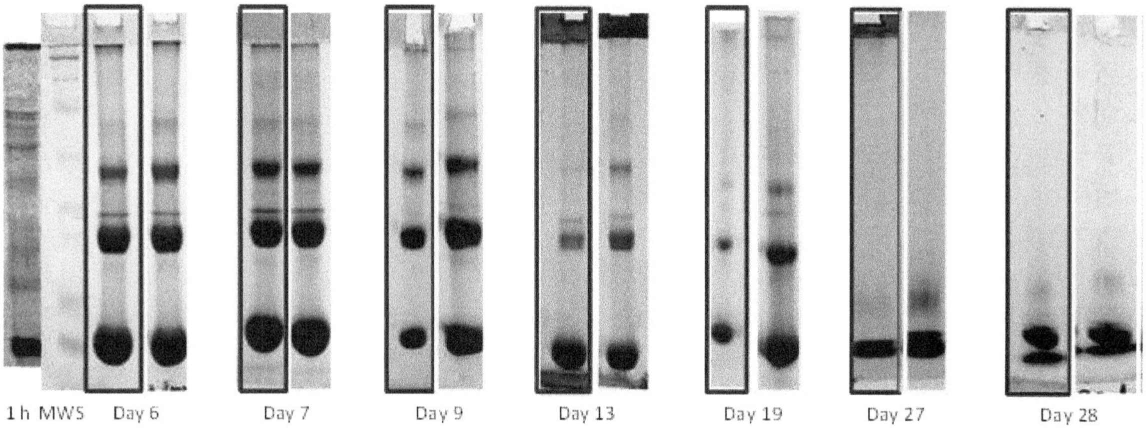

Figure 8. Band pattern of cytoskeletal proteins in RBCs at different days of incubation in the HPE from *Crataegus mexicana*, all lines at 100 mg/L. The protective effect of the extract on $FeSO_4$-induced protein damage is observed at different time periods of incubation. RBCs incubated in PBS (black margin) shown a decrease in size and band intensity as the incubation time is longer. RBCs incubated in HPE (white margin) shown that some bands are kept for a longer period of time. 10% SDS-PAGE. Experiments were run as triplicate: 1 μL per channel according to material and methods. MW= molecular weight standards.

Table 1. Scored distribution of RBCs protein bands with and without HPE at different times.

Molecule scored*	Time of incubation															
	1 h		day 6		day 7		day 9		day 13		day 19		day 27		Day28	
	C	E	C	E	C	E	C	E	C	E	C	E	C	E	C	E
band 1	+++	+++	+++	+++	+++	+++	+	+	+-	+-	+	+	+	-	-	-
New band								+	-	+	-	+-	-	-	-	-
band 2	+++	+++	+++	+++	+++	+++	+++	+++	+-	+	-	+-	-	-	-	-
band 3	+++	+++	+++	+++	+++	+++	+	+++	+-	++	-	+-	-	-	-	-
band 4	+++	+++	+++	+++	+++	+++	+	++	++	++	-	-	-	-	-	-
band 5	+++	+++	+++	+++	+++	+++	++	+++	+-	++	+	+++	-	-	-	-
New band									+	+	-	+-	-	-	-	-
New band												+-	-	-	-	-
band 6	+	+	+	+	+	+	-	+	-	-+	-	-+	-	+-	-	-
New band													+-	+	+-	+
band 7	+++	+++	+++	+++	+++	+++	++	+++	+++	+++	++	+++	++	++	++	++
New band													++	++	++	++

Values are the number of bands in RBCs indicated in each case. * Molecule scored according to the size and stain intensity as described in material and methods. C = control RBCs/PBS; E = RBCs/HPE. n = 3.

The banding pattern was compared between RBCs in PBS, and RBCs in HPE in an attempt to determine time dependent antioxidant protective effect on the cytoskeleton proteins. Meanwhile the banding pattern of some protein bands were observed in some cases up to the day 28 when the RBCs are incubated with HPE, the proteins in the control group were degraded, and some of them in a time point of incubation were absent (Figure 8). Because the band pattern was constant in each different concentration, it has been only shown the band pattern with HPE in a concentration at 100 mg/L. These observations were scored (Table 1).

Singh and Rajini (2008), in a similar experiment showed the antioxidant protective activity against ferrous- ascorbate using potato peel extract in different concentrations, but they did not determine time dependent changes during long periods of time. This makes clear that incubating RBCs in extracts containing polyphenols improve the antioxidant effect on the cytoskeletal proteins. Many pro-oxidants such as H_2O_2, Fe^{2+}-ascorbate and AAPH are capable to induce alterations in the erythrocyte membrane and a decrease in cytoskeletal proteins that lead a series of changes in the morphology and rheology properties of the cells (Singh and Rajini, 2008; Battistelli

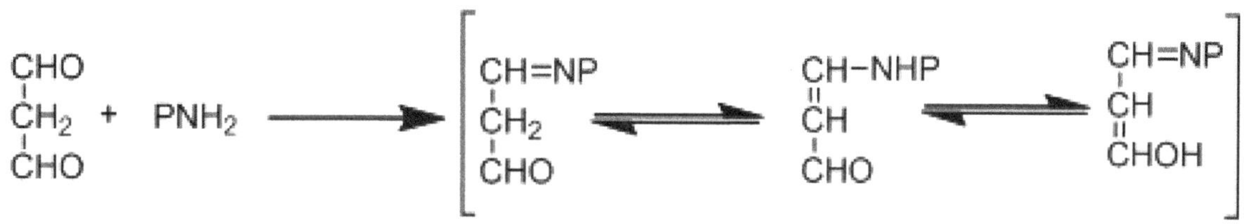

Figure 9. Formation of protein carbonyls by reaction of protein amino groups (PNH_2) with peroxidation products (MDA) of polyunsaturated fatty acids (PUFA).

et al., 2005; Srour et al., 2000). In these experiments, the use of the pro-oxidant $FeSO_4$ is responsible of the oxidative attack inducing this changes and subsequent non-enzymatic protein fragmentation and membrane deformation, but the HPE of *C. mexicana* is capable of protecting the RBCs of the oxidative process reducing the morphologic changes and retarding the hemolysis in the eryptosis process.

DISCUSSION

The most abundant cells in the blood are the RBCs with a 45% of its total volume, with an average life span of 120 days, and upon which main function is transportation of O_2 and CO_2. During their life span RBCs are exposed to many stressful situations through the capillaries, and suffering osmotic, energetic, and oxidative stress. Therefore, the high concentration of oxygen and haemoglobin promote the oxidative process and impaired RBCs functions and damage in the cell. The oxidative stress process is catalysed by Fe^{2+}, via the Fenton reaction to produce hydroxyl-radical; a process that has been well documented (Cheng and Li, 2007). The oxidation of the polypeptide backbone is $O_2^{\cdot-}$ dependent; is caused by the $^{\cdot}OH$-dependent abstraction of the α-hydrogen atom of an amino acid residue, this results in the formation of a carbon-centered radical that react with O_2 in a series of molecular reactions, forming alkoxyl radicals and aldehydes, the last one may participate in Schiff-base formation and Michael addition (Dean et al., 1997). Many of the steps in this pathway that are mediated by interactions with $HO_2^{\cdot}$ can also be catalysed also by Fe^{2+}. The intermediate radicals may undergo side chain reactions with other amino acid residues in the same or a different protein molecule to generate a new carbon-centered radical and the cleavage of the peptide bond of proteins. In addition, carbonyl groups may be introduced into proteins by reactions with aldehydes (MDA) produced during lipid peroxidation (Figure 9).

This carbonyl groups and MDA are associated with aging, oxidative stress, many chronic diseases, and human red blood cells (Berlett and Stadtman, 1997). Therefore, these oxidized products in the cell proteins of RBCs, can contribute in many morphologic changes and reduced life span of the erythrocyte. An increase in oxidation raises hydrophobicity, while a limited oxidation increases hydrophilicity. This change in turn produces inner and outer rheology alterations on the cell bilayer membrane, and such responses will further heighten the eryptosis process in RBCs.

Hawthorn is a source of natural polyphenol antioxidants (García-Mateos et al., 2012). Acetone extract from hawthorn *C. mexicana* inhibited oxidative stress induced by the pro-oxidant $FeSO_4$ in RBCs. It was demonstrated and confirmed by TBARS analysis, evidences in cell morphology, and protein electrophoretic pattern using different concentrations. Our results by TBARS analysis showed a relationship between low doses of HPE to inhibit MDA formation induced by $FeSO_4$ in the RBCs. These results are similar to those demonstrated in previous studies using homogenate from rat brain cells.

$FeSO_4$-induced lesion of the cell membrane; hemolysis and morphologic changes in the RBCs were inhibited in a concentration-dependent manner in the presence of HPE, and the life span of the erythrocytes were extended in the RBCs samples that contained the acetone extract. The morphologic changes associated to the eryptosis process were retarded during the time of the experiments. It was observed that the crenated cells, acanthocytes, and echinocyte were present for longer periods of time, before the hemolysis process and/or schistocyte formation took place. This protection supplied by the hawthorn extract was observed during the first 24 h of the experiment, as compared with the control group, where the eryptosis and life span of RBCs are reduced in a significant period of time. Zhu et al. (2002), determined the antioxidant protective effect of the cocoa feeding in rats, to inhibit erythrocyte hemolysis up to for 240 min in a dose dependent manner, suggesting that some polyphenols in HPE have the capacity of quenching free radicals in the cell membrane of the erythrocyte. Our study also showed a longer period of time on the inhibition of $FeSO_4$-induced hemolysis and morphologic changes in the erythrocytes, in a concentration-dependent manner up to for 28 days. This results support the fact that polyphenols in hawthorn fruits improve the antioxidant protective effects in RBCs, and may be working in an interactive way on their lipid bilayer of the cells.

Changes in the number, relative mobility and staining intensity of the protein bands observed in the electrophoretic gels, indicate that cytoskeletal proteins are susceptible to oxidation due the pro-oxidant $FeSO_4$ and during life span of the erythrocyte under the experimental condition. Furthermore, RBCs samples with the different hawthorn extract concentrations showed that many proteins bands are maintained during different times of the experiment and, also with a higher concentration of the extract the band pattern staining of some proteins was more intense. This may indicate that the extract contribute in some way in the inhibition of the protein oxidation for a longer and extended period of time, probably by protecting the –SH protein groups against oxidation. Also, it can inhibit the ability of proteases to degrade the oxidized forms of other proteins, and therefore to increase cellular resistance to oxidative stress (Asgary, et al., 2005). The delay time for protein oxidation and polymerization has a relationship on the stability and morphology of the erythrocytes. The lipid bilayer membrane and protein stability should be related to antioxidants concentration present in the HPE of *C. mexicana*, reducing cell dehydration, K loss and other osmotic cell changes (Wandersee et al., 2005). This protection in the cytoskeletal proteins is related to better form and shape of the erythrocyte as compared with those RBCs without the extract of *C. mexicana*. The observed reduction in the number and intensity of the band pattern could be related to increased proteolysis or instability of the absent bands, but is also unlikely in view of the finding that results were similar in all the experiments in RBCs comparing with or without the acetone extract. The pathophysiology of this changes, remains to be determined using specific and purified antioxidants.

The deformability of the red blood cells is an important factor in the rheology of the blood (Bronkhorst et al., 1995; Dupire et al., 2012), and it is related to many important cell properties, flow rate and blood diseases. The deformability is the result of several mechanical and geometrical properties of the RBCs, and is determinant in the stability of the membrane lipids and cytoskeletal proteins of the cell. Consequent disruption of the membrane-cytoskeleton by any oxidative process may be responsible for the bebbling formation in RBCs (Spangler, 2011; Liu et al., 2012). Thus, hydrodynamic stress and oxidation of proteins and phospholipids can also induce morphological and rheological changes of the cell. HPE of *C. mexicana*, was able to inhibit oxidative stress maintaining the structural and morphologic integrity of the cell, with less changes and damage in the morphology of the erythrocyte under the experimental conditions. HPE reduce the eryptosis process, retarding the formation of crenated cells and increasing the flow rate of the erythrocyte, contributing to the cell motion and avoid energetically a costly oxidation and deformation, to keep the cell elasticity and shape. This could be possible by the presence of antioxidants metabolites in the acetone HPE. This capacity of the hawthorn may be used to avoid erythrocyte adhesion and vasoocclusion in the blood stream in many blood diseases, and to inhibit peroxidation of biomembranes that can induce damage during ischemia/reperfusion, inflammation and aging.

In spite of this, is well known that echinocytes are formed after 3 to 4 days in whole blood kept at 4° C and within 24 h in blood incubated at 37° C (Brecher and Bessis, 1972), but with the acetone extract of *C. mexicana* the crenated cells, acanthocytes and echinocytes, were formed later compared with the control group and the shape of the cell membrane was kept for 28 days, showing inhibition during the time of the experiment of these morphological changes. Thus, may be, the oxidative process in the cells is inhibited by the antioxidants in the extracts, that maintain the morphology of the cell and its life span for a longer time, reducing the eryptosis process.

The attenuation of RBCs lesion was demonstrated in our study, and this is similar to those reported by Baek et al. (2012) in guinea pigs by haptoglobin therapy. However, the hawthorn extract is a natural product than can be consumed regularly by the human being in healthy and disease conditions. Wautier et al. (1981) mentioned the increased adhesion of erythrocytes to endothelial cells in diabetes mellitus and its relation to vascular complications, therefore, the nutraceutical application of hawthorn could inhibit the adhesion by keeping the RBC membrane from being oxidized by free radicals.

These results are in broad agreement with the published reports regarding anti-oxidant capacity of other fruits and vegetables extracts, such as potato peel extracts, cocoa extracts, and flaxseed oil. A large body of evidence implicates oxidative stress in the pathogenesis of red blood cell diseases, affecting their life span, shortening the eryptosis process, and increasing the hemolysis process. This process might also result in a systemic inflammatory response. Therefore, we favor the hypothesis that the observed protective effects on the morphology and life span of RBCs is due to the presence and collective action of the antioxidant molecules present in the acetone HPE. These results are related to the use of hawthorn in Mexico and other countries since ancient times in order to treat blood and other related systemic diseases. However many of this complex dynamics and interactions remain unexplored experimentally.

There is a need to identify more effective and novel compounds of fruits according to food consumption in traditional diets like the Mexican one, as well as how theirs constituents are acting against oxidative stress in cells, tissues and, in human diseases. The broad range of fruits and theirs parts to inhibit and remove ROS and intermediate radicals in many forms encourage continuing the purification of antioxidants in Mexican fruits and determining their mechanisms of action in lipids and proteins in red blood cells.

Conclusions

This study provides the first *in vitro* evidence for an active role of *C. mexicana* as antioxidant on erythrocytes. Hawthorn is a fruit with anti-oxidant potential, reduces morphological changes and extends life span of the red blood cells *in vitro*. Such protection could be due to the inhibition of membrane lipoperoxidation and oxidative damage in cytoskeletal proteins. The conjunction of antioxidants in this extract, and not just specific purified antioxidants, could be helpful to keep normal the rheology properties of the erythrocyte apparently without any pro-oxidative damage.

The antioxidants present in the acetone extract may contribute to protect the erythrocyte cell membrane against lipid peroxidation induced by oxidative stress, thereby participating in diminishing the morphological and many rheology changes that may happen in the pathogenesis of vasoocclusive diseases, such diabetes mellitus, metabolic syndrome, different types of anemia, and the inhibition of cell storage lesion in RBCs.

More studies are required to correctly identify the bioactive compounds of *C. mexicana* and how it contribute to the erythrocyte protection, and their mechanisms of action in the protection of the RBCs. This study give the scientific biochemical bases to use hawthorn extracts in a clinical setting, to prevent or treat many diseases and conditions associated to oxidative stress.

Conflict of interests

The authors did not declare any conflict of interest.

ACKNOWLEDGMENTS

This study was supported by grant 105663 from the Mexican Council of Science and Technology (CONACYT). We also acknowledge CONACYT for providing the financial support (54150) to José Antonio Banderas Tarabay during his Ph.D. studies.

REFERENCES

Aebi H, Bossi E, Cantz M, Matsubara S, Suter H (1968). A catalasemia in Switzerland. In: Hereditary Disorders of Erythrocyte Metabolism. Beutler E, editor. pp. 41-65.

Andrade-Cetto A, Heinrich M (2005). Mexican plants with hypoglycaemic effect used in the treatment of diabetes. J Ethnopharmacol. 99: 325-348.

Arrieta JD, Siles-Barrios J, García-Sánchez B, Reyes-Trejo and Sánchez-Mendoza ME (2010). Relaxant effect of the extracts of *Crataeugus mexicana* on guinea pig tracheal smooth muscle. Pharmacogn. J. 2: 40-46.

Asgary S, Naderi GH, Askari N (2005). Protective effect of flavonoids against red blood cell hemolysis by free radicals. Exp. Clin. Cardiol. 10: 88-90.

Baek JH, D´Agnillo F, Vallelian F, Pereira CP, Williams MC, Jia Y, Schaer DJ, Buehler PW (2012). Hemoglobin-driven pathophysiology is an in vivo consequence of the red blood cell storage lesion that can be attenuated in guinea pigs by haptoglobin therapy. J. Clin. Invest. 122:1444-1458.

Ballas SK (1977). Abnormal erythrocyte membrane protein pattern in severe megaloblastic anemia. Clin. Res. 25: 333A.

Battistelli M, De Sanctis R, De Bellis R, Cucchiarini L, Dachá M, Goobi P (2005). Rhodiola rosea as antioxidant in red blood cells: ultra-estructural and hemolytical behaviour. Eur. J. Histochem. 49:243-254.

Berlett BS, Stadtman ER (1997). Protein oxidation in aging, disease, and oxidative stress. J. Biol. Chem. 272: 20313-20316.

Brecher G, Bessis M (1972). Present status of speculated red cells and their relationship to the discocyte-echinocyte transformation. A critical review. Blood 40:333-343.

Bronkhorst PJH, Streekstra GJ, Grimbergen J, Nijhof EJ, Sixma JJ, Brakenhoff GJ (1995). A new method to study shape recovery of red blood cells using multiple optical trapping. Biophys. J. 69: 1666-1673.

Cai Y, Luo Q, Sun M, Corke H (2004). Antioxidant activity and phenolic compounds of 112 traditional Chinese medicinal plants associated with anticancer. Life Sci. 74:2157-2184.

Cañas P, Olivares M, Celedón C, Llaguno S, Valenzuela A (1988). Actividad del sistema de defensa antioxidante y agresión oxidativa en eritrocitos de recién nacidos de bajo de peso y distintas edades gestacionales. Rev. Chil. Pediatr. 59:83-88.

Cheng Z, Li Y (2007). What is responsive for the initiating chemistry of iron-mediated lipid peroxidation: An update. Chem. Rev. 107: 748-746.

Dean RT, Fu S, Stocker R, Davies MJ (1997). Biochemistry and pathology of radical-mediated protein oxidation. Biochem. J. 324:1-18.

Domínguez M, Nieto A, Marin JC, Jeffery E, Céspedes CL (2005). Antioxidant activities of extracts from *Barkleyanthus salicifolius* (*Asteraceae*) and *Penstemon gentianoides* (*Scrophulariaceae*). J. Agric. Food Chem. 53: 5889-5895.

Dupire J, Socol M, Viallat A (2012). Full dynamics of a red blood cell in shear flow. PNAS 109: 20808-20813.

Edwards JE, Brown PN, Talent N, Dickinson TA, Shipley PR (2012). A review of the chemistry of the genus *Crataegus*. Phytochemistry 79: 5-26.

Esterbauer H, Cheeseman KH (1990). Determination of aldehydic lipid peroxidation products. Methods Enzymol. 186: 407-421.

Folin C, Ciocalteau V (1927). Tyrosine and trypthopan determination in proteins. J. Biol. Chem. 73: 627-650.

García-Mateos R, Aguilar-Santelises L, Soto-Hernández M, Nieto-Angel R, Kite G (2012). Total phenolic compounds, flavonoids and antioxidant activity in the flowers of *Crataegus* spp. from Mexico. Agrociencia 46: 651-662.

Hebbel RP, Leung A, Mohandas N (1990). Oxidation-induced changes in microrhelogic properties of the red blood cell membrane. Blood 76: 1015-1020.

Hillery CA, Du MC, Montgomery RR, Scott JP (1996). Increased adhesion of erythrocytes to components of the extracellular matrix. Isolation and characterization of a red blood cell lipid that binds thrombospondin and laminin. Blood 87 (11): 4879-4886.

Jangde R (2011). An Overview of resealed erythrocyte for cancer therapy. Asian J. Res. Pharm. Sci. 1: 83-92.

Kempe DS, Lang PA, Duranton C, Akel A, Lang KS, Huber SM, Wieder T, Lang F (2006). Enhanced programmed cell death of iron-deficient erythrocytes. FASEB J. 20: 368-370.

Khattak KF, Simpson TJ, Ihasnullah (2008). Effect of gamma irradiation on the extraction yield, total phenolic content and free radical-scavenging activity of *Nigella staiva* seed. Food Chem. 110: 967-972.

Kibanova D, Nieto-Camacho A, Cervini-Silva J (2009). Lipid peroxidation induced by expandable clay minerals. Environ. Sci. Technol. 43: 7550-7555.

Kim HJ, Chen F, Wang X, Chung YH, Jin Z (2005). Evaluation of antioxidant activity of vetiver (*Vetiveria zizanioides* L) oil and identification of its antioxidant constituents. J. Agric. Food Chem. 53: 7691-7695.

Kowalczyk E, Kowalski J, Blaszczyk J, Gwoździński L, Cieéwierz J, Sienkiewicz M (2012). Estimation of cell membrane properties and erythrocyte red-ox balance in patients with metabolic syndrome. Mol.

Biol. Rep. 39: 11113-11118.

Krishnaveni G, Krishnamoorthy B, Muthukumaran M (2013). Future Prospectus in Targetting Drug Delivery: A Review on Resealed Erytrocytes - A Promising Carrier. Schol. Acad. J. Pharm. 2: 81-88.

Laemmli UK (1970). Cleavage of structural proteins during the assembly of the head of bacteriophage T4. Nature 277: 680-688.

Lam RYY, MPhil, Woo AYH, Leung Po-Sing, Cheng CHK (2007). Antioxidant actions of phenolic compounds found in dietary plants on low-density lipoprotein and erythrocytes *in vitro*. J. Am. Coll. Nutr. 26: 233-242.

Lang F, Lang KS, Lang PA, Huber SM, Wieder T (2006). Mechanisms and significance of eryptosis. Antioxid. Redox Signal. 8: 1183-1192.

Liu L, Yang W, Fu J, Yu M, Huang Q, Wang D, Xu J, Chun D, Yao P, Huang F (2012). Effects of flaxseed oil on anti-oxidative system and membrane deformation of human peripheral blood erythrocytes in high glucose level. Lipids Health Dis. 11: 88.

Lowry OH, Rosebrough NJ, Farr AL, Randall RJ (1951). Protein measurement with the Folin phenol reagent. J. Biol. Chem. 193: 265-275.

Lynch RE, Lee GR, Cartwright GE (1976). Inhibition of superoxide dismutase of methemoglobin formation from oxyhemoglobin. J. Biol. Chem. 251: 1015-1018.

Mackenzie SB (2000). El eritrocito. En: Hematología Clínica (2nd ed.) México: Ed. El Manual Moderno. pp. 39-60.

Matteucci E, Cocci F, Pellegrim L, Gregori G, Giampietro O (1995). Measurement of ATPase in red cells: setting up and validation of a highly reproducible method. Enzyme Protein 48: 105-119.

Méndez-Iturbide D, Banderas-Tarabay JA, Nieto-Camacho A, Rojas-Chávez A, García-Meza MG (2013). Antioxidant capacity of extracts from hawthorn (*Crtaegus mexicana*) skin. Afr. J. Food Sci. 7:150-158.

Middleton IJ, Kandaswami C, Theoharides TC (2000). The effects of plant flavonoids on mammalian cells: Implications for inflammation, heart disease, and cancer. Pharmacol. Rev. 52: 673-751.

Nagababu E, Gulyani S, Earley CJ, Cutler RG, Mattson MP, Rifkind JM (2008). Iron-deficiency anemia enhances red blood cell oxidative strees. Free. Radic. Res. 42:824–829.

Ng TB, Liu F, Wang ZT (2000). Antioxidative activity of natural products from plants. Life Sci. 66: 709-723.

Nugraheni M, Santoso U, Suparmo, Wuryastuti H (2011). *In vitro* antioxidant, antiproliferative and apoptosis effect of *Coleus tuberosa* L. Afr. J. Food Sci. 5: 232-241.

Ohkawa H, Ohishi N, Yagi K (1979). Assay for lipid peroxides in animal tissues by thiobarbituric acid reaction. Anal. Biochem. 95: 351-358.

Pandey KB, Rizvi SI (2011). Biomarkers of oxidative stress in red blood cells. Biomed. Pap. Med. Fac. Univ. Palacky Olomouc Czech. Repub. 155:131-136.

Salawu SO, Akindahunsi AA, Sanni DM, Decorti G, Cvorovic J, Tramer F, Passamonti S, Mulinacci N (2011). Cellular antioxidant activities and cytotoxic properties of ethanolic extracts of four tropical green leafy vegetables. Afr. J. Food Sci. 5: 267-275.

Singh N, Rajini PS (2008). Antioxidant-mediated protective effect of potato peel extract in erythrocytes against oxidative damage. Chem. Biol. Interact. 173:97-104.

Sorata Y, Takahama U, Kimura M (1984). Protective effect of quercetin and rutin on photosensitized lysis of human erythrocytes in the presence of hematoporphyrin. Biochim. Biophys. Acta 799: 313-317.

Spangler EJ (2011). Computer simulation of blebbing in erythrocyte membranes. Proccedings of the National Conference on Undergraduate Research (NCUR). Ithaca College, New York.

Srour T, Bilto YY, Juma M, Irhimeh MR (2000). Exposure of human erythrocytes to oxygen radicals causes loss of deformability, increased osmotic fragility, lipid peroxidation and protein degradation. Clin. Hemorrheol. Microcirc. 23:13-21.

Van Asbeck BS, Hoidal J, Vercellotti GM, Schwartz BA, Moldow CF, Jacob HS (1985). Protection against lethal hyperoxia by tracheal insufflation of erythrocytes: Role of red cell glutathione. Science 227: 756-759.

Wandersee NJ, Punzalan RC, Rettig MP, Kennedy MD, Pajewski NM, Sabina RL, Scott JP, Low PS, Hillery CA (2005). Erythrocyte adhesion is modified by alterations in cellular tonicity and volume. Br. J. Haematol. 131: 366-377.

Wautier JL, Paton RC, Wautier MP, Pintigny D, Abadie E, Passa P, Caen JP (1981). Increased adhesion of erythrocytes to endothelial cells in diabetes mellitus and its relation to vascular complications. N. Engl. J. Med. 305:237.

Willis MS, Wians FH (2003). The role of nutrition in preventing prostate cancer: a review of the proposed mechanism of action of various dietary substances. Clin. Chim. Acta 330:57-83.

Zhu QY, Holt RR, Lazarus SA, Orozco TJ, Keen CL (2002). Inhibitory effects of cocoa flavanols and procyanidin oligomers on free radical-induced erythrocyte hemolysis. Exp. Biol. Med. 227: 321-329.

Zhu QY, Schramm DD, Gross HB, Holt RR, Kim SH, Yamaguchi T, Kwik-Uribe CL, Keen CL (2005). Influence of cocoa flavanols and procyanidins on free radical-induced human erythrocyte hemolysis. Clin. Dev. Immunol. 12:27-33.

3

Identification of the lactic microorganism responsible for milk coagulation in Abeche (Chad)

Doutoum A. A.[1], Tidjani A.[2] , Abdelaziz Arada I.[3], Tidjani S. M. T.[3], Balde M.[1], Sylla K. S. B.[1], Seydi Mg[1] and Toguebaye B. S.[4]

[1]Laboratory of Hygiene and Food Industry of Animal Origin (HIDAOA). EISMV, P.O Box 5077, Dakar, Senegal.
[2]Faculty of Human Health, University of N'Djamena, P. O. Box: 1117, N'Djamena, Chad.
[3]Laboratory of Microbiology Immunology Infectious Pathology (MIPI). EISMV, P.O. Box 5077, Dakar, Senegal.
[4]Department of Animal Biology, Faculty of Science and Technology, University Cheikh Anta Diop (UCAD), Senegal.

Milk and dairy products are consumed by many people in its various forms in Chad. This product has a major share in the labor market. Milk is turned into curds of various types and other dairy products empirically, but there is not much information reported about the lactobacilli species responsible for curd production in Chad. Therefore we have chosen to locate and identify them. We started with the suburban area of Abeche. This research conducted aimed at identifying the lactic acid bacterial strains in 100 samples of milk curd (laban) obtained from the suburban area of Abeche, Chad. Identification of lactic acid bacteria was made using the API 20 STREP and API 50 CHL galleries. The bacteria were cultured in de Man, Rogosa and Sharpe (MRS) and M17 medium. In addition to the identification, confirmation was done at the Pasteur Institute of Dakar with the API MT web software. Results have identified *Lactococcus lactis* subsp. *lactis*, *Lactococcus lactis* subsp. *cremoris*, *Lactococcus lactis* subsp. *diacetylactis* as well, other contaminants were identified, including *Enterococcus durans* and *Enterococcus faecium*. *Lactobacillus plantarum* identified in all samples analyzed.

Key words: Laban, *Lactococcus*, *Lactobacillus*, Identification, Abeche.

INTRODUCTION

Milk is the staple food in a domestic economy. In Chad, the local artisanal dairy industry has developed considerably based on micro-enterprise collection and dairy processing (Koussou, 2008). Milk is a highly nutriatious food for its richness in carbohydrates, fats, vitamins and minerals (Aggad et al., 2009; Ahmed et al., 2010). Indeed, the dairy product comes in many forms depending on the region and varied tastes. The curd is especially appreciated by the Chadian consumers. Fermented milk has always been traditionally consumed in many countries (Nakasaki et al., 2008). Most studies on traditional fermented products showed that mesophilic lactic acid bacteria are the main cause of these fermentations group (Ben Amor et al., 1998). Lactic acid bacteria also through their lipase can decompose fat and free fatty acids in milk, resulting in the appearance of rancid odor in the dairy product. When lipolysis is not controlled or exaggerated, there is the appearance of strong and rancid odors (Vignola, 2002). Among the lactic acid bacteria with habitat as milk, we have *Streptococcus*, *Lactobacillus*, *Enterococcus*, *Leuconostoc* and *Aerococcus* (Luquet and Corrieu, 2005). In Chad, especially in Abeche, no study has been done about this subject. This lack of scientific information led us to seek and identify these bacteria in different varieties of "Laban", which is very well consumed by the population. The objective of this study was therefore

Table 1. Biochemical analysis of isolated strains with API 20 STREP.

Biochemical parameter	SLA1	SLA2	SLA3	SLA4	SLA5	SLA6	SLA7	SLA8	SLA9	SLA10
Hydrolysis hyppurate	+	+	+	+	+	+	+	+	+	+
Hydrolysis of esculin	+	+	+	+	+	+	+	+	+	+
Galactosidase	-	-	-	-	-	-	-	-	-	-
Glucuronidase	-	-	-	-	-	-	-	-	-	-
Phenylalanine deaminase	-	-	-	-	-	-	-	-	-	-
Leucinearyl aminase	+	+	+	+	+	+	+	+	+	+
Hydrolysis of arginine	+	+	+	+	+	+	+	+	+	+
Hemolysis	-	-	-	-	-	-	-	-	-	-
Profile fermentation										
Ribose	+	+	+	-	-	+	+	+	-	+
Arabinose	+	+	+	+	+	+	+	+	+	+
Mannitol	-	+	-	+	+	+	+	-	-	+
Sorbitol	-	-	-	-	-	-	-	-	-	-
Lactose	+	+	+	+	+	+	+	+	+	+
Trehalose	+	+	+	+	+	+	+	+	+	+
Inulin	+	+	+	+	+	-	+	+	+	+
Raffinose	-	-	-	-	-	-	-	-	-	-
Starch	-	-	-	-	+	+	-	+	-	-
AGlycogene	-	-	-	-	-	-	-	-	+	-

SLA, Strain lactic Abeche

to identify some species of lactic acid bacteria present in "Laban".

MATERIALS AND METHODS

Sampling

In the suburban area of Abeche, we collected aseptically and randomly 100 samples of 250 ml of milk curd in sterile plastic jars. Before introducing the curd in the jars, the technician washes hands first with soapy water and take precautions to avoid contact with the curds. After collection, the samples were quickly introduced into a cooler box containing bottles of dry ice and transported at a low temperature to the laboratory for analyses.

Isolation and Identification of lactic acid bacteria

de Man, Rogosa and Sharpe (MRS) and M17 as agar media and broth were used to isolate lactic acid bacteria after incubating at 30°C for 48 h. One milliliter of each milk sample was homogenized in 9 ml of sterile saline. Lactic acid bacteria were enumerated by serial dilutions in sterile saline that were placed on the appropriate agar medium. Gallery API 20 STREP and API 50 CH gallery-L with their reagents (Biomerieux, France) were used in the identification using software API Web TM (Biomerieux, France).

RESULTS

Isolation and identification of lactic flora

The Table 1 shows the results of biochemical analyses of isolated strains with API 20 STREP. The test showed that the positive and negative reactions tend to equilibrate.

The Table 2 shows the digital profile and identification of strains. It was observed that there was a predominance of L*acotococcus lactis* subsp *lactis*. The results of analyzes based on the biochemical and physiological criteria of isolated bacteria are shown in Table 3. Captured in Table 4 are the biochemical analyses of isolated strains with API 50 STREP.

DISCUSSION

The results of this study have identified six strains of *Lactococcus lactis* subsp. *lactis*, two strains of *Lactococcus lactis* subsp. *cremoris* and one flora contamination represented by a strain of *Enterococcus faecium* and *Enterococcus durans* strain. Ben Amor et al. (1998) found similar results of 12 isolates. They showed that the mesophilic lactic flora is represented by three species of *Lactococcus* flora associated with contamination of the *Enterococcus genus*. Moreover, Ngassam's studies (2007) on 10 samples of curled have identified a single species in all samples (100%): *Lactococcus lactis* subsp. *lactis* with a percentage of identification of 98%; while El-Baradei et al. (2007) have identified in the Egyptian traditional curd milk (Domiati) the species *Lactococcus lactis* subsp. *lactis*, *Lactococcus garvieae* and *Lactococcus raffinolactis*. Previously Lafarge et al. (2004) have also determined the curd pre- dominant presence of *Lactococcus lactis* subsp. *lactis* species by molecular methods. Also, further more studies (Katinan et al., 2012; Conte, 2008; Badis et al., 2005) reported the prevalence of *Lactococcus*

Table 2. Digital profile and identification of strains of *Lactococci* by the API 20 STREP galleries.

Souche	Identification			% Identification	Tests against
	Genus	Species	Subsp.		
SLA1	*Lactococcus*	*lactis*	*lactis*	43.8	2
SLA2	*Lactoccocus*	*lactis*	*cremoris*	82.2	1
SLA3	*Lactococcus*	*lactis*	*lactis*	95.3	1
SLA4	*Lactococcus*	*lactis*	*lactis*	97.5	2
SLA5	*Lactococcus*	*lactis*	*lactis*	99.8	0
SLA6	*Lactococcus*	*lactis*	*lactis*	59.9	2
SLA7	*Lactococcus*	*lactis*	*cremoris*	85.7	1
SLA8	*Lactococcus*	*lactis*	*lactis*	98.9	0
SLA9	*Enterococcus*	*faecium*		93.6	3
SLA10	*Enterococcus*	*durans*		52.8	3

SLA, Strain lactic Abeche.

Table 3. Biochemical analysis of isolated strains with API 50 STR.

Growth condition	LBA1	LBA2	LBA3	LBA4	LBA5	LBA6	LBA7	LBA8	LBA9	LBA10
15°C	+	+	+	+	+	+	+	+	+	+
45°C	-	-	-	-	-	-	-	-	-	-
6.5% NaCl	+	+	+	+	+	+	+	+	+	+
pH 9.6	-	-	-	-	-	-	-	-	-	-
Milk sunflower	ACr	ACr	ACr	ACr	ACr	ACr	ACr	ACr	ACr	ACr
ADH	-	-	-	-	-	-	-	-	-	-
Type of fermentation	Homo	Homo	Homo	Homo	Homo	Homo	Homo	Homo	Homo	Homo

DHA, Arginine dihydrolase

Table 4. Biochemical analysis of isolated strains with API 50 STREP.

Profile fermentation	LBA1	LBA2	LBA3	LBA4	LBA5	LBA6	LBA7	LBA8	LBA9	LBA10
D-Arabinose	-	-	-	-	-	-	-	-	-	-
L-Arabinose	+	-	+	+	+	+	+	-	-	+
Ribose	+	+	+	+	+	+	+	+	+	+
D-Xylose	-	-	-	-	-	-	-	-	-	-
L-Xylose	-	-	-	-	-	-	-	-	-	-
Adonitol	-	-	-	-	-	-	-	-	-	-
Galactose	+	+	+	+	+	+	+	+	+	+
D-Glucose	+	+	+	+	+	+	+	+	+	+
D-Fructose	+	+	+	+	+	+	+	+	+	+
D-Mannose	+	+	+	+	+	+	+	+	+	+
L-Sorbose	-	-	-	-	-	-	-	-	-	-
Rhamnose	+	-	+	+	+	+	+	-	-	+
Inositol	-	-	-	-	-	-	-	-	-	-
Mannitol	+	+	+	+	+	+	+	+	+	+
Sorbitol	+	-	+	+	+	+	+	-	-	+
N-Acetyl Glucosamine	+	+	+	+	+	+	+	+	+	+
Esculin	+	+	+	+	+	+	+	+	+	+
Maltose	+	+	+	+	+	+	+	+	+	+
Lactose	+	+	+	+	+	+	+	+	+	+
Sucrose	+	+	+	+	+	+	+	+	+	+
Trehalose	+	+	+	+	+	+	+	+	+	+
Gluconate	±	+	+	+	+	+	±	+	+	+

in milk. Ouadghiri et al. (2009) studies indicate that Morocco "Lben" is typically dominated by *Lactococcus lactis* species (41%), *Leuconostoc pseudomes enteroides* (36%) and *Lactobacillus plantarum* (15%). Our results are also consistent with those found by Biatcho (2006) with lactobacilli in large quantities in the craft milk. However, dairy technology, lactococci play a role of "bioconservator." It is recognized *Lactococcus lactis* having a role of initiator agent for curdling milk.

Compared to biochemical tests and API 50 CHL tests, our results show that the germs identified belong to the *Lactobacillus genus*, including *Lactobacillus plantarum*. The percentage of identification obtained by the API software web MT are the order of 99.9% for most strains except LBA2, LBA8 and LBA9 that are a little discriminating profile compared to *Cornobacterium pisicola* Identification with a percentage of 52.6% and 3 test against (ribose, melibiose and sorbitol). Our results are similar to those of Ben Amor et al. (1998). These authors on eight strains were able to identify seven strains of *Lactobacillus plantarum* and strain *Cornabacterium pisicola* with a percentage of 50.7% and identification of three tests against. *Lactobacillus plantarum* has also been isolated (16 isolates) in raw goat milk from two Algerian goat populations (Badis et al., 2005). Ngassam (2007) meanwhile isolated a variety of six species of the *Lactobacillus genus*. The differences in species and their diversity may be explained by the origins of milk.

Conclusion

This work has identified certain lactic acid bacteria in curd available in Abeche. The study identified some lactic acid bacteria in the "Laban" consumed in of Abeche town. These include *Lactoccocus lactis* subsp. *lactis Lactoccocus lactis* subsp. *cremoris* and *Lactobacillus plantarum*. But these results are far to be completed because the country is too large as well as the agro-ecological zones are varied. Knowledge of most lactic acid bacteria in Chad will help in the production of dairy products with longer shelf-life. Other studies, including the characterization of molecular biology of lactic acid bacteria is necessary to provide benefit information to improve the technological and sanitary quality of milk.

REFERENCES

Aggad H, F Mahouz, YA Ammar, Kihal M (2009). Evaluation of the hygienic quality of milk in western Algeria. Rev. Med. Vet. 160(12): 590-595.

Ahmed AI, Mohammed AA, Faye B, Blanchard L, Bakheit SA. (2010). Assessment of quality of camel milk and gariss, north Kordofan State, Sudan. Res. J. Anim. Vet. Sci. 5(1): 18 – 22.

Badis A, Laouabdia-Sellami N, Guetarni D, Kihal M, Ouzrout R (2005). Phenotypic characterization of isolated from raw goat's milk from two local goat populations "Arabia and Kabyle" lactic acid bacteria. Sci. Technol. C - No. 23, June (2005). pp. 30-37.

Ben Amor K, Cornelieus CA, Mahjoub Thonart (1998). Identification of the lactic flora of traditional Tunisian fermented milk (Lben) and evaluation of flavoring compounds. Microb. Hyg. Ali. 10 (27) 31 - 36.

Biatcho NS (2006). Assessing the implementation of hygiene in a traditional dairy Dakar "The DIRFEL" crop milk to its transformation into curd called "SOW PUR" Thesis: Med. Vet. Dakar 11.

Conte S (2008). Evolution organoleptic, physico-chemical and microbiological characteristics of traditional quail milk. Memory depth of livestock degree. Cheikh Anta Diop University, Senegal, pp. 45.

El-Baradei G, Delacroix-Buchet A, J-C Ogier (2007). Biodiversity of Bacterial Ecosystems in Traditional Egyptian Domiati cheese. Egypt: American Society for Microbiology. Appl. Environ. Microbiol. 73(4): 1248-1255.

Katinan Coulibaly Remi Sadat AW, CHATIGRE Kouamé Olivier Kouassi Bohoussou Maxime ASSIDJO Nogbou Emmanuel. (2012). Evaluation of chemical and microbiological quality of curds handicrafts produced and consumed in the city of Yamoussoukro, Cote d'Ivoire. J. Appl. Biosci. 55: 4020-4027.

Koussou MO (2008). Dynamics of changes in the sector of livestock in Chad: the case of the dairy industry of N'Djamena. Doctoral thesis, Institute of Sciences and Industries of the living and the environment (Agro-Paris-Tech). pp. 239.

Lafarge V, Ogier JC, Girard V, Maladen V, Leveau JY, Delacroix-Buchet A (2004). The potential of TTGE to study bacterial some raw milk and dairy products.169 –178 - In: Milk 84. - Paris: INRA, EDP Sciences.

Luquet F, Corrieu MG. (2005). Lactic acid bacteria and probiotics. Edition 8c Tec Doc Lavoisier. Paris. pp. 307.

Nakasaki K, Yanagisawa M, Kobayashi K (2008). Microbiological quality of fermented milk produced by repeated-batch culture. J. Biosci. Bioeng. 105(1):73 – 76.

Ngassam CT (2007). Contribution to the study of microbiological characteristics of lactic acid bacteria fermented milk crafts in Senegal: case of Niayes area. Dakar: Th Méd. Vet N°. 19.

Ouadghiri M, Vancanneyt M, Vandamme P, Naser S, Gevers G, Lefebvre K, Swings J, Amar M (2009). Identification of lactic acid bacteria in Moroccan raw milk and traditionally fermented skimmed milk _lben'. J. Appl. Microbiol. 106:486–495

Vignola CL (2002). Sscience and technology of the milk, processing of milk - Quebec: Foundation and dairy technology of the Quebec. pp. 600.

Analysis of quality attributes of banana drinks blended with aqueous extracts of ginger

Yéo Mohamed Anderson[1] , Chatigre O.[1], Elleingand E.[2], Koni Pascal[3] and Koffi Ernest[1]

[1]Laboratoire de Biochimie Sciences des Aliments, UFR Biosciences, Université Félix Houphouet Boigny de Cocody, Abidjan, Côte d'Ivoire.
[2]Laboratoire de Genie chimie et Agroalimentaire, Institut Nationale Polytechnique Félix Houphouet Boigny de Yamoussoukro, Côte d'Ivoire.
[3]Société Ivoirienne de Technologie Tropicale (I2T), Côte d'Ivoire.

This study aimed to investigate the effect of ginger (*Zingiber officinale*) on the microbiological, physicochemical and sensory quality of banana (*Musa* spp .) drink. Production of banana juice was achieved with hot and unheated water extraction. The various juices obtained were supplemented with aqueous extracts of ginger in concentrations of 2.5 and 10%, stored and analyzed periodically for microbiological and physicochemical (pH, vitamin C, proteins, total sugars, Mg, k, Ca, Fe, Na, Zn) characteristics. The results showed that ginger addition to either hot or unheated banana juice extraction treatment significantly reduced growth of microorganisms as reflected in total viable counts from day 4 of storage. Vitamin C, proteins and mineral contents increased as compared to control samples. The results of this study suggest that the incorporation of ginger extracts into banana drink could be an effective means of preserving banana drinks.

Key words: Quality, banana drink, blended, ginger.

INTRODUCTION

Plantain is one of the primary sources of food product intended for consumption in Côte d' Ivoire. It is the second food crop in Côte d' Ivoire with an annual production of 1.6 million tons. Its main production area is the midwest (ANADER, 1998).

Although it occupies a prominent place in the diet of Ivorian, production of plantain remains seasonal. This seasonality is characterized by a period of relative abundance (October to March) where post-harvest losses are high. During this period, the sales prices are low and another period of shortage in urban markets (April-September) marked by the high cost of banana and inaccessibility of the product for low income households.

Thirty to forty percent of this production is unfortunately lost before consumption, mainly due to poor harvest, transport and storage (Lépengué, 1999). Indeed, these losses are due to storage of bananas at room temperature (28-35°C) than the recommended 8-12°C (Varoquaux et al., 2002) temperature, the remoteness of production and lack of accessible roads to urban areas.

Moreover, not processing or very low level of trans-formation of plantain is an obstacle to the develo-pment of the plantain sector in Africa in general and in Côte d'Ivoire in particular.

Thus, "the production of plantain juice clarified and stabilized" will not only help reduce post-harvest losses but also offer a new product that will improve the income of the plantain industry. However, the juice of plantain like that of *Hibiscus sabdariffa* (zobo) (Omemu et al., 2006) has a shelf life of not more than 24 h if not refrigerated.

To solve this problem of conservation, Dougheri et al. (2007) have used benzoic acid thus extending the shelf life of zobo (a drink made from the calyces of *Hibiscus subdariffa*) to 14 days. They were able to preserve the same the organoleptic qualities beyond 14 days. However, the use of these chemicals has various negative effects on the health of consumers (Adesokan et al., 2010). In this regard, Kolapo et al. (2007) preferred natural plant extracts with antimicrobial properties. The aim of the present study was to determine the effect of ginger (*Zingiber officinale*) on microbiological, physicochemical and sensory quality of fruit juice of plantain (*Musa* spp.)

MATERIALS AND METHODS

Sampling

Plantain variety "Corne" and the roots of ginger (*Z. officinale*), were purchased at market Guro Adjamé (Abidjan) in Côte d'Ivoire. These samples were carefully sorted and manually washed with bleach, then rinsed thoroughly with distilled water before use.

Ginger juice extraction

One hundred grams of ginger roots were cut into small pieces with a sterile knife.

The ground material obtained is then mixed in 50 ml of distilled water using a "Kenwood blender" having cutting blades until a fine paste. The paste is dissolved in 100 ml of distilled water and filtered through a muslin cloth. The filtrate is stored in sterile bottles in a refrigerator at 5°C until use (Adesokan et al., 2010).

Banana juice production

A cold and hot water extraction methods were performed. For both methods, 500 g banana pulp were dispersed in 1 L of distilled water in a blender brand "Phillips" . The resulting mash was filtered using a sieve of 0.250 mm diameter. The filtrate was then centrifuged at a speed of 4000 rev/min for 30 min with a centrifuge brand Heraeus Christ, Bofuge III. Residues separated juice was placed in a sterile bottle and ginger juice added at different concentrations (2.5 and 10%). Juice of plantain non-supplemented ginger served as controls.

Physicochemical analyses

Measurement of pH at 25°C was done for each day for 15 days using a pH meter, according to AOAC (1990). Proteins were determined by Kjédhal method (BIPEA 1976). The dosage of vitamin C was produced according to the AFNOR method (1995). Mineral elements: iron, potassium, calcium, magnesium, sodium, copper, zinc, manganese were determined by atomic absorption spectrophotometer according to the method of AOAC (1990) and Abulude et al. (2007).

Microbiological analyses

The various banana juices were carefully diluted in sterile distilled water. Appropriate dilutions were poured onto agar PDA (potato dextrose agar) and MacConkey respectively for the enumeration of total coliforms and flora. Boxes cast in MacConkey medium were incubated at 35°C for 24 h while those of PDA medium were incubated at 28°C for 72 h and supplemented with streptomycin to prevent bacterial growth (Adesokan, 2005).

Sensory analysis

Different extracts of banana juice were analyzed by a panel of 17 people previously formed by the Ivorian society of Tropical Technology (I2t). The samples were evaluated on a scale of 0 to 10.

Statistical analysis

STATISTICA 10 software was used to perform statistical analyses. Comparisons between the dependent variables were determined by analysis of variance and Duncan's test and Newman Keuls. Statistical significance was defined at 5%. The results for sensory tests were subjected to analysis of variance using the software Tastel +

RESULTS

The physicochemical analyzes of the various plantain drinks are shown in Table 1. The results show that the vitamin C content (mg/100 g) increased from 102.61 to 135.01 in hot extracts samples (C) and from 108.01 to 264.46 in cold extracts samples (F). The lowest value, 102.61 is the ascorbic acid content of hut sample treatment with no ginger (C0) and the highest value, 264.46 is that of cold sample treatment with 10% ginger (F10). The pH values vary slightly from 4.75 to 4.99 in samples F0, F10 and C10. Total sugars in turn have values ranging between 51.4 mg/100 ml in the C0 sample and 67.2 mg/100ml in the F2,5.

Moreover the results indicate that the lowest protein content (0.25%) is that of the sample C0 while the highest content 0.65% is that of cold extraction treatment supplemented with 10% ginger. Plantain drinks contain varying concentrations of Mg^{2+}, K^{+}, Ca^{2+}, Fe^{2+}, Na^{+} and Zn. The sample F10, that is, the cold water extract treatment containing 10% ginger has a higher concentration than the other juice samples. The enumeration of the total flora, expressed in cfu/ml of various banana juice samples are summarized in Table 2.

These results show that there is a steady increase in the total flora with time of storage. However, the samples

Table 1. Contents of proteins, total sugars, minerals, vitamin C and pH of banana juice supplemented with ginger extracts.

Parameters	C0	C2,5	C10	F0	F2, 5	F10
Vitamine C (mg/100 ml)	102.61 ± 0.602^{d}	108.01 ± 0.13^{c}	135.01 ± 0.25^{b}	108.01 ± 0.01^{c}	135.01 ± 0.04^{b}	264.46 ± 1.25^{a}
pH	4.93 ± 0.01^{b}	4.96 ± 0.23^{a}	4.99 ± 0.15^{a}	4.75 ± 0.26^{c}	4.97 ± 0.20^{a}	4.99 ± 0.01^{a}
Total sugars (mg/100 ml)	51.4 ± 0.772^{f}	54.9 ± 1.01^{c}	57 ± 1.5^{b}	54.2 ± 0.54^{d}	67.2 ± 1.2^{a}	55.3 ± 0.02^{e}
Proteins ((%)	0.25 ± 1.03^{d}	0.56 ± 0.01^{b}	0.59 ± 0.56^{b}	0.53 ± 1.6^{c}	0.56 ± 0.83^{b}	0.65 ± 1.5^{a}
Mg^{2+} (mg/100 ml)	143.01 ± 0.25^{d}	152.56 ± 1.54^{b}	148.34 ± 1.02^{c}	137.11 ± 1.03^{e}	177.41f ± 2.03^{a}	130.29 ± 0.97^{f}
K^{+} (mg/100 ml)	155.33 ± 1.01^{b}	124.13 ± 0.21^{e}	115.83 ± 0.56^{f}	132.41 ± 0.40^{d}	138.33 ± 1.21^{c}	191.66 ± 1.01^{a}
Ca^{2+} (mg/100 ml)	411.85 ± 0.45^{e}	1397.32 ± 0.74^{d}	3954.2 ± 1.47^{a}	255.82 ± 0.98^{f}	1620.05 ± 0.46^{c}	3523.42 ± 1.51^{b}
Fe^{2+} (mg/100 ml)	10.98 ± 0.77^{f}	12.23 ± 0.68^{e}	14.01 ± 1.57^{c}	12.63 ± 0.01^{d}	14.18 ± 0.13^{b}	15.67 ± 0.58^{a}
Na^{+} (mg/100 ml)	25.8 ± 0.04^{f}	27.43 ± 0.01^{e}	27.6 ± 0.25^{d}	59.75 ± 0.77^{a}	38.3 ± 0.77^{b}	27.93 ± 1.24^{c}
Zn^{2+} (mg/100 ml)	0.112 ± 0.41^{c}	0.113 ± 0.21^{c}	0.167 ± 0.15^{a}	0.131 ± 0.05^{b}	0.132 ± 1.21^{b}	0.171 ± 0.65^{a}

Means with different letters in the same row indicate significant differences ($p \leq 0.05$) (n= 3). C0: juices extracted with hot water with 0% ginger; C2,5: extracted juice in hot water with 2.5% ginger; C10: juices extracted with hot water with 10% ginger; F0: juices extracted in cold water with 0% ginger; F2,5: juice extracted in cold water with 2.5% ginger; F10: juices extracted with cold water with 10% ginger.

Table 2. Total viable counts (cfu/ml) of the different banana drinks during storage.

	Day 1	Day 4	Day 7
C0	9×10^{4} ^{b}	2.02×10^{6}b	3×10^{7} ^{b}
C2,5	3.3×10^{3} ^{d}	2.78×10^{4}d	2.9×10^{6} ^{e}
C10	$7 7 \times 10^{2}$ ^{f}	1.1×10^{4}f	4.8×10^{5} ^{f}
F0	3×10^{5} ^{a}	4.9×10^{6} ^{a}	7×10^{7} ^{a}
F2,5	3.4×10^{3} ^{c}	4.66×10^{4} ^{c}	3.3×10^{7} ^{c}
F10	3.9×10^{3} ^{e}	3×10^{4} ^{e}	3.5×10^{6} ^{d}

Sample codes are as stated in Table 1. Values are means of three replicates.

Table 3. Sensory evaluation of different banana drinks.

Sample	Color		Texture		Taste		Appearance		Aroma		Acceptability	
	D1	D7	D1	D7	D1	D7	D1	D7	D1	D7	D1	D7
C0	7.2^{a}	6.6^{a}	3.4^{c}	3.7^{c}	4.4^{ba}	4.7^{a}	3.3^{d}	3.4^{c}	5.2^{a}	5.8^{b}	5.3^{a}	5.2^{a}
C2,5	6.6^{ba}	4.9^{b}	4.6^{cb}	4.1^{cb}	3.9^{cba}	3.4^{b}	5.3^{c}	5.6^{b}	3.7^{b}	3.9^{c}	4.9^{a}	4.4^{b}
C10	4.1^{c}	4.3^{cb}	4.4^{cb}	4.4^{cba}	3.3^{c}	2.2^{c}	5.9^{cb}	5.8^{ba}	2.5^{c}	1.9^{d}	4.7^{a}	3.4^{c}
F0	4.6^{c}	3.5^{c}	6.5^{ba}	5.4^{cba}	4.7^{a}	4.6^{a}	5.3^{c}	5.6^{b}	7^{a}	6.5^{a}	5.4^{a}	5.2^{ba}
F2,5	5.4^{cb}	5.0^{b}	6.1^{a}	5.7^{a}	4.5^{cba}	4.2^{a}	6.8^{ba}	7.3^{aba}	5.5^{a}	6,4^{ba}	4.8^{a}	4.7^{ba}
F10	5.6^{cb}	5.1^{ba}	5.4^{ba}	5.2^{ba}	3.8^{cb}	2.3^{cb}	6.4^{a}	7.9	3.3^{cb}	3.6^{c}	4.7^{a}	3.2^{c}

Values are means of three replicates.

fortified with ginger contain less germs than the controls. After seven days of storage, the sample F0 contain the highest flora (7.10 cfu/ml), while the lowest (4, 9.10 cfu/ml) quantity is contained in the C10 sample.

The organoleptic properties of plantain juice are shown in Table 3 and these results indicate that the cold extract F2 has the highest overall acceptability of 5.4, while the cold extract sample F10 has the lowest acceptability (3.2).

DISCUSSION

At the end of this study, it is clear that vitamin C content is higher in the different juices supplemented with 2.5 and 10% ginger extracts. This shows that the vitamin C level can be improved by ginger addition. This assertion seems justified by research conducted by Adesokan et al. in 2005 who indeed improved the ascorbic acid content of the juice "ZOBO" through the use of mixtures of ginger and garlic in fresh or powder forms. The pH value of banana juice was increased with increasing ginger concentration. This implies that plantain juice with more acid should not be used when the stomach is empty. Proteins content also increases with the content of ginger. These show that ginger improves protein content as in the study of Adesokan et al. (2005) and Olayemi et al. (2011). Different plantain drinks characterized in this study have varying contents of metal ions such as Mg^{2+}, K^{+}, Ca^{2+}, Fe^{2+}, Na^{+} and Zn^{2+}. A study by Olayemi et al. (2011) on three different "zobo" drinks showed variations in their mineral contents following spices incorporation. Similar results were revealed by Adesokan et al. (2005) study in the "zobo" drinks supplemented ginger.

In addition, we found that ginger extract can reduce the total flora of plantain juice during 7 days of storage. The presence of mono and sesquiterpenoids within the ginger extracts is considered main cause of their antimicrobial activity (Bajpal et al., 2009).The mode of action of these compounds involves cell walls and cell membranes attacks affecting their permeability and functions (electron transport, nutrient uptake and enzyme activity).

These results are in agreement with those of Adesokan et al. (2005) and Ogiehor et al. (2008) who were able to maintain the quality of "zobo" drink for 42 days by combining refrigeration and ginger (0.2%) addition.

Conclusion

From this work, it can be concluded that ginger provides mineral fortification and antimicrobial benefits to banana drink during storage for 7 days. Therefore, it is suggested that ginger could be used to extend the shelf life of fruit juices, which seems more healthful than synthetic additives. The ginger extracts showed an interesting antimicrobial activity and could therefore be used to preserve various food products. The extracts may find application in antibacterial and antifungal therapies. Additional research is needed to elucidate mechanisms of action of bioactive molecules in the extracts.

Conflict of Interests

The author(s) have not declared any conflict of interests.

ACKNOWLEDGEMENT

This research was financially supported by the WAAPP (West Africa Agriculture Productivity Program) fund for the World Bank.

REFERENCES

A.O.A.C. (1990). Official methods of Analysis (15th Ed.).Association of Official Analytical Chemist, Washington D.C.

Abulude FO, Ogunkoya MOAdesanya WO (2007). Description and performance evaluation of a juice extraction constructed in Nigeria. Res. J. Appl. Sci. 2:1-34

Adesokan IA (2005). Characterization of killer yeasts from Nigerian traditional fermented alcoholic beverages. M.Sc. Thesis, Department of Microbiology ,University of Ibadan , Nigeria. 174-177.

Adesokan IA, Abiola OP, Adigun MO, Anifowose OA (2013). Analysis of quality attributes of Hibiscus *sabdariffa* (Zobo) drinks blended with aqueous extract of ginger and garlic. Afr. J. Food Sci. 7(7):174-177.

Adesokan IA, Abiola IP, Ogundiya MO (2010). Influence of ginger on sensory properties and shelf – life of Ogi, a Nigerian traditional fermented food. Afr. J. Biotechnol. 9 (12):1803-1808.

AFNOR (1995). Recueil de normes françaises, contrôle de la qualité des produits alimentaires-analyses sensorielles.

ANADER (1998).Agence Nationale pour le Développement Rural. Enquète de base 1996-1997

Bajpal VK, Al-Reza SM, Choi UK, Lee JH, Kang SC (2009). Chemical composition, antibacterial and antioxidant activities of leaf essential oil and extracts of Metasoquioa glyptostroboides Miki ex HU. Food Chem. Toxicol. 47:1876-1993.

BIPEA (1976) Bureau International d'Etude Analytique. Recueil des méthodes des communautés Economiques Européennes 1ère édition, Bruxelle, pp. 67-114.

Dougheri JH , Alabi G ,Elmahmood AM (2007). Effect of some chemical preservatives on the shelf –life of zobo drink. Afr. J. Microbiol. Res. 2:037-041.

Kolapo AL, Popoola TOS, Sanni MO, Afolabi RO(2007). Preservation of soybean daddawa condiment with dichloromethane extract of ginger. Res. J. Microbiol. 6:13-18.

Lépengué AN (1999). Activités enzymatiques et troubles de maturation des bananes plantain entreposées aux températures élevées. Diplôme d'Etudes Approfondies, en Physiologie Végétale (Agrophysiologie), UFR Biosciences, Université de Cocody-Abidjan, 68 p.

Ogiehor IS, Nwafor OE, Owhe-Ureghe UB (2008). Changes in the quality of Zobo beverages produced from *Hibiscus sabdariffa* (Linn Roscelle) and the effects of extract of ginger alone or in combination with refrigeration. Afr. J. Biotechnol. 7(8):1176-1180.

Olayemi F, Adedayo R, Muhummad R, Bamishaiye E (2011). The nutritional quality of three varieties of zobo (*Hibiscus sabdariffa*) subjected to the same preparation condition. Plant Food Hum. Nutr. 6:10-15.

Omemu AM, Atayese AO, Obadina (2006). A Survey of the microflore of Hibiscus sabdariffa (Roselle) and resulting « zobo » juice. Afr. J. Biotechnol. 5(3):254-259.

Varoquaux P, Gouble B, Ducamp MN, Self G (2002). Méthode permettant d'optimiser l'emballage des fruits sous atmosphère modifiée. Fruits 57(5-6):313-322.

Toxicity of leaf powder of *Lepidagathis alopecuroides* to Nile tilapia *Oreochromis niloticus* Juveniles

Okey, I. B[1]*, Keremah, R. I[2], Ikpi, G. U[1] and Ebeniro, L. A [3]

[1]Cross River University of Science and Technology, Obubra Campus, Cross River State, Nigeria.
[2] Department of Fisheries/ Livestock Production Technology, Niger Delta University, Amasssoma, P. M.B. 071, Yenaogoa, Bayelsa State, Nigeria.
[3]Federal College of Agriculture, Ishiagu Ebonyi State, Nigeria.

Oreochromis niloticus **juveniles (mean body weight, 13.99±0.38 g and length, 8.32±0.93 cm) were exposed in a daily renewal bioassay to different concentrations of** ***Lepidagathis alopecuroides*** **(0.00, 1.0, 1.5, 2.0, 2.5 and 3.0 mg/l), for 96 h to assess its toxicity. The opercular (obf), tail beat (obf) frequencies and mortalities were monitored during the exposure period. The pattern of response of OBF/min and TBF/min decrease with time of exposure, but increased with increase in concentration of the toxicant. Opecurlar beat frequency (OBF) per minute was more variable from the 24th h, whereas TBF were less variable throughout the duration of the exposure. The cumulative mortality of the fish increased with time and concentrations of the toxicants. Less than 50% mortality was recorded for 1.0 and 1.50 mg/l by the 96th h, while the threshold concentrations that killed 100% of exposed fish were 3.0, 2.5 and 2.0 mg/l at the 48th and 72nd h, respectively. Wide variations were recorded in OBF and mortality at the various times of exposure and concentrations. The relationship between OBF, TBF and time of exposure was negative and significant at (p< 0.01) but positive with mortality and insignificant. The 24 and 96 hrLC_{50} were 2.27 and 0.88 mg/l and their safe concentration were 0.23 and 0.09 mg/l, respectively. The mean lethal time decreased with increase in the concentration of** ***L. alopecuroides.*** **The results suggest that aqueous extracts of the leaves of** ***L.alopecuroides*** **has piscicidal property and are highly toxic to** ***O. niloticus.*** **Hence, care should be exercised in the use of the toxicant under culture conditions except for pond cleansing of unwanted fauna before stocking.**

Key words: Opercular beat frequency, tail beat frequency, mortality, *Lepidagathis alopecuroides* and *Oreochromis niloticus*.

INTRODUCTION

Some plants contain compounds of various classes that have insecticidal, piscicidal and molluscicidal properties (Cagauan and Arce, 1992). Plants from different families have been applied for catching fish, control of predators and reduction of overpopulation in aquaculture ponds all over the world and are considered advantageous when viewed against the backdrop of using persistent chemical (Van Andel, 2000; Tiwari and Singh, 2003). Frequent application of high concentrations these ichthyotoxins in water may have adverse effects not only on fish species but also on other aquatic fauna. Several studies have shown that plant toxins at very low concentrations are very toxic to all groups of aquatic fauna (Goktepe et al., 2004; Gabriel and Okey, 2009).

L.alopecuroides (Vahl) is a tropical shrub belonging to the family Acanthaceae commonly found in the coastal countries of West Africa. The leaves are used to immobilize fish in many communities of Rivers and Cross River

*Corresponding author. Email: piusbass@yahoo.com.

States of Nigeria. In Rivers State it is used for quick kill of hardy fish like mudskippers (Obomanu et al., 2007) and in Cross River State it is normally applied in pools to kill mostly catfishes and tilapias constituting the major source of dietary protein (Ekanem et al., 2003). Phytochemical screening of the plant revealed the presence of alkaloids, tannins, saponins, glycosides and flavonoid (Obomanu et al., 2005) all of which have been proven to posses some form of toxic effect on organisms. The aqueous extract of the leaves of *L. alopecuroides* on the haematological, biochemical, behavioral (Opercular and tail beat frequencies) and mortality of African catfishes of various sizes have been reported (Gabriel and Okey, 2009: Keremah, et al., 2010).

Nile tilapia (*Oreochromis niloticus*) is important and commercially valued fish for the Nigerian fishery and aquaculture industry. They are widely cultured in ponds and occur naturally in freshwater and swamps where they are fished in large numbers at the onset of the dry season using plant poisons. Although, the toxicity level and lethal concentration of *L.alopecuroides* have been determine for African clarrids, no report on the toxicity of the plant to Nile tilapia. This study was there to determine the toxicity of leaf powder of *L. alopecuroides* to *Oreochromis niloticus* juveniles.

MATERIALS AND METHODS

Bioassay was conducted in the Fisheries Wet Laboratory, Cross Rivers University of Technology (CRUTECH). Two hundred and fifty *O. niloticus* juveniles (mean body weight, 13.99±0.38g and length, 8.32±0.93 cm) were obtained from CRUTECH fish Farm Obubra campus Cross River State, Nigeria and acclimated to Laboratory conditions using glass tanks (45 x 40 x 40cm) of 40 liters capacity, filled with 30 liters of tap water (characteristics: Temperature 26.52±1.21°C, DO_2 5.47± 0.56 mg/l, total alkalinity 16.19 ±0.97 mg/l as $CaCO_3$ and pH, 7.63± 2.11. The water in the containers was renewed daily and the fish were fed with commercial fish feed (Pellets) containing 35% crude protein at 1% of their body weight (FAO, 1986). Unconsumed feed and faecal wastes were removed and water replenished regularly as recommended by Oyelese and Faturoti (1995). Feeding was discontinued 24 h before the commencement of and during experiment, to minimize the contamination of the test aquaria. *L. alopecuroides* were procured from Ofumbongha, Obubra Cross Rivers State, where they are mostly use to kill fish in pools and streams. The leaves were sun dried, then pulverized with a sterile grinding machine and were then sieved with 100-micron sieve to obtain a fine powder. A range finding test was conducted according to OECD (2001) and Omitoyin, et al (2006), respectively to determine the concentrations of *L. alopecuroides* used in the definite test. The following concentrations 1.0, 1.5, 2.0, 2.5 and 3.0 mg/l were prepared in triplicates from a stock solution and introduced into the glass aquaria. Ten fish each were randomly placed in each of the aquaria holding 30l of the test solution. Test media was renewed daily for through out the duration of the experiment. The behaviour Obf and Tbf, mortality and other external changes in the body of the test fish were observed and recorded at 12, 24, 48, 72 and 96 h of exposure time as described by Odiete (1995), and Omitoyin et al. (2006). Dead fish were promptly removed to avoid contamination of the test media.

Data obtained from the experiments were subjected to ANOVA using statistical package for the social sciences, (SPSS) version 13. Where differences exist, Duncan multiple range test separated them and F- test used for significant differences ($p < 0.05$) between the various treatments (Wahua, 1999). Correlation and regression between OBF min^{-1}, TBF min^{-1} and cumulative mortality against concentrations and time for the different species was done according to Zar (1996). An analysis of the lethal concentration (LCs) and median lethal time (MLTs) with associated confidence interval were done with Probit analysis. Safe concentrations at the various time intervals were obtained by multiplying the lethal concentration (LC_{50}) by a factor of 0.1 (EIFAC, 1983; Koesoemadinate, 1980).

RESULTS

O. niloticus juveniles exposed to *L. alopecuroides* at different concentrations shows initial distress swimming movements, rapid opercular movements, loss of balance, incessant gulping of air, excessive mucus secretion, unusual lethargy and fish settling at the bottom motionless with slow opercular movement and erratic swimming before death.

The abnormal behaviour displayed by the fish increased with increasing concentration of *L. alopecuroides* in water but decreased with increase in time of exposure.

The pattern of response of OBF/min and TBF/min decrease with time of exposure, but increased with increase in concentration of the toxicant (Figures 1 and 2). Opecurlar beat frequency (OBF) per minute was more variable from the 24th h, whereas TBF were less variable throughout the duration of the exposure. The cumulative mortality of the fish increased with time and concentrations of the toxicants (Figure 3). Less than 50% mortality was recorded for 1.0 and 1.50 mg/l by the 96th h, while the threshold concentrations that killed 100% of exposed fish were 3.0, 2.5 and 2.0 mg/l at the 48th and 72nd h, respectively. No mortality was recorded in the control (0.00 mg/l) group throughout the exposure period. Wide variations were recorded in OBF and mortality at the various times of exposure and concentrations (Tables 1 and 2). Regression analysis showed that the relationship between OBF, TBF and time of exposure was negative and significant at ($p< 0.01$) but positive with mortality and insignificant (Table 3). The 24 and 96 hrLC_{50} were 2.27 and 0.88 mg/l and their safe concentration were 0.23 and 0.09 mg/l, respectively (Table 4). The mean lethal time decreased with increase in the concentration of *L. alopecuroides*.

DISCUSSIONS

The water quality parameters recorded were within acceptable ranges for toxicity test (APHA, 1998). They may not have acted synergistically with the toxicant to affect the behavior as well as mortalities recorded in this study. Similar observations were reported by Onusiriuka and Ofodike (1994) in acute test solutions of Akee apple, *Blighia sapida* and *Kigelia Africana* extracts and Gabriel and Okey (2009) to acute concentrations of *L. alopecuroides* to which *C. gariepinus* were exposed.

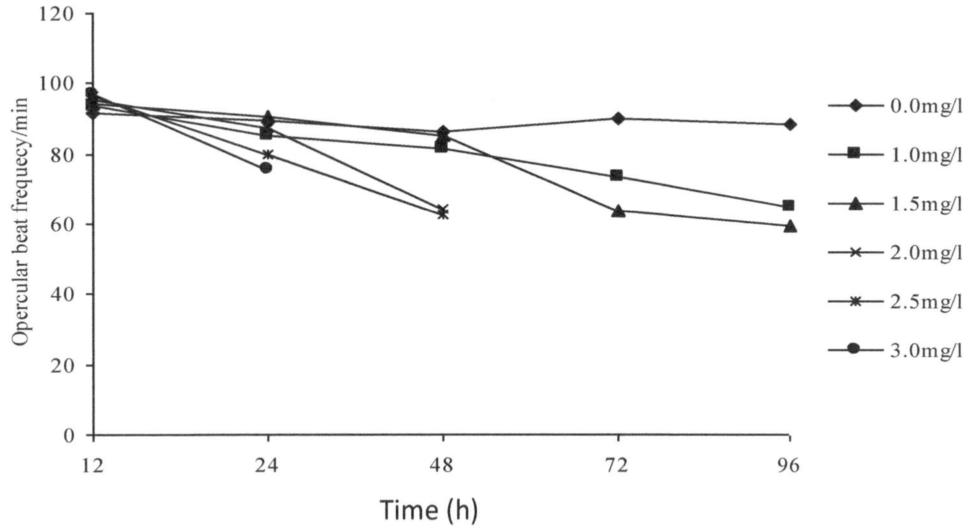

Figure 1. Changes in opercular beat frequency/min of *O. niloticus* juveniles exposed to aqueous extract of *L. alopecuroides*

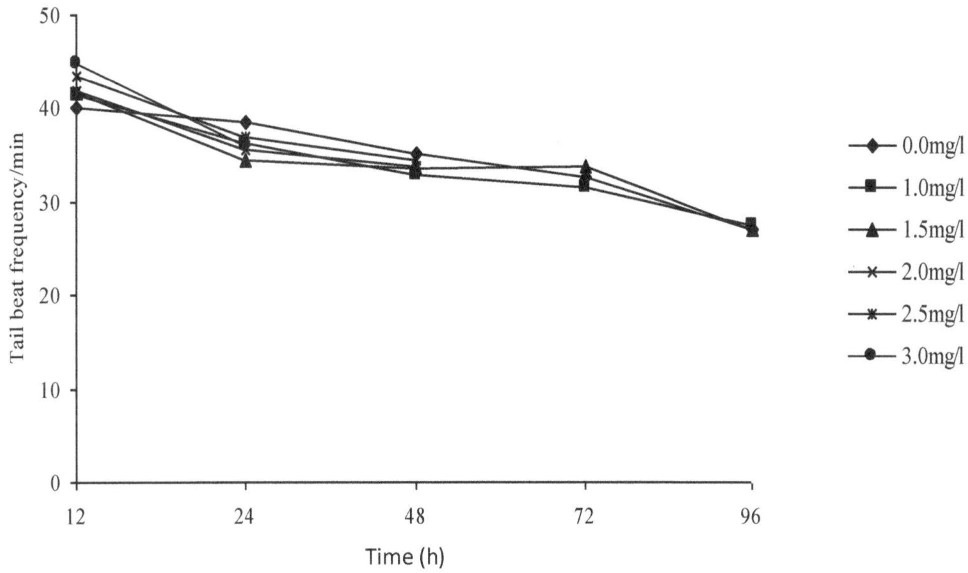

Figure 2. Changes in tail beat frequency/min of *O. niloticus* juveniles exposed to aqueous extract of *L. alopecuroides*.

Heath (1991) and Adeyemo et al., (2004) have reported changes in haemotological parameters and swimming activity of fish due to changes in the water quality parameters.

The behavioural pattern in this study revealed that fish exposed to toxicants usually exhibits some behavioral changes such as increased opercular and tail beat frequencies, mucus secretion and gulping for air (Nwanna et al., 2000). The behavioral changes observed in this study were similar to those reported for *O. mossambicus*, *T. niloticus* and *O. niloticus* exposed to leaf extracts of *Apodytes dimidiate* and *Thevitia nerifolia* (Sambasivam et al., 2003). This also compared favorably with those reported by Ayotunde and Offem (2005) and Gabriel and Okey (2009). Signs of agitated behaviour, respiratory distress and abnormal nervous behavours including eventual deaths were observed in *O. niloticus* exposed with ethanolic and aqueous extracts of *Ipomoea aquatica* leaves (Ayoola et al., 2011). The 96 hrLC_{50} (0.88 mg/l) recorded in this study was smaller than the 4.2 mg/l reported for aqueous extract of pawpaw seed powder (Ayotunde and Offem 2005) and 7.13 g/l reported for aqueous extract of *Euphorbia poisonii* leaves to tilapia *Oreochromis niloticus* fingerlings (Ayoola et al., 2011).

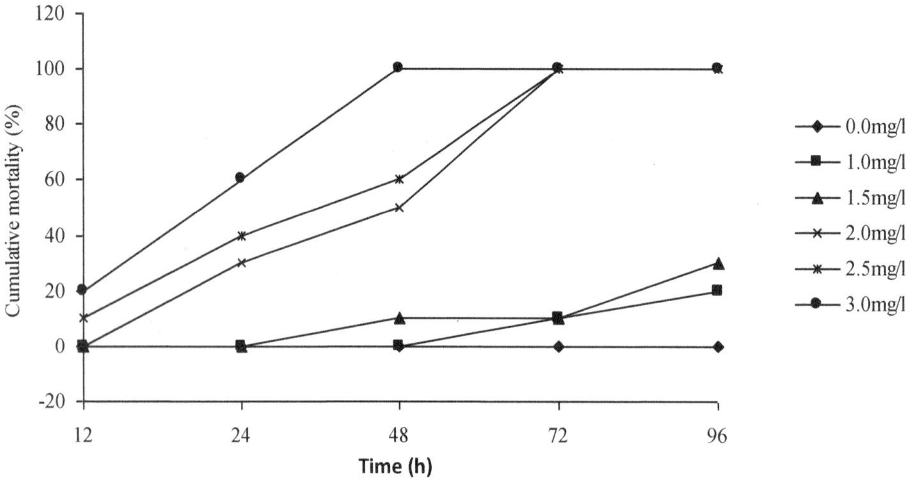

Figure 3. Changes in the cumulative mortality of *O. niloticus* juveniles exposed to aqueous extract of *L.alopecuroides.*

Table 1. Opercular and tail beat frequency and cumulative mortality (%) of *O. niloticus* juveniles exposed to aqueous extract of *L.alopecuroides.*

Variable	Concentration of *L. alopecuroides* (mg/l)					
	0.00	1.00	1.50	2.00	2.50	3.00
OBF/min	91.07±6.76^{a}	79.73±9.62b^{c}	78.71±9.62^{c}	81.84±5.57b^{c}	79.58±7.88b^{c}	86.41± 10.10^{b}
TBF/min	34.65±7.88^{b}	33.88± 7.25^{b}	34.13± 9.65^{b}	37.18± 4.77^{ab}	38.23± 6.39^{a}	40.35± 9.90^{a}
Mortality	0.00±0.00^{d}	6.33± 22.13^{d}	24.25 ±17.26^{c}	55.71± 27.13^{b}	84.29±24.50^{a}	98.35± 38.52^{a}

Means with the same superscript in the row are not significantly different (p>0.05).

Table 2. Opercular and tail beat frequency and cumulative mortality (%) of *O. niloticus* juveniles at various durations of exposure to aqueous extract of *L.alopecuroides.*

Variable	Duration of exposure (h)				
	12	24	48	72	96
OBF/min	94.73± 8.15^{a}	84.68± 6.09^{b}	76.05± 8.00^{c}	75.79± 10.13^{c}	70.82± 11.66^{c}
TBF/min	42.19± 6.42^{d}	36.28± 5.56^{b}	34.02±8.44^{b}	32.65± 6.33^{b}	27.19± 9.90^{c}
Mortality	2.56± 6.33^{d}	22.39± 14.57^{c}	41.73± 27.14^{b}	56.25± 32.44^{a}	48.99± 26.13^{b}

Means with the same superscript in the row are not significantly different (p>0.05)

Table 3. Regression lines f or the prediction of the values o f OBF/min., TBF/min. an d cumulative mortality of *O. niloticus* juveniles exposed to aqueous extract of *L.alopecuroides.*

Independent variable	Dependent variable	Regression equation	(r^2)	(r)	Significant level
Time	OBF	y=93.10-0.25x	0.92	0.82	0.05
Time.	TBF	y=42.09-0.15x	0.95	0.90	0.05
Time	Mort.	y=7.53+ 0.33x	0.55	0.30	ns
Conc.	OBF	y=85.53+1.50x	0.82	0.91	0.0001
Conc.	TBF.	y=32.97+5.00x	0.81	0.90	0.001
Conc.	Mort.	y=10.38+26.30x	0.90	0.81	0.05

Where y = independent variable (concentration, Time), x = dependent variable (OBFmin^{-1} , TBFmin^{-1} and mortality. r^2 = coefficient of determination, r = coefficient of correlation.

Table 4. Lethal (LCs) and safe concentrations and Mean Lethal Times (MLTs) and associated 95% confidence limit of *L. alopecuroides* on *O. niloticus* juveniles.

Exposure time (h)	Lethal Conc. and associated 95% C.L			Safe Conc.	T.F	Conc. (mg/l)	MLT and associated 95% C.L		R.T
	LC_5	LC_{50}	LC_{95}				MLT_{50}	MLT_{95}	
24	1.04 (0.15-0.28)	2.27 (40.95-70.06)	3.94 (1.94-7.31)	0.02	1	1.0	102.72 (87.15-96.47)	164.40 (46.72-65.03)	1
48	0.88 (12.45-43.70)	1.63 (0.10-1.67)	2.31 (0.00-0.00)	0.02	1.39	1.5	83.98 (66.70-99.05)	98.57 (43.59-56.08)	1.22
72	0.64 (65.08-78.45)	1.02 (0.00-0.00)	2.04 (23.34-56.89)	0.01	2.23	2.0	51.67 (11.95-13.13)	71.11 (79.60-98.74)	1.99
96	0.34 (23.90-78.65)	0.88 (0.46-0.70)	1.64 (1.44-6.30)	0.01	2.58	2.5	34.77 (36.08-79.09)	48.88 (41.58-69.81)	2.95
						3.0	19.94 (97.65-103.7)	35.93 (48.74-86.83)	5.15

LC= Lethal Concentration, T.F= Toxicity factor =LC50 value at 24hrs ÷ LC50 value of any other periods, C.L= Confidence MLT=Mean Lethal Time, R.T=Relative Time=MLT50value at 1mg/l ÷ MLT50 value at any other concentrations.

This indicates that *L. alopecuroides* is more toxic than pawpaw seed and *Euphorbia poisonii* to *O. niloticus*. The toxicity of *L. alopecuroides* to *O. niloticus* is higher than the result of Wade et al. (2002) who reported that the 96 hLc50 of 0.19 mg/l^{-1}, for the Nile tilapia *Oreochromis niloticus*, fingerlings exposed to effluent of cassava (*Manihot esculenta*) which may be attributed to the age of fish and the strength of the active ingredient in the toxicant.

Positive correlation between concentration (mg/l) and mortality agrees with observations of Tiwari and Singh (2003) in *C. carpio* exposed to ethanol extracts of *Nerium indicum*. Results of the regression of OBF on the concentration of the toxicant and time lend credence to the observation and claim by several authors that OBF is better tool for assessing acute toxicity of the plant extract than TBF based on the unit changed in each of these variables with that of the toxicant (Ekweozor et al., 2001;; Bobmanuel et al., 2006). This may explain why several authors prefer the use of obf to tbf in the study of stress responses of fish to the effects of toxicants, (Onusiriuka and Ufodike, 1994; Babatunde et al., 2000; Oti, 2002).

The lethality of the leave extract of *L. alopecuroides* to fish species had already been confirmed by previous studies (Obomanu et al., 2005, 2007; Gabriel and Okey, 2009). Since the toxicant was highly toxic to the exposed fishes, under field applications to catch fish by artisanal fishermen or clear ponds of unwanted tadpoles and amphibians which are major competitors and predators in fishponds (Nguenga et al., 2000), priced culture fishes like the test species may be adversely affected.

Conclusion

The results suggest that aqueous extracts of the leaves of *L. alopecuroides* has piscicidal property and are highly toxic to *O. niloticus*. From the toxicity tests *L. alopecuroides* powder concentration as low as 0.88 mgl^{-1} in the medium can be potentially hazardous to some fish species in freshwater. The high potency of the plant extract may possibly be due to the ability of the components to impair oxygen transfer through the gills leading to death which rate was proportional to the toxicant concentration. Hence, care should be exercised in the use of the toxicant under culture conditions except for pond cleansing of unwanted fauna before stocking. Therefore, acute toxicity data of the present study provide baseline information needed to develop models of *L. alopecuroides* powder effects on ecological systems.

REFERENCES

Adeyemo OK, Akintoye OA, Oghi ML (2004).Acute toxicity of chlorpyrifos (Dursban) to *Clarias gariepinus* juveniles. Trop. Vet. 22: (1) 4-8.

APHA (1998). American Public Health Association Standard methods for examination of water and wastewater. 13th Edition APHA Washington, D.C USA. pp.1254

Ayoola SO, Kuton MP, Idowu AA, Adelekun AB (2011).Acute toxicity of Nile tilapia (*Oreochromis niloticus*) juveniles exposed to aqueous and ethanolic extracts of *Ipomoea aquatica* leaf. Nat. Sci. 9(3): 91- 99.

Ayotunde EO, Ofem BO (2005). Acute and Chronic Toxicity of Pawpaw (*Carica papaya*) Seed Powder to Nile Tilapia *Oreochromis niloticus* (Llinne 1757), Fingerlings. J. Agric. Technol. Envinron. 1: 1- 4.

Babatunde MM, Oklademeji AA, Balogun TK (2000). Acute toxicity of gramozone to *Oreochromis niloticus* in Nigeria. Water Air Soil Pollut. 131:1-10.

Bobmanuel NOK; Gabriel UU, Ekweozor IKE (2006). Direct toxic assessment of treated fertilizer effluents to Oreochromis niloticus, and catfish hybrid (Heterobranchus bidorsalis ♂ x Clarias gariepinus ♀). Afri. J. Tech. 5 (8): 653-642.

Cagauan AG, Arce RG (1992). Overview of pesticides use in rice-fish farming in Southeast Asia. In: C. R. dela Cruz, C. Lightfoot, B. Costa-Pierce, V. R. Carangal and M. P. Bimbao (eds.) pp. 217-233..

EIFAC(1983). Revised report on fish toxicity testing procedures: EIFAC Tech. Paper 24, Rev.1: FAO, Rome. pp.37.

Ekanem AP, Meinelt T, Kloas W, Knopf K (2003). Effects of the extracts of two African fish poison plants *Asystasia vogeliana* and *Tephrosia vogelii* on embryos of zebrafish (*Danio rerio*). Bull. Environ. Contam. Toxicol. 71: 551–556.

FAO (1986). Manual of Methods in Aquatic Environment. Research. Part 10. Short-term static Bioassays. FAO Fisheries Technical Report Paper, pp: 247-6

Ekweozor IK, Bobmanuael NO, Gabriel UU (2001). Sublethal effects of ammoniacal fertilizer effluents on three commercial fish species from the Niger Delta, Nigeria. J. Appl. Sci. and Environ. Manag. 5 (1): 63-68.

Gabriel UU, Okey IB (2009). Effects of aqueous extracts of *Lepidagathis alopecuroides* on the behaviours and mortality of hybrid (*Heterobranchus bidorsalis* ♀ **x** *Clarias gariepinus* ♂) fingerlings. Res. J. Appl. Sc. Eng. Technol. 1 (3): 116-120.

Goktepe I, Portier R, Ahmedna M (2004). Ecological assaessment of neem based pesticides. J. Environ. Sci. Health B39: 311-320.

Heath AG (1991). Water pollution and fish physiology. Lewis publishers, Boca, Ranton, Florida. U.S.A.

Keremah RI, Okey IB, Gabriel UU(2010). Relative toxicity of aqueous leaf extracts of *Lepidagathis alopecuroides* (Vahl) R. Br. ex Griseb to the clariids, *Clarias gariepinus* and *Heterobranchus bidorsalis* fingerlings. Agric. Biol. J. North Am. 1(5): 834-840

Nguenga D, Forbin I, Teugels GG, Ollevier F (2000): Predation capacity of tadpole (*Bufo regularis*) using African catfish *Heterobranchus longifilis* larvae: impact of prey characteristics on vulnerability to predation. Aquac. Res. 31:931-936.

Nwanna LC, Fagbenro OA, Ogunlowo ET (2000). Toxicity of textile effluent to *Clarias gariepinus* and *Heterobrauchus bidorsalis* and Hybrid fingerlings. Responsible Aquaculture in the New millennium. Abstract of Inter. Conf. Aqua 2000, P. 510 (Flos, C.R. and Cesswell, L) C Euro Aqua Soc. Special Pub. No. 28 Ostaude, Belguim.

Obomanu FG, Fekarurhobo GK, Howard IC (2005). Antimicrobial activity of extracts of leaves of *Lepidagathis alopecuroides* (Vahl). J. Chem. Soc. Nig. 30:33-35.

Obomanu FG, Ogbalu OK, Gabriel UU, Fekarurhobo GK, Abadi SU (2007). Piscicidal effects of *Lepidagathis alopecuriodes* on mudskipper, *Periophthalmus papillio* from the Niger Delta. *Res.* J. Appl. Sci. 2(4): 282-387.

OECD (2001). Organisation of Economic Cooperation and Development. Guidance document on Aquatic Toxicity Testing of Difficult substances and mixtures, pp. 34-57.

Odiete WO (1999). Environmental physiology of animals and Pollutions. Published by Diversified Resources Limited. pp: 5, 44, 69, 74-79, 157-167, 187-189, 220 and 223.

Omitoyin BO, Ajani EK, Adesina BT, Okuagu CNF (2006). Toxicity of lindane (gamma hexachlorocyclohexane) to *Clarias gariepinus* (Burchell 1822). W. J. Zool. 1: 57-63.

Onusiriuka BC, Ufodike FB (1994). Acute toxicity of water extracts of Akee apple, *Bligha sapida* and Sausage plant, *Kigella africana* on African catfish, *Clarias gariepinus*. J. Aquat. Sci. 9:35-41.

Oti EE (2002). Acute toxicity of cassava mill effluent to the African catfish fingerlings. J. Environ. Biol. 20(3): 112- 117

Oyelese OA, Faturoti EO (1995). Growth and mortality estimates in *Clarias gariepinus* fed graded levels of processed cassava peels. J. Trop. For. Resour. 11: 71-81.

Sambasivam S, Karpagam G, Chandran R, Khan SA (2003). Toxicity of leaf extracts of yellow Oleander *Thevetia nerifolia* on tilapia *O. niloticus*. J. Environ. Biol. 24 (2): 201-204

Tiwari S, Singh A (2003). Control of common freshwater predatory fish, *Channa punctatus*, through *Nerium indicum* leaf extracts. Chemosphere 53:865-875.

Van Andel T (2000): The diverse uses of fish-poison plants in Northwest Guyana. Econ. Bot. 54: 500–512.

Wade JW, Omoregie E, Ezenwaka (2002). Toxicity of cassava (*Manihot esculenta* CRANTS) effluent on the Nile Catfish *Clarias gariepinus* (l) under laboratory condition. J. Aquat. Sci., 17: 2-34

Wahua TAT (1999). *Applied statistics for scientific studies*. Afrika Link Books. Aba, Nigeria. pp.365

Zar JH (1996). Biostatistical Analysis. 3rd Edition. Englewood Cliffs, New Jersey, Prentice Hall.

6

Evaluation of fruit quality traits of traditional varieties of tomato (*Solanum lycopersicum*) grown in Tunisia

Amira Ben Aoun, Belgacem Lechiheb, Leila Benyahya and Ali Ferchichi

Institut des Régions Arides, Laboratoire d'aridoculture et cultures oasiennes, 4119 Médenine, Tunisia.

Physicochemical, nutritional, agronomic and sensorial parameters, which define fruit quality of tomato (*Solanum lycopersicum*), were evaluated in 13 traditional varieties collected from several localities in Tunisia, using three commercial varieties as controls. Several varieties were identified as better score for total solid, soluble solid, sugars/acid ratio and vitamin C concentration than the commercial varieties. These varieties could be a resource of good quality in breeding programmes. The correlation between specific parameter seen in this study showed that the local varieties can be differentiated, not just as a function of their morphological attributes but also as a function of their organoleptic, nutritional and sensorial quality.

Key words: Tomato, quality, selection, breeding programme.

INTRODUCTION

Major fruit quality of interest to both fresh market and industrial tomato include a number of agronomic, organoleptic and nutritional qualities. Organoleptic quality is evidenced via physicochemical parameters that make product to be acceptance to consumers (Stevens, 1972; Kader et al., 1977). Total fruit solids content is particularly important to the processing industry and has probably received more attention than any other fruit trait. Total solids content of cultivated tomato amount to 4.5 to 8.5% of its fresh weight (André et al., 2005), though this percentage can be much higher in some wild species (Bertin et al., 2000).

Total solids comprise all fruit components such as aromas except water and volatiles. In cultivated tomato, the soluble (SS) and insoluble solids (ISS) account for about 75 and 25%, respectively, of total solids (Majid, 2007). Reducing sugars (glucose and fructose) are the major components of the SS; sucrose is also present but in very small quantities (Malundo et al., 1995). Remaining soluble solids consist of organic acids, lipids, minerals and pigments. ISS include proteins, cellulose, hemicellulose, pectins and polysaccharides, which determine fruit juice. Estimates of SS contents of the commercial cultivars of tomato range between 4.6 (mostly fresh tomato) and 6.3% (mostly for processing) of fresh weight. However, accessions have been identified within related wild species of tomato, including *Lycopersicon pimpinellifolium*, *Lycopersicon chmielewskii* and *Lycopersicon cheesmanii*, with much higher concentrations (9 to 15%) of SS (Chen and Foolad, 1999). Titratable acidity, pH, fruit firmness are important fruit quality characteristics of tomato. Organic acids give the fruits sourness, and affect flavour by acting on the perception of sweetness (Fisher et al., 1997). The major organic acids in tomato are citric and malic acid, with citric acid predominating (Davies and Hobson, 1981). Acidity influences storability of processed tomato. Lower pH reduces the risk of pathogen growth in tomato products, such as *Bacillus coagulans*, which is found to be completely inhibited by a pH below 4.1 (Majid, 2007). Titratable acidity has no significant effect on tomato flavour unless pH is low. For this reason, a pH below 4.5 and citric acid content of above 0.35 g/100 g of fruit fresh weight are desirable. Firmness is one of the major factors contributing to shelf quality of tomato fruit. Consumers judge the quality of fresh tomatoes by their firmness, colour and taste (Rosenfeld et al., 1994).

Organoleptic quality is correlated with sensorial parameters. Flavour is a function of both taste (e.g., sugars and acids), and aroma (e.g., volatile compounds) components. Sugars, acids and sugar/acid ratios have been defined as good indicators of tomato flavour (Stevens et al., 1977; Kader et al., 1978).

In addition, tomatoes may provide a convenient matrix by which nutrients and other health-related food components can be supplied to humans (Sanchez-Moreno et al., 2006). Thus, tomatoes contribute significantly to the dietary intake of vitamin C, lycopene, beta-carotene, folate, potassium, flavonoids and vitamin E (Willcox et al., 2003). Fresh and processed tomatoes are rich sources of a natural antioxidant which can replace synthetic antioxidants and thus protect cells from oxidants that have been linked to cancers of the stomach, oesophagus, lung, pharynx, endometrium, pancreas, and colon (Hounsome et al., 2008).

Tomatoes can provide a significant proportion of total antioxidants in the diet, in the form of carotenes and phenolic compounds. Lycopene predominates among carotenoids and is mainly responsible for the red colour of tomato fruits and their derived products (Valverde et al., 2002).

In recent years, with the arrival of new commercial varieties, fruit external attributes such as colour, size and shape have a priority in improvement programs, whereas organoleptic and nutritional attributes have been regarded to be of secondary importance in food production. However, these parameters alone do not guarantee correct flavour and texture quality of a product. The sum of sugars, organic acids and the amount of volatile compounds, as well as colour, shape and texture determine the sensory properties of tomatoes (Azodanlou et al., 2003). One possibility for improving the quality of numerous horticultural crops might be to obtain new cultivars from hybridisations between the present day improved varieties and traditional local lines.

The objective of this work was to select varieties with excellent fruits characteristic and that can easily compete with commercial varieties and thus could considerably ease future breeding programmes.

MATERIALS AND METHODS

Plant material and crop conditions

Thirteen traditional varieties of tomato, collected from the traditional agrosystem of Tunisia, were grown in winter cycle in covered greenhouse at the Institut des Régions Arides (Southern Tunisia, Latitude 33°35' N Longitude 10°48'3" E Altitude 105 m). The climate in this region can be defined as arid inferior, with mild winter. Measured air temperature in the greenhouse corresponding to the harvest period is in average of 31.65°C.

The experimental design consisted of randomized blocks replicated four times. Each replicate contained five plants per variety. These varieties were called IRA, followed by a number to identify their origin. The commercial varieties "Marmande VF", "Rio Grande" and "Ventura", the three most cultivated varieties in Tunisia during the last decade, were used as controls. Tomatoes were harvested at the optimum ripeness stage between 9:00 and 11:00 am and immediately used for analysis.

Tomato samples were analysed for their agronomic traits (fruit weight, diameter), physicochemical traits (total solid percentage, citric acid concentration, pH and total soluble solid content), nutritional traits (vitamin C, potassium content) and sensorial traits (intensity of red colour, firm texture, elasticity, intensity of aroma and taste).

Physicochemical and nutritional analysis

Soluble solid content (°Brix) was determined using a hand refractometer (Reichert, Scientific Instruments), pH was measured with a pH-meter (pH-200L; Neo Met). For determination of total acidity, 10 g of tomato puree was titrated to 8.0 using 0.1 mol/L KOH. The titrated volume (mL) corresponds directly to total acidity expressed as g/L citric acid (Azodanlou et al., 2003). To determine potassium content, tomato homogenates were dried at 80°C for 48 h and were subsequently digested in a 60% nitric and 60% perchloric acid mixture (85:15, v/v) at 200°C for 6 h (Anza et al., 2006). Cationic mineral ($K+$) was determined by flame atomic absorption spectrophotometry (Sherwood, Model 410).

Vitamin C (Ascorbic Acid) was determined as described by Serrano et al. (2007). A portion of 25 g of fruit was added to 25 ml of 4.5% ortho-phospohric solution. The mixture was homogenized and centrifuged at 9000 *g* for 25 mn at 4°C. The supernatant was filtered through Whatman No.1. Then 10 ml of the filtered sample were passed through a Millipore 0.45 micrometer membrane and were thus ready to be injected in the HPLC system (Knauer; UV and RI detecteur; Pump K501).

Sensory analysis

Sensory analyses were performed by a trained panel of 10 judges. For each variety, four different attributes were revealed: one related to appearance (red colour intensity), one to flavour (intensity of aroma), two to texture (firm texture, juiciness, and elasticity), and one to taste (tomato taste). Each panellist received 10 samples, then the panel rated the different parameters on 1-6 scale (e.g. 1 = very weak aroma intensity and 6 = very strong aroma intensity). The same liking scale was used for the overall appreciation.

Statistical evaluation

Statistical analysis was performed using SPSS program (version 15). Analysis of variance (ANOVA) and Duncan's multiple range test ($P<0.05$) was used to establish possible significant variation among varieties of instrumental parameter (agronomic, organoleptic and nutritional parameters) analyzed. To assess key relationships which exist among characteristics involved in tomato fruit organoleptic quality, principal agronomic, nutritional and sensory parameters, Pearson's correlations between all traits pairs were calculated and the significance of their associations was tested with t-test at a significance level of 0.05 and 0.01.

RESULTS AND DISCUSSION

Comparison between different tomatoes cultivar by average of all instrument traits calculated is presented in Table 1. Traditional tomato variety IRA17 had the highest

Table 1. Mean value of agronomic, physicochemical and nutritional parameters evaluated in tomato fruits of traditional varieties (IRA) and controls.

Variety	TS	TSS	pH	TA (g/l citric acid)	K+	Vitamin C (mg/100 g)	Weight (g)	Diameter	TSS/TA
Marmande VF	7.86^{bc}	4.5^{cd}	4.45^{c}	7.9^{c}	2.72^{b}	8.05^{abc}	150.25^{c}	90.96^{g}	0.56^{ab}
IRA 61	7.11^{ab}	3.3^{abcd}	4.31^{abc}	3.97^{ab}	2.1^{ab}	10.89^{cd}	108.25^{abc}	57.69^{cdef}	0.85^{abc}
IRA 622	7.11^{ab}	3.67^{abcd}	4.49^{c}	4.27^{ab}	1.86^{ab}	9.39^{abc}	116.25^{bc}	64.37^{def}	0.88^{bc}
IRA 103	7.30^{b}	2.35^{ab}	4.39^{bc}	2.52^{a}	1.81^{a}	9.45^{bc}	15.5^{a}	27.02^{a}	0.91^{bc}
IRA 17	10.32^{d}	2.67^{abcd}	4.46^{c}	3.25^{ab}	2.19^{ab}	10.55^{bcd}	25.25^{ab}	30.99^{ab}	0.83^{abc}
IRA 9	7.19^{ab}	4.22^{bcd}	4.41^{c}	9.05^{c}	2.43^{ab}	10.91^{cd}	324.25^{d}	91.03^{g}	0.47^{a}
IRA 162	8.82^{c}	3.95^{abcd}	4.35^{abc}	5.72^{b}	1.9^{ab}	10.53^{bcd}	126.25^{c}	92.57^{g}	0.69^{ab}
Rio Grande	6.82^{ab}	2.02^{a}	4.41^{c}	3.12^{ab}	2.72^{b}	6.01^{a}	97.25^{abc}	42.41^{abc}	0.65^{ab}
IRA 22	6.70^{ab}	3.85^{abcd}	4.36^{abc}	3.15^{ab}	2.14^{ab}	10.16^{bcd}	87^{abc}	46.02^{abcd}	1.13^{c}
IRA 23 A	7.58^{bc}	2.40^{abc}	4.41^{c}	4.85^{ab}	2.19^{ab}	8.57^{abc}	145.75^{c}	68.97^{ef}	0.45^{a}
IRA 23 B	7.57^{bc}	4.57^{d}	4.19^{a}	5.12^{ab}	2.38^{ab}	10.56^{bcd}	150.25^{c}	77.52^{fg}	0.95^{bc}
IRA 5	5.88^{a}	3.02^{abcd}	4.37^{abc}	3.45^{ab}	2^{ab}	9.15^{abc}	125^{c}	65.18^{def}	0.89^{bc}
IRA 21	7.02^{ab}	3.17^{abcd}	4.38^{abc}	3.47^{ab}	2.1^{ab}	8.80^{abc}	122^{c}	50.21^{bcde}	0.92^{bc}
IRA 2	7.83^{bc}	2.5^{abcd}	4.3^{abc}	4.57^{ab}	2.19^{ab}	9.59^{bcd}	120^{c}	63.51^{def}	0.58^{ab}
Ventura	6.74^{ab}	2.65^{abcd}	4.38^{abc}	2.9^{a}	2.29^{ab}	7.30^{ab}	80.25^{abc}	41.16^{abc}	0.92^{bc}
IRA 3	7.12^{ab}	2.47^{abcd}	4.21^{ab}	3.6^{ab}	2^{ab}	**12.94^{d}**	90.5^{abc}	60.44^{cdef}	0.73^{abc}

Means for groups in homogeneous subsets are displayed, alpha = 0.05, harmonic mean sample size used = 4,000.

total solid percentage. Both controls, Rio Grande and Ventura had significant lowest total solid content. In the industry, this parameter dictates the factory yield (Pedro et al., 2005), in which the highest tomato total solids content amount to less tomato to be used to produce processed tomato product. Higher value in TSS content (4.57° Brix) was found in traditional variety IRA23B. The lowest value was recorded in Rio Grande (2.02° Brix). Higher value in TA (9.05 g/L citric acid) was observed in IRA 9.

Traditional variety IRA 103 had the lowest value (2.52 g/L citric acid). It was reported that higher fruit acidity is an advantage, as it causes a lower incidence for fungal infection (Mohammed et al., 1999). The higher value in SS/TA ratio (1.13) was found in IRA 22. The lowest value (0.47) was recorded in IRA 9, which had the lowest titratable acidity. Some researchers have defined sugars, acids and sugars/acids ratio as good indicators of tomato flavour (Kader et al., 1978), and sugars/acids ratio as important parameter in differentiating tomato flavour among varieties (Stevens, 1972). All varieties had pH values equal or bellow 4.49, which is considered to be ideal for correct fruit sourness.

Higher vitamin C value was recorded in IRA3, whereas the three controls had significant lowest vitamin C concentrations. Potassium concentrations had the smallest range of variation in chemical parameters analysed in this study. Agronomic parameters, fruit diameter and weight, had larger range of variation in instrumental parameters.

Relationships among traits

Results of correlation analysis between instrumental traits are summarised in Table 2. Controls were not considered in correlation analysis because of their higher heterozygous characteristics (Gomez et al., 2001). Thus, only correlations between tomato traditional varieties traits were calculated. Fruit weight and diameter were positively correlated with TA on the one hand (0.689, $P<0.01$; 0.662, $P<0.01$, respectively) and with TSS on the other hand (0.358, $P<0.01$; 0.387, $P<0.01$, respectively). Whereas, correlations between these agronomics parameters and SS/TA ratio were negatives.

Among chemical traits, the strongest positive correlation was observed between sugars content and titratable acidity (0.602, $P<0.01$). Some recent studies, for example, Georgelis (2002), Colombani et al. (2001) and Getinet et al. (2008) had confirmed this correlation and had shown that positive correlation between sugars and titratable acidity means that, generally, plant with high sugars have more free organic acids than plants with low sugars.

Whereas many other studies support the hypothesis that organic acids are produced within the fruit from stored carbohydrate material although a proportion may also be translocated from leaves and roots to fruit (Sakiyama et al., 1976; Davies and Maw, 1972; Getinet et al., 2008).

Among the sensory traits (Table 3), red colour intensity was positively correlated with aroma intensity (0.619, $P<0.01$), taste (0.834, $P<0.01$) and juiciness (0.619, $P<0.01$).

Taste and aroma, which describe close flavour components of fruit, were strongly correlated (0.924, $P<0.01$), while, fruit elasticity was negatively correlated to most of sensory traits.

Table 2. Spearman's correlation coefficients and level of significance among instrumental traits analysed in fruits of traditional varieties.

	TA	TSS	TS	TSS/TA	K+	Vit C	pH	Weight	Diameter
TA	1.000								
TSS	**0.602****	1.000							
TS	0.088	0.020	1.000						
TSS/TA	**-0.342***	**0.452****	0.017	1.000					
K+	0.190	0.205	-0.020	0.074	1.000				
Vit C	0.088	0.027	0.201	-0.054	0.130	1.000			
pH	-0.133	-0.129	0.126	0.012	-0.021	-0.030	1.000		
Weight	**0.689****	**0.358****	-0.107	**-0.277***	0.142	0.041	0.016	1.000	
Diameter	**0.662****	**0.387****	-0.56	**-0.311***	0.207	0.188	-0.94	**0.698****	1.000

**Correlation is significant at the 0.01 level (2-tailed), *Correlation is significant at the 0.05 level (2-tailed).

Table 3. Spearman's correlation coefficients and level of significance among sensorial traits analysed in traditional varieties fruits.

	Firm texture	Juiciness	Red colour intensity	Aroma intensity	Taste	Elasticity
Firm texture	1.000					
juiciness	0.040	1.000				
Red colour intensity	-0.028	0.619**	1.000			
Aroma intensity	0.008	0.491**	0.626**	1.000		
taste	-0.028	0.748**	0.834**	0.904**	1.000	
elasticity	0.162	-0.486**	-0.508**	-0.853**	-0.795**	1.000

**Correlation is significant at the 0.01 level, *Correlation is significant at the 0.05 level.

Table 4. Spearman's correlation coefficients and level of significance between instrumental and sensorial traits analysed in fruits of traditional varieties of tomato.

	Firm texture	Juiciness	Red colour intensity	Aroma intensity	Taste	Elasticity
TA	**0.311***	**0.385****	Ns	ns	ns	ns
TSS	ns	ns	Ns	ns	ns	ns
TS	ns	ns	Ns	0.553**	0.357**	-0.425**
TSS/TA	ns	-0.417**	Ns	ns	-0.318*	ns
K	ns	ns	Ns	ns	ns	ns
Vit C	ns	ns	Ns	ns	ns	ns
pH	ns	-0.280*	-0.308*	ns	ns	ns
weight	0.426**	0.349*	Ns	ns	ns	ns
diameter	0.366**	0.492**	Ns	ns	ns	ns

**Correlation is significant at the 0.01 level (2-tailed), *Correlation is significant at the 0.05 level (2-tailed), ns = correlation not significant.

Correlation between sensory and instrumental traits

Correlations between sensory and instrumental traits are shown in Table 4. Total acidity had a positive correlation with firm texture and juiciness. This is to be expected when we look at physiological process, which take place during fruit maturation, when a decrease in acidity and firmness is observed.

Positive correlation was recorded between firm texture and agronomic traits. This correlation can be investigated as association of selection traits, which can offer easy hybridisation possibilities. Red colour intensity was negatively correlated to pH. Total solid content had positive correlation with aroma intensity and taste on one hand and negative correlation with elasticity on other hand. These correlations confirm the importance of this parameter in tomato flavour.

Correlations analysis shows that there is no correlation between vitamin C and potassium contents and all parameters analysed in this study. They can be correlated to physiological parameter such as flowering time, growth and harvest date or environmental factors.

Conclusions

To assess traits that contribute to defining the target quality to design strategies to improve it is a long-term objective of tomato breeding programmes.

In general, results of this study show that traditional varieties had several specific parameters, with score better than the commercial varieties; these varieties could thus be used for improvement programs as a source material for optimal quality characteristics. Some local varieties, which had the same range of some quality characteristics as commercial varieties, have the advantage of being pure line and the transfer of such a character in a breeding programme may be less difficult.

REFERENCES

Anza M, Rigal P, Garbisu C (2006). Effects of variety and growth season on the organoleptic and nutritional quality of hydroponically grown tomato. J. Food Qual. 29:16-37.

Azodanlou R, Darbellayb C, Luisierc JL, Villettazc JC, Amadoa R (2003). Development of a model for quality assessment of tomatoes and apricots. Elseiver Lebensm.-Wiss. U.-Technol. 36:223-233.

Bertin N, Guichard S, Leonardi C, Longenesse JJ, Langlois D, Navez B (2000). Seasonal evolution of the quality of fresh greenhouse tomatoes under Mediterranean conditions, affected by air vapour pressure deficit and plant fruit load. Ann. Bot. 85:741-750.

Chen FQ, Foolad MR (1999). A molecular linkage map of tomato based on a cross between *Lycopersicon esculentum* and *L. pimpinellifolium* and its comparison with other molecular maps of tomato. Genome 42(1):94-103.

Colombani V, Causse M, Langlois D (2001). Genetic analysis of organoleptic quality in fresh market tomato. 1. Mapping QTLs physical and chemical traits. Theor. Appl. Genet. 102:259- 272.

Davies JN, Hobson GE (1981). The constituents of tomato fruit - the influence of environment, nutrition, and genotype. Crit. Rev. Food Sci. Technol. 15:205-280.

Davies JN, Maw GA (1972). Metabolism of citric and malic acids during ripening of tomato fruit. J. Sci. Food Agric. 23:969-979.

Georgelis N (2002). High fruit sugar characterization, inheritance and linkage of molecular markers in tomato. An M.Sc. Thesis. Presented to the School of Graduate Studies of Florida University. p. 81.

Getinet H, Seyoum T, Woldetsadik K (2008). The effect of cultivar, maturity stage and storage environment on quality of tomatoes. Elsevier. J. Food Eng. 87:467-478.

Hounsome N, Hounsome B, Tomos D, Edwards-Jones G (2008). Plant metabolites and nutritional quality of vegetables JFS R: Concise Reviews/Hypotheses in Food Science 1 -18.

Kader AA, Morris LL, Stevens MA, Albright-Holton M (1978). Composition and flavour quality of fresh market tomatoes as influenced by some postharvest handling procedures. J. Am. Soc. Hortic. Sci. 103:6-13.

Kader AA, Stevens MA, Albright-Holton M, Morris LL, Algazi M (1977). Effect of fruit ripeness when picked on flavor and composition in fresh market tomatoes. J. Am. Soc. Hortic. Sci. *102*:724-731.

Majid R (2007). Genome Mapping and Molecular Breeding of Tomato. Int. J. Plant Genomics p. 52.

Malundo MM, Shewfelt RL, Scott JW (1995). Flavor quality of fresh tomato *(Lycopersicon esculentum* Mill.) as affected by sugar and acid levels. Postharvest Biol. Technol. 6:103-110.

Mohammed M, Wilson LA, Gomes PL (1999). Postharvest sensory and physiochemical attributes of processing and non-processing tomato cultivar. J. Food Qual. 22:167-182.

Rosenfeld D, Shmulevich I, Galili N (1994). Measuring firmness through mechanical acoustic excitation for quality control of tomatoes, Food Automation Congress, February 1994.

Sakiyama R, Stevens A (1976). Organic acid accumulation in attached and detached tomato fruits. J. Am. Soc. Hortic. Sci. 101:394-396.

Serrano IO, Hernandez-Jover T, Martin-Belloso O (2007). Comparative evaluation of UV-HPLC methods and reducing agents to determine vitamin C in fruits. Food Chem. 105:1151-1158.

Stevens MA (1972). Relationships between components contributing to quality variation among tomato lines. J. Am. Soc. Hortic. Sci. 97:70-73.

Stevens MA, Kader AA, Albright-Holten MA, Algazi M (1977). Effect of fruit ripeness when picked on flavor and composition in fresh market tomatoes. J. Am. Soc. Hortic. Sci. 102:724-731.

Valverde IM, Periago MJ, Provan G, Chesson A (2002). Phenolic compounds, lycopene and antioxidant activity in commercial varieties of tomato (*Lycopersicum esculentum*). J. Sci. Food Agric. 82:323-330.

Willcox JK, George L, Catignani GL, Lazarus S (2003). Tomatoes and cardiovascular health. Food Sci. Nutr. 43:1-18.

7

An evaluation of milk quality in and around Rawalakot Azad Kashmir

Saiqa Bashir[1], Muhammad Siddique Awan[1], Shahzad Akbar Khan[2] , Habib Ahmed Rathore[1], Muhammad Akhtar Qureshi[2] and Zulfiqar Hussain Kathu[3]

[1]Department of Food Technology, The University of Poonch Rawalakot, Pakistan.
[2]Faculty of Veterinary and Animal Sciences, The University of Poonch Rawalakot, Pakistan.
[3]Department of Animal Husbandry, Government of AJK, Pakistan.

The present study was conducted to evaluate the quality of milk from Rawalakot and its surroundings. Milk samples were collected from house hold animal, milkman and restaurant milk in order to assess the hygienic status of raw milk, which was collected from three different parts of the city, Rawalakot Azad Jammu and Kashmir. It was observed that the milk samples which were collected from house hold milk showed better quality compared to the other two sources. For the assessment of adulterants, clot-on-boiling (COB), alcohol perception and methyleneblue tests were conducted. In COB test, the result was 29% positive in house hold animal milk, 38% in milkman and 38% in restaurant milk; whereas, in COB test, results of milkman and restaurant milk were similar. The results of alcohol tests showed variation among the three sources of milk: in household animal milk (APT) it was 34%; 44% in milkman and 34% in restaurant milk. Methyleneblue test results show variation among the three sources; the mean values of A+, B+, C+ and D+ for HH were 29, 44, 17, 11 and 14%; 62, 11, 14% were recorded for MM; whereas 14, 33, 33 and 11%, in RM, respectively. The quality of milk was different in all three sources. Milk composition was assessed by specific gravity, fat % and acidity test. The result of Gerber test (fat% age) shows slight variation among the three sources of milk (7.00%, 6.31% and 5.55), respectively. The results of specific gravity were 0.128 g/ml for house hold, 1.027 g/ml for milkman and 0.125 g/ml for restaurant milk, respectively. All three sources showed low specific gravity than standards. The result of acidity test shows slight difference in all three sources of milk. The acidity of milk samples was recorded as 0.18 in house hold milk, 0.21 for milkman milk and 0.19 in restaurant milk. The acidity was higher than the standard values in both milkman and restaurant milk. In organoleptic test, the result was based on sensory evaluation and scoring system. The lowest value was 5.5 and highest value was 7.4. As the results show the HH animal milk was graded as good compared to the other two sources of milk.

Key words: Organoleptic, adulterants, milk, acidity.

INTRODUCTION

Milk is the lacteal secretion which is free from colostrums during the period of 15 days before parturition and five days after; it contains: 8.5% solids-not-fat, 87.20% water, 3.50% protein, 3.70% fat, 4.90% milk sugar, 0.70% ash and 12.80% dry matter (Tenvir, 2007). Thus milk is a balance diet which contains all basic components to fulfill the needs of an individual. Milk protein is considered as high quality protein with all essential and non essential amino acid, highest biological value and easily digested to the extent of 97 to 98 % (Tenvir, 2007). Milk and milk products contribute ⅓rd of world's intake of animal protein. Milk plays a vital role in building a healthy society and can be used as vehicle for rural development. Fresh or raw milk as diet contributes to infants and adults all over the world (Cousin, 1982).

Pakistan is in 3rd position in milk production with annual contribution of 43.29 million tons (Ministry of Livestock and Dairy Development Pakistan, 2009-10). Buffalo and cow account for 96% of the total milk production in Pakistan, of which buffalo is in first position and cattle, 2nd position (Khan et al., 2008). Livestock sector contributes 11% of total GDP and provides main work force for 30-35 million people of rural population (Anonymous, 2008). Milk and milk products represent 27% of the total house hold expenditure on different food items (Zia, 2006).

There are several factors which decrease the quality of raw milk. These include poor animal health, delivery services, poor house hygiene, unhygienic milking practices, udder infection and zoonotic infection. Other factors such as presence of antibiotic residues and animal diseases (mastitis) also affect milk quality. On the other hand, environmental changes such as seasonal and nutritional change also affect milk composition.

Absence of cooling rooms and hygiene facilities may affect the quality standards of milk (Blowey and Edmondson, 2000; Grange, 1998; Hogeven et al., 2000). Milk quality can also be decreased by some other factors like milk adulteration during and after milking, udder disorder and inflammation of mammary glands of different milking animal cattle, sheep and camels (Molla and Bekele, 2000; Edmondson, 2001; Ali, 2001).

Milk differs in compositional variations in different species due to factors such as species of animal, breed, individuality, stage of lactation, frequency of milking, age, seasonal variations, feed, milking interval, use of drugs and hormonal change (Ensminger, 1993). All these factors cause economic losses in dairy industry throughout the world (Kossabati et al., 1998).

In Rawalakot (AJK), majority of the people use raw milk but no proper guidelines are available for milking, handling and storing at house hold level restaurants and shops. In AJ&K very limited work has been under taken to assess the quality standards of milk.

The objective of this study was to evaluate the hygienic quality and level of microbial contamination in raw milk which was collected from house hold animals, milk man and restaurants. The present study was therefore planned to investigate the quality and safety standards.

MATERIALS AND METHODS

The present study was carried out in the Food Technology Research Laboratory of Department of Food Sciences and Technology Faculty of Agriculture Rawalakot University of Azad Jammu Kashmir, to study the hygienic quality of raw milk from Rawalakot and its surroundings.

Collection of samples

Milk samples were collected from house hold animals directly, from the milkers/gawala and from the respondents of restaurants in bottle from Rawalakot Azad Jammu and Kashmir, and its surroundings. They were collected in aseptic plastic bags with the help of sampling spatula. They were labeled, placed in ice packing and then transported to the laboratory for further analysis. These samples were transferred in to the refrigerator immediately at 4°C and culturing was conducted within 24 h. The sample was subjected to the following quality tests.

Quality tests of milk

Clot on boiling test

This test was performed according to the method described by Marshall (1992) for acid milk having pH less than 5.8 or abnormal milk (e.g. colostral or mastitis milk) to assess milk acidity. The alcohol test was done by using a 68% ethanol solution. Tests were done immediately after the samples were delivered to the laboratory within four hours. 2 ml of the raw milk sample was taken and mixed with 68 percent ethanol solution in a sterile test tube. The solution was prepared from 68 ml 96 percent alcohol and with 28 ml of distilled water.

Alcohol perception test

This test was also performed according to method of Marshall (1992); the test is based on the proteins (instability) and concentration of acid or rennet is increased. Test assesses the increased levels of albumen colostrum milk and salt concentrates (mastitis).

Methylene blue test (MBRT)

Methylene blue test for the assessment of mastitis was performed according to the procedure described by Awan and Rahman (2005). The test is used to diagnose mastitis, the ability of bacteria to reduce the color of methylene blue dye from the milk sample. Dye reduction time is inversely proportional to the presence of total number of bacteria in sample; hence the greater the bacterial population, the shorter is the dye reduction time.

Acidity test

Acidity in different milk samples has been calculated by the method described by Marshall (1992). The natural acidity of milk is 0.16 - 0.18%. Figure higher than this signifies developed acidity due to the action of bacteria on milk sugar.

Table 1. Value for clot-on-boiling of different milk samples.

Sources	Positive (%)	Negative (%)
House hold	29^{b}	71^{a}
Milkman	38^{a}	62^{b}
Restaurant	38^{a}	62^{b}

Table 2. Values for alcohol precipitation test of different milk samples.

Sources	Positive (+ive)%age	Negative (-ive) %age
House hold	34^{b}	66^{a}
Milkman	44^{a}	56^{b}
Restaurant	35^{b}	65^{a}

Specific gravity

The specific gravity of milk was determined by Lactometer according to Marshall (1992).

Gerber test

Gerber test is used for the determination of fat percentage. For this purpose butyrometer is used. Fat % in different milk samples was determined by the method of Marshall (1992).

Organoleptic tests

The rapid segregation of low quality milk samples at milk receiving platform was based on Marshall (1992). Milk grade should have good sight, smell and taste.

Statistical analysis

The data obtained were subjected to statistical analysis by using CRD (Steel, 19 97).

RESULTS AND DISCUSSION

The present study was carried out to assess the hygienic quality of raw milk from Rawalakot and its surroundings. The research project was conducted in Food Technology Research Laboratory Faculty of Agriculture Rawalakot, University of Azad Jammu and Kashmir. About 102 milk samples were collected from different sources mainly, the house hold (HH), milkman (MM) and restaurant milk (RM). These samples were analysed for different parameters like adulteration (clot on boiling, alcohol precipitation test, methylene blue test), gross milk composition (specific gravity, acidity, fat) and organoleptic characteristics.

Adulterations

The adulterants such as chemicals, starch, oils, salts, soaps and water are commonly added to milk by milk sellers which may adversely affect the milk quality. So it is necessary to evaluate the milk quality with regard to adulterants that are added to the milk to increase the quantity and raising extra money. Therefore, the quality of milk affected by these adulterants was evaluated in the laboratory by clot on boiling (COB) and alcohol perception test (APT), respectively.

Clot-on-boiling

The results pertaining to the test clot on boiling (COB) are presented in Table 1. The results obtained from different sources differ significantly from each other. The positive results for COB (29, 38 and 38%) were obtained from house hold, milkman and restaurant milk respectively. Higher difference in COB values for house hold animal milk and milkman might be due to the management practices during the handling of raw milk. It was obvious from the results that HH milk was in better condition as compared to other sources. There was no significant difference observed in COB values of milkman and restaurant that might be due to similar practices in handling, keeping, transportation and storage practices. Variations in COB values are the main reasons of quality deterioration of milk. Furthermore it is revealed that deterioration increases as the steps increase towards marketing. Similar trend in deterioration was investigated previously by Grimaud et al. (2004).

Alcohol perception test (APT)

The results pertaining to the APT test are shown in Table 2. The result of the APT test shows the significant differences among all three sources of milk. Positive cases recorded 34% in HH milk whereas in MM, 44% and RM, 35% values. The result shows high level of acid and excess of salt concentration in the milk samples that indicate low quality milk. Milk coagulated only when the acidity of milk reached 0.21-0.23%. According to the results the positive test was slightly lower in RM and HH but a big difference was observed in MM milk. The result of APT for house hold animal's milk showed better quality compared to the other milk sources. The milkman frequently used the adulterants (chemicals, salts, starch, etc) in milk before marketing; therefore, such adulterants can cause the increase of acidity of milk. The results are in agreement with Grimaud et al. (2004), who observed

Table 3a. Values for methylene blue test of different milk samples.

Sources	A+ (%)	B+ (%)	C+ (%)	D+ (%)
House hold	29[e]	44[b]	17[f]	11[h]
Milkman	14[g]	62[a]	11[h]	14[g]
Restaurant	14[g]	33[c]	33[d]	11[h]

A + 8 h, B + 6 h, C + 4 h, D + 2 h.

Table 3b. Grading of samples on the basis of (MBRT) in different milk sources.

Quality of milk	Decolonization time
Excellent	More than 8 h
Good	About 6 h
Fair	Between 2 and 6 h
Poor	Less than 2 h

Table 4. Mean values for acidity (%) of different milk samples.

Source	Mean
House hold	0.18[b]
Milkman	0.21[a]
Restaurant	0.19[b]

Table 5. Mean values for specific gravity (g/ml) of different milk samples.

Source	Mean
House hold	1.028[a]
Milkman	1.027[a]
Restaurant	1.025[b]

similar results in milk samples taken from market.

Methylene blue reduction test (MBRT)

The data regarding to methylene blue reduction test (MBRT) are given in Table 3a and b. The results obtained from the three different sources are found highly significant. The mean values of A+, B+, C+, and D+ for HH were 29, 44, 17 and 11, 14, 62, 11%;14% was recorded for MM and 14, 33, 33 and 11% in RM, respectively. In all three sources of milk grading was different; milkman milk contains high percentage (14% D+) of poor quality milk. However, restaurants milk contains high percentage (33% C+) of fair milk quality; whereas milk quality of HH milk was comparatively better than the other two sources. The poor quality of milk for restaurants and milkman may be due to the following reasons.

The milkman collected the milk from infected animals and the casual organism of mastitis may easily be transferred from one animal to another animal; even by milking practices. In case of mastitis positive animals, they need time to be cured and to recover, but the common people of the area are not well aware about mastitis and its cure. The quality of milk depends on the degree of diseases invasion and the rate of contamination; as the number of organisms (*Staphylococcus aureus*) was more in milk, the quality of milk was lower. If the microbial load is low milk is considered of good quality. Furthermore increase in microbes decreases the quality of milk graded as B, C, D or poor quality of milk. According to the results, the high quality of milk was observed in house hold animal whereas milkman milk and restaurant milk percentage decreased. The result of the test is in agreement with Chatterjees et al. (2006).

Acidity test

Data pertaining to acidity of milk samples have been shown in Table 4; while its statistical evaluation is given in Table 8. The mean values (Table 8) are found to be significant. It is revealed from the work that acidity of MM (0.21) was higher than that of HH and RM. The acidity of milk increased by increasing temperature of storage room.

In MM and RM acidity of milk was higher than normal. It has two main reasons: the milkman has surplus store of milk after buying, so he used it next day by mixing it with fresh milk. Unsterilized utensils also enhance milk acidity. Milk is transported from far off places, which is another factor of high acidity because almost all the restaurants have to purchase the raw milk from local milkman or house hold animals. In restaurant milk, the raw milk is collected in big utensil/container in which milk of different sources was mixed, and this kind of mixing may increase the acidity of milk because some of them are already acidic and can be the source of high acidity in whole milk container.

Specific gravity

The result pertaining to specific gravity is presented in Table 5 and its statistical analysis is given in Table 9. The analysis of variance shows that the mean values for

Table 6. Mean values for fat percentage of different milk samples.

Source	Mean
House hold	7.00[a]
Milkman	6.31[a]
Restaurant	5.55[b]

Table 7. Mean values for the sensory evaluation of different milk samples collected from house hold, milk man and restaurants.

Source	Color	Flavor	Taste
House hold	7.4a	6.9a	6.0[b]
Milkman	6.9b	6.5b	6.6[a]
Restaurant	5.9c	5.7c	5.5[c]

Table 8. Mean values for acidity (%) of different milk samples.

Source	Mean
House hold	0.18[b]
Milkman	0.21[a]
Restaurant	0.19[b]

Table 9. Analysis of variance for specific gravity.

Source	df	SS	MS	F-Value	P
Source	2	0.0001088	0.000544	1.63*	0.025
Error	6	0.00200	0.0033		
Total	8	0.0020			

*Significant.

Table 10. Analysis of variance for fat percentage.

Source	df	SS	MS	F-V	p
source	2	2.97787	1.48893	14.6**	0.0050
Error	6	0.61193	0.10199		
Total	8	3.58980			

**Highly significant.

specific gravity were found to be significant. The mean values of the HH milk (1.028 g/ml) and MM (1.027 g/ml) are at par whereas RM (1.025g/ml) is higher than the other two sources. Hence there is slight difference in all the three sources. The specific gravity of HH animal milk was almost similar with MM milk; whereas the specific gravity of RM was found lower than that of the other sources. The recorded specific gravity was lower than the standard value of specific gravity that was 1.033 g/ml for buffalo and 0132 g/ml for cows, respectively.

Since the common practice is to add water to milk in every platform/collection point, the addition of water causes the decrease in specific gravity of milk as specific gravity of milk is slightly higher than the water. Thus the change in composition changes the specific gravity of milk. Due to removal of fat, the specific gravity of milk may increase because the weight of fat is much lower than the water. The fat content has low density so it increases the specific gravity of milk and reduces the milk nutritive contents; finally makes it low standard milk quality. The adulteration rate of water was almost common besides that some chemicals and powders might be commonly added to increase the specific gravity of milk. The widely used adulterant was water due to its cheapness and easy availability and it becomes the source of serious problem. This practice cannot be stopped because the people and MM used it as weapon of earning and it becomes a challenge for hygiene and safety measure of raw milk. The specific gravity of MM contains the lowest values compared to the other two sources. However, the results of specific gravity of milk were lower in all three sources as compared to standard and it may be an indication of adulteration, which was water. The results of the test were in line with findings of Grimaud et al. (2004) and similar trends were found in findings of Khan et al. (2008).

Gerber test

The data regarding the fat percentage test are presented in Table 6. The statistical evaluation (Table 10) represented that all three sources have significant differences. The average value of HH animal was 7.00% fat; in MM it was 6.31%, whereas means value for RM was 5.55%. Hence the fat percentage found varied in different sources of milk.

Raw milk received from HH animal was very pure and adulteration was lower than the other sources. The variation in the fat % may be due to the species or breed difference, as all the animal species have different fat percentage. In MM and RM milk the difference among the percentage was slightly lower as compared to HH; this might be due to mixing of milk from different sources. Mixing of milk is most common practice in MM milk and RM, because milk purchased from different sources may show variation in fat percentage. Fat proportions also deviate by the skimming practices of raw milk which is also common in milkman and other milk sellers. The fat percentage of MM and RM was lower than that of HH animal milk. In RM the fat percentage was recorded as 5.55%; it was not objectionable but comparatively lower than that of the other two sources.

Organoleptic evaluation

The results regarding organoleptic characteristics test are presented in Table 7. It is observed that samples from house hold, milkman and restaurants are significantly different in color, flavor and taste.

Color

The mean values of color for different milk sources (Table 7) showed grade A for HH, grade B for MM and grade C for RM respectively. The results show that HH milk has normal (yellowish white) color as compared to the other two sources. However, the color of MM sources is slightly different from HH milk source and shows more yellowish pale color than normal milk color. However, the milk color for the RM source shows abnormal milk color; the abnormality of color might be due to adulteration factors like water, chemical (starch, formaldehyde, olive oil, etc.). The color difference in milk was also observed due to change in feeding habits of animal. The animal which eats more consolidated feed has more yellowish or pale color of milk compared to pasture feeding animals.

Flavor

The comparison of mean values for the flavor of different milk sources (Table 7) showed A grade for HH milk, B grade for MM milk and C grade for RM samples. The results showed that the flavor for all three milk sources was highly significant; all the three sources show variation among them. The flavor for HH milk was almost equal to normal flavor of milk but in case of MM the flavor was slightly bitter than normal taste of milk. However, the flavor for the RM sources was not good and has off flavor different from normal milk flavor. This flavor differences among the milk sources are due to addition of chemicals, long term storage and lack of storage facilities and also due eating habits of animals. Commonly, milk flavor is based on animals' feed; those animals that eat herbs or bushes have unpleasant smell that gives the off flavor milk. Some medicinal plants also produce bad smell and off flavor, which is objectionable; this leads to the devalue of milk. The results are in line with the findings of Islam et al. (1984).

Taste

The results regarding taste of different milk sources (Table 7) showed that the taste for HH milk was B grade, A grade for MM and C grade for RM, respectively. The taste for HH and RM has poor results as compared to the MM sources of milk. The variation of taste in different milk samples depends on feedings habits of animal to some extent; the nutritive feeding of animals given the good results in all aspect of milk characteristics such as flavor taste and color. There were few samples in which abnormalities seemed to be sweet less taste of milk (containing saltish taste, bitter taste); due to all these abnormalities, milk was graded as low quality. Overall evaluation of HH milk samples was good and better than the other two sources of milk.

Conclusion

The milk samples collected from house hold, milkman and restaurant represented considerable differences among them with regard to quality. Adulteration test (COB, APT and MBR) showed that milk sample from house hold milk has superior quality to milkman and restaurants milk. Milk composition (fat percentage, specific gravity, acidity) in HH and MM is almost similar but lower in RM. Specific gravity in all three sources was lowered. House hold milk due to lower acidity was better than the other two sources. Organoleptic test (color, flavor, taste) result regarding taste and color is also better in house hold milk sample. In case of flavor, house hold milk and milkman's man have quite better results than the restaurants milk. However, the contamination status in milkman and restaurants was higher and hence represented poor quality than the house hold milk samples. The low quality of raw milk in milkman and restaurant milk is due to mis-handling, contamination originating from farm level and lack of storage facilities.

REFERENCES

Ali Q (2001). Goats-undervalued assets in Asia.The regional work shop on Goat production system and markets, Luang Prabang 24-25:47-49.

Anonymous (2008). Pakistan Economic Survey. Government of Pakistan. Ministry of Finance, Economic Advisor Wing Islamabad.

Awan JA, U-Rehman S (2005). Elements food and nutrition. Published by NUI-TECH Communications, Faisalabad-Pakistan. ISBN:969-8543-03-1.

Blowey R, Edmondson P (2000). Mastitis controls in dairy herds an illustrated practical guide farming. 6:1-138.

Chatterjee SN, Bhattacharjee I, Chatterjee SK, Chandra G (2006). Microbiological examination of milk in Tarakeswar India with special references to coli forms. Afr. J. Biotechnol. 5:1383-1385.

Cousin MA (1982). Presence and activity of psychrotrophic microorganisms in milk and dairy products. J. Food Prot. 45:172-207.

Edmondson P (2001). Mastitis control in dairy herds, an illustrated practical guide farming press. 6:1-138.

Ensminger ME (1993). Milk Secretion and its handling. J. Dairy Cattle Sci. 3:416-422.

Grange JM (1998). The genus mycobacterium and species of mycobacterium in grange mycobacterium and human diseases.

Edward Arnold London, 25-102. FAO. J. Food Agric. 6: 57-59.

Grimaud P, Serunjogi ML, Grillet N (2004). An evaluation of milk quality in Uganda. Afr. J. Food Agric. Nutr. Dev. 7:1-16.

Hogeven H, Vorst V, Younas M (2000). Automatic milking system and milk quality In Netherlands. Proceeding of the international Symposium Robotic Milking, Lelystad 17-19 August 73-82.

Islam MN, Hossain SMI, Mannan AM (1984). Studies on the physical parameters and chemical qualities of market milk in Mymensingh town Bangladesh. Bangladesh J. Anim. Sci. 13:52-56.

Khan TG, Zinnah MA, Siddique MP, Rashid MH, Islam MA, Choudhury KA (2008). Physiology and Microbiological qualities of raw milk collected from Bangladesh Agriculture University dairy farms and its surrounding villages. Faculty of Veterinary Science Bangladesh. J. Vet. Med. 6:217-221.

Kossabati MA, Hovi M, Esslemont RJ (1998). Incidence of clinical mastitis is in dairy Herds in England. J. Vet. Res. 143: 649-653.

Molla B, Bekele T (2000). Mastitis in lactating camels (Camelus dromedaries) in African region north-eastern Ethiopia. Berl Munch Tierarztl Wochenschr, Afr. J. Agric. Res. 114:169-172.

Tenvir A (2007). A case study of milk production and marketing on small and medium scale contract Forming of Haleeb Foods Limited. Pakistan.

Zia U (2006). Analysis of milk marketing China and Pakistan FAO food Laws Manual 2010. Factor inflecting and variation in bulk milk of sheep flocks in Spain, Home Journal 93:1587-1597.

Physicochemical and bacteriological analyses of drinking water from wash boreholes in Maiduguri Metropolis, Borno State, Nigeria

Mustafa Alhaji Isa*, Ibrahim Alkali Allamin, Haruna Yahaya Ismail and Abubakar Shettima

Department of Microbiology, University of Maiduguri, P. O. Box 1069, Maiduguri, Borno State, Nigeria.

Physicochemical and bacteriological analyses of water samples were carried out from five wash borehole used for drinking purpose in Maiduguri Metropolis, Nigeria. The bacteriological analysis was carried out using multiple tube (most probable number) technique for enumeration of both total coliform count and differential *Escherichia coli* count. The results obtained were compared with World Health Organization (WHO), National Agency for Food and Drug Administration and Control (NAFDAC) and Nigeria Standard of Drinking Water Quality (NSDWQ) standard for drinking water. None of the samples complied with the bacteriological standard as the total coliform count ranges between 6 x 10^3 and 145 x 10^3 MPN/ml. Also, with the exception of samples from Hausari and Bulabulin-ngarannam, the other samples did not comply with the pH standard recommended by WHO, NAFDAC and NSDWQ. Total dissolved solid of the water samples from Hausari also exceeded the standard requirements. Although no *E. coli* was detected in the entire samples, there is need to create awareness about the present situation of the wash boreholes and the necessity for further treatment by consumers, before it can be used for both drinking and domestic purposes.

Key word: World Health Organization (WHO), National Agency for Food and Drug Administration and Control (NAFDAC), Nigeria Standard of Drinking Water Quality (NSDWQ), water, wash boreholes.

INTRODUCTION

The quality of drinking water is a powerful environmental determinant of health (WHO, 2010). Water plays an indispensable role in sustenance of life and it is a key pillar of health determinant, since 80% of diseases in developing countries are due to lack of good quality water (Cheesbrough, 2006). Drinking water quality management has been a key pillar of primary prevention for over one and half centuries and it continues to be the foundation for the prevention and control of water borne diseases (WHO, 2010). Contaminated water is a global public health threat placing people at risk of a host of diarrhoeal and other illness as well as chemical intoxication (Okonko et al., 2009). The major risk to human health is faecal contamination of water supplies.

Serious ill health can be caused by water contaminated from faeces being passed or washed into river, stream, pool or being allowed to seep into well or borehole (Cheesbrough, 2006). There has been a report of borehole water contamination through many domestic waste water and livestock manure especially if there is a puncture in a layer of soil (Obi and Okacha, 2007). These waste and sewage when deposited near the boreholes may travel with percolating rain water directly into the boreholes or may travel along the well-wall or surrounding material of the drill-holes (Obi and Okacha, 2007). There are several variants of the faecal-oral pathway of water borne disease transmission. These include contamination of drinking water catchments (example, human or animal faeces), water within the distribution system (such as leaky pipe or obsolete infrastructure) or of stored household water as a result of unhygienic handling (WHO, 2010). Increase in human

*Corresponding author. E-mail: mustafaalhajiisa@gmail.com.

population pose a great pressure on provision of safe drinking water especially in developing countries (Okonko et al., 2009). Consequently, water borne diseases such as cholera and typhoid often have their outbreak especially during dry season (Banu and Menakuru, 2010; Adenkunle et al., 2004). High prevalence of diarrhoea among children and infants can be due to the use of unsafe water and unhygienic practice (Oladipo et al., 2009; Tortora et al., 2002). Thus, many infectious diseases are transmitted by water through faecal oral contamination. Diseases due to drinking of contaminated water leads to the death of five million children annually and make 1/6 of the world population sick (Shittu et al., 2008). Also, water may contain toxic inorganic chemicals which may cause either acute or chronic health effect. Acute effects include nausea, lung irritation, skin rash, vomiting and dizziness, sometime death usually occurred. Chronic effect, like cancer, birth defects, organs damage, disorder of the nervous system and damage to the immune system are usually more common (Erah et al., 2002). Inorganic chemicals like lead may produce adverse health effect which include interference with red blood cell chemistry, delay in normal physical and mental development in babies and young children, slit deficit in attention span, hearing and learning abilities of children and slight increase in blood pressure in some adults. Also, presence of chromium in drinking water had been shown to result in chronic toxic effect (including liver and kidney damage, internal haemorrhage and respiratory disorders) in animal and human by ingestion. Although, the sources of metal contaminants of the underground water are uncertain, it may likely be due to natural process and anthropogenic activities (Erah et al., 2002). In addition, rural water also have excessive amount of nitrite from microbial action on agricultural fertilizer, when ingested nitrite compete for oxygen in the blood (Oladipo et al., 2009).

The use of groundwater is of great importance to Maiduguri inhabitants, as superficial water sources are lacking in the vicinity. Although, Lake Alau has been dammed up for that reason and serve as a reservoir for drinking water since the late 1980s (Rudiger, 2002). Many boreholes have been drilled with a depth of more than 600 m all over Maiduguri metropolis (Rudiger, 2002). Although, due to increase in population, public water supply from these boreholes is very unstable and unpredictable as supplies are often irregular. As a result, most of the people drilled their boreholes at a depth less than 30.48 m which are often called wash boreholes. These boreholes are widely recognized within the vicinity and it becomes an important source of drinking water in the area. The population congestion and consequent indiscriminate dumping of polluted water may enhance the infiltration of harmful compound into the ground water. Thus, the possibility of these contaminations may justify the purpose of this research. The objective of the present study was to analyze and determine the overall physical, chemical and bacteriological quality of drinking water from wash boreholes in Maiduguri Metropolis and the results was compared with the standard guidelines set by WHO, NAFDAC and NSDWQ. It should be noted, however, that there are no available literatures or researches that have been conducted on wash boreholes in the identified area upon which the authors can lay their hands. This research would also provide an insight into the water quality of the boreholes in Maiduguri, Nigeria.

MATERIALS AND METHODS

Sample site

Maiduguri Metropolitan area is the capital of Borno state, located in North eastern Nigeria. It covers more than 3000 km^2 of different land units. Maiduguri and its immediate environs is known for its dryness, with semi-arid climate, savannah or tropical grasslands vegetation, light annual rainfall of about 300 to 500 mm and the average daily temperature ranging from 22 to 35°C, with mean of the daily maximum temperature exceeding 40°C between March and June before the onset of the rains in July to September. It has mainly sandy loam soils (Arku et al., 2011). It is divided into twenty-one different geographical zones (Rudiger, 2002). In this study, five geographical zones were randomly selected; these include Bulabulin-ngarannam, Hausari, Moduganari, Gwange and Mashamari. Sampling sites were selected based on their closeness to refuse and waste disposal site. Water samples were collected from five different wash boreholes in the month of February and used for both physicochemical and bacteriological analyses.

Collection of water samples

Water samples were collected from five wash boreholes located at different geographical zones within Maiduguri Metropolis using two sterile 250 ml plastic bottles for each sample. To ensure the sterility of the samples, the borehole taps were sterilized by means of cigarette lighters, after which the taps were opened to flow for 1 to 2 min. Then, the plastic bottles were filled with water up to 200 ml leaving some space to allow shaking before analysis. The collected samples were delivered to the laboratory within 30 min of collection.

Physicochemical analysis

The water from samples were analysed for temperature, colour, turbidity, pH, total dissolve solid, conductivity, total alkalinity, manganese, iron, chromium, nitrite, copper and fluoride as describe by FAO (1997a).

Bacteriological analysis

Bacteriological characteristics of the water samples were determined using multiple tube fermentation method (most probable number) for enumeration of both total coliform count and differential *Escherichia coli* count. Lauryl Tryptose Broth (LTB) along with fermentation tubes (Durham tubes) was used. A serial dilution of the water sample to be tested was made and inoculated into LTB growth media. Samples were then incubated at 35°C for 48 h for the presumptive test for total coliform count. After the positive tubes were transferred to Brilliant green lactose bile broth (confirmation test) and incubated for 48 h at 35°C, the growth or gas production confirmed the presences of coliform (Nollet Leo, 2007).

Table 1. Physicochemical analyses of water samples from wash boreholes.

S/N	Parameter	Gwange	Mashamari	Hausari	Moduganari	Bulabullin Ngarannam	WHO		NAFDAC	NSDWQ
							Highest desirable level	Maximum permissible level		
1	Temperature at the time of analysis (°C)	26.6	27.5	29.2	28.3	30.0	-	40	-	Ambient
2	Colour (TCU)	Not determined	Not determined	Not determined	Not determined	Not determined	6		15	15
3	Turbidity (NTU)	0.77	0.85	0.78	0.85	0.98	5	25	-	5
4	Ph	6.04	6.50	5.72	6.28	6.60	7.0-8.5	6.5-9.2	6.5-8.5	6.5-8.5
5	Total dissolve solid (mg/L) (TDS)	374	332	547	94	247	500	1500	500	500
6	Conductivity (µs/cm)	747	664	1092	189	496	NS		NS	1000
7	Total Alkalinity (mg/L)	60	100	60	60	100	80	120	100	-
8	Iron (mg/L)	0	0	0	0	0	-	-	-	0.3
9	Chromium (mg/L)	0	0	0	0	0	-	-	-	0.05
10	Nitrite (mg/L)	0	0	0	0	0	-	-	-	0.2
11	Copper (mg/L)	0.11	0.32	0.47	0.05	0.09	-	-	-	1
12	Fluoride (mg/L)	0.33	0.15	0.63	0.92	0.13	-	-	-	1.5

NS: No standard; NAFDAC (2001), WHO (2001), NSDWQ (2007).

RESULTS

Physicochemical parameters of water samples from five wash boreholes within the Maiduguri Metropolis were analysed. The result is presented in Table 1 and indicated that the temperature of the water samples at the time of the analysis ranged from 26.6 to 30°C, with water samples from Bulabulin-ngarannam having the highest temperature of 30°C, while Gwange had the lowest temperature of 26.6°C (Table 1).

Water sample from Bulabulin-ngarannam had the highest turbidity of 0.98 NTU, while Gwange had 0.77 NTU, none of the samples had objectionable colour. The pH of the water samples ranged from 6.0 to 6.6, while the total dissolve solid ranged from 94 to 574 mg/L and conductivity measured at µs/cm also ranged from 189 to 1098. Total alkalinity ranged from 60 to 100, while the copper content ranged from 0.05 to 0.47 mg/L, with the water sample from Hausari having the highest copper content of 0.47 mg/L and that of Moduganari having the lowest copper content of 0.05 mg/L. Fluoride also ranged from 0.13 to 0.92 mg/L, but nitrite, chromium and iron were not detected in each of the samples (Table 1).

The most probable number for presumptive total coliform count of the water samples ranged from 6000 to 145,000 MPN per 100 ml. Water sample from Gwange had the highest coliform count of 145,000 MPN per 100 ml, followed by that of Bulumkutu, Bulabulin-ngarannam and Hausari with 65,000, 36,000 and 35,000 MPN per 100 ml, respectively. However, lowest coliform count of 6000 MPN per 100 ml was recorded in water sample from Mashamari (Table 2).

DISCUSSION

The result of physicochemical analysis of water shows that the pH of the water samples from Gwange, Hausari and Moduganari do not comply with standard requirements. Their values are less than the lower limits of the pH (6.5) recommended by WHO, NAFDAC and NSDWQ, except for those from Mashamari and Bulabulin-ngarannam. Even though pH has no direct effect on human health,

Table 2. Bacteriological analysis of water.

Site	Total coliform count	Differential *Escherichia coli* count (cfu)
Mashamari	6.00 x 10^3	Not detected
Gwange	145.00 x 10^3	Not detected
Hausari	35.00 x 10^3	Not detected
Bulabulin ngarannam	36.00 x 10^3	Not detected
Bulumkutu	65.00 x 10^3	Not detected
WHO standard	0 per 100 ml	0 per 100 ml
NSDWQ standard	10 per 100 ml	0

its indirect action on physiological process cannot be over emphasized (Adenkunle et al., 2004; NSDWQ, 2007). Also total dissolve solid (TSD) of the water sample from Hausari exceeded the standard recommended by WHO, NAFDAC and NSDWQ (500 mg/l). The TDS is the term used to describe the inorganic salt and small amount of organic matter present in solution or water. The principal constituents are usually calcium, magnesium, sodium and potassium cation, carbonate, hydrogen carbonate, chloride, sulphate and nitrate anion (WHO, 1996). The presence of TDS in water may affect its taste (WHO, 1996). It has been reported that drinking water with extremely low concentration of TDS may be unacceptable because of its flat insipid taste (WHO, 1996; Bruvold and Ongerth, 1969). The turbidity of all water samples used in this study is in agreement with both WHO and NWDSQ standard. Water turbidity is very important because high turbidity is often associated with higher level of disease causing microorganism, such as bacteria and other parasites (Shittu et al., 2008). The total alkalinity of all water samples are in agreement with both WHO (80 to 120 mg/l) and NAFDAC (100 mg/l) standard. Also, fluoride content of all water samples fell within the standard limit of NSDWQ (1.5 mg/l), likewise the copper content of all the water samples used in this study which is in agreement with NSDWQ standard of 1 mg/l. Although, presence of copper above the standard set by NSDWQ may cause gastrointestinal distress with a shorter term exposure, while a long term exposure may experience liver or kidney damage (EPA, 2012).

All the water samples analysed in this study have unobjectionable colour which is in agreement with the standard colour of 6 TCU by WHO and 5 TCU by both NAFDAC and NSDWQ (NAFDAC, 2001: WHO, 2001; NSDWQ, 2007). Higher conductivity of 1092 µs/cm was observed in the water sample collected from Hausari, although there is no disease or disorder associated with conductivity of drinking water (NSDWQ, 2007). Iron, chromium and nitrite were not detected in each of the samples.

The total coliform counts of all the water samples were generally high. They exceeded the standard requirement of 10 total coliform count per 100 ml for NSDWQ and zero total coliform count per 100 ml for WHO (NSDWQ, 2007: WHO, 2002). High total coliform counts vividly indicate that the water from the wash boreholes is faecally contaminated. The water sample from Gwange had the highest total coliform count (145,000 MPN per 100 ml). This finding is not surprising considering the high population and close proximity of the wash borehole to pit latrines. The sewage could seep slowly into underground water, thereby polluting it. Also, long term usage of boreholes may lead to deterioration of the water quality, because the pipeline may become corroded with random cracks and in most cases clogged with sediment (Onemano and Otun, 2008). This will allow the passage of inorganic metals and bacteria. The implication of this finding is the possibility of the presence of pathogens that may cause acute intestinal illness, which are generally considered discomfort to health and could become fatal for some susceptible groups (such as infants, elderly and those who are sick) (Addo et al., 2009; Olowe et al., 2005; NSDWQ, 2007). In addition to human and animal waste contamination, parasitic organism such as *Giardia* and *Cryptosporidium* may be present (EPA, 2003; Shittu et al., 2008). Generally, underground water is often considered as the purest form of water (Shittu et al., 2008), although it's vulnerability to contamination could be due to improper construction, animal waste, proximity to toilet facilities, sewage, refuse dump site and various human activities surrounding it (Bilton, 1994; Shittu et al., 2008). However, no *E. coli* were detected in all the water samples, which indicate that all the water samples are free from recent faecal contamination. The ability to detect faecal contamination in drinking water is necessary, as pathogenic microorganisms from human and animal faeces in drinking water pose the greatest danger to public health.

Conclusion

The principal aims of monitoring drinking water are to prevent the spread of water borne diseases and to protect the health of the community. The importance of access to good quality water cannot be overemphasized. Increase in population in Maiduguri Metropolis coupled with the rise in human activity pose a great pressure on provision of safe drinking water. This necessitates large number of people to consume water from wash boreholes which constitute a major health problem due to close proximity of wash boreholes to pit latrines. This study recorded high number of coliform counts in water samples analyzed, thus making it unsafe for drinking and require further treatment. Therefore, there is an urgent need for awareness to be created about the present situation of these wash boreholes, to enlighten the people on the necessity for further treatment of this water before they can be used for drinking and domestic purposes.

REFERENCES

Adenkunle LV, Sridhar MKC, Ajayi AA, Oluwade PA, Olawuyi JF (2004) Assessment of the health and socio-economic implication of sachet water in Ibadan, Nigeria. Afri. J. Biomed. Res. 7(1):5-8.

Arku AY, Musa SM, Mofoke ALE, Dibal JM (2011). Re-examining raw effluents from Nigerian Bottling Company Maiduguri for crop irrigation. J. Appl. Phytotechnol. Environ. Sanit. 1 (1):43-49.

Addo KK, Mensah GI, Bekoe M, Bonsu C, Akyen ML (2009) Bactrioelogical quality of sachet water produced and sold in Teshie-Nungua, Surburbs of Accra, Ghana. Afr. J. Food Agric. Nutr. Dev. 9(4):1019-1030.

Banu N, Menakuru H (2010) Enumeration of Microbial Contamination in School Water. A Public Health Challenge. Health 2(6): 582-588. doi: 10.4236/health.2010.26086.

Bilton G (1994). Waste Water Microbiology. Gainesville, New-York. Wiley IBS. pp. 118.

Bruvold WH, Ongerth HJ (1969). Taste quality of mineralized water. J. Am. Water Works Assoc. pp. 61-170.

Cheesbrough M (2006). District laboratory Practice in Tropical Countries. Part 2. Cambridge University Press. pp. 143-157.

EPA (2003). Environmental Protection Agency, Safe Drinking Water Act. EPA 816-F-03 016.

EPA (2012). Environmental Protection Agency Basic information about copper in drinking water. Available on http:// water.epa.gov/drinking/contaminants/basicinformation/copper.cfm. Last update on 6 march 2012.

Erah PO, Akujieze CN, Oteze GE (2002). A quality of ground water in Benin City: A baseline study on inorganic chemicals and microbial contaminants of health importance in boreholes and open wells. Trop. J. Pharm. Res. 1(2): 75-82.

FAO (1997a). Chemical Analysis Manual for Food and Water, 5th ed. FAO Rome I. P:20-26.

NAFDAC (2001). National Agency for Food and Drug Administration and Control, Ministry Safety bulletin, Volume 1. Recommendation, National agency for food, Drug, Administration and control. Lagos, Nigeria.

NSDWQ (2007). Nigeria Standard for Drinking Water Quality, Nigeria Industrial Standard, Approve by Standard Organization of Nigeria Governing Council. ICS 13. 060. 20: 15-19.

Nollet Leo ML (2007) Hand Book of Water Analysis. CRC Press Taylor and Francis Group.

Obi CN, Okocha CO (2007) Microbiological and Physico-Chemical of Selected Bore-hole Waters in World bank Housing Estate, Umuahia, Abia State, Nigeria. J. Eng. Applied Sci. 2(5):920-929.

Okonko IO, Ogunjobi AA, Kolawale OO, Babatunde S, Oluwole I, Ogunnusi TA, Adejoyi OD, Fajobi EA (2009). Comparative Studies and Microbial Risk Assessment of a Water Samples Used for Processing Frozen Sea foods in Ijora- Olopa, Lagos State, Nigeria. EJEAFChe. 8(6): 408-415.

Oladipo C, Onyenike IC, Adebiyi AO (2009) Microbiological analysis of some vended Sachet water in Ogbomoso, Nigeria. Afr. J. Food Sci. 3(12):406-412.

Olowe OA, Ojurongbe O, Opaleye OO, Adedosu OT, Oluwe RA, Eniola KIT (2005) Bacteriological Quality of Water Samples in Osogbo Metropolis. Afr. J. Clin. Exper. Microbiol. 6(3): 219-222.

Onemano JI, Otun JA (2003). Problems of Water quality standard and monitoring in Nigeria. Paper presented at the 29th WEDC International Conference at Abuja Sheraton Hotel and Tower, Nigeria on 22-26 September 2003.

Rudiger KV (2002). To the memory of Dr. Ursula (Uschi) Greinert-Byer. Published Drukerei Franz Hansen, Bergisan Gladbach. 70-90.

Shittu OB, Olaitan JO, Amusa TS (2008) Physico-Chemical and Bacteriological Analysis of Water Used for Drinking and Swimming Purpose. Afr. J. Biochem. Res. 11:285-290.

Tortora, JG, Funke RB Case LC (2002) Microbiology an introduction. Media update of seven edition. Including Bibliography and index Publisher. Daryl fox. pp. 258-260.

World Health Organization (WHO) (1996). Total dissolve solid in drinking water. Guideline for drinking water quality. 2:0-1.

World Health Organization (WHO) (2001). Guidelines for Drinking Water Quality, Recommendation, Geneva. p. 130.

World Health Organization (WHO) (2010). Guidelines for Drinking-water Quality. Recommendation, Geneva, p: 1-6. Retrieve from http://www.who.int/water_sanitation_health/WHS_WWD2010_guidelines_2010_6_en.

9

Contamination of cowpea and by-products by organophosphorous pesticide residues in Ngaoundere markets: Dietary risk estimation and degradation study

Sonchieu Jean[1]*, Ngassoum Martin Benoit[2], Tchatchueng Jean Bosco[2], Ashutosh Kumar Srivastava[3] and Laxman Prasad Srivastava[3]

[1]Department of Social Economy and Family Management, Higher Technical Teachers' Training College (HTTTC), University of Bamenda, Cameroon.
[2]Department of Applied Chemistry, ENSAI, University of Ngaoundere, P. O. Box: 455, Ngaoundere, Cameroon.
[3]Pesticides Toxicology Laboratory, Indian Institute of Toxicology Research, (CSIR), P B No 80, M G Marg, 226001, Lucknow, India.

The study was carried out to monitor some organophosphorous (OP) pesticide residues in stored cowpea in Ngaoundere's markets and two by-products (koki and fritter), to assess the dietary risk exposure through quotient risk (RQ) and to degrade some pesticides using kanwa (rock salt) solution. The levels of nine OP pesticides were determined by gas chromatography equipped with nitrogen photometric detector (GC-NPD) and GC-MS for confirmation. Six OP pesticides (dichlorvos, methyl-parathion, malathion, profenofos, diazinon and chlorpyrifos) were found in concentrations ranging from 0.02 to 5.4 mg/kg in peripheral zone while only five OP pesticides (dichlorvos, methyl-parathion, malathion, profenofos and chlorpyrifos) were found in the urban area (0.02 to 4.62 mg/kg). High amounts of these compounds were found in koki and fritter. Malathion, methyl-parathion and dichlorvos were the most frequent (27 to 89%) and some levels exceeded the maximum residue limits (MRLs) or the acceptable daily intake (ADI) per FAO; high values of RQ were found between February and May for all foodstuffs showing high risk for consumer at this period; malathion, dichlorvos, and profenofos were denatured by kanwa solution at the rate of reduction varying from 65% (malathion) to 98% (dichlorvos) with the production of malaoxon while methyl-parathion was not.

Key words: Organophosphorous, pesticide residues, monitoring, cowpea, dietary-risk, degradation.

INTRODUCTION

The use of pesticides in agriculture becomes more and more intensive in the Northern part of Cameroon. Small scale farmers, who are the main users, abusively spray those chemicals to protect crop and harvested products from pests. Since 1989, because of the financial crisis that hit the country, the government decided to set up a program of liberalizing and privatizing the pesticide sub-sector (Souop, 2005). The objective was to make farmers responsible for the purchase and use of pesticides. This decision fundamentally changed the policy of crop protection and increases risk of poisoning in Cameroon since users were not trained for this purpose. Therefore, selling and use became most rampant among the population and users. Their ignorance about pesticide toxicity led to misuse: abusive application, inadequate usage, etc. (Nguegang et al., 2005). Organochlorine pesticides are continuously banned (NRCPPCS, 2007). In the opposite, an addition of organophosphorous compounds is increasing and their association with fungicides is rising. Then organophosphorous pesticides are widely applied in agriculture because their degradation elsewhere is known to be rapid (Kavi and Reetha, 2012).

The general public is mainly exposed to organophosphorous pesticides through the intake of contaminated foods (such as grains, vegetables, and fruits), which are directly used with during cropping and storage (Matthews et al., 2003 and Gimou et al., 2008). These OP pesticides cause two major toxic effects. The first is the well known acute toxicity initiated by inhibition of acetylcholinesterase with subsequent accumulation of acetylcholine at nerve endings.

The second effect is organophosphate-induced delayed polyneuropathy (OPIDP; ataxia and paralysis) associated with inhibition of at least 70% neuropathy target esterase activity sometimes followed by the irreversible transformation of inhibited enzyme to its non-reactivatable form (Milan , 2001). They act by binding to the enzyme acetylcholinesterase, disrupting nerve function, resulting in paralysis and death (Liu et al., 1999). Both of these effects are related to chemical structure of OP that can be largely influenced by metabolism of these compounds that may form potent inhibitors of the enzymes.

Kicth and Ntoukam (1990) show that cowpea is the most threatened grain produced among maize and millet in the Northern part of Cameroon. The production of the cowpea in this area represents 98.5% of the total national production and estimated at 130101 tons in 2010, produced in 115 374 ha (AGRI-STAT, 2011). This production is locally consumed, sold in the southern Cameroon and about 40% is exported to Gabon and Nigeria. The unit price per kilogramme is estimated at 1.1 USD (MINADER, 2012). The protection of this cowpea using white powder organophosphorous pesticides is much popularized as; to differentiate it from red powder (organochlorine and some carbamates) supposed to be used only for seed treatment. However, the misuse of those compounds is still harmful for consumers and users. Since cowpea is produced once a year, the protection during storage becomes necessary to preserve grain quality throughout the year.

In Ngaoundere, cowpea is consumed in two main forms (koki and fitters) sold in restaurant and popular places (schools, markets, etc.). The use of some ingredients such as rock-salt during cooking is very common to avoid flatulence, to tenderize or to improve aroma (Efiuvwevwere and Chidi, 1996). The commercial transaction between producers and consumers is generally done by sellers (wholesalers and retailers). They store grains to make most economic benefits within the duration since the price depends on the period of the year (Langyintuo, 1999). Therefore, contamination of this foodstuff by pesticide residues becomes a dramatically topical issue in the area because of multiple treatments.

In developed countries, pesticide residues control in market food is routinous, but in developing countries like Cameroon, few databases are available (Sonchieu et al., 2010 and Yanhong et al., 2006). Many intoxication cases due to OP pesticides through food contamination have been reported by Apa (2010) in Ngaoundere and elsewhere in the country. Many methods of degradation have been developed to reduce amounts in water. This advanced oxidative process (AOP) has been used by authors to degrade pesticide residues in food (Randwan et al., 2005 and Alvaro et al., 2011), but products used for this degradation are laboratory chemicals (acetic acid and bicarbonate of sodium) or microorganisms, which can also damage some functions of the human body (Luisa, 2012). However, those products are found in some house materials used to cook or to flavor: such as the rock-salt commonly known as kanwa.

Objectives of this study were foremost to carry out the monitoring of the organophosphorous pesticide residues in whole cowpea grains and into two by-products (koki and fritters) during twelve months (November 2010 to October 2011) both in Ngaoundere and peripheral areas; and secondly to assess dietary risk and a method of degradation.

MATERIALS AND METHODS

Sampling

Samples of cowpea grains were collected from 10 sites during 12 months in the Ngaoundere central markets: Petit Marché [(&ll= 7.322372, 13.586655), (4 sites)] and Grand Marché [(&ll= 7.326661, 13.582073), (1 site)]; and Ngaoundere peripheral markets: Bamyanga [(&ll= 7.293068, 13.576682), (2 sites), Bantaï [(&ll= 7.310454, 13.597641), (2 sites)], and Dang [(&ll= 7.42273, 13.551979), (1 site)], following the guidance given by Sharma et al. (2007). While those of koki and fritters were collected from restaurants during the same period. Approximately 1 kg of samples were separately collected, put in pharmaceutical bags and directly transported to the laboratory for immediate preparation and extraction. The extract was stored at -20°C until final processing.

Source of materials and sample preparation

Cowpea grains, koki and fritter samples were collected directly from ten local identified sellers who were randomly chosen, according to the availability of foods from them.

Samples of cowpea grains were properly homogenized and 500 g picked for comminution using electrical blinder (ZENPLIN®) or were triturated using laboratory mortar (koki and fritters). After each operation, all the materials were washed and dried before the next grinding. No sample was sieved. The comminuted samples were mixed and homogenized before withdrawing 25 g for analysis. The rock-salt (kanwa) was purchased at the Ngaoundere market, ground and sieved (Ø=01 mm).

Test of matrix quality

The matrices (cowpea, *koki* and fritter) used for recovery were provided from Mokolo (Northern Cameroon area) on November 2010. Five blanks were taken for each type of food and collected directly from farmers or restaurants during harvest, before any spraying. Analysis was initially carried out for each blank to determine if there were any detectable contaminants. All contaminated samples were discarded. Any interference chromatographic peaks with investigated insecticides were assessed.

Materials and reagents

All organic solvents were products of Spectrochem or Fisher Scientific, high performance liquid chromatography (HPLC) grade, supplied by Supleco of Lucknow, India. Standard OP pesticides of purity 99.0 to 99.9% were provided by the Bureau of Indian Standards (BIS) and supplied by the Core Laboratory, Certified Reference Materials (CRMs) from BCKV (Bidhan Chandra Krishi Vishridhyalya) Centre of India. Stock standard solutions of OP pesticides were prepared in n-hexane at about 100 mg/L. An intermediate solution of each pesticide at 10 mg/L was prepared in the same solvent. The mixture of all pesticides was also made up in n-hexane at 0.5 mg/L and used as working standard. This working solution contained 8 OP pesticides which are all OP pesticides authorized and commonly used in the country: dichlorvos, dimethoate, diazinon, methyl-parathion, edifenfos, malathion, chlorpyrifos and profenofos.

Extraction

Extraction was carried with 100 mL of mixed HLPC grade solvents (acetone and n-hexane) (1:1, v/v) in 250 mL flasks and were mechanically shaken for 1 h using a reciprocating shaker (Sambros®) at 180 translations per min. After filtration, 250 mL of distilled water and 20 ml of saturated sodium chloride solution were added. After 5 min of vigorously shaking of the mixture and 15 min of rest, the two phases were separated. The aqueous phase was retrieved and 75 mL of dichloromethane (DCM) added and mechanically shaken for 10 min. The two organic phases were then mixed and dehydrated with a convenient quantity of activated anhydrous sodium sulphate (anhy. Na_2SO_4). The samples were then concentrated in a rotary evaporator until dryness at 35±10°C. The residue was dissolved in 10 mL of an iso-octane/acetone mixture (9:1, v/v).

Clean-up

A 2.5 mL aliquot of the above extract was loaded on the cartridge and eluted with 60 mL of n-hexane saturated with acetonitrile through a chromatographic column containing 2 g of activated florisil. Elution took about 1 hour. The cartridge provided by Supelco (Lucknow, India) was the Florisil-Sigma, F9127-5006 mesh: 60-100/PR (pesticide residues) with 1 cm of activated anhydrous Na_2SO_4 loaded on top. The eluate was concentrated under the same conditions as described above and the residue dissolved in 5mL of an iso-octane/acetone mixture (9:1, v/v). GC determination was either carried out immediately or the extract was kept refrigerated at 4°C for a maximum of 48 h.

Removal of fat/oil

Since fat is detrimental to the column, extract obtained above was treated to remove fat from samples of koki and fritter. 30 mL of petroleum ether were added to extract and mixed with acetonitrile (20 mL) and shake vigorously: repeated three times. The acetonitrile phase was kept while the other was discarded. 20 mL of petroleum ether was also added to the volume of acetonitrile phase to remove other trace of fat. 30 mL of n-hexane were added to the acetonitrile phase and the mixture was vigorously shaken: three times repeated. The n-hexane phase was washed with 250 mL of bi-distilled water and 20 mL of saturated NaCl. The operation has been repeated once and all n-hexane phases were mixed. Dehydration of this mixture was done using a convenient quantity of activated anhydrous sodium sulphate (anhy. Na_2SO_4) and following by filtration. The filtrate was concentrated in a rotary evaporator until dryness at 35±10°C. The residue was dissolved in 10 mL of pure hexane for clean-up.

Risk estimation

The dietary risk was calculated using the quotient method based on the Risk Quotient (RQ) for each combination of contaminants found in a given foodstuff. The assessment of this risk parameter was calculated using the formula described below (EAWAG, 2003 and EC, 2003).

$$RQ = \frac{Concentration}{\mathrm{TR}V / NOAEL} \quad \text{and } RQ = RQ_1 + RQ_2 + \ldots\ldots\ldots RQ_n$$

Where, 1, 2, ….n represent a specific pesticide found in the same sample; TRV is the toxicity reference value which is the no observed adverse effect level (NOAEL). The RQ reference value is “1”, any value above describes a situation of risk and vice-versa.

Scheme of Kanwa solution application on cowpea grains

The ground sieved kanwa was dissolved in bi-distilled water and the pH fitted at 9.52. One kilogram of grain of cowpea was separately spiked with formulations of pesticides found to present dietary risks: Dichlorvos (DDVP) 125 g/L + malathion 100 g/L (Digrain 4), methyl-parathion (Paratox) and profenofos (Plantofos 75 EC), and stored for three months to evaluate the effect of kanwa during this period of storage. Analyses were carried-out once a month. The experimentation was first led on water contaminated with studied pesticides. Mixtures are shown in Table 1, where conditions are the following: time (30 min), temperature (25°C), pH (9.52) and no shaking. The weight used during soaking refers to whole grains or flour.

Apparatus and operating conditions

Gas chromatography: conditions in NPD detector

Gas chromatography in this case was performed using a Carlo Erba gas chromatography equipped with a nitrogen phosphorus detector (NPD). An Optima 5-MS60® (30mx25 mm I.D., 0.32 nm film thickness) capillary column was used. Nitrogen was used as carrier gas at a flow-rate of 2 ml min^{-1}. The oven temperature conditions were: initial temperature isothermal at 90°C for 1 min, increased by 10°C.min^{-1} to 340°C and held for 1 min. Injector and detector temperatures were 200 and 300°C, respectively.

Conditions in GC/MS

A Shimadzu GC-MS QP-2010 was used to identify target compounds, equipped with a split/splitless injector and a DB-5MS® (30mx0.25mm I.D., 0.25 mm) J&W column. EI mass spectra were generated using electron energy of 70 eV, monitoring for ions m/z

Table 1. Levels of spiking and concentration in grains and water.

Pesticides	Spiked amount (mg/kg or mg/L)	Water	Concentration	
	Stored grain		Weight/Vol. of *kanwa*	Vol. of *kanwa* /Vol. spiked water
Dichlorvos	0.1	1.4	5 g/10 mL	1 mL/5 mL
Methyl- parathion	0.1	2.1	5 g/10 mL	1 mL/5 mL
Malathion	0.1	1.3	5 g/10 mL	1 mL/5 mL
profenofos	0.1	1.0	5 g/10 mL	1 mL/5 mL

Table 2. Data of the method validation for active ingredients and metabolite.

Pesticide	n	mLOD (mg kg^{-1})	mLOQ (mg kg^{-1})	spiking level (mg Kg^{-1})	Recovery ±RSD (%)		
					koki	fritters	cowpea
Parent products							
Dichlorvos	3	0.002	0.006	B+0.5	97±7	95±6	89±10
Dimethoate	3	0.002	0.003	B+0.5	79±16	77±14	73±3
Diazinon	3	0.002	0.006	B+0.5	100±12	98±14	78±2
Methyl Parathion	3	0.002	0.006	B+0.5	83±7	81±9	103±11
Chlorpyrifos methyl	3	0.001	0.003	B+0.5	71±3	73±2	88±5
Malathion	3	0.002	0.006	B+0.5	85±10	100±16	106±1
Chlorpyrifos	3	0.003	0.009	B+0.5	74±11	75±9	85±4
Profenofos	3	0.002	0.006	B+0.5	109±16	105±10	103±6
Edinofos	3	0.002	0.006	B+0.5	75±12	101±16	78±6
Degradation product							
Malaoxon	3	0.002	0.006	B+0.5	99±16	77±12	102±1

LOD, Limit of detection; LOQ, limit of quantification.

50→450 in full-scan and selected ion recording (SIR) modes. The scan interval was 0.3 s (1425 amu.s^{-1}), injection volume was 2 ml and the source temperature employed for the ionization technique was 200°C. Injector temperature and detector temperature were set at 250°C. The column was held at 80°C for 1 min, then increased at a rate of 6°C min^{-1} to 150°C, followed by heating to 250°C at a rate of 20°C min^{-1}, and finally held constant at 250°C for 2 min.

Standardization of the method and calculation

Quantification was performed by comparing the peak areas to those of the calibration curve standard while identification was carried out by comparing the retention times with the standard. The precision of the methods expressed as repeatability (% RSD), was evaluated by analyzing samples in triplicate fortified at 0.015; 0.15; 0.2 and 0.25 mg kg^{-1}, depending on the sensitivity of the detector to the compound. OJ/EU (2004), determination coefficients were used to estimate linearity. Peak areas were measured relative to standards. Blanks used to estimate recoveries were provided as non-treated grains and directly acquired from local farmers during harvest. They were used as controls and showed no interference from the pesticides being monitored. The method limits of detection (mLOD) and quantification (mLOQ) were calculated experimentally from signal-to-noise ratios of 3.0 and 10.0, respectively, by spiking samples at low concentrations and subjecting them to sample preparation and clean-up. Blank extracts were used for estimation of background noise in chromatographic analysis. Identification of active ingredients was confirmed by GC–MS on more contaminated samples and on some formulations used for powered grains. Following the conditions described above, a standard curve was used to calculate sample concentrations (x) according to the linear equation: $y = ax + b$.

RESULTS AND DISCUSSION

Method validation

Table 2 presents data of the method validation. Data rule the acceptability for applicability performance of analysis, since repeatability is still conformed to the European Commission (EC) directives (EC, 2010).

Rate of contamination of foodstuffs

Analysis of cowpea samples and by-products from Ngaoundere revealed the presence of seven OP pesticides at rate ranging from 66 to 87% (Table 3). Malathion, methyl-parathion and dichlorvos are found in all types of foodstuffs but with predominance of the former two compounds (Malathion and methyl-parathion) in all contaminated samples. Although, grain samples

Table 3. Rate of cowpea and by-products contamination from Ngaoundere compared to the MRL/ADI.

OP pesticide	Grains (Petit Marché) n	Grains (Petit Marché) Rate of contamination (%) [%>MRL]	Grains (peripheral zone) Rate of contamination (%) [%>MRL]	*Koki* n	*Koki* Rate of contamination (%) [%>ADI]	Fritter n	Fritter Rate of contamination (%) [%>ADI]
	120	66 [40]	86 [50]	60	87 [60]	60	87 [70]
Dichlorvos	120	25 [0]	20 [0]	60	15 [15]	60	20 [20]
Dimethoate	120	-	2 [0]	60	-	60	-
Diazinon	120	-	4 [0]	60	3 [0]	60	-
Methyl Parathion	120	55 [35]	60 [40]	60	55 [55]	60	35 [35]
Chlorpyrifos-methyl	120	-	-	60	-	60	-
∑Malathion	120	75 [0]	90 [5]	60	85 [60]	60	90 [70]
Chlorpyrifos	120	5 [0]	4 [0]	60	-	60	5 [0]
Profenofos	120	5 [5]	10 [5]	60	5 [5]	60	5 [0]
Edinofos	120	-	-	60	-	60	-

n, Total number of samples; MRL, maximal residues limit; ADI, acceptable daily intake; [...] = % samples higher than ADI/MRL.

from the peripheral area and by-products are most contaminated (ranging from 15% (dichorvos) to 90% (malathion)) and present levels higher than norms [maximum residue limits (MRL) and acceptable daily intake (ADI)] comprised between 15% (dichlorvos, koki) and 70% (malathion, fritter). However the methyl-parathion in contaminated samples might go up to 100% higher than norms (koki and fritter) despite its low frequency. The edinifos and chlorpyrifos-methyl were not found in any sample.

Levels of contamination

Analysis of cowpea samples and by-products in Ngaoundere showed contamination throughout the year at various levels. The mean of amounts are presented in Figure 1, methyl-parathion presents higher levels than those of other compounds in grains of Petit Marché and fritter (varying from 4.5 to 5.5 mg/kg), while the malathion presents the same profile in koki and grains of peripheral zone (varying between 5.4 and 8.6 mg/kg). Levels of all OP pesticides in each type of food fluctuate and show maxima during February, March, April and May, but remain low during other months.

These concentrations maybe higher than the limit of report (LR) which is MRL (cowpea grains) and ADI (koki and fritter) or not as shown in Table 4; which also presents levels less than the limit of report varying from 40% (methyl-parathion in koki) to 100% (malathion and dichlorvos in grains). Amounts of pesticides found to be greater than the limit of report were 10, 100, 1000 or 10000 times higher than these norms. The methyl-parathion is the only pesticide showing amounts 10000x higher than the Limit of report in samples of koki (5%) and fritter (2%). The majority of amounts of OP pesticide residues in grains were ranging from LR to LRx10 even those of peripheral area were highest, while those in koki and fritter were varying from LRx10 to LRx100. Although, the methyl-parathion presented higher amounts among compounds, following by dichorvos and malathion. Other OP pesticides residues such as diazinon, chorpyrifos and dimethoate were present at lower concentrations.

Dietary risk exposure

This parameter assessed according to the risk quotient (RQ) refers to Table 5 showing variations of calculated RQ values for each type of food throughout the period of study. Risks of intoxication (exposure) are given for 12 months. All calculated RQ values were lower than the critical level ("1") except koki of the month of May (RQ=1.77). The RQ calculated values grouped two to three pesticides but this combination did not influence the RQ. Higher calculated values of RQ were ranging between months of the February and May, showing high exposure risk.

Degradation study

The use of kanwa solution degrades OP pesticides such as malathion, dichlorvos and pro-

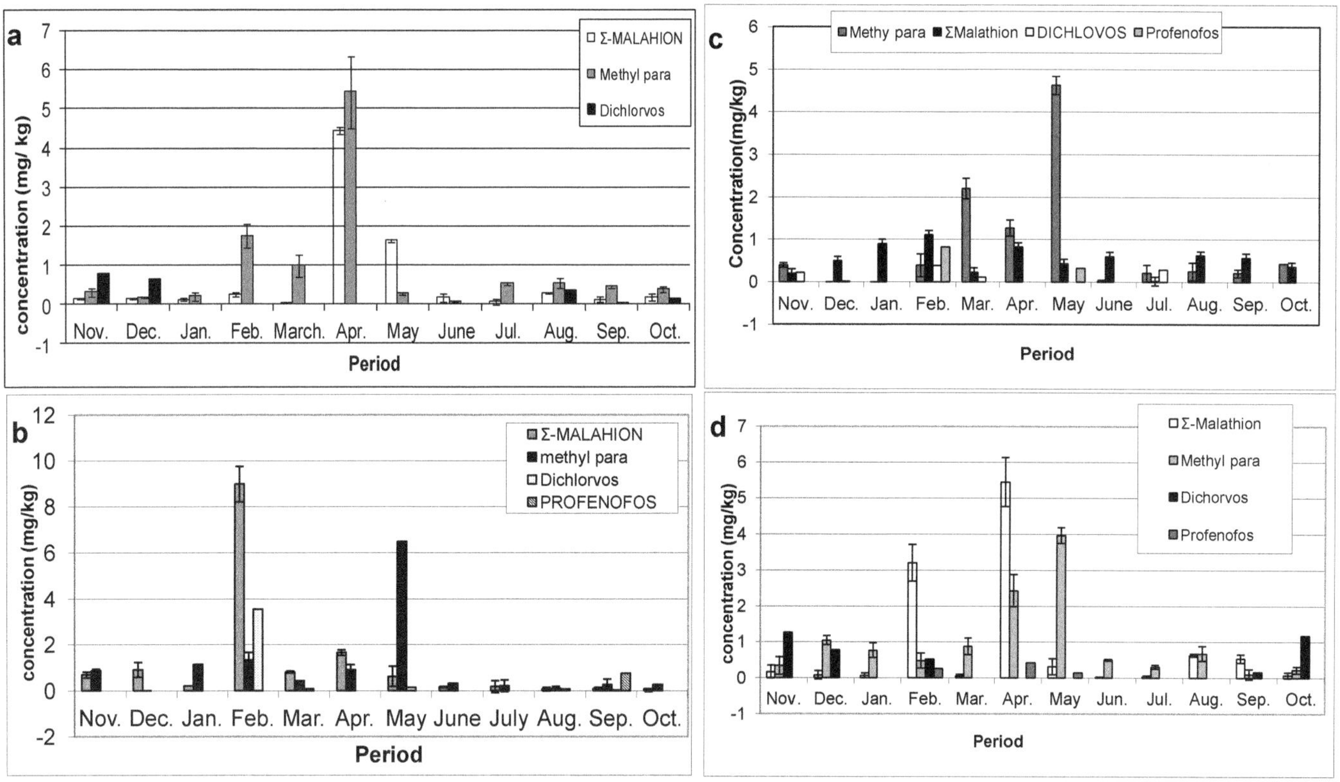

Figure 1. Evolution of OP pesticides amounts in cowpea and by-products from Ngaoundere: November 2010 - October 2011. **a,** Fritter; **b,** *koki*; **c.** grains from Petit Marché (Ngaoundere); **d,** grains from peripheral zone (Ngaoundere).

fenofos (Table 6). After 5 min of mixture, only dichlorvos was degraded, but from 15 to 60 min, all OP pesticides were denaturized; except methyl-parathion. Values of recovery (without treatment) are significantly different from those obtained after application. The reduction of levels of pesticides studied within that time showed significant difference between values (Table 6). The production (formation) of malaoxon followed the same principle but values were not significantly different. Then OP pesticides were reduced at the rate varying from 20 to 98%, starting from 15 min, except methyl-parathion (Figure 2). The duration did not significantly influence the rate of reduction after 15 min.

The results of kanwa solution's application on cowpea grains and flour are detailed in Table 7. The level of recovery are not significantly different ($p<0.01$) between samplings. However, significant differences are observed between each value of recovery and subsequent value after application of kanwa solution. Figure 3 indicates levels of reduction with production of malathion's metabolite (malaoxon). This rate of reduction did not change during the three months of storage. Amounts of malaoxon produced were not significantly different between months.

DISCUSSION

This study is in line with Sonchieu et al. (2010) which confirm the work done on vegetable and fruits by Gimou et al. (2008) because cowpea is a staple food to be eaten with other secondary foods. Then, a dietary risk may be accessed through the complementarity of that data. High concentrations of malathion were also observed in April (Sonchieu et al., 2010). Those high concentrations and frequencies confirm observations of Kicth and Ntoukam (1990) showing that usage of synthetic pesticides is practiced by 73% of those manipulating grains during storage. The methyl-parathion and malathion are still most used since they are most popularized by sellers. Their presence in many formulations sold in the country in many cases as powder, facilitates their accessibility and spraying. The most common formulation is the "poudrox" which had shown it effectiveness on pests, maybe more than "Caiman Rouge" which is a mixture of malathion and dichlorvos. Other OP pesticides (dimethoate, chlorpyrifos, chlorpyrifos-methyl, profenofos and diazinon) determined during this study are commonly used during cropping and mainly in vegetables since they are in liquid form (Apa, 2010 and MINARDE, 2010). Their presence in grains is an indicator of inadequate use or persistence.

Highest amounts observed between February and May are in agreement with observation of Langyintuo et al. (2003), who showed that the main pests (*Callosobruchus maculatus*) damages happen three months after harvest. This period corresponds to that of best exchanges

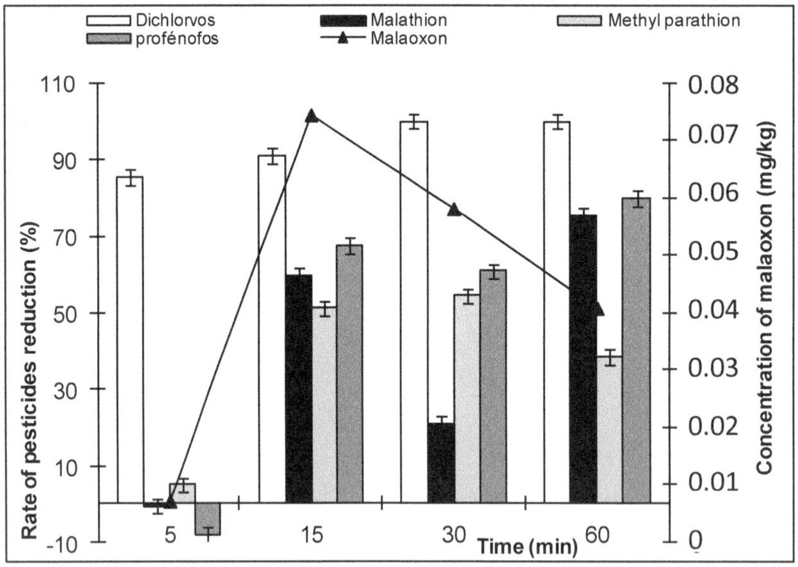

Figure 2. Rate of OP pesticides reduction and metabolite production in distilled water after soaking in *kanwa* solution.

Table 4. Variation of concentrations of main pesticides found in *koki*, fritter and cowpea grains in Ngaoundere's markets.

Foodstuff	Pesticide	Concentration (mg/kg)					Total (n)
		<LR (5)[	**[5-50]**	**[50-500]**	**[500-5000]**	**[5000<]**	
Petit *Marché* grains	ΣMalathion: n(%)	48 (100)	0(0)	0(0)	0(0)	0(0)	48 (100)
	Dichlorvos: n(%)	48(100)	0(0)	0(0)	0(0)	0(0)	48(100)
	Methyl-Parathion	<LR (0.1)[	[0.1-1.0]	[1.0-10.0]	[10.0-100.0]	[100.0<]	
	n (%)	30 (63)	14(29)	3 (6)	1(2)	0(0)	48(100)
Peripheral area grains	ΣMalathion: n(%)	70 (97)	2 (3)	0(0)	0(0)	0(0)	72(100)
	Dichlorvos: n(%)	72 (100)	0(0)	0(0)	0(0)	0(0)	72(100)
	Profenofos	<LR (0.05)]	[0.05-0.5]	[0.5-5.0]	[5.0-50.0]	[50.0<]	
	n(%)	71 (99)	1 (1)	0(0)	0(0)	0(0)	72(100)
	Methyl-Parathion	<LR (0.1)	[0.1-1.0]	[1.0-10.0]	[10.0-100.0]	[100.0<]	
	n(%)	42 (58)	20 (28)	5 (7)	5(7)	0(0)	72(100)
Koki	ΣMalathion	<LR (0.3)	[0.3-3.0]	[0.3-30.0]	[30.0-300.0]	[300.0<]	
	n(%)	45(75)	14(23)	0(0)	1 (2)	0(0)	60(100)
	Profenofos	<LR (0.03)	[0.03-0.3]	[0.3-3.0]	[3.0-30.0]	[30.0<]	
	n(%)	59(98)	1 (2)	0(0)	0(0)	0(0)	60(100)
	Methyl-Parathion	<LR (0.003)	[0.003-0.03]	[0.03-0.3]	[0.3-3.0]	[3.0<]	
	n(%)	24 (40)	0(0)	18 (28)	16 (27)	2 (5)	60(100)
	Dichlorvos	<LR (0.004)	[0.004-0.04]	[0.04-0.4]	[0.4-4.0]	[4.0<]	
	n(%)	52(87)	1 (2)	6(9)	1(2)	0(0)	60(100)
Fritter	ΣMalathion	<LR (0.3)	[0.3-3.0]	[0.3-30.0]	[30.0-300.0]	[300.0<]	
	n(%)	45(75)	13(22)	2(3)	0(0)	0(0)	60(100)
	Methyl-Parathion	<LR (0.003)	[0.003-0.03]	[0.03-0.3]	[0.3-3.0]	[3.0<]	
	n(%)	38(63)	0(0)	9(15)	13(21)	1(1)	60(100)
	Dichlorvos	<LR (0.004)	[0.004-0.04]	[0.04-0.4]	[0.4-4.0]	[4.0<]	
	n(%)	49(82)	1(2)	6(9)	4(7)	0(0)	60(100)

LR, Limit of report (MRL/ADI); n, number of sample.

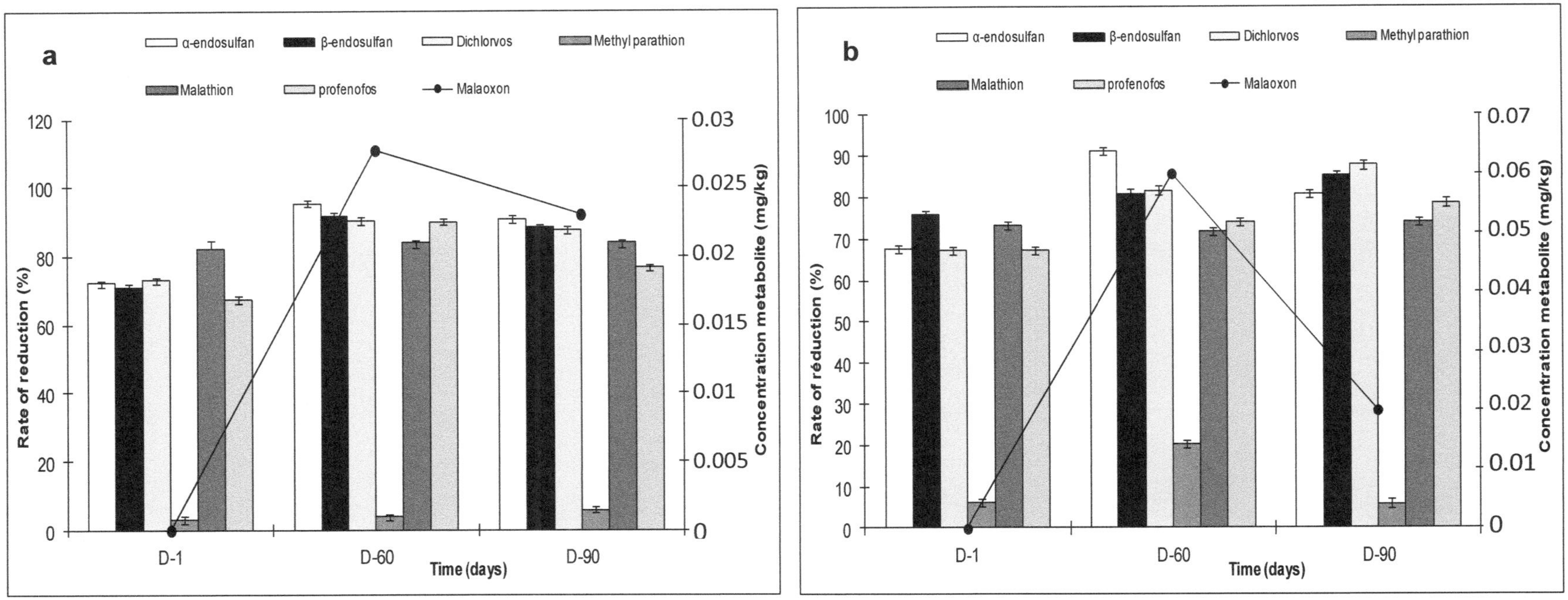

Figure 3. Rate of OP pesticides reduction in stored cowpea and metabolite production after soaking in *kanwa* solution. **a)** Rate of OP pesticides reduction in cowpea flour; b) rate of OP pesticides reduction in cowpea whole grains. D-, Days after spraying.

for cowpea commercialization in Cameroon; subsequently, the appropriate period of multiple treatment for greatest preservation (langyintuo, 1999). This is the main reason why grains from peripheral areas have shown higher amounts than those from Ngaoundere central markets (urban area): wholesalers in the city supply to retailers at the peripheral area, who constitute another step of grain treatment. This observation demands further investigation.

The high pollution of by-products issuing cowpea grains is due to the addition of some ingredients (palm oil, cotton oil, pepper, etc.) which can also contain pesticide residues (Mbofung et al., 1999; Sougnabe et al., 2010). The effect of heat did not effectively affect the amount of these pesticides during cooking (WHO, 2002; Lalah and Shem, 2002).

The level of risk calculated according to a man of 60 kg is relatively low, but may become dangerous for children and young infants who are not neglected consumers (EC, 2003). They are exposed to pesticide daily intake through koki or fritter, since they consume more food per kilogram of body weight than do adults (Noriko et al., 1998). Many symptoms related to pesticides effects maybe found within the local population as controlled in Indian population by Ajith et al. (2006). However, observed effects are correlated to the health of the consumers (EC, 2003). The high risks observed in February, March, April and May are likely related to the levels of contamination observed during this period. Results of this investigation show concordance to those observed by Gimou et al. (2008) in Yaounde between June and August in general food consumption.

The degradation of OP pesticides in the kanwa solution which is an alkaline medium is attributed to the pH effects, since Perihan et al. (2007) demonstrated that the alkalinity affects pesticides. After five minutes of soaking, there was no reduction characterizing the negative chart part. This effect is great as Peggy (2002) advised pesticides users to control the acidity of the water

Table 5. Variation of risk quotient (RQ) in cowpea and by-products from Ngaoundere's markets.

Period	Grain (Petit Marché)		Grains (peripheral zone)		Fritter		*Koki*	
	Number of pesticides	RQ	Number of pesticides	RQ	Number of pesticides	RQ	Number of pesticides	RQ
November	3	0.03	3	0.03	3	0.03	2	0.01
December	3	0.01	2	0.08	3	0.02	2	0.01
January	2	0.01	1	0.05	3	0.03	2	0.03
Febuary	3	0.04	4	0.04	2	0.12	3	0.12
March	2	0.15	3	0.06	2	0.07	3	0.08
April	3	0.1	2	0.18	2	0.37	2	0.06
May	3	0.32	3	0.28	2	0.03	3	1.77
June	2	0.01	2	0.04	2	0.01	2	0.01
July	2	0.02	3	0.03	2	0.04	2	0.03
August	2	0.02	2	0.05	3	0.05	3	0.01
September	3	0.02	2	0.02	3	0.03	3	0.02
October	3	0.08	2	0.03	3	0.03	2	0.02

Table 6. Effects of *kanwa* on distilled water pesticides fortified.

Pesticide	Variation of concentration (mg/mL)							
	5 min		15 min		30 min		60 min	
	Recovery	After application	Recovery	After application	Recovery	After application	Recovery	After application
Dichlorvos	0.98^{1}	0.14 ± 0.04^{a2}	1.06^{1}	0.09 ± 0^{b2}	0.94^{1}	ND	1.03^{1}	ND
Malathion	1.46^{1}	1.47 ± 0.1^{a1}	1.11^{1}	0.45 ± 0.1^{b2}	1.2^{1}	0.35 ± 0.01^{c2}	1.16^{1}	0.28 ± 0.1^{d2}
Methyl-parathion	2.02^{1}	1.92 ± 0.04^{a}	2.05^{1}	1.71 ± 0.01^{b1}	2.08^{1}	0.95 ± 0.01^{c2}	2.00^{1}	1.23 ± 0.03^{d2}
Profenofos	0.94^{1}	1.02 ± 0.2^{a}	0.98^{1}	0.32 ± 0.01^{b2}	0.82^{1}	0.32 ± 0.01^{b2}	0.86^{1}	0.17 ± 0.02^{c2}
Malaoxon	0.0	0.01 ± 0^{2}	0.0	0.07 ± 0.05^{2}	0.0	0.06 ± 0.01^{2}	0.0	0.04 ± 0.01^{2}

Values of the same rows with similar digits or letters are not significantly different at P>0.01.

Table 7. Effects of *kanwa* solution on stored cowpea pesticides fortified.

Pesticide	Variation of concentration (mg/kg)					
	Cowpea flour					
	Day 1		Day 60		Day 90	
	Recovery	After application	Recovery	After application	Recovery	After application
Dichlorvos	0.74^{1}	0.17 ± 0.02^{a2}	0.64^{1}	0.06 ± 0.01^{a2}	0.72^{1}	0.15 ± 0.02^{a2}
Methy-parathion	0.50^{1}	0.43 ± 0.02^{a1}	0.42^{1}	0.4 ± 0^{a1}	0.50^{1}	0.44 ± 0.02^{a1}
Malathion	0.50^{1}	0.08 ± 0.01^{a2}	0.77^{2}	0.12 ± 0.01^{a2}	0.60^{1}	0.07 ± 0.01^{a2}
profenofos	0.69^{1}	0.22 ± 0.01^{a2}	0.27^{2}	nd	0.34^{1}	0.03 ± 0.01^{a2}
Malaoxon	0.0	nd	0.0	0.03 ± 0.01^{2}	0.0	0.02 ± 0.01^{2}

Values of the same rows with similar digits or letters are not significantly different at P>0.01.

before any dilution for spraying. Radwan et al. (2005) and Alvero et al. (2001) observed this same effect using solutions of sodium carbonate (Na_2CO_3). The prompt reaction was also observed on organochlorine (lindane and heptan) and some organophosphates (diazinon) (Nyakundi et al., 2011 and Xiangyu et al., 2013). This degradation is attributed to oxidation or hydrolysis reactions by desulfuration (P-S→P-O) pathway (Eto, 1974). The production of the malaoxon is then due to the attack of P=S and P-O bonds by the OH^- ions. The formation of this degradation product follows the reaction (NRCPPS, 2007; Eto, 1974) as follows:

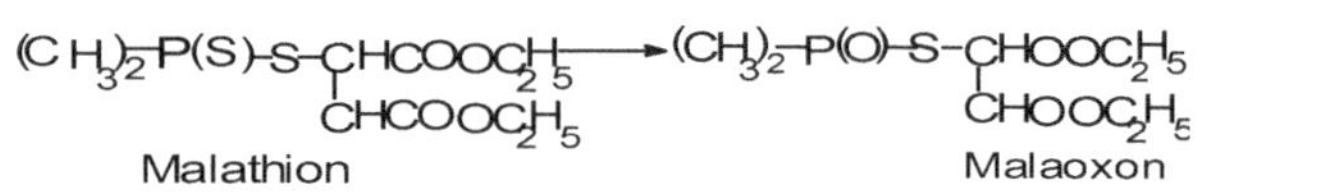

Table 7. Continued.

Pesticide	Variation of concentrations (mg/kg)					
	Whole grain					
	Day 1		Day 60		Day 90	
	Recovery	After application	Recovery	After application	Recovery	After application
Dichlorvos	0.74[1]	0.24±0.02[b2]	0.29[2]	0.05±0.01[a2]	0.29[2]	0.04±0.02[a2]
Methy-parathion	0.39[1]	0.27±0.06[a1]	0.29[1]	0.25±0.1[a1]	0.24[1]	0.24±0.2[a1]
Malathion	0.40[1]	0.1±0.01[a2]	0.71[2]	0.41±0.1[a2]	0.68[2]	0.38±0.2[a2]
profenofos	0.80[1]	0.26±0.01[a2]	0.27[2]	0.06±0 [a2]	0.22[2]	0.04±1[a2]
Malaoxon	0.0	nd	0.0	0.03±0.01[2]	0.0	0.02±0.01[2]

Conclusions

These results provide an important information on the current contamination status of a key agricultural product in Cameroon since cowpea produced in Northern Cameroon is sold in the whole country or in other neighboring countries such as Nigeria and Gabon. This study points to the need for urgent action to control the use and management of some excessively applied OP pesticides; since recent literature reveals that the largest proportion of human acute toxicity data is related to OP pesticide intoxications.

ACKNOWLEDGEMENT

The authors express their thanks to the Academy of Science for the Developing World (TWAS) and the Council of Scientific and Industrial Research (CSIR) for the funding and support provided. They also thank the Director of IITR, Lucknow, for his research interest. IITR manuscript number: 2869

REFERENCES

AGRI-STAT (2011).Cereals and legumes production. Yearbook of statistics of agricultural sector, campaign 2008 & 2009. DESA, MINADER, Yaoundé, Cameroon. 16:24-25.

Ajith JJ, Alex P, Anooj TV, Bipin N, Dheerej D, Kishore T, Vipin MG (2006). Effects of endosulfan on human beings. Report submitted to National Institute of Technology, Calcut, India. 40 pp.

Alvaro S, Geolar F, Villafuerte-Castrejon ME, Tejeda-Cruz A, Bosch M (2011). Two fold role of calcined hydrotalcites in the degradation of methyl parathion pesticide. Beilstein J. Nanotechnol. 2:99-103.

Apa (2010). North Cameroun: a food contamination made six dead. website:www.bonaberi.com/ar,nord_cameroun_une_intoxication_alimentaire_fait_six_morts,5346.html. (Accessed Oct 2010).

EAWAG (Swiss Federal Institute for Environmental Science and Technology) (2003). Agriculture and water quality. Publ. EAWAG news 59:20-23.

EC (European Commission) (2003). Technical guidance document on risk assessment. TGD Part I, Institute for Health and Consumer Protection, European Chemicals Bureau, European Commission, Ispra, Italy, 311 pp.

EC (European Commission) (2010). Method validation and quality control procedures for pesticide residues analysis in food and feed. Document No. SANCO/10684/2009 Supersedes Document No. SANCO/3131/2007 Implemented by 01/01/2010: 1-55.

Efiuvwevwere BJ, Chidi FE (1996). Influence of Fermentation Time and an 'Indigenous Tenderiser' (Kanwa) on the Microbial Profile, Chemical Attributes and Shelf-Life of Rice Masa (a Nigerian Fermented Product). J. Sci. Food Agric. 71:442-448.

Eto M (1974). Organophosphorous Pesticides. Org. Biol. Chem. 1:143-147.

Gimou MM, Charrondiere UR, Leblanc JC, Pouillot R (2008). Dietary exposure to pesticide residues in Yaoundé: the Cameroonian total diet study. Food Addit. Contam. (part A). 25(4):458-471.

Kavi KS, Reetha D (2012). Efficiency of Bacterial Isolates in the Degradation of Malathion and Parathion. Int. J. Pharm. Biol. Arch. 3(3):659-665.

Kicth LW, Ntoukam G (1990). Storage of cowpea in ash. In: Tech. bull. Bean/Cowpea; CRSP, East Lasing, MI, USA. 56 pp.

Lalah J, Shem OW (2002). The effect of boiling on the removal of persistent malathion residues from stored grains. P. Appl. Chem. 73:1147-1155.

Langyintuo AS (1999). Hedonic price analysis as a multidisciplinary activity: the case of cowpea in Cameroon. In: Bean/Cowpea CRSP West Africa Regional Report. 1:36-56.

Langyintuo AS, Lowenberg-DeBoer J, Faye M, Lambert D, Ibro G, Moussa B, Kergna A, Kushwaha S, Musa S, Ntoukam G (2003). Cowpea supply and demand in West and Central Africa Field. Crop Res. 82:215-231.

Liu J, Olivier K, Pope CN (1999). Comparative neurochemical effects of repeated methyl parathion or chlorpyrifos exposures in neonatal and adult rats. Toxicol. Appl. Pharm. 158:186-196.

Luisa B (2012). Biological degradation of pesticides: a solution to the problem of environmental contamination. Agrotec 2:78-81.

Matthews G, Wiles T, Baleguel P (2003). A survey of pesticides application in Cameroon. Crop Prot. 22:707-714.

Mbofung CMF, Ribgy N, Waldron KW (1999). Use of two varieties of hard-to-cook beans (*Phaseolus vulgaris*) and cowpea (*Vigna unguiculata*) in the processing of koki (a steamed legume product). Plant Food Hum. Nutr. 54:131-150.

Milan J (2001). Biotransformation of organophosphorous compounds. J. Toxicol. 166:139-160.

MINADER (2012). INFO PRIX. Publication trimestrielle des prix des produits et intrants agricoles. Cellule des informations et de l'alerte rapide. Secrétariat Général. Direction des enquêtes et statistiques des produits agricoles. Ministère de l'Agriculture et du Développement Rural. Cameroun. (007):1-58

MINARDE (2010). List of products homologated on October 2010. Commission of national homologation of phytosanitary products and apparatus certification for Spraying. Secretary of the commission. Yaounde, Cameroun. pp. 1-11.

Nguegang P, Parrot L, Lejoly, Joiris VJ (2005). Yaounde Dregs exploitation: production system, traditional ability and urban and peri-urban agricultural potentialities for local development. In: Agriculture and urban development in West and Central Africa. L. Parrot and R. Kahane, eds. Monpellier, CIRAD, France. 40 pp.

Noriko Y, Akiko K, Hiroko N, Makoto N, Eiko M, Keiko M, Masanori S, Yoshio I (1998). Effect of Cooking on Pesticide Residues Including Captan in Domestic Crops I. Jpn. J. Food Chem. 1:61-68.

NRCPPCS (National Registration Commission of Phytosanitary Products and Certification of Sprayers) (2007). List of pesticides authorized in Cameroon. Ministry of Agriculture and Rural Development Yaoundé, Cameroon. pp. 1-11.

Nyakundi WO, Magoma j, Ochora, Nyende AB (2011). Biodegradation of diazinon and methomyl pesticides by white rot fungi from selected horticultural farms in rift valley and central Kenya. J. Appl. Technol. Environ. Sanit. 1(2):107-124.

OJ/EU (Official Journal/ European Union) (2004). Recommendation of the commission of 9 January 2004 about coordinated community's programme for control in perspective to insure the respect of MRL of pesticide residues on/in cereals and others crops products. C (2003) 5400:60- 64.

Peggy M (2002). Water pH and its Effect on Pesticide Stability. Fact Sheet Fs-02-36 University of Nevada Reno. 56 pp.

Perihan A, Arpad A, Steven J, Lehotay, Andrew C (2007). Validation of an efficient method for the determination of pesticides residues in fruits and vegetables using ethyl acetate for extraction. J. Environ. Sci. Heal. (Part B). 42:481-490.

Randwan MA, Abu-Elamayem MH, Shiboob, Abdel-Aal A (2005). Residual behaviour of profenofos on some field-grown vegetables and its removal using various washing solutions and household processing. Food Chem. Toxicol. 43(4):553-557.

Sharma KK, Trivedi TP, Sharma RP, Shashi AV, Aruna Kumar T, Virender KB, Kui BG (2007). Sampling techniques for pesticides residues analysis. Pesticide residues Analysis Manuel. Indian Council of Agricultural Research, New Delhi.110 020. India. pp. 56-60.

Sonchieu J, Ngassomu MB, Tchatchueng JB, Ashutosh KS, Laxman PS (2010). Survey of pesticide residues in maize, cowpea and millet from the Northern Cameroon. Food Addit. Contam. (part B). 3(3):178-184.

Sougnabe SP, Yandia A, Acheleke J, Brevault T, Vaissayre M, Ngartoubam LT (2010). Pratiques phytosanitaires paysannes dans les savanes d'Afrique centrale. In : *Actes du colloque « Savanes africaines en développement : innover pour durer »*, 20-23 avril, Garoua, Cameroun 2009 ; L. SEINY-BOUKAR, P. BOUMARD , éds. Prasac, N'Djaména, Tchad ; CIRAD, Montpellier, France, cédérom. 136:7-14.

Souop D (2000). Obsolete pesticides stocks in Cameroon. In: *OECD-FAO-UNEP workshop in obsolete pesticide.* 13-15 September 2000; *Ms Anne LINDSAY,* Alexandra,Virginia, USA. 67 pp.

WHO (World Health organization) (2002). Children's health and environment: A review of evidence. A joint report from the European Environment Agency and the WHO Regional Office for Europe Environmental issue. Report No 29 Luxembourg: Office for Official Publications of the European Communities. 223 pp.

Xiangyu C, Chao Y, Ruihua L, Qiang L, Wei Z, Jianli L, Cunjiang S, Chuanling Q, Ashok M (2013). Simultaneous degradation of organophosphate and organochlorine pesticides by *Sphingobium japonicum* UT26 with surface-displayed organophosphorus hydrolase. Biodegradation 24(2):295-303.

Yanhong B, Ling Z, Jiang W (2006). Organophosphorous pesticide residues in market foods in Shaanxi area, China. Food Chem. 98:240–242.

Development of nutritionally enriched instant weaning food and its safety aspects

Mohammed A. Satter[1], Syeda Absha Jabin[1], Nusrat Abedin[1], Taslima Arzu[1], Kanika Mitra[1], A. M. Abdullah[2] and D. K. Paul[3]

[1]Institute of Food Science and Technology (IFST), Bangladesh Council of Scientific and Industrial Research (BCSIR), Dr. Qudrat- E-. Khuda Road, Dhanmondi, Dhaka -1205. Bangladesh.
[2]BCSIR Laboratories, Dhaka, Bangladesh Council of Scientific and Industrial Research (BCSIR), Dr. Qudrat- E-. Khuda Road, Dhanmondi, Dhaka -1205. Bangladesh.
[3]Islamic University, Kushtia, Bangladesh.

Malnutrition is persisting among children in Bangladesh due to lack of proper weaning foods. A highly nutritive instant weaning food was prepared in our laboratory by using locally available food resources with the aim to ensure the availability of low cost weaning food in Bangladesh. The developed food was evaluated for their nutritional characteristics and microbiological quality. The food contained the major nutrients like moisture, ash, fat, protein, fiber, carbohydrate and energy 2.43%, 2.26%, 11.32%, 15.98%, 1.06%, 75.35% and 456.6 kcal/ 100 g, respectively which were comparable to those of the three good quality imported commercial weaning foods F-1, F-2 and F-3. The vitamin A, iron and calcium contents were significantly different ($p<0.05$) than the commercial foods. The highest protein efficiency ratio and feed efficiency ratio were shown in the rats feed on the prepared weaning food than the imported commercial weaning foods. The overall bacteriological status of the prepared and the imported commercial weaning foods were observed to be satisfactory. The costs of the developed weaning food is considerably cheaper than the three imported commercial weaning foods of same quality and suitable for low income people of Bangladesh.

Key words: Instant, low cost, weaning food, cereals, rice.

INTRODUCTION

Inappropriate feeding practices of infant and young children are the most serious obstacles to maintaining adequate nutritional status and contribute to levels of malnutrition in Bangladesh that are amongst the highest in the world. Poverty, malnutrition and disease are inter-linked with each other. Malnutrition in children is the consequence of a range of factors, which are often related to poor food quality, insufficient food intake, severe and repeated infectious diseases; or frequently it involves some combination of the three (deOnis et al., 1993). According to Institute of Public Health and Nutrition (IPHN), Bangladesh (2007), almost one-half of children under five years are underweight and 42% are stunted. Only 42% of infants aged less than six months are exclusively breastfed and almost one-third (29%) of children aged 6 to 9 months do not receive any solid or semi-solid foods. The most common complementary foods include khichuri, bhaat dal, suji and muri (IPHN, 2007).

The introduction of supplementation in terms of weaning foods prepared from easily available and low cost ingredients is of vital importance to meet the require-ments of the growing children (Saeeda et al., 2009). In most developing countries, commercial weaning foods of excellent quality either imported or locally produced are generally 10 to 15 times higher than the cost of the common staple foods due to sophisticate processing, expensive packing, extensive promotion and solid profit margins (Bahlol et al., 2007). Therefore, development of supplementary foods based on locally available cereals and legumes has been suggested by the Integrated Child Development Scheme (ICDS) and Food and Agriculture Organization (FAO) to combat malnutrition among mothers and children of low socio-economic groups (Imtiaz et al., 2011).

Rice alone constitutes 92% of the total food grains produced annually in Bangladesh. It provides about 80% of the people's energy intake, the majority of their protein intake, and a considerable proportion of several micro-nutrients in their average daily diet (Provash and Mahendran, 2012). At present, there are no instant weaning foods manufactured in Bangladesh from locally available food resources and all the weaning foods available in the local markets are imported commercially (Imtiaz et al., 2011). Bangladesh is a very densely populated poor country. Hence, these imported commercial foods are too expensive for low income families. Thus, in this study, a cereal based weaning food is prepared to ensure the availability of low cost weaning food which is microbiologically safe to reduce malnutrition and mortality rate in infants. We also compare the nutritive value of the prepared weaning food with the available imported commercial weaning foods in Bangladesh.

MATERIALS AND METHODS

Collection of imported commercial weaning foods and raw materials

The three imported commercial weaning foods (Cerelac wheat – milk as F-1, Cerelac rice- milk as F-2 and Biomil rice-milk as F-3) were selected on the basis of its popularity and market availability. F-1 and F-2 both are marketed by Nestle Bangladesh Limited. F-3 is marketed by Fasska Food Industries Limited, Bangladesh. These three commercial weaning foods are recommended for children aged 6 months and above. These weaning foods were purchased from the super markets in Dhaka, Bangladesh.

Rice, sugar, skim milk powder, butter etc. were purchased from a local market in Dhaka city, Bangladesh and transported to the laboratory in clean polyethylene bags for later use.

Processing of rice

Rice powder was blended with enzyme and incubated at 70°C for 90 min with occasional stirring. Emulsifier was blended and mixed again with the butter thoroughly. The mass was then passed through the roller dryer to make flakes. The white flakes were grounded and passed through 1 μm mesh sieve.

Nutritional criteria for developing instant weaning food

The weaning food was developed taking into consideration the:

i) Codex Alimentarius Standards for foods for special dietary uses (including foods for infants and children) (WHO, 1994).
ii) Recommended daily allowances (RDA) for energy, protein and other nutrients for 7 to 12 months old infant by Food and Nutrition, Institute of Medicine (IOM, 2005).

Formulation of instant weaning food

Rice flakes, skim milk powder, butter, vitamin premix and sugar were blended for the preparation of weaning food. The complete processing steps of weaning food are shown in Figure 1.

Chemical analysis of weaning foods

Nutrient analysis

The nutrient compositions of the imported commercial weaning foods and as well as the prepared weaning food that is prepared in our laboratory were carried out as follows:

Determination of moisture content: Moisture content was determined by oven-dry method as the loss in weight due to evaporation from sample at a temperature of 105°C. The weight loss in each case represented the amount of moisture present in the sample:

$$\text{Moisture (\%)} = \frac{\text{Weight of original sample} - \text{weight of dried sample}}{\text{Weight of original sample}} \times 100$$

Determination of crude protein: The crude protein content was determined following the micro Kjeldahl method (AOAC, 2005). Percentage of nitrogen (N) was calculated using the following equation:

$$\text{Nitrogen (\%)} = \frac{(S-B) \times N \times 0.014 \times D}{W \times V} \times 100$$

Where D = Dilution factor, T = titre value = (S-B), W = weight of sample, 0.014 = constant value.

Crude protein was obtained by multiplying the corresponding total nitrogen content by a conventional factor of 6.25. Thus, crude protein (%) = % of N × 6.25.

Determination of crude fat: Crude fat was determined by the soxhlet extraction technique followed by AOAC (2005). Fat content of the dried samples was easily extracted into organic solvent (petroleum ether) at 60 to 80°C and followed to reflux for 6 h. Percentage of fat content was calculated using the following formula:

$$\text{Crude fat (\%)} = \frac{\text{Weight of fat in sample}}{\text{Weight of dry sample}} \times 100$$

Determination of ash: Ash content was determined by combusting the samples in a muffle furnace at 600°C for 8 h according to the method of AOAC (2005):

$$\text{Ash content (\%)} = \frac{\text{Weight of ash}}{\text{Weight of sample}} \times 100$$

Determination of crude fiber: The bulk of roughage in food is referred to as the fiber and is called crude fiber. Milled sample was dried, defatted with ethanol acetone mixture and then the experiment

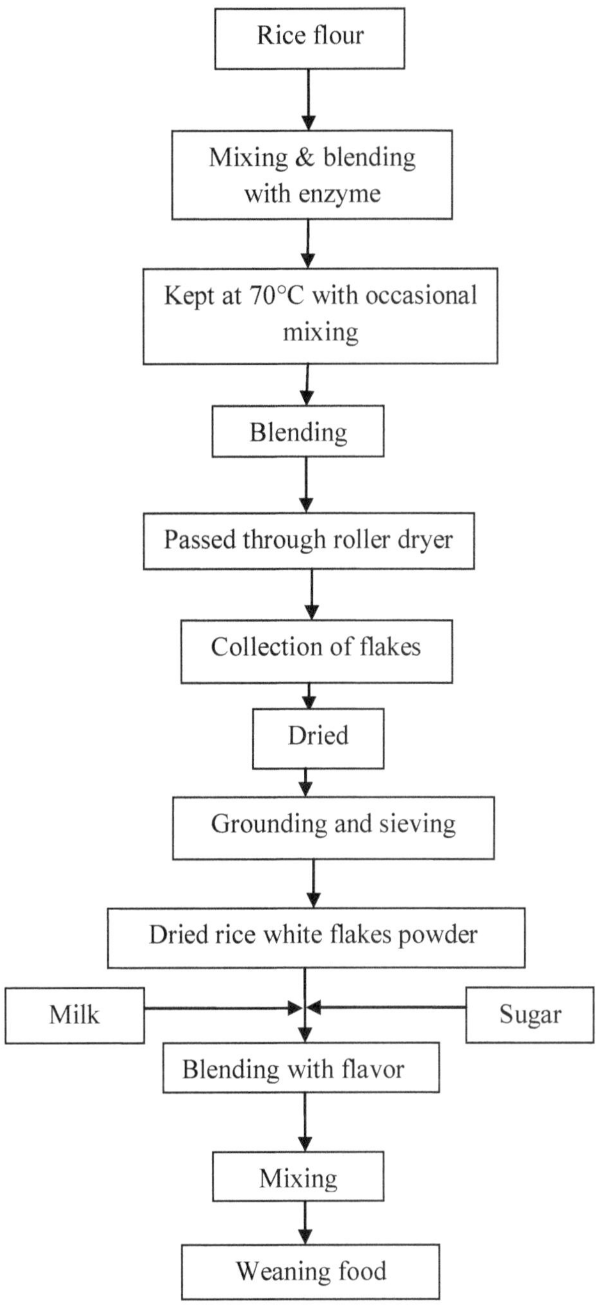

Figure 1. Flow diagram of weaning food formulation from rice, skimmed milk powder and sugar.

was carried out using the standard method as described in AOAC (2005):

$$\text{Crude fiber (\%)} = \frac{\text{Weight of residue} - \text{Weight of ash}}{\text{Weight of sample}} \times 100$$

Determination of carbohydrate: The carbohydrate content was estimated by the difference method. It was calculated by subtracting the sum of percentage of moisture, fat, protein and ash contents from 100% according to AOAC (2005):

Carbohydrate (%) = 100 – (moisture% + fat% + protein% + ash%)

Determination of total energy: The total energy value of the food formulation was calculated according to the method of Mahgoub (1999) using the formula as shown in the following equation:

Total energy (kcal/100 g) = [(% available carbohydrates × 4) + (% protein × 4) + (% fat × 9)]

Determination of vitamin A: About 10 g of sample was homogenized, weighed and transferred into a ground bottom flask, 30 ml of extraction solution, 0.1% antioxidant and few drops of KOH were added and reflux for 30 min at 70°C. The sample was cool down, vitamin A was extracted into hexane, and the combined hexane extract was washed with water and then dried the hexane layer to about 2 ml on a water bath or rotary evaporator. The final volume was made up to 50 ml with the mobile phase. The mobile phase, standard and sample were filtered through 0.45 µ membrane filter and were degassed before injection. Calibration curve was made by a standard in mobile phase with five point calibrations and analyzed independently by HPLC. A standard curve was plotted between the concentration of vitamin A and peak area obtained. For HPLC analysis, an Eelipse × BD – C_{18} column (4.6 × 250 mm 5 µm) was used with a linear gradient of methanol: water (95:5) at constant flow rate of 1 ml /min by using a binary pump with column temperature 40°C. A multiple wavelength detector was employed for the detection of pecks using a wavelength of 325 nm and a bandwidth of 8 nm.

Determination of minerals: The mineral contents were determined after the ash content determination. The ash residue of each formulation was digested with perchloric acid and nitric acid (1:4) solution. The samples were left to cool and contents were filtered through Whatman filter paper 42. Each sample solution was made up to a final volume of 25 ml with distilled water. The aliquot was used separately to determine the mineral contents of Fe and Ca by using an Atomic Absorption Spectrometer (Spectra AA 220, USA Varian).

Determination of amino acids: Amino acid composition of samples was measured on hydrolysates using an amino acid analyzer based on HPLC system. Sample hydrolysates were prepared following the method of Moore and Sterin (1963).

Determination of microorganism: Microbiological examination of the weaning foods was performed to assess bacterial, fungal and yeast load under laboratory condition. Standard Plate Count (SPC), fungal and yeast count and enumeration of total coliform and *Salmonella* of the weaning foods were examined according to BAM (1998). Plate count method was employed for the examination of total number of viable microbes present in the sample. Standard plate count (SPC) was estimated by decimal dilution technique followed by pour plate method and spread method for fungus and yeast. Streak plate method was used to isolate the specific microorganism. Isolation and enumeration of total coliform were performed by most probable number (MPN) method using MacConkey broth (Harrigen and MacCance, 1976).

Rat bioassay

To assess the nutritional quality of the weaning foods, a standard rat bioassay procedure was done in the animal house section, Institute of Food Science and Technology (IFST), BCSIR, Dhaka. Long-Evan (28 to 30) days old rats were used to study the growth after feeding of the specific weaning foods. The rats were divided into four groups containing 4 to 5 male rats in each cage, main-

Table 1. Nutritive Value of the Prepared Weaning Food as compared with Imported Commercial Weaning Foods

Nutrient	Prepared weaning food	F1	F2	F3
Moisture (%)	2.43 ± 0.076^{a}	3.35±0.33^{b}	2.79 ± 0.26^{ac}	3.85 ± 0.30^{d}
Protein (%)	15.98 ± 0.48^{a}	14.42 ± 0.33^{b}	13.79 ± 0.22^{bc}	13.00 ± 0.45^{cd}
Fat (%)	11.32 ± 0.25^{a}	8.08 ± 0.38^{b}	10.16 ± 0.36^{c}	6.78 ± 0.28^{d}
Carbohydrate (%)	75.35 ± 0.61^{a}	66.60 ± 0.65^{b}	69.56 ± 0.94^{c}	64.43±0.52^{d}
Ash (%)	2.26 ± 0.35^{a}	3.15 ± 0.10^{b}	2.85 ± 0.20^{ac}	2.33 ± 0.48^{ac}
Fiber (%)	1.06 ± 0.08^{a}	2.31 ± 0.25^{b}	1.89 ± 0.21^{bc}	1.55 ± 0.28^{ac}
Energy(kcal/100 g)	456.67 ± 7.64^{a}	423.33 ± 10.40^{b}	427.67 ±11.24^{bc}	391.33 ± 3.21^{d}
Vitamin A (µg/100 g)	289 ± 3.60^{a}	260 ± 5.00^{b}	257.67 ± 2.57^{bc}	241.67 ± 7.64^{d}
Vitamin B1 (mg/100 g)	0.55 ± 0.03^{a}	0.67 ± 0.03^{b}	0.59 ± 0.03^{ac}	0.42±0.04^{d}
Iron(mg/100 g)	7.32 ±0.24^{a}	7.57 ± 0.18^{a}	7.46 ± 0.37^{a}	6.58 ± 0.19^{b}
Calcium (mg/100 g)	446.00 ± 5.29^{a}	330.00 ± 5.00^{b}	436.67 ± 2.89^{ac}	495.00 ± 5.00^{d}

Mean values by different superscripts within columns are significantly different at ($p<0.05$).

tained at room temperature with 12 h light/dark cycle. They were subjected to a 7 days acclimatization period and the individual groups were given same amount of commercial laboratory chow diet. After 7 days, each rat group acclimatized on average 55 g of different weaning foods and was feed water *ad labitum* for 4 weeks. The control group was feed on casein diet containing 75% starch, 10% oil, 4% salt, 1% vitamin mix and 10% casein protein. Food consumption, spoilage and weight gain were recorded in every 4 days. Protein efficiency ratio (PER) and feed efficiency ratio (FER) was calculated using the following formula:

$$PER = \frac{\text{Weight gain of test group (g)}}{\text{Total protein consumed (g)}}$$

$$FER = \frac{\text{Gain in body weight (g)}}{\text{Food consumed (g)}}$$

Estimation of cost

The cost of developed weaning food was determined according to the price of the materials used in preparing the food, gas, electricity, jars, water and others. The cost of the prepared food was compared with market price of the three imported commercial baby foods at the same time.

Statistical analysis

The data were analyzed using SPSS version 17.0. The mean and standard deviations of the triplicate analyses were calculated. The analysis of variance (ANOVA) was performed to determine significant differences between the means using Dunnett's T3 tests.

RESULTS AND DISCUSSION

Nutritional composition

The nutritional compositions (g/100 g dry weight) of the prepared weaning food and imported commercial weaning foods were summarized in Table 1. The energy content of the prepared weaning food was 456.67 kcal/100 g. For the imported commercial weaning foods, the energy content ranged from 391.33 to 427.67 kcal/100 g which was significantly lower than the prepared weaning food. For all the weaning foods both prepared and commercial, the energy density per 100 g of the dry food was lower than the minimum energy (483.9 kcal/100 g) recommended in the Codex Alimentarius Standards for weaning/follow up foods (FAO/WHO, 1994). Protein is one of the most important nutrients required in weaning foods. The prepared weaning food had a protein content of 15.98% which was higher than the minimum amount (14.52%) specified in Codex Alimentarius standards. The protein levels of the commercial weaning foods ranged from 13.00 to 14.42% which were significantly lower than the prepared weaning food and slightly below the Codex Alimentarius standards lower limit. According to FAO/WHO Codex Alimentarius Standards for follow – up/ weaning foods, the protein content should range from 14.52 to 37.70 g/100 g (FAO/WHO, 1994). The prepared weaning food contained 11.32% of fat which was significantly higher ($p<0.05$) than the commercial weaning foods ranged from 6.78 to 10.16% but lower than the specified amount in the Codex Alimentarius Standards (range 14.52- to 41.13%). For the commercial weaning foods, F-2 had a fat content of 10.16% which is significantly higher ($p<0.05$) than the F-1 and F-3.

The lower fat content had contributed to lower energy value of the prepared weaning food. The lower fat content may also have contributed to the increase in the shelf - life of the formulation by decreasing the chances of rancidity (Onuorach and Akijede, 2004). Ash content is an important nutritional indicator of mineral content and an important quality parameter for contamination, particularly with foreign matter (for example, pebbles) (Fennema, 1996). The ash content of the prepared weaning food was 2.26%; this result was significantly lower than F-1 which contained 3.15% (Table 1); but the ash content of the prepared food was comparatively within the same range as that of F-2 (2.85%) and F-3 (2.33%). No standard for ash concentration has been specified for weaning/follow–up foods in the Codex Alimentarius Standards (FAO/WHO, 1994). The moisture content of the prepared weaning food was 2.43%. This result was significantly different from the commercial weaning foods F-1 and F-3 which were 3.35% and 3.85%, respectively (Table 1). The commercial weaning food F-2 contained 2.79% moisture which has no significant difference from the prepared food. The percent crude fiber of the prepared weaning food was 1.06 (Table 1). This value was significantly lower than the value of commercial weaning foods which ranged from 1.55 to 2.31%. Low fiber concentration in the prepared weaning food could be due to use of highly refined rice flours.

Fiber is an important dietary component in preventing overweight, constipation, cardiovascular disease, and diabetes and colon cancer (Mosha et al., 2000). High dietary fiber content has been reported to impair protein and mineral digestion and absorption in human subjects (Whitney et al., 1990). Some fiber related fractions such as polyphenols and non-starch polysaccharides, bind minerals such as Ca, Zn and Fe, making them unavailable for human nutrition (Fairweather-Tait and Hurrell, 1996). The carbohydrate content in the prepared weaning food was 75.35% (Table 1). This result was significantly ($p<0.05$) higher than the commercial weaning foods which ranged from 64.43 to 66.60%. The carbohydrate levels of the prepared weaning food and all the imported commercial weaning foods were higher than the lower limit (41.13 to 73.79 g/100 g) of the Codex Alimentarius Standards (FAO/WHO, 1994). Vitamins are substances which are indispensible for the growth and maintenance of good health. According to Codex Alimentarius Standards (FAO/WHO, 1994), the recommended levels (µg/100 kcal) of vitamin A for processed cereal based foods for infants and young children ranged from 60 to 180 µg/100 kcal. On comparing with this standard, the prepared weaning food had vitamin A content of 63.28 µg/100 kcal and the imported commercial weaning foods F-1, F-2, and F-3 had vitamin A content of 61.42, 60.25 and 61.76 µg/100 kcal, respectively which were above the minimum limit (60 µg/100 kcal) specified in the Codex Alimentarius standards. The vitamin A content of the prepared weaning food was significantly higher ($p<0.05$) than the commercial weaning foods (Table 1).

The prepared weaning food had vitamin B1 content of 120.44 µg/100 kcal (Table 1). For the imported commercial weaning foods, F-1, F-2 and F-3 had vitamin B1 content of 158.27, 137.96 and 107.33 µg/100 kcal, respectively. The vitamin B1 contents of the locally prepared weaning food and the commercial weaning foods were higher than the minimum amount (60 µg/100 kcal) specified in the Codex Alimentarius Standards (FAO/WHO, 1994). Calcium is an essential element in infants and young children for building bones and teeth, functioning of muscles and nerves, blood clotting and for immune defense (Whitney et al., 1990). The concentration of Ca of the prepared weaning food was 446.0 mg/100 g which was significantly different from the imported commercial weaning foods F-1 and F-3 by 330.00 and 495.00 mg/100 g, respectively (Table 1). According to the Codex Alimentarius standards, Ca concentrations in weaning foods should not be less than 435.51 mg/100 g of the dry food. On the basis of this standard, the prepared weaning food and all the imported products, except F-1 had the Ca concentrations above the minimum amount (435.51 mg/100 g) specified in the Codex Alimentarius Standards (FAO/WHO, 1994). Iron is an essential micronutrient for the synthesis of hemoglobin (an oxygen carrier in the red blood cells), myoglobin (used for muscle contraction) and enzymes/coenzymes (used in various metabolic pathways). Iron also enhances the body's immune system thus, reducing infections and fostering proper functioning of other organs of the body (Whitney et al., 1990).

Iron concentrations in the prepared weaning foods were 7.32 mg/100 g (Table 1); while in the imported commercial weaning foods, the concentration ranged between 6.50 to 7.57 mg/100 g. The prepared weaning food and all the imported products had the iron concentrations above the minimum amount (4.8 mg/100 g) specified in the Codex Alimentarius Standards (FAO/WHO, 1994). According to Food and Nutrition, Institute of Medicine (IOM, 2005), the recommended daily allowances (RDA) for energy, carbohydrate, protein, fat, vitamin A, vitamin B1, Ca and Fe for 7 to 12 months old infant are 743 kcal, 95 g, 11 g, 30 g, 500 µg, 0.3 mg, 270 mg and 11 mg, respectively. A 100 g portion of prepared weaning food assayed in this study could meet the daily requirements (% of RDA) of energy (61.46%), carbohydrate (79.32%), protein (145.27%), fat (37.67%), vitamin A (57.8%), vitamin B1 (183.33%), Ca (165.19%) and Fe (66.55%). For the imported commercial weaning foods, a 100 g portion of the food could meet the daily requirements (% of RDA) of energy (52.67 to 56.98%), carbohydrate (67.82 to 73.22%), protein (118.18 to 131.09%), fat (22.60 to 33.87%), vitamin A (48.33 to 52.00%), vitamin B1 (140.0 to 223.33%), Ca (122.22 to 183.33%) and Fe (59.82 to 68.82%).

These findings suggest that, both prepared and commercial weaning foods have the potential to provide the recommended daily requirements for energy and other nutrients for infants and young children. These foods, however, may be unable to meet the RDAs for energy and macro/micronutrients as the children grow older due to increased demand for nutrients to support growth.

Quality evaluation

In a well - balanced weaning food, 13% of the total energy is usually derived from protein, 26% from fat and 61% from carbohydrates, respectively (Royal Tropical Institute, Amsterdam Weaning Food, 1987). The pre-

Table 2. Percent calories from different nutrients of weaning foods

Weaning foods	Percent calories		
	Protein	Carbohydrate	Fat
Prepared food	14.00	66.00	22.30
F-1	13.63	62.93	17.18
F-2	12.89	65.05	21.38
F-3	13.29	65.85	15.41

% Protein calories = Protein% X 4/ Total energy of formulations; % Carbohydrate calories = Carbohydrate % X 4/ Total energy of formulations; % Fat calories = Fat % X 9 / Total energy of formulations

Table 3. Amino acids composition (g/100g crude protein) of the prepared weaning food and percentage of RDA of amino acids met per 100g of prepared weaning food.

Amino acid	Prepared food	*RDA	% of RDA met
Aspartic Acid	0.62	-	-
Threonine	0.42	2.6	16
Serine	0.54	-	-
Glutamic Acid	1.43	-	-
Glycine	0.43	-	-
Alanine	0.94	-	-
Valine	0.30	4.2	7
Methionine	0.54	2.2	24
Isoleucine	0.96	4.2	22
Leucine	0.20	4.8	4
Tyrosine	0.60	-	-
Lysine	1.34	4.2	31
Arginine	1.31	2.0	65

* Source of RDA: FAO/WHO (1973).

pared weaning food had higher concentration of calories from carbohydrate and protein than the recommended values (Table 2). The prepared weaning food and the imported commercial weaning foods F-1, F-2, F-3 provided 22.30 and 17.18, 21.38 and 15.41% of the total calories from fat, indicating low energy density in these baby foods. Such an imbalance of calories from different nutrients may affect the quality of the diet (Khan, 1989).

Amino acid composition

The amino acid compositions of the prepared weaning food were shown in Table 3. The results showed that the prepared weaning food contained thirteen amino acids including seven essential amino acids. The percentage of recommended dietary allowance (RDA) of essential amino acids of the prepared food was also shown in Table 3. The results showed that the prepared weaning food met 16, 7, 24, 22, 4, 31 and 65% of the RDA of essential amino acids which was threonine, valine, methionine, isoleucine, leucine, lysine and arginine, respectively. The low percentage of the essential amino acids of the prepared weaning food compared with the RDA of essential amino acids for children could be ascribed to the fact that the weaning food was prepared from plant based food materials (rice). A number of studies have shown that the protein content of plant- based food materials is inadequate to meet the protein requirements of individuals compared with food material produced from animal sources (Ijarotimi and Olopade, 2009).

Microbiological quality of weaning foods

The overall bacteriological status of the prepared and the imported commercial weaning foods was observed to be satisfactory (Table 4). The obtained results revealed that the total viable bacterial count, total yeast and mold count, total coliform count and presence of *Salmonella* per gram were absolutely nil/g in the all weaning foods analyzed, when packets were opened. The low counts of the examined foods indicated adequate thermal process, good quality of raw materials and as a result of the good

Table 4. Comparative Microbiological Load of Weaning Foods

Baby food sample	Total viable count (CFU/g)	Yeast and mold count (CFU/g)	Total Coliform count (CFU/g)	*Salmonella*/g
Prepared food	Less than 10	Less than 10	Nil	Nil
F-1	Nil	Nil	Nil	Nil
F-2	Nil	Nil	Nil	Nil
F-3	Nil	Nil	Nil	Nil

Table 5. Food intake, protein intake and body weight gain of rats, for assessment of PER and FER

Weaning foods	Food intake (g)	Protein intake (g)	Weight gain up (g)	PER	FER
Casein diet	545.6	53.05	90.50	1.71	0.17
Prepared food	640.33	62.05	130.65	2.11	0.20
F-1	615.45	60.99	115.56	1.89	0.19
F-2	620.65	61.65	105.60	1.71	0.17
F-3	555.5	55.55	98.65	1.78	0.18

Table 6. Cost of prepared weaning food and imported commercial weaning foods

Cost/ year (900000 packets)	Developed weaning food	F-1	F-2	F-3
Raw materials	71963256 BDT	-	-	-
Manpower	1,159,500 BDT	-	-	-
Utilities	210,000 BDT	-	-	-
Depreciation and other expenditures	1003600 BDT	-	-	-
Other financial expenses	10000000 BDT	-	-	-
Revenue from sales	152019984 BDT	-	-	-
*Price per packet (500g/packet)	263 BDT	350 BDT	385 BDT	450 BDT

*1USD= 77.88 BDT; BDT= Bangladesh Taka; USD= United States Dollar.

different processing conditions under which the production of foods was carried out.

Nutritional quality of weaning foods

In a 21-days study of food intake, the prepared weaning food had the highest value (640.33 g per rat) and it differed from other foods where food intake ranged from 545.6 to 620.65 g per rat (Table 5). Group of rats feed on the prepared weaning food gained the highest body weight (130.65 g per rat) than the other groups of rats feed on the three imported commercial weaning foods where body weight gain ranged from 98.65 to 115.56 g per rat. Also, protein intake was high in the rats feed on of the prepared weaning food. PER and FER were the highest for those rats feed on the prepared weaning food (2.11 and 0.2). The PER of commercial weaning foods were ranged from 1.71 to 1.89 which were similar to casein diet. The imported commercial weaning foods F-1, F-2 and F-3 had 0.19, 0.17 and 0.18 of FER, respectively; and were lower than the prepared weaning food but similar to casein diet. PER was high in the prepared weaning food than the other imported commercial weaning foods (Table 5). So the growth rate was also higher of those rats feed on the prepared weaning food than those rats feed on the commercial foods. Based on earlier studies, it has been recommended that the PER of the weaning foods should not be less than 2.1 and preferably 2.3 (PAG, 1971). Thus, the prepared weaning foods would meet the required standards for PER and adequately satisfied the recommended nutritional quality.

Cost of weaning foods

The cost of the developed weaning food and other commercial products (F-1, F-2 and F-3) were calculated as shown in Table 6. The cost of the prepared weaning food is suitable for the low national income in Bangladesh and cheaper than those of commercial foods.

Conclusion

A highly nutritive weaning food prepared from rice and milk in Cereal Technology section, IFST, BCSIR, Dhaka and safety aspects of it was compared to imported commercial weaning foods. From the nutritional analysis in various constituents like moisture, protein, fat, ash, carbohydrate, crude fiber and energy among four (one prepared and three commercial) weaning foods; the prepared weaning food has shown the most satisfactory result. The protein (an essential nutrient for rapid growth in infants) was highest in the prepared weaning food than imported commercial weaning foods. From the vitamins (vitamin-A and -B1) and minerals (iron and calcium) analysis of those weaning foods, prepared weaning food has also given the most satisfactory result. From the microbial point of view, the prepared weaning food showed satisfactory results with regards to microbial food safety. In rat bioassay, the highest PER and FER were shown in the rats feed on the prepared instant weaning food. Thus, the highly nutritive low cost instant weaning food in our laboratory was satisfactory to meet food safety requirement, ensure the rapid growth of infants and finally, help to reduce malnutrition situation and give extra support to nourishment for the children of Bangladesh.

ACKNOWLEDGEMENTS

The authors wish to express their thanks to the Institute of Food Science and Technology (IFST), Bangladesh Council of Scientific and Industrial Research (BCSIR), Dhaka, Bangladesh, for supporting to do this research.

REFERENCES

AOAC (2005). Official Methods of Analysis. Association of Official Analytical Chemists, 18th ed. Washington, D.C.

BAM (1998). Bacteriological Analytical Manual, 8th Ed. Revision A, Chapter 3.

Bahlol HEM, Sharoba AM, EL-Desouky AI (2007). Production and evaluation of some food formulas as complementary food for babies using some fruits and vegetables. Annals Agric. Sci. Moshtohor 45(1): 147-168.

DeOnis M, Monteiro C, Akre J, Clugston G (1993). The Worldwide Magnitude of Protein-Energy Malnutrition: An Overview from the WHO Global Database on Child Growth. Bull. World Health Org. 71 (6): 703-712.

Fairweather-Tait S, Hurrell RF (1996). Bioavailability of minerals and trace elements. Nutr. Res. Rev. 9: 295–324.

Fennema OR (1996). Food Chemistry, 3rd edn. New York.

Harrigan WF, McCance ME (1976). Laboratory Methods in Food and Dairy Microbiology. Academic Press: London.

Ijarotimi OS, Olopade AJ (2009). Determination of Amino Acid Content and Protein Quality of Complementary Food Produced from Locally Available Food Materials in Ondo State, Nig. Mal J. Nutr.15(1):87-95.

Imtiaz H, Burhanuddin M, Gulzarm A (2011). Evaluation of weaning foods formulated from germinated wheat and mungbean from Bangladesh. Afr. J. Food Sci. 5(17): 897-903.

IOM (2005). Dietary Reference Intakes (DRIs): Recommended Intakes for Individuals. Food and Nutrition Board, Institute of Medicine, Washington, D.C: National Academies.

IPHN (2007). National Strategy for Infant and Young Children Feeding in Bangladesh. Institute of Public Health and Nutrition (2007). Directorate General of Health Services. Ministry of Health and Family Welfare, Dhaka, Bangladesh.

Khan MA (1989). Desirable dietary patterns for Pakistan. Proceedings of FAO/PARC Workshop on Dietary Guidelines for Food and Agriculture Planning. Islamabad: National Agricultural Research Centre.

Mahgoub SEO (1999). Production and evaluation of weaning foods based on sorghum and legumes. Plants Foods Hum Nutr. 54: 29-42.

Moore S, Stein WH (1963). Chromatographic amino acids determination by the use of automatic recording equipment. Methods Enzymol. 6: 819-831.

Onuorach CE, Akijede FA (2004). Comparative Evaluation of Four Formulated Weaning Foods and A Commercial Product. Nig. Food J. 22: 48-53.

PAG (1971). Protein Advisory Group of the United Nations, PAG Guideline No. 8, 'Protein-rich mixtures for use as weaning foods. New York: FAO/WHO/UNICEF.

Provash KK, Mahendran S (2012). Trend models for analyzing sustainable production of rice in Bangladesh. Sci. Res. Essays. 7(6): 669-678.

Royal Tropical Institute, Amsterdam (1987). Weaning Food: A New Approach to Small Scale Weaning Food Production from Indigenous Raw Materials in Tropical Countries, 2nd edn. Amsterdam: RTI, Rural Development Programme, Nutrition and Agrotechnology Section.

Saeeda R, Muhammad NS, Amer M, Nouman S, Khalid N, Muhammad A (2009). Preparation and quality evaluation of nutritious instant baby food from indigenous sources. Pakistan J. Agric. Res. 22: 1-2.

Mosha TCE, Laswai HS, Tetens I (2000). Nutritional composition and micronutrient status of home made and commercial weaning foods consumed in Tanzania. Plant Foods Human Nutr. 55(3): 185-205.

Whitney EN, Hamilton EMN, Rolfes SR (1990). Understanding Nutrition, 5th edn. NewYork: West Publishing Company.

WHO (1994). Codex Alimentarius Standards for Foods for Special Dietary Uses (including foods for infants and children, Vol. 4. Joint FAO/WHO Food Standards Program Rome: WHO, Codex Alimentarius Commission.

WHO (1973). Ad Hoc Expert Committee, Energy and Protein requirements. WHO Technical Report Series, No.522; FAO Nutrition Meetings Report Series, No. 52; WHO, Geneva; FAO, Rome, Italy.

11

The effect of thermal processing on fumonisin B_1 (FB_1) levels in maize-based foods

Mohanlall R., Odhav B. and Mohanlall V.

Department of Biotechnology and Food Technology, Faculty of Applied Science, Durban University of Technology, P. O. Box 1334, Durban 4001, South Africa.

Fumonisin B_1 (FB_1) is a mycotoxin from *Fusarium verticillioides* that is frequently associated with maize. Fumonisins have been implicated as the causal agents of a variety of animal diseases and are epidemiologically linked to the high incidence of human esophageal cancer in some regions of the world. Thermal treatments are used in many processes involving grain and its derivatives, but little is known about the effects of common processing methods on the fumonisin content of food. The objectives of this study were to determine the thermostability of this toxin in contaminated maize, at different time/temperature combinations, as well as to determine the effect of baking and frying on the stability of FB_1 spiked into maize-based foods. The identity of FB_1 in extracts before and after heat treatments was confirmed by high-performance liquid chromatography. For each thermal process, the fumonisin content was inversely proportional to the processing temperatures. An initial FB_1 concentration of 217 mg/g in the control, was reduced to 184 mg/g when treated at 100°C for 2 h. Oven temperature of 220°C for 30 min showed extensive reduction of FB_1 to a concentration of 1.1 mg/g as compared to 94 mg/g in the control. Baking maize muffins spiked with 1.25 µg/g (dry weight) FB_1 at 200°C for 20 min resulted in an average FB_1 loss of 70%. Frying of maize chips spiked with 5 µg/g (dry weight) FB_1 at 190 to 210°C for 5 to 10 min resulted in an average FB_1 loss of 67%. The results of this study indicate that boiling temperatures are ineffective in producing any significant reduction in FB_1 levels. Thus, there might be correlation between the under processing of fumonisin-contaminated foods and the high incidence of esophageal cancer in certain regions of South Africa.

Key words: Mycotoxins, *Fusarium verticillioides,* maize-based food products, esophageal cancer.

INTRODUCTION

Fungal growth and subsequent contamination of animal feeds and human foodstuffs by toxins (mycotoxins) constitute both an economic and a health risk. The mycotoxins of major concern for human health are produced by three main genera of fungi *(Aspergilius, Fusarium* and *Penicillium)* (Nelson et al., 1994). The fumonisins produced by *Fusarium verticillioides* and *Fusarium proliferation,* is one family of mycotoxins, which contaminate feeds and foodstuffs, predominantly maize-based products throughout the world. This fungus has been suspected of being involved in human and animal diseases since its original description in 1904 (Sheldon,

1904).

They can produce high amounts of fumonisins in tropical and subtropical regions (Marasas, 2001; Reddy et al., 2009; Shephard et al., 1996), whereas in colder regions, the fumonisin contamination is much lower (Logrieco et al., 2002; Miller, 2008). In addition to maize-derived products, fumonisins have also been detected in rice (Yazar and Omurtag, 2008), black tea leaves (Martins et al., 2001), asparagus (Logrieco et al., 1998) and pine nuts (Marin et al., 2007).

Maize is the staple cereal food grown and consumed by the rural farming communities of Africa, especially in the Transkei region of South Africa (Shephard et al., 2002), where the effects of fumonisins were first discovered in 1988. The Transkei region has one of the highest incidences of esophageal cancer in the world, which seems to be associated with the fumonisin intake (Makaula et al., 1996; Shephard et al., 2002). A positive correlation in the number of ecological studies between dietary fumonisins and human esophageal cancer rates has been reported in Africa and China (Sydenham et al., 1990; Chu and Li, 1994). In Southern Africa, the highest rate of human esophageal cancer occurred in the southwestern districts of Transkei where maize is the main diet, and these populations can be chronically exposed to highly contaminated food (Marasas et al., 1981). Several strains of *F. verticillioides* isolated from maize produced in these districts have been found to be acutely toxic to ducklings. When culture material of these isolates grown on autoclaved maize, was fed to experimental animals, the lesions induced included cirrhosis and nodular hyperplasia of the liver and intraventricular cardiac thrombosis in rats, leukoencephalomalacia (LEM) and toxic hepatosis in horses, pulmonary edema in pigs, nephrosis and hepatosis in sheep and acute congestive heart failure in baboons (Kriek et al., 1981).

On the basis of the toxicological evidence, the International Agency for Research on Cancer has declared *F. verticillioides* toxins as potentially carcinogenic to humans as class 2B carcinogens (Shephard et al., 1996). In view of these toxic effects and the ubiquitous nature of *F. verticillioides* contamination of maize, control measures and potential decontamination procedures are required to reduce the health risks posed by contaminated maize consumption in various areas of the world. Also, South Africa cannot afford to dump imported grain or export grain if it is contaminated. Thus, contamination of maize by *F. verticillioides* and fumonisins is a growing concern that may threaten animal and human health and is detrimental to the marketability of maize throughout the world.

Studies on the occurrence of fumonisins in maize and maize-based foods have shown that the highest levels of the toxin are in the whole grain and those maize products that undergo the least/mildest forms of processing such as maize meal, maize flour and grits. Those maize products that are more highly processed such as corn flakes and cereals tend to have either no detectable, or very low levels of fumonisins (Stack and Eppley. 1992). Industrial processes such as milling, extrusion, baking and frying are used by the food industry to convert raw cereal grains into a multitude of consumer products. Little is known about the effects of thermal processing on fumonisin levels in maize-based foods. Thus, the purpose of this study was to determine the thermal stability of FB_1 in maize-based foods.

MATERIALS AND METHODS

Growth and maintenance of *F. verticillioides*

A toxigenic strain of *F. verticillioides* (MRC 826, Pretoria, South Africa) was used throughout this study. The strain was subcultured on malt extract agar (Oxoid) and Potato Dextrose Agar (Oxoid), for 7 to 12 days at 25°C until profuse sporulation was evident. Mycelial plugs from a sporulating region were excised from the cultures using a scalpel blade and used for inoculation of maize. Stock cultures were maintained by subculturing onto malt extract agar and Sabouraud dextrose agar slants fortnightly.

Inoculation of maize patty cultures

Two 250-ml Erlenmeyer flasks were filled with 100 g of dry maize meal (Nyala, Premier Milling, Newtown), plus 100 ml distilled water and autoclaved at 121°C for 15 min. The flasks were inoculated with mycelial plugs from a 15 day old culture of *F. verticillioides* and incubated at 25°C for 4 weeks. The flasks were then dehydrated in a vacuum oven for 48 h at 45°C. Dried maize cultures were ground (Warring blender) to the consistency of flour. The FB_1 concentration in this contaminated maize was determined by high performance liquid chromatography analysis and was considered the initial concentration.

Processing treatments

Boiling

The inoculated maize cultures as previously prepared were subsequently used to determine the effects of thermal treatments. Two procedures were used for treatment. Firstly, boiling water, a moist heat source was used. The maize cultures were processed by the method of Dupuy et al. (1993) with slight modifications. A 500 mg sample of contaminated maize was placed into stoppered test tubes and immersed in boiling water at 100°C for 30 min, 1 and 2 h. *F. verticillioides* inoculated maize cultures which were not thermally treated were used as the controls to determine the FB_1 concentration prior to processing. Temperature was measured using calibrated thermocouples (Omega Engineering). Secondly, an oven was used to provide a dry heat source. A 500 mg sample of contaminated maize was placed into stoppered test tubes and processed at 100°C for 30 min, 1 and 2 h in an oven. This was achieved by placing the test tubes in beakers filled with sand. All experiments were performed and analyzed in triplicate.

Baking

The baking and frying studies were modified from the method described by Jackson et al. (1997) with a few changes as follows. A commercial maize muffin mix (Golden Cloud, Tiger Milling and

Baking, Germiston) was used for baking studies. The muffin mix was spiked with 1.25 µg/g FB_1 (w/w). Fresh milk, eggs and FB_1 standard solution were blended together and mixed with the dry muffin mix. Muffin batter (50 g) was placed onto each of the six cups in a nonstick muffin pan and baked in a laboratory convection oven at 220°C for 20 min. Calibrated thermocouples (Omega Engineering) were used to monitor the oven temperature. Inoculated maize cultures were processed at temperatures of 240 to 300°C for 20 min in a muffle furnace. The controls were treated the same except that they were not baked. All experiments were performed and analyzed in triplicate.

Frying

Maize meal spiked at 5 µg/g FB_1 (w/w) was used for frying experiments. This was achieved by mixing a 100 g maize meal with an equal weight of water containing the FB_1 solution. The resulting dough was formed into circles of approximately 10 cm in diameter and 1.5 mm in thickness with a rolling pin. The circles were cut into four equal pieces and fried in pure sunflower oil at 190 to 210°C for 5 to 10 min. In this study, the control used to determine the FB_1 concentration prior to processing was FB_1 spiked maize chips that were not fried. Calibrated thermometers were used to monitor the oil temperatures during frying. For all the frying runs, oil temperatures dropped by 10to 20°C after the chips were added to the oil. However, the required temperatures were reached within 2 min of starting the frying run. All experiments were performed and analyzed in triplicate.

Extraction

Fumonisin B_1 (98% pure) for standards and processing studies were obtained from Sigma Chemical Co; Midrand South Africa. Standards were prepared in 1:1 acetonitrile - water. The standards allowed qualitative and quantitative validation, by TLC and HPLC respectively. δ-Phthaldialdehyde (OPA) was purchased from Sigma Chemical Co. All reagents were of analytical grade and solvents were of HPLC grade.

Loss of FB_1 in maize cultures, muffins and chips was measured according to the method of Sydenham et al. (1992). Contaminated processed and unprocessed samples were extracted with methanol-water (3:1) (v/v), in a 1:2 (v/v) ratio and placed onto an orbital shaker for 1 h. The resulting slurries were centrifuged (500 xg, 10 min at 4°C). The supernatants were then filtered through Whatman No. 4 paper. The filtered extracts before and after processing were purified using the Bond Elut SAX cartridges (Varian, Harbor City, USA). Cartridges were fitted onto the Vac-Elut manifold (Varian) and were conditioned by washing first with 5 ml methanol followed by 5 ml of methanol-water (3:1) (v/v). Cartridges were not allowed to run dry. A 5 ml aliquot of the filtered extract was applied to the cartridge, thereafter the cartridge was washed with 5 ml methanol-water (3:1) followed by 3 ml methanol to remove contaminants. A flow rate of 2 ml/min through the cartridge was maintained. Fumonisins were eluted and collected with 10 ml of 0.5% acetic acid in methanol at a flow rate not higher than 1 ml/min in a suitable vial. The eluate was evaporated to dryness under a stream of nitrogen at 50°C. Dried samples were stored at 4°C until HPLC analysis.

Thin layer chromatography

Detection of FB_1 in extracts before and after processing were done by the method of Dupuy et al. (1993) with modifications, dried extracts and standards were reconstituted in 20 µl of acetonitrile-water (1:1), by vortexing for 5 min. 5 µl of the corn extracts and standards was spotted on Silica Gel 60 plates (10 x 10 cm, Merck). The plate was developed in 1-butanol-acetic acid-water (20:10:10) (v/v/v). After solvent evaporation, the plate was sprayed with 0.5% ρ-anisaldehyde and heated at 100°C for 10 min. For each plate, extracts and standards were spotted in duplicate.

High-performance liquid chromatography

Loss of FB_1 in inoculated maize cultures, muffins and chips were measured according to the method of Sydenham et al. (1992) with modifications. Briefly, residues were dissolved in 200 ul of acetonitrile-water (1:1). A 50 µl volume of the purified extracts or standards were transferred to the bottom of a small test tube and 450 µl of -phthalaldehyde reagent was added to each test tube. The solution was vortexed for 20 s, and 20 µl was injected into the HPLC system within 2 min of adding the OPA reagent. The HPLC column used was a Lichrosorb 5-µm C_{18} reversed phase column (Merck). A liquid chromatography M45 pump and injector with a fluorescence detector (335 nm excitation wavelength and 440 nm emission wavelength) was used to determine the identity and concentrations of FB_1 in the sample extracts. The mobile phase was methanol : 0.1 M sodium dihydrogen phosphate (68:32) adjusted to pH 3.35 with ortho-phosphoric acid. The flow rate was 1.0 ml/min. The fumonisin content of the maize sample was calculated from the chromatographic peak areas according to Shepard (2000).

RESULTS AND DISCUSSION

Boiling water

Migration rates of crude extracts of inoculated maize patty cultures processed at a 100°C in boiling water for 30 min, 1 and 2h can be seen in Figure 1A. Trailing and co-elution were present in the sample extracts, indicating that the fumonisins and the dirt were bound, thus affecting the migration rates (Rf) values of the sample extracts. Positive identification of fumonisins was achieved after SAX cleanup in the processed and unprocessed samples (Figure 1A). Purification of extracts resulted in no dirt beneath the sample extracts and thus equivalent (Rf values) between samples and FB_1 standards were obtained. Positive identification was achieved after SAX clean up in processed and unprocessed *F. verticillioides* inoculated maize cultures by TLC. In Figure 1B, sample extracts migration rates (Rf values) are equivalent to those of the standards. Unprocessed inoculated maize patty cultures used for the boiling water experiments contained an initial concentration of 217.11 mg/g. Thermal treatments resulted in small decreases of fumonisin levels. Inoculated maize cultures processed at 100°C for 30 min contained 205.95 mg/g of FB_1. Thermal treatments at 100°C for 1 h contained a FB_1 concentration of 194.80 mg/g. Processing at 100°C for 2 h resulted in a FB_1 concentration of 183.65 mg/g (Table 1). Unprocessed inoculated maize cultures used for the oven experiments contained 1706.80 µg/g FB_1. Thermal treatment in an

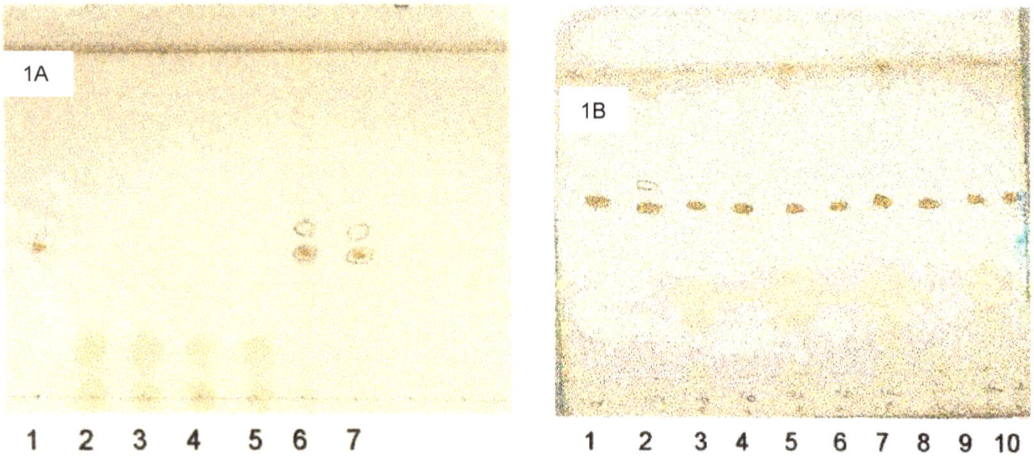

Figure 1. A: TLC of crude processed and unprocessed *F. moniliforme* inoculated maize cultures. Lane 1, 6, 7– FB_1 standards, lane 2 represents the control, lane 3- samples processed at 100°C for 30 min, lanes 4- samples processed at 100°C for 1 h and lanes 5- samples processed at 100°C for 2 h. B: TLC of thermally processed purified extracts in boiling water (100°C). Lanes 1 and 2– FB_1 standards, lanes 3 and 4- control, lanes 5 and 6- sample extracts treated at 100°C for 30 min, lanes 7 and 8- sample extracts treated at 100°C for 1 h and lanes 9 and 10- sample extracts treated at 100°C for 2 h.

Table 1. Thermal treatments of inoculated maize cultures at 100°C in boiling water and in oven.

Sample (n=3)	Boiling water (100°C)		Oven (100°C)	
	Concentration (mg/g)	Loss (%)	Concentration (µg/g)	Loss (%)
Control	217.11	0 ± 2	1706.80	0 ± 2
100°/30min	205.95	5 ± 1	1085.51	36 ± 1
100°/1h	194.80	10 ± 2	709.65	58 ± 1
100°/2h	183.65	15 ± 2.5	272.50	84 ± 5

Results are the mean and standard deviation of three determinations.

oven resulted in significantly higher FB_1 percentage losses as compared to boiling water. Processed inoculated maize cultures at 100°C for 30 min contained 1085.51 µg/g of FB_1. Processed maize cultures at 100°C for 1 h contained 709.65 µg/g of FB_1 and treatment at 100°C for 2 h contained a FB_1 content of 272.50 µg/g (Table 1).

Oven treatments

A convection oven was used to process *F. verticillioides* maize patty cultures at a higher range of temperatures from 140 to 220°C for 30 min. None of the above temperatures appeared to affect fumonisin levels at this stage. Equivalent migration rates (Rf values) between sample extracts and the FB_1 standards achieved positive identification. The control refers to an unprocessed inoculated maize patty culture sample extract which contained an initial concentration of 93.73 mg/g. Thermal treatments showed increase in FB_1 reduction as temperatures increased. Inoculated maize cultures processed at 140°C for 30 min contained 75.11 mg/g of FB_1. Processing at 160°C resulted in a FB_1 content of 48.59 mg/g, treatment at 180°C resulted in 34.75 mg/g FB_1. While processing at 200 and 220°C for 30 min resulted in FB_1 concentrations of 22.54 and 1.14 mg/g, respectively (Table 2).

These results support the data obtained by Alberts et al. (1990), who reported that boiling culture material of *F. verticillioides* for 30 min had no effect on the FB_1 concentration. It highlights the fact that FB_1 might be found in foods cooked at 100°C for a few minutes. Dupuy et al. (1993) found similar reductions in FB_1 levels when dry maize was heated at temperatures <150°C.

Muffle furnace

In an effort to determine the temperature at which fumonisins are eliminated, inoculated maize cultures were processed at temperatures of 240 to 300°C for 20

Table 2. Effects of thermal treatments on *F. moniliforme* inoculated maize cultures processed for 30 min in an oven.

Sample (n=3) (°C)	Concentration (mg/g)	Loss (%)
Control	93.73	0
140	75.11	20
160	48.59	49
180	34.75	63
200	22.54	76
220	1.14	99

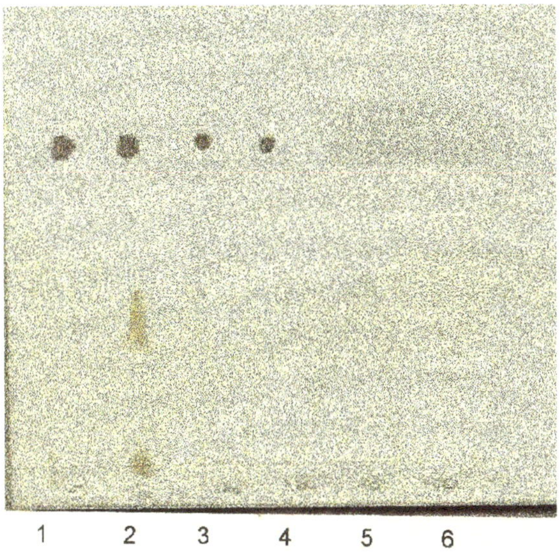

Figure 2. Thermal treatments of *F. moniliforme* inoculated maize cultures treated at 240 to 300°C in a muffle furnace. Lane 1- fumonisin standard, lane 2- control, lanes 3, 4, 5 and 6- samples processed at 240, 260, 280 and 300°C for 20 min.

min in a muffle furnace (Figure 2). Fumonisins were detected at 240 and 260°C, but not detected at 280 and 300°C.

Evaluation of processing temperatures

Baking

Evaluation of processing temperatures involved baking and frying studies. Maize muffins spiked at 1.25 µg/g FB_1 standard were baked at 220°C for 20 min in a convection oven. Fumonisins were positively identified by comparison of the migration rates (Rf values) of the sample extracts and the FB_1 standard indicating that the baking process does not sufficiently degrade fumonisins. The control for baking studies was an unbaked spiked muffin, which had an initial concentration of 0.96 µg/g.

Table 3. Effects of baking on fumonisin levels in maize muffins.

Sample (n = 3)	Concentration (µg/g)	Loss (%)
Control	0.96	0
Muffin 1	0.37	62
Muffin 2	0.28	71
Muffin 3	0.23	76

Baking of spiked maize muffins resulted in an average FB_1 percentage loss of 70% (Table 3).

The results of these baking experiments are in accordance with those by Scott and Lawrence (1994). FB_1 and FB_2 retentions of 36 and 28% were reported for muffins baked at 220°C for 25 min. Baking of maize meal muffins at 175 and 200°C by Jackson et al. (1997) resulted in small but statistically significant ($p<0.005$) losses of FB_1. The percent of FB_1 retained in muffins baked at 175°C (83.7 ± 3.5%) was slightly greater than that in muffins baked at 200°C (72.4% ± 5.9%). As compared to the data reported here, the higher FB_1 retention obtained by Jackson (1997) could be due to the lower baking temperatures used.

Results of previous studies on the thermal stability of FB_1 in maize compares well to those obtained here. The mechanism for fumonisin loss, however, may differ in each study depending on the matrix and temperature used. Jackson et al. (1996a, b) showed that when solution reached temperatures >150°C, partial and fully hydrolyzed fumonisins were the major decomposition products. The loss of FB_1 at temperatures <150°C and the lack of hydrolysis products suggest that hydrolysis may not be the major mechanism for decomposition in the baked maize muffins studied here. However, another possible explanation is that HFB_1 formed during heating reacted with the other components of the muffin mix. The loss of FB_1 during baking may be due to the non-enzymatic browning reaction described by Murphy et al. (1996). The reaction occurs when the primary amine group of FB_1 reacts with free aldehyde ketone groups in reducing sugars such as glucose or fructose. Murphy et al. (1996) found that when 100 mM fructose or glucose was heated with 5 µm FB_1 at 80°C, the reaction followed apparent first order kinetics. The presence of reducing sugars in the muffin mix and added milk (lactose) suggests that losses of FB_1 observed in the baked muffins may have been due to non-enzymatic browning. The rate and extent of nonenzymatic browning increase in food with increasing heating time and temperature (Mottram, 1994).

Frying

TLC results of FB_1 spiked fried chips, showed equivalent

Table 4. Effects of frying temperatures on fumonisin levels in maize chips.

Sample (n=3)	Concentration (µg/g)	Loss (%)
Control	1.98	
Fried 1	0.66	67
Fried 2	0.68	66
Fried 3	0.64	67

Rf values for the samples and the FB_1 standards. Thus, indicating that frying temperatures does not degrade fumonisins completely. A sample of spiked maize chips that was not fried was used as the control, which contained a FB_1 concentration of 1.98 µg/g (Table 4). Processing of maize chips at frying temperatures of 190 to 210°C for 5 to 10 min resulted in an average FB_1 percentage loss of 67%.

The results shown here for the thermal processing of fumonisin contaminated maize based-foods are generally in agreement with those of previous studies on the thermal stability of FB_1 in maize. Alberts et al. (1990) reported no loss of fumonisins when culture material of *F. verticillioides* was boiled for 30 min. When raw milk spiked with FB_1 and FB_2 was pasteurized (62°C for 30 min), there was no loss of either toxin (Maragos ad Richard, 1994). In contrast, Scott and Lawrence (1994) observed losses of FB_1 and FB_2 exceeding 70% in maize meal heated to 190°C for 60 min and about 100% in maize meal heated to 220°C for 25 min. Dupuy et al. (1993) reported that the decomposition of fumonisin in dry maize heated at 100 to 150°C followed first order kinetics but at a rate indicative of high thermal stability. Jackson et al. (1996a, b) studied the effects of heating of FB_1 and FB_2 in an aqeous buffer. They found that the rate and extent of fumonisins loss increased with increasing temperature, and that hydrolysed fumonisins were formed.

REFERENCES

Alberts JF, Gelderblom WCA, Thiel PG, Marasas WFO, Van Schalkwyk DJ, Behrend Y (1990). Effects of temperature and incubation period on production of fumonisins B_1 by *Fusarium verticillioides*. Appl. Environ. Microbiol. 6:1729-1733.

Brown NL, Rheeder JP, Shephard GS, Marasas WFO. The occurrence of Fumonisins and *Fusarium verticillioides* in maize collected from rural areas in Africa. Presented at the SAAFoST/IPSA Congress, Cape Town, SA, 27-29 September 1999.

Chu FS, Li GY (1994). Simultaneous occurrence of fumonisin B_1, and other mycotoxins in moldy corn collected from the People's Republic of China in regions with high incidence of esophageal cancer. Appl. Environ. Microbiol. 60:847-852.

Dupuy J, Le Bars P, Boudra, Le Bars J (1993). Thermostability of fumonisin B_1, a mycotoxin from *Fusarium verticillioides* in corn. Appl. Environ. Microbiol. 59:2864-2867.

Jackson LS, Hlywka JJ, Senthil KR, Bullerman LB (1996b). Effects of thermal processing on the stability of fumonisin B_1 in an aqueous system. J. Agric. Food Chem. 44:1984-1987.

Jackson LS, Hlywka JJ, Senthil KR, Bullerman LB, Musser SM (1996a). Effects of time, temperature, and pH on the stability of fumonisin B_1 in an aqueous model system. J. Agric. Food Chem. 44:906-912.

Jackson LS, Katta SK, Fingerhut DD, Devries JW, Bullerman LS (1997). Effects of baking and frying on the fumonisin B_1 content of corn-based foods. J. Agric. Food Chem. 45:4800-4805.

Kriek TS, Kellerman TS, Marasas WFO (1981). A comparative study of the toxicity of *Fusarium verticilliodes* (*F. verticillioides*) to horses, primates, pigs, sheep and rats. Onderstepoort. J. Vet. Res. 48:129-131.

Logrieco A, Doko B, Moretti A, Frisullo S, Visconti A (1998). Occurrence of fumonisin B1 and B2 in *Fusarium proliferatum* infected asparagus plants. J. Agric. Food Chem. 46:5201-5204.

Logrieco A, Mule G, Moretti A, Bottalico A (2002). Toxigenic *Fusarium* species and mycotoxins associated with maize ear rot in Europe. Eur. J. Plant Pathol. 108:597-609.

Makaula AN, Marasas WF, Venter FS, Badenhorst CJ, Bradshaw D, Swanevelder S (1996). Oesophageal and other cancer patterns in four selected districts of the Transkei, Southern Africa: 1985-1990. Afr. J. Health Sci. 3:11-15.

Maragos CM, Richard JL (1994). Quantitation and stability of fumonisin B_1 and B_2 in milk. J. AOAC Int. 77:1162-1167.

Marasas WFO (2001). Discovery and occurrence of the fumonisins: A historical perspective. Environ. Health Perspect. 109:239-243.

Marasas WFO, Wehner FC, Van Rensburg SJ, Van Schallkwyk DJ (1981). Mycoflora of corn produced in human esophageal cancer areas in Transkei, Southern Africa. Phytopathology 71:792-796.

Marin S, Ramos AJ, Vazquez C, Sanchis V (2007). Contamination of pine nuts by fumonisin produced by strains of *Fusarium proliferatum* isolated from *Pinus pinea*. Lett. Appl. Microbiol. 44:68-72.

Martins ML, Martins HM, Bernardo F (2001). Fumonisins B1 and B2 in black tea and medicinal plants. J. Food Prot. 64:1268-1270.

Miller JD (2008). Mycotoxins in small grains and maize: Old problems, new challenges. Food Addit. Contam A. 25:219-230.

Mottram DS (1994). Flavour compounds formed during the Maillard reaction. In Thermally Generated Flavors. Parliament TH, Morello MJ, McGorrin RJ, Eds., American Chemical Society: Washington, DC, pp. 104-126.

Murphy PA, Hendrich S, Hopmans EC, Hauck CC, Lu Z, Buseman G, Munkvold G (1996). Effect of Processing on fumonisin content of com. In Fumonisins in food. Jackson LS, DeVries JW, Bullerman LB. Eds., Plenum Publishing: New York. pp. 223-234.

Nelson PE, Dignami MC, Anaissie EJ (1994). Taxonomy, Biology and Clinical aspects of *Fusarium* species. Clin. Microbiol. Rev. pp. 479-504

Reddy KRN, Abbas HK, Abel CA, Shier WT, Oliveira CAF, Raghavender CR (2009). Mycotoxin contamination of commercially important agricultural commodities. Toxin Rev. 28:154-168.

Scott PM, Lawrence GA (1994). Stability and problems in recovery of fumonisins added to corn-based foods. J. AOAC Int. 77:541-545.

Sheldon LA (1904). A corn mold (*Fusarium verticillioides* n. sp.). 17th Ann. Rep. Agric. Exp. Stn.: Nebraska, USA.

Shephard GS, Leggott NI, Stockenstrom S, Somdyala NIM, Marasas WFO (2002). Preparation of South African maize porridge: effect on fumonisin mycotoxin levels. S. Afr. J. Sci. 98:393-396.

Shephard GS, Thiel PG, Stoekenstrom S, Sydenham EW (1996). Worldwide survey of fumonisin contamination of maize and maize-based products. J. AOAC Int. 79:671-687.

Stack ME, Eppley RM (1992). Liquid chromatographic determinations of Fumonisins B_1 and B_2 in corn and corn products. J. AOAC Int. 75:834-337.

Sydenham EW, Gelderblom WCA, Thiel PG, Marasas WFO (1990). Evidence for the natural occurrence of fumonisin B_1, a mycotoxin produced by *Fusarium verticillioides*, in corn. J. Agric. Food Chem. 38:285-290.

Sydenham EW, Shephard GS, Thiel PG, Marasas WFO, Stockenstrom S (1991). Fumonisin contamination of commercial corn-based human foodstuffs. J. Agric. Food Chem. 39:2014-2018.

Turner PC, Nikieme P, Wild CP (1999). Fumonisin contamination of food: progress in development of biomarkers to better assess human health risks. Elsevier 443:81-93.

Yazar S, Omurtag GZ (2008). Fumonisins, trichothecenes and zearalenone in cereals. Int. J. Mol Sci. 9:2062-2090.

12

Hygienic and sanitary evaluation of minimally processed vegetables sold in public fairs in the Western Region of Paraná State, Brazil

Alexandre Carvalho de MOURA[1], Fabiana Gisele da Silva PINTO[2], Eliana Almeida Mira de BONA[3], Luciana Pagliosa Carvalho GUEDES[4] and Izabel Aparecida SOARES[5]

[1]University of the Federal Border South, Campus Royalty Pr. Agricultural Engineering doctoral program, Brazil.
[2]State University of West Paraná - UNIOESTE, Cascavel Pr Campus Dr in Microbiology, Brazil.
[3]Technique in the Itaipu Technological Park FoundationUNIOESTE, Campus Pastor Rattlesnake Master in Conservation And Management of Natural Resources, Brazil.
[4]State University of West Paraná - UNIOESTE, Cascavel Pr Campus Dr in Statistics, Brazil.
[5]Federal University of South Border Royalty Campus Brazil.

The microbiological quality of minimally processed vegetables sold in the public fairs of the numerous towns in the Western Region of Paraná State, Brazil, was studied. Lettuce, salad rocket, cabbage and chicory randomly collected were sent to the Food Analysis Laboratory (Lanali) in Cascavel, Brazil. The vegetables were analyzed for total and thermotolerant coliforms, aerobic mesophilic bacteria and *Salmonella* sp. Total coliforms were detected at different rates in all samples of vegetables analyzed. Cabbages from town 5 had the highest contamination rate (3.6×10^6 UFC/g) and the highest mean contamination rate (4.1×10^4 UFC/g) in terms of thermotolerant coliforms in 147 samples, at different count rates. In all the samples collected and analyzed, *Salmonella* sp. occurred in 10. Results show that vegetables sold in the public fairs in the Western Region of Paraná, Brazil were not good for human consumption due to feces-originated bacteria or high mesophilic counts. In fact, they are the possible sources of toxin and infections caused by pathogens.

Key words: Coliforms, *Salmonella*, vegetables.

INTRODUCTION

The consumption of vegetables by humans is highly relevant for a balanced diet as they contain vitamins, have health benefits and lead subsequently to healthier life style. In fact, their intake does not only reduce the development of disease, but also their fibers regulate the digestive functions of the human body (Germano and Germano, 2003). Vegetables are also a source of vitamins, especially vitamin C, B complex and A (β-carotene); while minerals, such as iron, calcium, potassium and magnesium are highly important components found in them (Philippi , 2003).

The consumption of minimally processed vegetables is highly relevant for the transmission of several infectious diseases especially those caused by irrigation with contaminated water or through the farm workers' lack of hygiene (Takayanagui et al., 2001). Since the culture of vegetables demands a moist environment, the constant irrigation of vegetables mainly leads to the formation and development of pathogenic microorganisms (Germano and Germano, 2003).

Food-transmitted diseases are mainly due to inadequate handling, lack of personal hygiene, dirtiness of the person who is supposed to clean the places where vegetables are prepared and sold and improper cleaning of the equipment and utensils before and after food manipulation (Vitti et al., 2010).

Since most vegetables may be consumed raw up to seven days after harvest, they may also cause many diseases. In fact, microorganism may be viable for a period between 7 and 40 days and may thus become a source of public health concern (Mogharbel, 2007).

Microbiological evaluations of minimally processed vegetables are required to verify the hygiene and sanitary conditions of food consumed by people. Verifying the presence of certain micro-organisms in vegetables is mandatory within the context of public health (Franco and Landgraf, 2005; Mogharbel, 2007).

MATERIALS AND METHODS

Current investigation was conducted between March and June 2012 at all fair stalls which sold vegetables in five different towns in the western region of Paraná. Sixty samples of all available vegetables, namely, lettuce, salad rocket, cabbage and chicory were collected in triplicates. Specific fairs were selected owing to the population´s interest in consuming products originating from these markets. The vegetables were conditioned individually in sterilized plastic bags, identified and sent to the Food Microbiology Laboratory (LANALI) of Cascavel PR, Brazil, for microbiological analysis.

Further, 25 g aliquots of each vegetable collected were retrieved and microbiologically evaluated to determine total and thermotolerant coliform counts (UFC/g), following methodology of Norm 62 of August 2003 (Brasil, 2003).

Total coliforms were counted in Violet Red Bile Agar and the counting of suspected colonies was done at a later phase, according to legislation. They were then inoculated in Brilliant Green Bile Lactose Broth at 2% and incubated (48h/36±1ºC) to confirm total coliforms. Thermotolerant coliforms were inoculated in EC broth and tubes were incubated (48h/44.5°C) in a warm bath under stirring (68rpm/min) for confirmatory proof.

In the case of *Salmonella* sp., the materials were pre-heated in buffered peptonated water (24h/35°C) and selectively enriched in Rappaport-Vassiliadis and Selenito-Cistina broth (24h/42°C); they were then isolated in agar XLD and brilliant-green phenol-red lactose sucrose (BPLS). Biochemical identification was confirmed by tests for urease, fermentation of glucose, sucrose and lactose in TSI medium, de-carboxylation of lysine in LIA medium, H2S production, motility and production of indol by SIM medium.

Total counting of mesophilic micro-organisms was undertaken by serial dilutions in plate count agar (PCA) with plates incubated at 35ºC for 24-48 h for reading and counting of viable colonies.

Statistic analysis of variance of data evaluated was done by ANOVA and media comparison by Tukey´s test, both at 5% significance. All variables at 5% significance did not have a normal distribution but variance heterogeneity among treatments.

Box-Cox transformation was conducted with these variables to obtain normal distribution and homogeneity of variances. Results were compared with rates given by current ANVISA[1] legislation, according to RDC n. 12 of 2nd January, 2001.

RESULTS AND DISCUSSION

Whereas total coliforms at different rates were reported in ll samples, thermotolerant coliforms were registered in 147 samples of all the samples, at different rates. *Salmonella* sp. occurred in 10 samples of the total samples evaluated

Cabbages from town 5 registered the highest contamination rate (3.6x10^6 UFC/g), followed by lettuce, chicory and salad rocket, when total coliform rates in the different vegetables were analyzed.

Analysis of variance for total coliforms at 5% significance showed that interaction was not significant with an influence of vegetables according to their origin, except in the case of cabbages. Results were obtained by Tukey´s test of mean multiple comparison (Table 1) with a significant difference of 5% in town 4 compared to the others.

Further, significant differences (5%) were detected among the vegetables from towns 1 and 4. Tukey´s test at 5% significance (Table 1) showed that only lettuce in towns 1 and 2 differed from the others. The vegetable group of cabbage and lettuce differed from chicory and salad rocket group in town 3; only lettuce and salad rocket from towns 4 and 5 were statistically different.

Similar results were reported by Silva et al. (2007) when they analyzed the microbiological quality of minimally processed vegetables sold in Porto Alegre RS Brazil. High coliform counts were registered. Only two samples had low total coliform rates. In fact, total coliforms may be very common in food and may originate from several contamination sources such as soil and irrigation water (Mogharbel, 2007).

Analysis of thermotolerant coliforms in different vegetables revealed that cabbages had the highest mean contamination rate (4.1x10^4 UFC/g) for town 5, followed by lettuces. Contamination data mainly occurred as a probable result of lack in personal hygiene, handling, transport and inadequate storing of the vegetables.

In fact, there were significant differences between the towns for all the vegetable types under analysis. Results were also given by Tukey´s mean multiple comparison test, especially with regard to cabbages from town 4 when compared to the others. A statistical difference for lettuce was reported between towns 3 and 5; whereas for cabbages significant difference was reported in town 4 when compared to towns 2, 3 and 5 (Table 2).

According to Mogharbel (2007), poor hygiene conditions during food processing, production, storage and handling, coupled with lack of treatment of irrigation water from streams and rivers and transport to the consumer market are factors for the occurrence of food contamination.

Similar results were reported by Takayanigui et al. (2001) who observed high concentrations of thermotolerant coliforms in vegetables sold without any sort of treatment and related to lack of hygiene during the collection process and food selling.

Results from the microbiological evaluation of minimally processed lettuces (*Lactuca sativa* L.) and cabbages

Table 1. Comparison for the presence *Escherichia coli* (in 10,000 units) in minimally processed vegetables between the cities of origin.

Vegetable	Town				
	1	2	3	4	5
1	55.80^{aA}	84.70^{aA}	230.00^{aA}	0.19^{bAB}	41.2^{abA}
2	0.83^{aB}	0.24^{aB}	5.90^{aA}	1.28^{aA}	1.63^{aAB}
3	15.00^{aB}	0.06^{aB}	0.05^{aB}	0.03^{aAB}	0.25^{aAB}
4	0.03^{aB}	0.02^{aB}	0.03^{aB}	0.005^{aB}	0.04^{aB}

Small letters represent comparison between lines; capital letters represent comparison between columns at 5% probability.

Table 2. Comparison for the presence of *Escherichia coli* (in 10,000 units) in minimally processed vegetables between the cities of origin.

Vegetable	Town				
	1	2	3	4	5
1	55.80^{aA}	84.70^{aA}	230.00^{aA}	0.19^{bAB}	41.2^{abA}
2	0.83^{Aab}	0.24^{aB}	5.90^{aAB}	1.28^{aA}	1.63^{aA}
3	15.00^{Aab}	0.06^{aB}	0.05^{aBC}	0.03^{aAB}	0.25^{aA}
4	0.03^{aB}	0.02^{aB}	0.03^{aC}	0.005^{aB}	0.04^{aA}

Small letters represent comparison between lines; capital letters represent comparison between columns at 1% probability

(*Brassica oleracea* L.) sold in Brasília DF Brazil identified thermotolerant coliforms in some samples analyzed. In fact, 19% of the 36 lettuce samples were contaminated by thermotolerant coliforms, whereas 81% were free from any contamination, according to current legislation. Further, 22% of 60 raw lettuces and cabbages analyzed were contaminated by thermotolerant coliforms and 78% were free from any contamination.

However, a study undertaken in the Brazilian Federal District revealed that all minimally processed lettuce, cabbage and turnip samples were contaminated by thermotolerant coliforms (Almeida et al., 2012). The same result was verified in another study in Uberlândia MG Brazil, where all raw food samples were contaminated at rates above those allowed by current legislation (Bonnas et al., 2005). *Escherichia coli* was reported in 8 out of 56 vegetable samples in Porto Alegre RS Brazil (Silva et al., 2007).

Vegetables actually free from thermotolerant coliforms indicated good sanitary conditions during processing and compliance with the microbiological standards of ANVISA (Sasaki et al., 2006). The contaminated vegetables were not fit for consumption due to fecal contamination.

Table 3 shows a descriptive analysis for data on *Salmonella* sp. Cabbages and lettuces had the highest contamination (approximately 66.67%), especially those from towns 2 and 3.

Contamination by *Salmonella* sp. is a public health concern due to the possibility of infection by toxins. However, its low contamination rate when compared to that by coliforms may be related to competition by other bacteria in the environment as well as by producers´ hygiene conditions when dealing with their products (Marques et al., 2006). Takayanagui et al. (2001) reported similar results, with only 9% of vegetables contaminated by *Salmonella* sp.

Mean mesophile counts in UFC / g varied between 2.5x10 and 3.6x106. The farmer from town 5 not only had the highest mesophile rates in harvested vegetables corresponding to data on total and thermotolerant coliforms but also showed contamination related to lack of hygiene. Perhaps due to their greater leaf surface, cabbages had the highest mesophilic count when compared to rates of the other vegetables.

Interaction was not significant at 5% for mesophile counting (Table 4). Consequently, the behavior of the variable may be analyzed separately with regard to each factor (town and vegetable type) significant at 1% probability. Mesophile count at 1% significance is influenced by town and vegetable type. Tukey´s test of mean multiple comparison for mesophiles (Table 5) shows that, at 1% significance, there were significant differences among all types of vegetables and between town 4 and towns 1 and 5.

Even after minimal processing, raw vegetables may still retain most of their previous microbiota. This is a serious health issue since pathogens may be included in the microbiota. In fact, they should be stored at proper

Table 3. Comparison of minimally processed vegetables as presence of *Escherichia coli* (in 10,000 units), the cities of origin.

Vegetable	Town				
	1	2	3	4	5
1	6.73^{aA}	1.09^{aA}	1.22^{aA}	0.02^{bA}	1.03^{aA}
2	0.61^{abA}	0.05^{bB}	0.03^{bB}	0.36^{abA}	2.38^{aA}
3	0.004^{abB}	0.009^{aB}	0.02^{aB}	0.00^{bB}	0.05^{aB}
4	0.001^{abB}	0.007^{aB}	0.03^{abB}	0.00^{bB}	0.003^{abB}

Small letters represent comparison between lines; capital letters represent comparison between columns at at 5% probability.

Table 4. Comparison of minimally processed vegetables as presence of *Escherichia coli* (in 10,000 units), the cities of origin.

Vegetable	Town				
	1	2	3	4	5
1	6.73^{aA}	1.09^{aA}	1.22^{aA}	0.02^{bA}	1.03^{abAB}
2	0.61^{aA}	0.05^{aAB}	0.03^{aAB}	0.36^{aA}	2.38^{aA}
3	0.004^{abB}	0.009^{abB}	0.02^{aB}	0.00^{bB}	0.05^{abBC}
4	0.001^{aB}	0.007^{aB}	0.03^{aB}	0.00^{aB}	0.003^{aC}

Small letters represent comparison between lines; capital letters represent comparison between columns at 1% probability.

Table 5. Comparison of minimally processed vegetables as the presence of Salmonella in the cities of origin.

Vegetable	Town					Total (%)
	1	2	3	4	5	
Cabbages	1	2	2	0	1	6 (54.55)
Lettuce	0	2	2	0	1	5 (45.45)
Chicory	0	0	0	0	0	0 (0.00)
Salad rocket	0	0	0	0	0	0 (0.00)
Total (%)	1 (0.09)	4 (36.36)	4 (36.36)	0	2 (18.18)	11

temperatures so that the growth of pathogens could be inhibited.

Conclusion

Vegetables sold in the public fairs of the main towns of the western region of Paraná are improper for natural human consumption since they contain fecal bacteria or high levels of mesophiles and certain quantity of *Salmonella;* and have the possibility of causing toxin and infections.

Cabbages had the highest contamination rate compared to the other vegetables, due to their larger leaf surface.

Vegetables from town 5 had the highest contamination, with the least hygiene and sanitary care.

Effective supervision by the health authorities should be warranted to ensure that agricultural products that are consumed raw could be sold in good conditions without the addition of any special treatment.

REFERENCES

Almeida AG de Almeida (2012). Resende, a. Análise microbiólogica em alfaces (*lactuca sativa* l.) E couves (*brassica oleracea* l.) Minimamente processadas e comercializadas em brasília – df. Sabios: Rev. Saúde e Biol. 7(3):52-59.

ANVISA - Agência Nacional de Vigilância Sanitária (2001). Resolução – rdc n° 12, de 02 de janeiro de 2001: aprova o regulamento técnico sobre padrões microbiológicos para alimentos. Diário oficial da união, brasília, df, 10 de janeiro de 2001.

Bonnas DS, Silva CC, Silva SA, Ferreira IM (2005). Qualidade higiênico sanitária de vegetais minimamente processados, comercializados no município de Uberlândia, MG. Revista Higiene Alimentar. 24:184-185.

Brasil (2003). Ministério Da Agricultura, Pecuária E Abastecimento. Secretaria de Defesa Agrope-cuária (DISPOA). Instrução Normativa nº 62, 26 de agosto de 2003. Oficializa os Métodos Analíticos Oficiais para Análises Microbiológicas para Controle de Produtos de origem Animal e água. Diário Oficial da União, Brasília, 26 de agosto de 2003.

Franco BDGM, Landgraf M (2005). Microbiologia dos alimentos. São Paulo: São Paulo.

Germano PML, Germano MIS (2003). Higiene e vigilância sanitária de alimentos. 2. ed. São Paulo: São Paulo.

Marques SC, Boari CA, Brcko CC, Nascimento AR, Piccoli RH, Brcko CC, Nascimento AR, Picccoli RH (2006). Avaliação higiênico sanitária de linguiças tipo frescal comercializadas nos municípios de Três Corações e Lavras MG. Ciência E Agrotcnologia, Minas Gerais, Nov/dec. 2006. Available at http://www.scielo.br/pdf/cagro/v30n6/a12v30n6.pdf.

Mogharbel AI (2007). Validação do emprego de instrumentos de coleta dedados, alface e manipuladores como indicadores de boas práticas em lanchonetes, 2007. 121.f. Doctoral thesis in Food technology. Universidade Federal do Paraná, Curitiba.

Philippi ST (2003). Nutrição e Técnica Dietética. São Paulo: Barueri, 1ª edição.

Sasaki FF, Aguila JS, Gallo CR, Ortega EMM, J Acomino AP, Kluge RA (2006). Alterações fisiológicas, qualitativas e microbiológicas durante o armazenamento de abóbora minimamente processada em diferentes tipos de corte. Horticultura Brasileira, São Paulo, abr./jun. 2006. Available at http://www.scielo.br/pdf/hb/v24n2/09.pdf.

Silva MA, Neves TRM, Frazatto C (2007). Avaliação da presença de formas evolutivas de parasitas de origem intestinal, em alfaces de restaurantes self-service. Revista Higiene Alimentar 21(152):76-78.

Silva SRP, Verdin SEF, Pereir ADC, Schatkoski AM, Corção MBR (2007). Microbiological quality of minimally processed vegetables sold in Porto Alegre, São Paulo. Braz. J. Microbiol. 38:4.

Takayanagui OM, Oliveira CD, Bergamine AMM, Oklno MHT, Febrônio IHP, Silva AAMC, Oliveira MA, Ribeiro EGA, Takayanigui AMM (2001). Fiscalização de verduras comercializadas no município de Ribeirão Preto. Revista da Sociedade Brasileira de Medicina Tropical, São Paulo, Jan./Feb. 2001. Available at http://www.scielo.br/scielo.php?script=sci_arttextandpid=S003786822 001000100006andlng=enandnrm=iso on 23 Sept. 2011.

Vitti MCD, Sasaki FF, Angelo PM, Jacomino P, Moretti CL, Kluger RA (2010). Atividade respiratória e aspectos microbiológicos de cultivares de batatas minimamente processadas e armazenadas em diferentes temperaturas. Ciência Rural.

13

Influence of gamma irradiation on the nutritional and functional properties of pigeon pea (*Cajanus cajan*) flour

O. P. Bamidele and C. T. Akanbi

Department of Food Science and Technology, Obafemi Awolowo University, Ile-Ife, Nigeria.

Effects of gamma irradiation on pigeon pea flour at various doses (0, 5, 10, 15 and 20 kGy) on the proximate composition and functional properties were investigated. Gamma irradiation resulted in a slight increase of crude protein at all doses and the crude lipid. Crude fibre and ash showed no apparent effects of irradiation between the non-irradiated sample and the irradiated. A significant increase of *in vitro* protein digestibility was observed in pigeon pea flour irradiated at 20 kGy. Phytic acid, a major anti-nutritional factor, was reduced significantly on irradiation, indicating the positive effect of gamma irradiation on pigeon pea flour. Significant enhancement in the water absorption and swelling capacities, bulk density of the flour was also recorded after irradiation. Results of the present investigation showed that application of gamma irradiation does not have major effect on the overall nutritional composition and functional properties of the flour and can be used as an effective method of preservation of pigeon pea flour and their products.

Key words: Pigeon pea flour, gamma irradiation, functional properties, anti-nutritional factors.

INTRODUCTION

Legumes play a major role in overcoming protein-energy malnutrition in developing countries, where scarcity of animal proteins prevails. Edible legumes fulfil the basic nutritional requirements as they are rich in proteins and other nutritional components such as essential minerals, unsaturated fatty acids and vitamins. However, the extent of production of legumes has failed to keep pace with the demands of ever-increasing populations (Ali and Kumar, 2000). The scarcity of fertile land and an over dependence on cereal-based food products have also aggravated the protein deficiency problems in humans (Sadik, 1991). To meet the ever-increasing protein demand, the exploitation of nonconventional seeds has become inevitable (Bhat et al., 2009). Such explorations may assuage the problems of food security, agricultural development, self-dependence and enhancement of the economy of developing countries. The little known pigeon pea seed (*Cajanus cajan*) and its flour might significantly contribute to world food production due to its wide distribution and adaptability to adverse environmental conditions. Nutritionally, pigeon pea seeds have been reported to be rich in proteins and the important amino acids such as methionine, lysine and tryptophan. The protein content is comparable with those in other legumes like cowpea and groundnut which have been used in complementing maize. It is rich in mineral quality and fibre content (Nene et al., 1990; El-Tabey, 1992). Pigeon pea grows well in Nigeria but is easily attacked by weevils; hard-to-cook phenomenon and the presence of anti-nutrients have limited its utilization. Irradiation processing of plant product has become an important physical preservation method to overcome the international quarantine barriers and to increase the safety and shelf life of the product, mainly by the elimination of spoilage micro flora (Diehl, 1995). Radiation processing has proved to be an effective means of decontamination and disinfestations of food and agricultural products (Anonymous, 1991; Loaharanu, 1994). Irradiation treatment itself or in combination with other processing methods has been shown to reduce or eliminate some of the anti-nutrients in cereals and legumes (Farag, 1989; Sattar et al., 1990).

Ionizing radiations have also been proved effective in improving the overall nutritional attributes, including some desired changes in functional properties of seed flours (Rahma and Mostafa, 1988; Dario and Salgado, 1994; Dogbevi et al., 2000). A joint expert committee convened by the FAO/IAEA/WHO stated that irradiation of any food commodity up to 10 kGy presents no toxicological hazard (Anonymous, 1991). Subsequently, a joint FAO/IAEA/WHO study group reviewed the toxicological, nutritional and radiation-induced chemical and physical aspects of irradiated foods above 10 kGy and concluded that application of ionizing radiation at 10 kGy or higher doses will be safe and nutritionally adequate (WHO, 1999). To fill the existing gap in the knowledge on the effects of γ-rays (low and high dose) on the nutritional potential of pigeon pea seeds, the current study was aimed to investigate the nutritional, anti-nutritional compositions and the functional properties of raw and irradiated pigeon pea flour. The results of this study will be useful to popularize pigeon pea flour for the successful exploitation of its nutritional value in the production of safe and inexpensive food products.

MATERIALS AND METHODS

Pigeon pea was purchased at Odori market in Iwo, Osun State and gamma irradiated at Energy Centre in Abuja at different doses (5, 10, 15 and 20 kGy). The samples were exposed to gamma rays generated by a cobalt-60 source (Gammacell 220, MDS Nordion, Ottawa, Canada) following the procedures described by Helinski et al. (2008) with a dose rate of ca. 2 Gy/min at 25°C and normal relative humidity. Double side irradiation (exposure to both sides) was performed for uniform dose delivery. A dosimetry system was used to measure the dose received by the batch based on the Gafchromic HD-810 film (International Specialty Products, NJ, USA; FAO/IAEA/USDA, 2003). Three dosimeters were included with each batch of pigeon pea seeds and read after irradiation with a radiachromics reader (Far West Technology Inc., CA, USA). All experiments were repeated 3 times and 3 replicates of each sample type were irradiated. The irradiated samples were milled with hammer mill and various analyses were carried out on it.

Proximate composition

The level of moisture, crude protein, crude fat, ash, crude fiber and carbohydrates of the irradiated (5, 10, 15, and 20 kGy) and non-irradiated pigeon pea flours were determined by the use of manual methods as outlined by AOAC (2000).

Determination of *in vitro* protein digestibility (IVPD)

IVPD was carried out according to the method described by Monjula and John (1991) with a minor modification. A known weight of the sample containing 16 mg nitrogen was taken in triplicate and digested with 1 mg pepsin in 15 ml of 0.1 N HCl at 37°C for 2 h. The reaction was stopped by the addition of 15 ml 10% trichloroacetic acid (TCA). The mixture was then filtered quantitatively, through Whatman No. 1 filter paper. The TCA soluble fraction was assayed for nitrogen using the micro-Kjeldahl method (AOAC, 2000). Digestibility was estimated by using the following equation:

$$\text{IVPD}\,\% = \frac{\text{N in Supernatant} - \text{Enzyme N}}{\text{Nin sample}} \times 100$$

Anti-nutrition determination

Determination of tannin content

The tannin content of the irradiated and non-irradiated samples was determined using the International Standardisation Organisation method (ISO, 1988). Tannic acid was used as a standard and the results were calculated as percentage tannic acid equivalents.

Determination of phytic acid content

The method reported by Reddy et al. (1982) was used for phytic acid and phytate-phosphorus determinations. Phytic acid was extracted from each 3 g flour sample with 3,5 trichloroacetic acid by shaking at room temperature followed by high-speed centrifugation (30, 000 xg for 5 min). The phytic acid in the supernatant was precipitated as ferric phytate, and iron in the sample was estimated. Phytate-phosphorus was calculated from the iron results assuming a 4:6 iron: phosphorus molecular ratio according to method 970.39 (AOAC, 2000). The phytic acid was estimated by multiplying the amount of phytate-phosphorus by the factor 3.55 based on the empirical formula $C_6P_6O_{24}H_{18}$

Determination of trypsin inhibitor

Trypsin inhibitor was determined using the method described by Arntfield et al. (1985). The samples were extracted by weighing 1.000 g of the sample and dissolving it in 50 ml of 0.5 M NaCl solution. The mixture was stirred for 30 min at room temperature and then centrifuged. The supernatant was filtered through Whatman No. 41 filter paper. The filtrate (extract) was used for the determination. Trypsin solution (20 ml) was added to 10 ml of the substrate in a test tube or beaker. A blank (10 ml) of distilled water was prepared. The content of the test tube was allowed to stand for at least 5 min after which its absorbance was measured at a wavelength of 410 nm. Trypsin activity was expressed as number of trypsin unit inhibited (TUI) per unit weight (g) of the sample analysed. Thus,

$$\text{TUI/mg} = \frac{\text{absorbance} \times 0.01\text{F}}{\text{Absorbance of standard}}$$

$$\text{TUI/mg} = \frac{b - a \times F}{0.01}$$

Where b = absorbance of sample, a = absorbance of blank, F = experimental factor.

$F = 1/w \times V_f / Va \times D$

Where w = weight of sample, V_f = total volume of extract, Va =

Table 1. Proximate composition of irradiated pigeon pea flour prior to storage (dry weight basis).

Proximate composition (g/100 g)	Irradiation dose (kGy)				
	0	5	10	15	20
Moisture	7.99 ± 0.120[d]	7.95±0.220[cd]	7.54±0.130[c]	6.84±0.160[b]	6.22±0.120[a]
Protein	19.37±0.590[a]	20.49±0.500[ab]	21.01±1.010[bc]	21.96±0.460[cd]	22.99±0.650[d]
Crude fat	1.88±0.010[a]	2.04±0.010[b]	1.95±0.010[b]	3.90±0.010[e]	3.87±0.010[d]
Crude fibre	5.59±0.153[e]	3.52±0.012[a]	3.70±0.010[b]	3.90±0.044[d]	3.77±0.010[c]
Ash	3.62±0.006[b]	3.47±0.012[a]	3.59±0.015[d]	3.46±0.010[a]	3.55±0.015[c]
Carbohydrate	61.55±0.015[b]	62.53±0.010[d]	62.21±0.010[c]	59.94±0.015[a]	59.60±0.025[a]

Values are means and standard deviation of triplicate of three samples (n = 9); Means followed by the same letter within the same row are not significantly different (p>0.05).

volume of extract used, D = dilution factor (if any).

Functional properties determination

Water absorption determination

The water absorption was determined by modified centrifuge method of Lin and Zayas (1987). Each sample (2.0 g) was transferred into lagged 50 ml centrifuge tube and weighed (W_1). Exactly 30 ml of hot distilled water (70°C) was added to each sample and at the same time washing down the inside of the centrifuge tubes using a glass stirring rod, the sample and the water was mixed for 30 min. The suspension was allowed to rest for 10 min; the flour adhering to the side of the centrifuge was scrubbed down with glass rod to prevent it from drying. Additional 10 ml of hot distilled water was used to wash the sample adhering to the stirring rod into the sample. The suspension was centrifuged at 1165 g for 25 min at 50°C. The tube was cooled in desiccator and weighed (W_2).

$$\text{Water absorption} = \frac{W_2 - W_1\,(\text{ml})}{\text{Weight of the sample (g)}}$$

Bulk density determination

Bulk density was determined using the gravimetric method as described by Okaka and Potter (1976). The sample (10 g) was weighed into a 25 ml graduated cylinder. The cylinder was gently tapped ten times against the palm of the hand. The bulk density was expressed as the sample per volume occupied by the sample.

Swelling capacity determination

Swelling capacity of the samples were determined by modification of Lin and Zayas (1987) method. Each sample (2 g) was dispersed in 40 ml distilled water. The resultant slurry was heated at a temperature of 70°C for 30 min in a water bath, cooled to room temperature and centrifuged at 598 xg for 30 min. The supernatant liquid was decanted and the centrifuge tube was placed in a hot air oven and dried for 25 min at 50°C. The residue was weighed (W_2). Centrifuge tube containing sample alone was weighed prior to adding distilled water (W_1).

$$\text{Swelling capacity} = \frac{W_2 - W_1\,(\text{ml})}{\text{Weight of the sample (g)}}$$

Statistical analysis

All experiments were carried out in triplicate using three samples at each replicate. Mean and standard deviation were calculated for each treatment. Data obtained from the phyisco-chemical results were subjected to analysis of variance (ANOVA) and the means were separated by lowest standard deviation test (SPSS version 16, 2008). Significant level was accepted at 5%.

RESULTS AND DISCUSSION

Proximate composition of irradiated pigeon pea flour (dry weight basis)

The proximate composition of raw and gamma-irradiated pigeon pea flour is presented in Table 1. The moisture content of pigeon pea flour was not substantially affected by gamma irradiation treatment. Similar findings were observed by Rady et al. (2002) who reported that gamma irradiation has no real effect on moisture content of oil seeds. Crude protein and carbohydrate constitute the major chemical constituents of the seeds. Gamma irradiation resulted in a moderate increase (p>0.05) in crude protein at all the irradiated doses (control, 19.37%; 20 kGy, 22.99%). The protein content is comparable with the values reported by Faris et al. (1987). The carbohydrates constitute over 50% of the pigeon pea flour constituents, probably due to low lipid content. Irradiation did not significantly alter the carbohydrates (p>0.05) and was slightly decreased at all doses except for 5 and 10 kGy. Stability of carbohydrates in pigeon pea flour on irradiation is advantageous as carbohydrates contribute to the calorific value as well as acting as an anti-marasmus in infant nutrition (Vadivel and Janardhanan, 2002). The crude fat on irradiation showed a dose-dependent increase which was significant at doses 15 and 20 kGy and this may be due to higher dose of irradiation. These results are in agreement with previous finding of Abu et al. (2005) who reported that at low dose irradiation (2 kGy); the oil absorption capacity of cow pea flour was not affected. However, an increase in oil absorption capacity was observed at higher doses (10 and 50 KGys); this may

Table 2. Functional properties of irradiated pigeon pea flour prior to storage (dry weight basis).

Functional properties	Irradiation dose (kGy)				
	0	5	10	15	20
Water absorption	1.88±0.020[a]	1.89±0.150[a]	1.96±0.020[b]	2.04±0.010[c]	2.06±0.010[c]
Swelling capacity	2.31±0.010[b]	2.60±0.010[c]	2.67±0.010[d]	2.29±0.020[a]	2.29±0.020[a]
Bulk density	50.31±0.015[ab]	50.20±0.010[a]	50.22±0.015[ab]	50.34±0.025[b]	50.27±0.015[ab]

Values are means and standard deviation of triplicate of three samples (n = 9); Means followed by the same letter within the same rows are not significantly different (p>0.05).

Table 3. Effect of gamma irradiation on anti-nutritional factors and *in-vitro* protein digestibility of pigeon pea flour.

Anti-nutritional factors (mg/100 g)	Irradiation dose (kGy)				
	0	5	10	15	20
Phytic acid	94.67±1.522[e]	83.33±0.575[d]	50.67±5.125[c]	41.00±1.725[b]	25.00±0.000[a]
Tannin	2.47±0.050[d]	2.27±0.020[c]	1.43±0.055[b]	0.83±0.055[a]	0.50±0.000[a]
Trypsin inhibitor	47.67±2.520[d]	32.33±2.520[d]	23.67±1.525[c]	16.33±1.525[b]	13.00±1.725[a]
IVPD (%)	60.67±1.160[a]	64.00±1.730[b]	69.00±1.000[c]	72.33±1.530[d]	75.00±2.000[e]

Values are means and standard deviation of triplicate of three samples (n = 9); Means followed by the same letter within the same row are not significantly different (p>0.05).

be due to exposure to non-polar protein site.

Crude fibre and ash on irradiation showed no apparent effect except that between non- irradiated sample and this was not in line with the report of Bhat et al. (2008) who reported a decrease in crude fibre and ash content of velvet bean seed. Low crude fibre is nutritionally valued as it traps less proteins and carbohydrates (Balogun and Fetuga, 1986). The dose-dependent decrease in fibre on irradiation has been attributed to depolymerization and delignification of the plant matrix (Bhat et al., 2008).

Functional properties of irradiated pigeon pea flour (dry weight basis)

Table 2 shows the result of functional properties of irradiated pigeon pea flour. Irradiation at 10 kGy and above showed a slight increase in water absorption capacity. Similar result was reported by A/Azim et al. (2009) which reported no apparent effect of gamma irradiation on water absorption capacity of groundnut flour. Seed flours with high water absorption capacity are suitable for production of soups and gravies (Oshodi and Adeladun, 1993). As polar amino acids are responsible for binding of water molecules by proteins, differences in water absorption capacity suggest variations in the polar amino acid composition in pigeon pea flour. Also, irradiation at 5 kGy and above showed a significant increase and a better swelling capacity except that of 15 kGy which showed a slight decrease. There was no significant difference in bulk density. This result was in line with the report of A/Azim et al. (2009) which reported that, the bulk density of ground nut flour was not affected significantly by gamma irradiation.

Functional properties of seed flours assume importance for the development of food products. Acceptability of legume flour depends entirely on the nutritional value as well as functional properties (Pour-El, 1981). Physicochemical changes in proteins influence functional properties such as the water absorption capacity, oil absorption capacity, protein solubility and foaming properties in cowpea products (Krer et al., 2000).

Effect of gamma irradiation on anti-nutritional factors

Table 3 present the result of effect of gamma irradiation on anti-nutritional factors in pigeon pea. Phytate, tannin and trypsin inhibitor in the samples decrease with increase in irradiation dose. Similar trend was reported by El-Niely (2007) that radiation processing significantly ($P \leq 0.05$) reduced the levels of phytic acid and tannins of legumes. Also, Bhat et al. (2009) who irradiated lotus seed at gamma dose range of 0 to 30 kGy. A decrease in phytic acid has been assigned to the production of free radicals during irradiation, thus leading to breaking of a chemical bond of phytic acid (Bhat et al., 2009). Phytic acid in legumes has been reported to lower the nutritional value by limiting the bioavailability of minerals and essential elements (Gustafsson and Sandbreg, 1995). Tannin concentration of raw pigeon pea is low compared to trypsin inhibitor and phytic acid concentration. Tannin is said to be present in the seed coat of legumes and pre-processing operation of the seed can result into reduction in tannin content (Gurumoorthi et al., 2003).

Effect of gamma irradiation on *in-vitro* protein digestibility

The *in-vitro* protein digestibility of pigeon pea flour shows a dose dependent increase (Table 3). The result is similar to that of Rehman and Shah (2001) that the removal of tannin and phytate create a large space within the matrix of flour which increase the susceptibility to enzymatic attack and consequently improve the digestibility of protein after radiation treatment. Bhat et al. (2008) also reported an increase in *in-vitro* protein digestibility of velvet beans seed due to decrease in some of the antinutritional factors such as phytates, trypsin and chymotrypsin inhibitors present in the seed; the analysis showed a significant difference ($p<0.05$).

Conclusion

The results of the present investigation showed that pigeon pea flour is a valuable source of nutrition due to high proteins and carbohydrates. The seed flour exhibited good functional properties, which will be of immense value in food industries in production of quality food formulations. Application of gamma irradiation as a method of preservation of pigeon pea flour quality did not show any adverse effect on the nutritional composition. However, it led to improvement of some of the functional properties (for example, water absorption and swelling of the flour) which is advantageous in food manufacturing/production.

REFERENCES

A/Azim AM, Nour Shireen EA, Hmed Gammaa AM Osman (2009). Effect of gamma irradiation on the physico-chemical characterization of groundnut (*Arachis Hypogaea*). Aust. J. Basic Appl. Sci. 3(3): 2856-2860,

Abu JO, Klaus M, Kwaku GD, Duodu G, Amanda M (2005) Functional properties of cowpea (*Vigna unguiculata* L. Walp) flours and pastes as affected by ã-irradiation. Food Chem. 93: 103-111.

Ali M, Kumar S (2000). Problems and prospects of pulses research in India. Indian Farm 50:4_13. and physiological aspects of dry bean carbohydrate. J. Food Chem. 13:25-68

Anonymous (1991). Analytical detection methods for irradiated foods. A review of the current litreature. IAEA-TECDOC-587. Vienna: IAEA.

AOAC (2000). Official methods of analysis international, 17th ed. Washington, DC: Association of Official Analytical Chemists. pp.1230.

Arntfield SD, Ismond MAH, Murray ED (1985). Stability of vicilin, a legume seed storage protein, with step-wise electrostatic modification Int. J. Pept. Protein Res. 26(6):584–590.

Balogun A, Fetuga BL (1986). Chemical composition of some under exploited leguminous crop seeds in Nigeria. J. Agric. Food Chem. 34: 189–192.

Bhat R, Sridhar KR, Seena A (2008). Nutritional quality evaluation of velvet bean seed (Mucuna pruriens) exposed to gamma irradiation. Int. J. Food Sci. Nutr. 54(4): 261-278

Bhat R, Sridhar KR, Alias AK, Young CC, Arun BA (2009). Influence of gamma radiation on the nutritional and Functional qualities of Lotus seed flour. J. Agric. Food Chem. 57(20):9524–9531

Bhat R, Sridhar KR, Alias AK, Young CC, Arun BA (2009). Influence of gamma radiation on the nutritional and Functional qualities of Lotus seed flour. J. Agric. Food Chem. Article 57:9524–9531.

Dario AC, Salgado JM (1994). Effect of thermal treatments on the chemical and biological value of irradiated and non-irradiated cowpea bean (Vigna unguiculata L.Walp.) flours. Plant Foods Hum. Nutr. 46:181-186.

Diehl JF (1995). Safety of Irradiated Foods, 2nd ed.; Marcel Dekkre Inc.: New York. pp. 189.

Dogbevi MK, Vachon C, Lacroix M (2000). Effect of gamma irradiation on the microbiological quality and on the functional propreties of proteins in dry red kidney beans (Phaseolus vulgaris). Radiat. Phys. Chem. 57:265-268.

El-Niely HFG (2007). Effect of radiation processing on antinutrients, *in vitro* protein digestibility and protein efficiency ratio bioassay of legume seeds. Radiat. Phys. Chem. 76: 1050–1057.

El-Tabey Shebata AM (1992). Hard-to-cook phenomenon in legumes. Food Rev. Inter. 8:191-221.

Farag MDEH (1989). Radiation deactivation of anti-nutritional factors: trypsin inhibitor and hemagglutinin in soybeans. Egypt J. Radiat. Sci. Appl. 6:207-215.

Faris DG, Saxena KB, Marumadar S, Singh U (1987). Vegetable Pigeon pea: a promising crop in India. ICRISAT, Patanchreu, A.P. 502324, India: International Crop Research Institute for Semi-Arid Tropics

Gurumoorthi P, Pugalenthi M, Janardhanan K (2003). Nutritional potential of five accessions of a South Indian tribal pulse Mucuna pruriens var. utilis: II. Investigations on total phenolics, tannins, trypsin and chymotrypsin inhibitors, phytohaemagglutinins, andin vitro protein digestibility. Tropical and Subtropical Agroecosystems, 1, 153–158.

Gustafsson EL, Sandbreg AS (1995). Phytate reduction in brown beans (Phaseolus vulgaris L.). J. Food Sci. 60:149–152.

Helinski MEH, Hassan MM, El-Motasim WM, Malcolm CA, Knols BGJ, El-Sayed B (2008). Methodology towards a sterile insect technique field release of *Anopheles arabiensis* mosquitoes in Sudan: irradiation, transportation, and field cage experimentation. Malar. J. 7: 1–10.

International Standards Organisation (ISO) (1988). Sorghum, Determination of Tannin Content. International standards Organisation for Standardisation, ISO 9648

Krer WL, Ward CDW, McWattres KH, Resurrection AVA (2000). Milling and particle size of cowpea flour and snack chip quality. Food Resour. Int. 34:39-45.

Lin CS, Zayas JF (1987) Protein solubility, emulsifying stability and capacity of two defatted corn germ proteins J. Food Sci. 52:1615.

Loaharanu P (1994). Food irradiation in developing countries: A practical alternative. IAEA Bull. 36:30-35.

Monjula S, John E (1991). Biochemical changes and *in vitro* protein digestibility of the endodermis of germinating Dolchos lablab. J. Sci. Food Agric. 55: 229-238.

Nene YL, Hall SD, Sheila VK (1990). Pigeon pea. CAB International, Wallinfork, University Press, Cambridge. pp. 490.

Okaka JC, Potter NN (1976). Cowpea powders dried with methionine. Preparation, storage, stability, organoleptic properties and nutritional quality. J. Food Sci. 41:48-53.

Oshodi AA, Adeladun MOA (1993). Proximate composition, some nutritionally valuable minreals and functional properties of three varieties of Lima beans (Phaseolus lunatus Linn.) flour. Int. J. Food Sci. Nutr. 43:181-185.

Pour-El A (1981). Protein functionality: Classification, definition and methodology. In: Chrery JP, editor. Protein functionality in foods. American Chemical Society Symposium Series 147. Washington, DC: American Chemical Society

Rady AH, Abdel Hady SM, Elnashabi FM, Afifi EA, Salam EM (2002). Influence of Gamma rays and Microwave h e a t in g on the Quality of Olive fruits and their virgin oil. Isotope Rad. Res. 34: 369-380.

Rahma EH, Mostafa MM (1988). Functional properties of peanut flour as affected by different heat treatments. J. Food Sci. Technol. 25:11-15.

Reddy NP, Pierson MD, Sathe SK, Salunkhe DK (1984). Chemical, nutritional, and physiological aspects of dry bean carbohydrates. Food Chem. 13(1):25-68.

Rehman ZU, Shah WH (2001). Tannin contents and protein digestibility of black grams (*Vigna mungo*)after soaking and cooking. Plant Food Hum. Nutr. 56: 265–273.

Sadik N (1991). Population growth and the food crisis: food, nutrition and agriculture/alimentation. Food Nutr. Agric. 1: 3–6.

Sattar A, Neelofar X, Akhtar MA (1990). Effect of radiation and soaking on phytate content of soybean. Acta Aliment 19:331-336.

SPSS 16.0 (2008). Statistical Package for the social science. Command syntax Reference. Inc. Chicago III.

Vadivel V, Janardhanan K (2002). Agro-botanical traits and chemical composition of Cassia obtusifolia L.: A lesser-known legume of the Western Ghats region of south India. Plant Foods Hum. Nutr. 57:151-164.

WHO (1999). High-dose irradiation: Wholesomeness of food irradiated with doses above 10 kGy. Report of a Joint FAO/IAEA/WHO Study Group. WHO Tech Rep Ser. 890:i_vi, 1_197.

14

Protective effects of garlic (*Allium sativum*) and ginger (*Zingiber officinale*) on physicochemical and microbial attributes of liquid smoked silver carp (*Hypophthalmichthys molitrix*) wrapped in aluminium foil during chilled storage

Fijelu Frank, Yanshun Xu, Qixing jiang and Wenshui Xia

School of Food Science and Technology, Jiangnan University, Wuxi, Jiangsu 214122, China.

The antioxidant and antimicrobial effects of equivalent concentrations of both garlic and ginger separately in fresh and powder form on quality attributes of liquid smoked silver carp (*Hypophthalmichthys molitrix*) during chilled storage were investigated for a period of 24 days. The control and the treated fish samples were analysed periodically for microbiological (total viable count) and physicochemical (pH, colour, water activity, free fatty acid, total volatile basic nitrogen; TVB-N, 2-thiobarbituric acid-reactive substances; TBARS) characteristics. The results showed that each spice in either fresh or powder form treatment significantly ($P < 0.05$) reduced growth of microorganisms as reflected in total viable counts (TVC) from day 6 of storage, lipid oxidation decrease as displayed in TBARS, chemical spoilage decrease as reflected in TVB-N after day 6 of storage as compared to the control samples. The results obtained from this study suggests that garlic and ginger in fresh and powder form, through their combined antioxidant and antimicrobial effects, are potentially useful in preserving liquid smoked silver carp at 4 ± 1°C.

Key words: Liquid smoke, silver carp, spices, shelf life, chilled storage.

INTRODUCTION

Silver carp (*Hypophthalmichthys molitrix*) is a prevalent freshwater fish, it has been one of the most widely cultured species all over the world due to its fast growth rate, easy cultivation and high feed efficiency ratio (Hu et al., 2008). This species is also known to provide a high content of important constituents of the human diet such as nutritional and readily digestible proteins, lipid, soluble vitamins, microelements and polyunsaturated fatty acids (Erkan and Ozden, 2007). However, distributors face challenges of deterioration during postmortem storage and processing as a result of damage mechanisms such as autolytic degradation, microbiological spoilage and lipid oxidation (Naseri and Rezaei, 2012).

Fish smoking is one of the oldest method of preservation, giving a characteristic flavour and colour to the product and increasing its shelf life. The preservative effect is due to the presence of some antimicrobial compounds in smoke such as phenols and formaldehyde (Muratore and Licciardello, 2005). Nevertheless, the conventional smoking process is commonly substituted by the use of liquid smoke (Hattula et al., 2001). Liquid smoke contains the same functional components such as phenols, carbonyls and acids that are found in vaporous smoke. Liquid smoke is free of harmful compounds such as polycyclic aromatic hydrocarbons (PAHs) which are ubiquitous environmental contaminants; these are commonly found in convectional smoke, also considered carcinogenic and mutagenic molecules (Alcicek, 2013). Additional advantages of liquid smoke are environmental friendliness, lower cost and easy application such as direct addition, drenching or dipping, impregnated (smoked) casings and atomization (Varlet et al., 2007). The smoking process is often coupled with other treatments, such as salting, addition of spices, packaging techniques and chilled storage, to produce synergistic effects on spoilage microorganisms and to increase shelf life (Visciano et al., 2008).

Ginger and garlic are spices, in addition to contributing taste and aroma to foods, also contain a variety of bioactive substances which are of considerable use from the standpoint of food science and technology. These may be used singly or in combination, and some act synergistically to control spoilage of foods, these have made it to be used as bio preservative (Yanishlieva et al., 2006). Recently, use of plant extracts as natural antioxidants has gained increasing interest because of the global trend of restriction in use of synthetic substances, also antioxidant rich plant extracts have potential benefits in food preservations (Uhart et al., 2006).

The aim of this study was to investigate the antioxidant as well as the antimicrobial effect of equivalent concentrations of two garlic and ginger preparations (fresh and powder form) on shelf life of liquid smoked silver carp wrapped in aluminum foil and stored at 4 ± 1°C by determining microbiological and physicochemical differences.

MATERIALS AND METHODS

Fresh garlic bulbs (*Allium sativum*, Chinese white garlic) and fresh ginger (*Zingiber officinale*, Chinese yellow ginger) were purchased from a local market. The dry skin of both fresh garlic and ginger were removed before use; they were peeled and crushed finely by using a kitchen hand held grater (Evasolo). Garlic powder and ginger powder were purchased from a supermarket. 4 kg of produced garlic powder was the weight equivalent of 16 kg of fresh garlic, also 2 kg of produced ginger powder was the weight equivalent of 10 kg of fresh ginger, according to the manufacturer details.

Preparation of fish

fifteen silver carp *(H. molitrix)* with average weight of 815.12 ± 14.55 g were purchased from a local market (Wuxi, Jiangsu, China), transported within 1 h in sealed polystyrene foam boxes containing ice to the Food Processing Technology Laboratory at Jiangnan University, where processing was carried out, then randomly divided into 5 groups for further treatments after they were gutted, eviscerated, deboned, the scale, skin, pin bones, debris were removed, and filleted.

Sample treatment and brining with liquid smoke

C: The samples without ginger and garlic treatment; T1: 10 g/kg ginger powder was added to brine; T2: 50 g/kg of fresh ginger applied on fish for 4 h at 4 ± 1°C after brining; T3: 12.5 g/kg garlic powder was added to the brine;T4: 50 g/kg of fresh garlic was applied on the fish for 4 h at 4 ± 1°C after brining.

All the samples above were immersed in brine containing 10% sodium chloride and 1% liquid smoke solution at a ratio of 1:1(w/w) for 4 h at 4 ± 1°C.

Drying, heating and cooling of samples

All samples were air dried at 20°C for 40 min. The samples were then heated in an oven at 75°C for 80 min. After cooling at 20°C for 30 min, the smoked products were wrapped in aluminium foil, and stored at 4 ± 1°C until analysis performed in replicates on day 0, 6, 12, 18 and 24 of storage.

pH value measurement

A 1 g sample of the fish flesh was homogenized in 10 mL of distilled water and the mixture was filtered. The pH of the filtrate was measured according to the method of Vyncke (1981) using a pH meter (Mettler Toledo 320-s, Shanghai, China).

Water activity

An adequate amount of ground sample was placed in a holding cup (about one-half full) and the water activity was read (AQUA LAB CX-2, Decagon Devices Inc, Pullman, WA, USA) after equilibration at 25°C.

Color evaluation

The color of fish was measured using the L*, a*, b* mode of CIE using color meter (Model TC-Pα G, Beijing optical Instruments Factory, Beijing, China). The measurement of L* (indicator of lightness), a* (indicator of redness), b* (indicator of yellowness) was performed in two replications (Lanier et al., 1991).

Determination of total volatile basic nitrogen (TVB-N)

TVB-N was determined according to the method of Antonacopoulos and Vyncke (1989). For total TVB-N, fish muscle (10 g) was homogenized with 6% perchloric acid (90 mL) for 1 min in an IKA T25 basic Ultra-Turrax (Janke & Kunkel, IKA Labor-technik, Selangor, Malaysia). The homogenates were filtered through a filter paper (Whatman no. 1) and filtrates alkalized by NaOH (20%). The distillate was titrated with 0.01 N HCl.

Determination of free fatty acid (FFA) content

Free fatty acid content of the fish samples was determined according to the method of Kirk and Sawyer (1991). A mixture of diethyl ether (25 mL), ethanol (70% v/v) (25 mL) and 1% phenolphthalein solution (1 mL) was prepared then neutralized with 0.1 M NaOH solution. Two grams quantities of fish samples were blended in the neutral solvent prepared above for about 20 min, and then titrated with 0.1 M NaOH with constant shaking until a pink color was formed which persisted for about 15 s. All samples were analyzed in triplicate and the free fatty acid content was expressed

Table 1. pH.

Storage time(days)	pH				
	C	T1	T2	T3	T4
0	6.07 ± 0.54^{aA}	6.1 ± 0.18^{aA}	6.15 ±0.07^{aA}	6.18 ± 0.25^{aA}	6.07 ±0.09^{aA}
6	6.25 ± 0.26^{aA}	6.23 ± 0.04^{aAB}	6.19 ±0.44^{aA}	6.2 ± 0.12^{aA}	6.13 ±0.13^{aA}
12	6.35 ± 0.08^{aA}	6.3 ± 0.14^{aAB}	6.26 ± 0.01^{aA}	6.3 ± 0.45^{aA}	6.15 ±0.21^{aA}
18	6.39 ± 0.13^{aA}	6.34 ± 0.06^{aAB}	6.29 ± 0.11^{aA}	6.32 ± 0.03^{aA}	6.2 ±0.07^{aA}
24	6.52 ± 0.12^{aA}	6.45 ±0.03^{abB}	6.33 ±0.1^{abA}	6.36 ±0.08^{abA}	6.27 ±0.16^{bA}

$^{A–B}$Means with different superscript letters in the same column indicate significant differences (p<0.05). $^{a–b}$Means with different superscript letters in the same row indicate significant differences (p < 0.05). Data are expressed as means ± standard deviation (n = 2).

as oleic acid equivalent.

Determination of 2-thiobarbituric acid-reactive substances (TBARS)

2-Thiobarbituric acid reactive substances (TBARS) were determined according to the method of Buege and Aust (1978). 0.5 g of fish fillet was homogenized in 10 ml of the mixture containing TBA (0.375 g/100 ml), TCA (15 g/100 ml) and HCl (0.25mol/l). The mixture was heated in the boiling water for 10 min, followed by cooling with the running tap water. The mixture was centrifuged at 3600 ×*g* for 20 min and the absorbance was measured at 532 nm (Model 135 WFZ-UV-2100, UNICOTM, Shanghai, China). The TBARS value was calculated from the standard curve of malonaldehyde and expressed as mg malonaldehyde/kg sample.

Total viable counts

10 g of each sample homogenate was diluted with 100 ml of 1% BPW at pH 7.5. Serial dilutions were made until 10^{-7} g/ml samples were obtained. A 1 ml aliquot of each dilution was placed in a Petri dish and approximately 15 ml plate count agar (PCA, Oxoid, CM325) was added. Appropriate decimal dilutions of the samples were prepared with the same diluents and 0.1 ml aliquots of appropriate dilution were spread on the agar plates. Each Petri dish was carefully shaken, in order to achieve a homogeneous distribution of the sample. After several minutes all Petri dishes were inverted and placed in an oven at 37°C for 48 h. After 48 h of incubation, all colonies were counted by hand, following the rules reported by Gilliland et al. (1976).

Statistical analysis

For data analysis, analysis of variance (ANOVA) was used. Significant differences were defined at $P < 0.05$. Comparisons of group means were obtained using Duncan's multiple range tests. All statistical analysis was performed using SPSS version 19.0 for windows software (SPSS Inc., Chicago, IL).

RESULTS AND DISCUSSION

Changes of pH

The pH value of smoked silver carp fillet was increased (P<0.05) with storage period (Table 1). Its values ranged between 6.07 and 6.52 for both control and treated samples throughout the entire storage period. The limit of acceptability is usually 6.8 to 7.0 (Erkan et al., 2011). This phenomenon is presumably because of the production of basic nitrogen compounds such as ammonia, TMA, and/or other basic nitrogenous compounds caused by microbial activity or proteolytic enzymes in a food material (Fan et al., 2009).

Water activity

Changes in water activity (Aw) are shown in Table 2. Initially, the Aw for all samples was in the range of 0.84 to 0.93, which increased significantly (P<0.05) with storage time to a final range of 0.92 to 0.99. The control samples showed higher water activity (Aw) value than the counter-part samples during time of storage. Water activity (aw) has its most useful application in predicting the growth of bacteria, yeast and mold. The knowledge of water activity is very important factor to guarantee the required stability towards microbial spoilage of the product and to ensure safety by avoiding any threat to the health of the consumer, this is because micro-organisms generally grow best between Aw values of 0.995-0.980, while most microbes cease growth at Aw <0.900. It is also necessary for the transport of nutrients and the removal of waste materials, to carry out enzymatic reactions, to synthesize cellular materials, and to take part in other biochemical reactions (Goksoy, et al., 2009).

Color

Color is one of the most important organoleptic characteristics. It influences the acceptability of the product and plays a major role in the purchase decision and evaluation of meat quality (O'Sullivan, et al., 2003).

L* values (lightness) decreased (P < 0.05) with the increase of storage time. Coordinates decreased from 84.36 on day 0 to 72.23 at the end of the experiment on

Table 2. Water activity.

Storage time (days)	Water activity				
	C	T1	T2	T3	T4
0	0.93±0^{bA}	0.93±0.01^{bA}	0.91±0^{bA}	0.91±0.02^{bA}	0.84±0^{aA}
6	0.95±0.01^{bBA}	0.94±0.01^{bB}	0.92±0.03^{bA}	0.93±0.01^{bB}	0.85±0.03^{aBA}
12	0.97±0.02^{cCB}	0.96±0.02^{cCB}	0.92±0^{bA}	0.94±0^{cbB}	0.87±0^{aBA}
18	0.98±0.01^{dDC}	0.96±0.04^{cC}	0.93±0^{bA}	0.95±0.01^{bB}	0.89±0.028^{aCB}
24	0.99±0.02^{aD}	0.97±0.03^{aC}	0.93±0.04^{aA}	0.95±0.05^{aB}	0.92±0^{aC}

$^{A–D}$Means with different superscript letters in the same column indicate significant differences (p<0.05). $^{a–d}$Means with different superscript letters in the same row indicate significant differences (p < 0.05). Data are expressed as means ± standard deviation (n = 2).

Table 3. Colour measurements.

Parameter/treatment	Storage period (days)				
	0	6	12	18	24
L*-value(lightness)					
Control	84.36±0.51^{dE}	82.04±0.001^{dD}	79.28±0.03^{eC}	76.51±0.3^{eB}	72.23±0.13^{eA}
T1	83.16±0.06^{dcE}	80±0.85^{cD}	77.44±0.62^{dC}	74.73±0.57^{dB}	70.68±0.45^{dA}
T2	79.28±1.81^{baD}	75.35±0.92^{bC}	74.22±0.03^{bC}	68.43±0.14^{bB}	63.95±0.03^{bA}
T3	81.31±0.98^{cbE}	78.53±0.75^{cD}	75.74±0.48^{cC}	71.82±0.00^{cB}	68.15±0.21^{cA}
T4	77.75±0.21^{aE}	72.83±0.52^{aD}	71.47±0.75^{aC}	66.94±0.03^{aBC}	61.95±0.55^{aA}
a*-value (redness)					
Control	4.3±0.28^{dA}	5.62±0.91^{cB}	6.65±0.03^{eB}	8.12±0.25^{dC}	9.49±0.01^{eD}
T1	3.58±0.11^{cA}	5.16±0.23^{cB}	6.42±0.03^{dC}	7.69±0.13^{dD}	8.98±0.25^{dE}
T2	0.33±0.01^{aA}	2.09±0.14^{bB}	3.63±0.13^{bC}	4.3±0.28^{bD}	5.91±0.001^{bE}
T3	2.24±0.08^{bA}	3±0.16^{bB}	4.37±0.11^{cC}	5.52±0.03^{cD}	6.93±0.04^{cE}
T4	0.35±0.01^{aA}	0.98±0.02^{aB}	2.01±0.01^{aC}	3.31±0.13^{aD}	4.92±0.04^{aE}
b*-value (yellowness)					
Control	27.12±1.24^{dC}	25.38±0.099^{dC}	22.63±1.26^{dB}	21.25±0.14^{dBA}	19.45±0.11^{eA}
T1	24.84±1.75^{dcD}	22.67±1.23^{cDC}	21.51±0.57^{dcCB}	19.51±0.51^{cBA}	18.23±0.17^{dA}
T2	21.45±0.21^{bD}	20.01±0.16^{bC}	18.74±0.13^{baB}	16.66±0.92^{baA}	15.77±0.06^{bA}
T3	23.12±0.96^{cbC}	21.4±0.14^{cbB}	19.99±0.41^{cbB}	17.93±0.68^{bA}	16.59±0.01^{cA}
T4	18.5±0.42^{aD}	17.69±0.58^{aDC}	17.49±0.28^{aC}	16.07±0.23^{aB}	15.12±0.001^{aA}

$^{a–e}$Means with different superscript letters in the same column indicate significant differences (p<0.05). $^{A–D}$Means with different superscript letters in the same row indicate significant differences (p < 0.05). Data are expressed as means ± standard deviation (n = 2).

the control samples, counterpart samples decreased from 77.75-83.16 at day 0 to 61.95-70.68 at the end of the experiment, depending on the treatment. Moreover, control samples showed higher values as compared to their counterpart samples (Table 3) but that difference was small and irrelevant. Many factors may contribute to more lightness on control samples than their counterpart samples, these include the concentration and type of pigments present (Viuda-Martos et al., 2010), water content, hygroscopicity of the material dissolved in the water matrix and surface water availability, but the pigment content and the fractions of MetMb and OxyMb were the most important factors for the variation in the L* value (Karlsson et al., 2006).

The a*-value was found to increase (P < 0.05) in all samples as duration increased. Again the values in control samples were higher than counterpart samples with increasing storage time. The a*-value in the control samples rose from 4.3 on day 1 to 9.49 at the end of the experiment. Treated samples rose from 0.33-3.58 to 4.92-8.98, depending on the treatment (Table 3). The increase of the a*value indicated that fish meat were redder, this

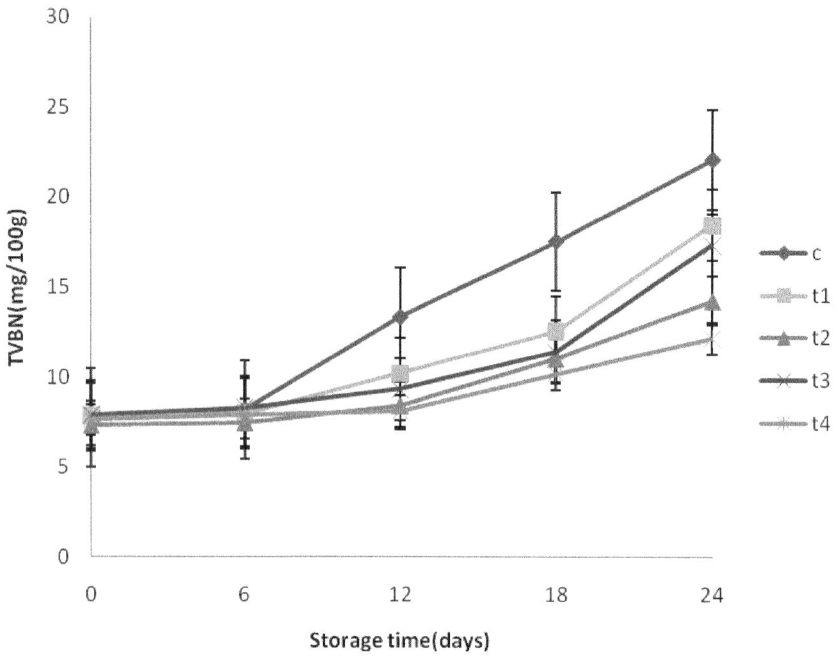

Figure 1. Total volatile basic nitrogen (mg/100 g).

could be due to less met-myoglobin formation (Faustman et al., 2010; Biswas et al., 2012). It is postulated that, spices are effective antioxidants that improve red color stability by inhibiting met-myoglobin formation in fish fillets (Allen and Cornforth, 2010). In the present study, the results do not establish the effectiveness of garlic and ginger as antioxidants and antimicrobials on red color stability.

For the yellow coordinate (b*), the value decreased ($P < 0.05$) in all samples during time of storage. Coordinates decreased from 27.12 on day 0 to 19.45 at the end of the experiment on the control samples, counterpart samples decreased from 18.5-24.84 at day zero to 15.12-18.23 at the end of the experiment, depending on the treatment (Table 3). Even though the magnitude of the differences was very small on all samples, the control samples showed higher yellowness than their counterpart samples. The differences for the yellow coordinate (b*), may be due to pH, oxidation extent, water activity, etc. in the matrix which have the greatest influence on this coordinate in many foods (Cofrades et al., 2004).

Total volatile base nitrogen (TVB-N)

The TVB-N of fish is an indicator of the freshness of the raw material (Zhou et al., 2011). Changes in TVB-N of different samples during the entire storage period are shown in Figure 1. The initial TVB-N value of silver carp was in the range of 7.31 to 7.90 mg/100 g. Similar results were also reported by Fan et al. (2009). TVB-N values of control samples showed significant ($P < 0.05$) differences from other samples on storage; as storage increased also its TVB-N value was higher than those other samples which increased more slowly than control samples after day 6 of storage. The TVB-N value of the control samples had already reached 12.32 mg/100 g on the day 12 of storage, while the values of counterpart samples did not reach this figure but was in the range of 8.08 to 10.2 mg/100 g. TVB-N values of samples T1, T2, T3 and T4 on the 24th day of storage were 18.48, 14.27, 17.36 and 12.16mg/100 g, respectively, while TVB-N value of control samples on the 24th day was 22.8mg/100 g. TVB-N values of all samples were lower than 25mg/100 g which was considered as the threshold for a good-quality fish product, high TVB-N values are unacceptable and are associated with unpleasant smell in the meat (Limbo et al., 2009). Assumably, this is because of the impact of the various treatments of TVB-N, which primarily includes nitrogen from ammonia, TMA, and dimethylamine which reflects the extent of degradation of proteins and non protein nitrogenous compounds which can be explained by proteolysis, due to enzymatic and microbial activities in the samples on storage (Erkan and Ozden, 2008). In the present study, the results establish the effectiveness of garlic and ginger as antioxidants and antimicrobials due to reduction in TVB-N on treated samples as observed, it is supposed that the spices used to treat silver carp were involved in TVB-N reduction.

Free fatty acids

The initial FFA values were in the range of 7.19 to 7.9

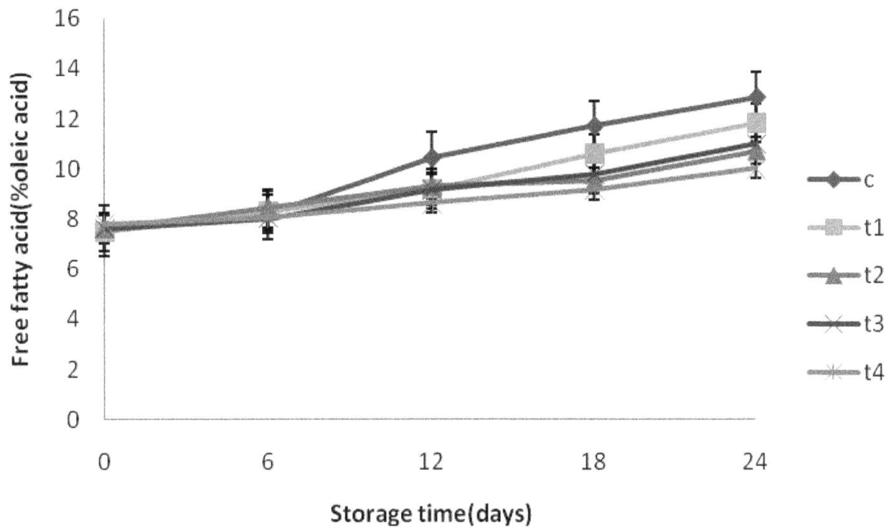

Figure 2. Free fatty acids (percentage oleic acid).

(oleic acid percentage) for all samples. FFA values increased with storage time (Figure 2); however the values in the control samples were higher than other samples during storage. Significant statistical differences were found between the treated samples and control samples ($P < 0.05$) after day 12 of storage. At the end of the storage period (day 24), values of FFA on control samples were found to be 12.83 (oleic acid percentage) which is higher than counterpart samples which were found to be 11.82, 10.72, 10.98 and 10.01 (oleic acid percentage) for T1, T2, T3 and T4, respectively. Accumulation of FFA is said to contribute to off flavor of the product and cause textural alterations by complexing with protein (Siddaiah et al., 2001). The results established the effectiveness of garlic and ginger as antioxidants which were greater in activities to inhibit the synthesis of free fatty acid in the treated samples than control samples on 24 days of storage.

Thiobarbituric acid reactive substances (TBARS)

Fish meat is particularly susceptible to oxidative changes because of the processing conditions, exposure of unsaturated fat and proteins to molecular oxygen. 2-Thiobarbituric acid (TBA) is widely used as an indicator of degree of lipid oxidation, and the presence of TBA reactive substances is due to the second stage auto-oxidation (Rezaei and Hosseini, 2008) during which peroxides are oxidized to aldehydes and ketones (Lindsay, 1991). The TBARS values of all silver carp samples increased as storage time increased, the increase of TBARS values in fish meat with increasing storage time is normal (Rong et al., 2009) (Figure 3). Significant differences were found between control samples and counterpart samples ($P < 0.05$) after day 6 of storage. The mean TBARS values of control were found to be 1.56 mgMA/kg of fish flesh which is higher than those of other samples which were found to be 1.2, 1.02, 1.1 and 0.93 mgMA/kg for T1, T2, T3 and T4, respectively, at day 24 of storage. There are two possible reasons for this phenomenon in the effectiveness of this product: 1, reduction in TBARS using garlic and ginger is related to peroxide-scavenging enzyme activity, which could reduce unsaturated fatty acid and total unsaturated fatty acid oxidation and 2, some active components in the garlic and ginger may involve desaturase and elongase activities (Mariutti et al., 2008).

Total viable counts

As shown, initial total viable counts (TVC) in all samples was in the range of 2.20 to 2.28 $\log_{10}$ CFU/g indicating very good fish quality, the value was found to increase ($p < 0.05$) but remained below 7 $\log_{10}$ CFU/g in all treatments for 24 days of storage, which is the Maximal Permissible Limit (MPL) for TVC recommended (ICMSF, 1986; Ojagh et al., 2010) in all samples. The values in the control samples were higher than their counterpart samples with storage time (Table 4). The increase of TVC in fish flesh during storage has been demonstrated by Bahmani et al. (2011). The high levels of microorganisms shown on control samples from day 6 of storage, resulted into significant differences ($P<0.05$) in TVC when compared with counterpart samples due to the strong antimicrobial activity of the organ sulfur compounds and other active components contained in garlic and ginger (Lu et al., 2011). TVC is the most common microbiological method aimed to detect and enumerate

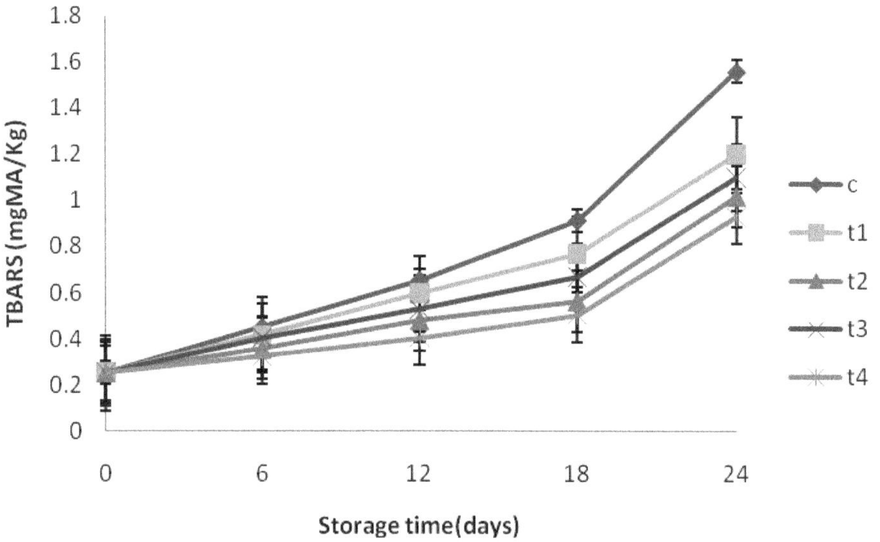

Figure 3. 2-Thiobarbituric acid-reactive substances (mgMA/kg).

Table 4. Total viable counts ($\log_{10}$ CFU/g).

Parameter	Storage time (days)				
	0	6	12	18	24
C	2.26±0.04^{aA}	3.53±0.45^{dB}	5.57±0.38^{cC}	6.12±0.11^{eD}	6.91±0.27^{cE}
T1	2.23±0.01^{aA}	3.08±0.16^{cB}	3.49±0.06^{bC}	4.16±0.18^{dD}	4.42±0.03^{bE}
T2	2.26±0.11^{aA}	2.71±0.14^{bB}	3.18±0.1^{aC}	3.72±0.07^{bD}	4.24±0.04^{baE}
T3	2.20±0.09^{aA}	2.86±0.03^{bB}	3.47±0^{bC}	3.91±0.01^{cD}	4.38±0.07^{bE}
T4	2.28±0^{aA}	2.49±0.09^{aA}	2.9±0.04^{aB}	3.44±0.17^{aC}	4±0.01^{aD}

$^{a–e}$Means with different superscript letters in the same column indicate significant differences ($p<0.05$). $^{A–E}$Means with different superscript letters in the same row indicate significant differences ($p < 0.05$). Data are expressed as means ± standard deviation ($n = 2$).

high proportion of the microbial population as possible. In practice, this usually means mesophilic, aerobic or facultatively anaerobic bacteria, which account for the major part of the microflora in fish. A TVC method can only provide an estimate of the microbial population based on those cells that are recoverable under the test conditions. Some viable cells may not be recoverable by any existing method, while others may only grow at low temperatures, in the presence of specific growth factors, or in the absence of oxygen (Fulford et al., 2004)

Conclusion

From the above results, it can be concluded that both garlic and ginger in fresh and powder form provide antioxidant and antimicrobial benefits to liquid smoked silver carp during chilled storage for 24 days. Therefore, it is suggested that garlic and ginger, as a natural herb, could be used to extend the shelf life of meat products, providing the consumer with food containing natural additives, which might be seen more healthful than those of synthetic origin. Further research is required to focus on understanding the mechanisms of action, in particular concentrations of active ingredients of both ginger and garlic in either powder or fresh form which applied to liquid smoked fish products.

ACKNOWLEDGEMENT

This research was financially supported by the earmarked fund for China Agriculture Research System (CARS-46).

REFERENCES

Alcicek Z (2013). Effects of Different Liquid Smoke Flavor Levels on the

Shelf Life of Venus Clam (Chamelea Gallina, L 1758) Meat. J Food Process Preserv. doi: 10.1111/jfpp.12052

Allen K, Cornforth D (2010). Comparison of spice-derived antioxidants and metal chelators on fresh beef color stability. Meat Science 85(4): 613-619.

Antonacopoulos N, Vyncke W (1989). Determination of volatile basic nitrogen in fish: A third collaborative study by the West European Fish Technologists' Association (WEFTA). Zeitschrift für Lebensmittel-Untersuchung und Forschung 189(4): 309-316.

Bahmani ZA, Rezai M, Hosseini SV, Regenstein JM, Bohme K, Alishahi A, Yadollahi F (2011). Chilled storage of golden gray mullet (Liza aurata). LWT - Food Sci. Technol. 44(9): 1894-1900.

Biswas AK, Chatli MK, Sahoo J (2012). Antioxidant potential of curry (Murraya koenigii L.) and mint (Mentha spicata) leaf extracts and their effect on colour and oxidative stability of raw ground pork meat during refrigeration storage. Food Chem. 133(2): 467-472.

Buege JA, Aust SD (1978). Microsomal lipid peroxidation. Methods in Enzymology. F. Sidney and P. Lester, Academic Press. Volume 52: 302-310.

Cofrades S, Serrano A, Ayo J, Solas MT, Carballo J, Colmenero FJ (2004). Restructured beef with different proportions of walnut as affected by meat particle size. Eur. Food Res. Technol. 218(3): 230-236.

Erkan N, Ozden O (2007). Proximate composition and mineral contents in aqua cultured sea bass (Dicentrarchus labrax), sea bream (Sparus aurata) analyzed by ICP-MS. Food Chem. 102(3): 721-725.

Erkan N, Ozden O (2008). Quality assessment of whole and gutted sardines (Sardina pilchardus) stored in ice. Int. J. Food Sci. Technol. 43(9): 1549-1559.

Erkan N, Tosun SY, Ulusoy S, Uretener G (2011). The use of thyme and laurel essential oil treatments to extend the shelf life of bluefish (Pomatomus saltatrix) during storage in ice. Journal für Verbraucherschutz und Lebensmittelsicherheit 6(1): 39-48.

Faustman C, Sun Q, Mancini R, Suman SP (2010). Myoglobin and lipid oxidation interactions: Mechanistic bases and control. Meat Sci. 86(1): 86-94.

Fulford MR, Walker JT, Martin MV, Marsh PD (2004).Total viable counts, ATP, and endotoxin levels as potential markers of microbial contamination of dental unit water systems. Br. Dent. J. 196:157 - 159.

Gilliland SE, Busta FF, Brinda JJ, Campbell JE (1976). Aerobic Plate Count M.L. Speck (Ed.), Compendium of Methods for the Microbiological Examination of Foods, America Public Health Association, Inc., Washington, pp. 107–131

Hattula T, Elfving K, Mroueh UM, Luoma T (2001). Use of Liquid Smoke Flavouring as an Alternative to Traditional Flue Gas Smoking of Rainbow Trout Fillets (Oncorhynchus mykiss). LWT - Food Sci. Technol. 34(8): 521-525.

Hu Y, Xia W, Ge C (2008). Characterization of fermented silver carp sausages inoculated with mixed starter culture. LWT - Food Sci. Technol. 41(4): 730-738.

ICMSF (1986). International commission on microbiological specifications for foods. Sampling plans for fish and shellfish. In ICMSF (Ed.), ICMSF, Microorganisms in foods. Sampling for microbiological analysis: Principles and scientific applications (Vol. 2, 2nd ed.). Toronto, Canada: University of Toronto Press.

Kirk S, Sawyer R (1991). Pearson's composition and analysis of foods, Longman Group Ltd.

Lanier TC, Hart K, Martin RE (1991). A manual of standard methods for measuring and specifying the properties of surimi, National Fisheries Institute, Washington, DC.

Limbo S, Sinelli N, Torri L, Riva M. (2009). Freshness decay and shelf life predictive modelling of European sea bass (Dicentrarchus labrax) applying chemical methods and electronic nose. LWT - Food Sci. Technol. 42(5): 977-984.

Lindsay RC (1991). Flavour of fish. Paper presented at 8th World Congress of Food Science & Technology, 29th September–4th October, Toronto, Canada.

Lu X, Rasco BA, Jabal JMF, Aston DE, Lin M, Konkel ME (2011). Investigating Antibacterial Effects of Garlic (Allium sativum) Concentrate and Garlic-Derived Organosulfur Compounds on Campylobacter jejuni by Using Fourier Transform Infrared Spectroscopy, Raman Spectroscopy, and Electron Microscopy. Appl. Environ. Microbiol. 77(15): 5257-5269.

Mariutti LB, Orlien V, Bragagnolo N, Skibsted LH (2008). Effect of sage and garlic on lipid oxidation in high-pressure processed chicken meat. Eur. Food Res. Technol. 227(2): 337-344.

Muratore G, Licciardello F (2005). Effect of Vacuum and Modified Atmosphere Packaging on the Shelf-life of Liquid-smoked Swordfish (Xiphias gladius) Slices. J. Food Sci. 70(5): C359-C363.

Naseri M, Rezaei M (2012). Lipid Changes During Long-Term Storage of Canned Sprat. J. Aquat. Food Prod. Technol. 21(1): 48-58.

Ojagh SM, Rezaei M, Razavi SH, Hosseini SMH (2010). Effect of chitosan coatings enriched with cinnamon oil on the quality of refrigerated rainbow trout. Food Chem. 120(1):193-198.

O'Sullivan MG, Byrne DV, Martens H, Gidskehaug LH, Andersen HJ, Martens M (2003). Evaluation of pork colour: prediction of visual sensory quality of meat from instrumental and computer vision methods of colour analysis. Meat Sci. 65(2): 909-918.

Rezaei M, Hosseini FS (2008). Quality Assessment of Farmed Rainbow Trout (Oncorhynchus mykiss) during Chilled Storage. J. Food Sci. 73(6): H93-H96.

Rong CX, Chang-hu X, Qi L, Bang-zhong Y (2009). Microbiological, chemical and sensory assessment of (I) whole ungutted, (II) whole gutted and (III) filleted tilapia (Oreochromis niloticus) during refrigerated storage. Int. J. Food Sci. Technol. 44(11): 2243-2248.

Uhart M, Maks N, Ravishankar S (2006). Effect of spices on growth and survival of salmonella typhimurium dt 104 in ground beef stored at 4 and 8c. J. Food Saf. 26(2):115-125.

Varlet V, Serot T, Knockaert C, Cornet J, Cardinal M, Monteau F, Bizec B, Prost C (2007). Organoleptic characterization and PAH content of salmon (Salmo salar) fillets smoked according to four industrial smoking techniques. J. Sci. Food Agric. 87(5): 847-854.

Visciano P, Perugini M, Conte F, Amorena M (2008). Polycyclic aromatic hydrocarbons in farmed rainbow trout (Oncorhynchus mykiss) processed by traditional flue gas smoking and by liquid smoke flavourings. Food Chem. Toxicol. 46(5):1409-1413.

Viuda-Martos M, Ruiz-Navajas Y, Fernandez-Lopez J, Perez-Alvarez JA (2010). Effect of added citrus fibre and spice essential oils on quality characteristics and shelf-life of mortadella. Meat Sci. 85(3): 568-576.

Vyncke W (1981). pH of fish muscle comparison of methods, Western European Fish Technologists' Association (WEFTA), Copenhagen, Denmark.

Yanishlieva NV, Marinova E, Pokorny J (2006). Natural antioxidants from herbs and spices. European Journal of Lipid Science and Technology **108**(9): 776-793.

Zhou R, Liu Y, Xie J, Wang X (2011). Effects of combined treatment of electrolysed water and chitosan on the quality attributes and myofibril degradation in farmed obscure puffer fish (Takifugu obscurus) during refrigerated storage. Food Chem. 129(4): 1660-1666.

15

Presence of *Listeria monocytogenes* in raw milk and traditional dairy products marketed in the north-central region of Morocco

Boujemaa El Marnissi[1] , Laila Bennani[2], Nezha Cohen[3], Abdelhakim El Ouali Lalami[2] and Rajae Belkhou[1]

[1]Laboratoire Agroalimentaire et Sécurité Sanitaire des Aliments (LASSA), Université Sidi Mohamed Ben Abdallah, Ecole Supérieure de Technologie Fès, Route d'Imouzzer B.P : 2427 - Fez, Morocco.
[2]Laboratoire Régional de Diagnostic Epidémiologique et d'Hygiène du Milieu, Direction Régionale de la Santé, Hôpital EL Ghassani-Fez, Morocco.
[3]Division de Microbiologie et Hygiène des Aliments et de l'Environnement Institut Pasteur du Maroc, 1 Place Louis Pasteur, Casablanca 20360, Morocco.

The objective of this study was to determine the prevalence of *Listeria monocytogenes* in raw milk and its two traditional dairy derivatives, the *Lben* (traditionally fermented skimmed milk) and *Jben* (traditional fresh cheese) marketed in the city of Fez in the center northern of Morocco. A total of 288 samples from three dairy products including one third of raw milk, one third of "Lben" and one third of "Jben" were collected from eight traditional dairies belonging to four sectors of the city of Fez between October 2009 and September 2010. Isolation and identification of *L. monocytogenes* were carried out according to official standard ISO procedure 11290-1 and the acidity of the various samples was performed according to the Moroccan standard. The overall prevalence of *L. monocytogenes* contamination was 5.90%. It was present in eight raw milk samples, five of *Lben* samples and in four *Jben* samples. The results reveal a variation of contamination from one sector to another with a higher contamination in the samples collected in the fall and winter. This suggests a link between management practices feed, hygienic conditions and *L. monocytogenes* contamination. The contamination levels achieved justify the control of the feeding cattle, milk pasteurization and the enforcement of the general principles of food hygiene in order to reduce consumer's exposure to *L. monocytogenes*.

Key words: *Listeria monocytogenes*, raw milk, traditional dairy products, food safety, Morocco.

INTRODUCTION

Currently, *Listeria monocytogenes* is considered one of the most important pathogens responsible for food-borne infection. It is often incriminated in outbreaks of human listeriosis (Ryser and Marth, 2007).

Pregnant women, infants, immunocompromised and the elderly people are at greatest risk for listeriosis (Gillespie et al., 2010). This infection is regularly monitored and reported in Europe and North America, but in Africa and other developing countries, only a few sporadic cases have been reported (Tazi, 1981; Boukadidda et al., 1994). In Morocco and other countries of North Africa, the studies on the incidence of human listeriosis are rare.

Evidence is accumulating to vindicate that the organism

is a food-borne pathogen, and dairy products have been implicated. In dairy industry, *Listeria* can contaminate directly or indirectly the products and the environment through contaminated raw milk, resulting in huge losses both in terms of public health and economy. This sensitizes the scientific and medical communities to focus on the safety of these products.

Due to the interest given to *L. monocytogenes* as emerging pathogens and to assess its possible role as agents of foodborne listeriosis in Morocco, several studies have been conducted on the prevalence of these bacteria in meat products (Kriem et al., 1998; Cohen et al., 2007; Ennaji et al., 2008) and in the marine environment (Bou-M'handi and El Marrakchi, 2003). On the other hand, little information is available related to the prevalence of *L. monocytogenes* in milk and dairy products available in Morocco.

The pasteurization of milk has been recognized to ensure effective consumer safety against *L. monocytogenes* (World Health Organization, 1988). However, Morocco has a wide range of traditional dairy products based on raw milk that have been known and highly appreciated by consumers for centuries. These foods have a loyal clientele, despite some apparent alterations and incidents in which consumers may be victims.

Accordingly, the aim of the present study was to investigate the prevalence of *L. monocytogenes* in raw milk and its two traditional derivatives, the *Lben* (traditionally fermented skimmed milk) and *Jben* (traditional soft white cheese) commercialized in the city of Fez in the central north of Morocco.

MATERIALS AND METHODS

Sample collection

In the period October 2009 to September 2010, a total of 288 samples of which 96 samples were raw milk, 96 *Lben* and 96 *Jben* were collected in Fez, located in the northern center of Morocco. Dairy products samples were collected from eight traditional dairies, belonging to four administrative sectors of the city (S1, S2, S3 and S4). Each dairy receives about 80 to 100 liter per day L day^{-1} from small farms located in four regions around the city (*Ras El Mae, Sidi Harazem, Boughioul/Sefrou and Al Gharb*). In order to collect information on farm characteristics, feeding practices and milk production, visits were organized to some farms. In this study, we randomly chose two dairies by sectors that were monthly sampled for a whole year. Milk sold in the traditional dairies is consumed without any heat treatment either in the production of traditional juice (fruits + raw milk) or for use in the manufacturing of traditional dairy derivatives (*Lben*, *Jben*, etc.). The samples were transported in iceboxes to Regional Laboratory of Epidemiological Diagnosis and Environmental Hygiene in Fez and analyzed immediately upon arrival.

Isolation and identification of *L. monocytogenes*

Isolation and identification of *L. monocytogenes* were carried out according to official standard ISO procedure 11290-1 (ISO, 2004).

Twenty five grams (25 g) of each sample were weighed into sterile stomacher bags, diluted with 225 ml of *Listeria* pre-enrichment Fraser broth (Biokar Diagnostics, BK115HA) homogenized and incubated at 30°C for 24 h. A 1 ml portion from this pre-enrichment culture was transferred to 9 ml enrichment broth (Complete Fraser broth, BIO-RAD), and incubated at 37°C for 24 h. A loopful of the enrichment culture was plated onto *Listeria* Ottaviani et Agosti agar [ALOA] (AES CHEMUNEX, AEB620088, Rennes, France), and PALCAM (Biokar Diagnostics, BK145F). Plates were incubated at 37°C for 24 to 48 h. Morphologically in ALOA *Listeria* grow as blue-green regular round colonies (detection of β-glucosidase by using a specific chromogenic substrate). *L. monocytogenes* also show an opaque halo which helps to easily differentiate them from other species of *Listeria*. The halo is due to the activity of a phospholipase involved in the infection process of pathogenic *Listeria*. PALCAM colonies *Listeria* are green with grayish reflections, or green with a black halo (esculin hydrolysis).

Typical colonies were taken as possible *Listeria* spp. Isolates from five suspect colonies on ALOA and PALCAM, respectively per sample were purified on tryptic soy agar with 0.6% yeast extract (Oxoid) (TSEYA) and identified using the following tests: Gram's staining, catalase reaction with 3% H_2O_2, oxidase test, motility at 25°C, test for glucose/lactose fermentation-iron utilization-gas production, fermentation of sugars (rhamnose, xyloseand mannitol), CAMP tests, methyl red-Voges Proskauer (MR-VP) reactions, nitrate reduction, indole production, urease production and β-haemolysis. Identified isolates of *L. monocytogenes* were confirmed in Institut Pasteur of Morocco with API *Listeria* (Biomerieux, 10300, Lyon, France).

Chemical analysis

The pH of the different samples was determined using a pH meter (Adwa" model "AD100"). The titratable acidity was measured by titration with 1 N NaOH in the presence of phenolphthalein and expressed in degrees Dornic according to Moroccan standard (NM 08.4.005, 2007).

Statistical analysis

Data were analyzed using statistical software package Epi Info version 6. The percentages of positive results were compared using the chi-square and Statistical significance was indicated by $P \leq 0.05$.

RESULTS

The results of the occurrence of *L. monocytogenes* and physicochemical parameters in analyzed samples of raw milk and traditional dairy derivatives are summarized in Table 1. Out of 288 samples of dairy products examined, 17 (5.90%) were found to be contaminated with *L. monocytogenes*. The prevalence of *L. monocytogenes* in raw milk, *Lben* and *Jben* samples was found to be 8.33, 5.20 and 4.16%, respectively. The average pH of analyzed samples was 6.6 for raw milk, 4.5 for *Lben* and 4.3 for *Jben*. The average titratable acidity measured in this study was 16, 80 and 95°D for raw milk, *Lben* and *Jben* respectively.

On the other hand, the prevalence of *L. monocytogenes* in samples from S1, S2, S3 and S4 was 4.16, 11.11, 2.77 and 5.55%, respectively. The highest prevalence of *L. monocytogenes* contamination was observed in autumn (11.11%) and winter (11.11%), while no positive

Table 1. Prevalence of *L. monocytogenes* isolated from raw milk and dairy derivatives.

Parameter		Raw milk	*Lben*	*Jben*	All
Number of analyzed samples		96	96	96	288
Number of positive samples	n	8	5	4	17
Incidence of *L. monocytogenes*	%	8.33	5.20	4.16	5.90
Average pH		6.6	4.5	4.3	-
Average titratable acidity (°D)		16	80	95	-

Table 2. Prevalence of *L. monocytogenes* in all samples of dairy derivatives sold in different sectors of Fez city between October 2009 and September 2010.

Sector	Collection season	Number of samples	Number of *L. monocytogenes* positive sample (%)
Sector 1	Autumn	18	2 (11.11%)
	Winter	18	1 (5.55%)
	Spring	18	0
	Summer	18	0
	Subtotal	**72**	**3 (4.16%)**
Sector 2	Autumn	18	4 (22.22%)
	Winter	18	3 (16.66%)
	Spring	18	0
	Summer	18	1 (5.55%)
	Subtotal	**72**	**8 (11.11%)**
Sector 3	Autumn	18	0
	Winter	18	2 (11.11%)
	Spring	18	0
	Summer	18	0
	Subtotal	**72**	**2 (2.77%)**
Sector 4	Autumn	18	2 (11.11%)
	Winter	18	2 (11.11%)
	Spring	18	0
	Summer	18	0
	Subtotal	**72**	**4 (5.55%)**
All		**288**	**17 (5.90%)**

samples were detected in spring (Table 2). Between seasons isolation rates of *L. monocytogenes* were statistically significant (χ^2 = 14.19; $P < 0.05$). However, the difference between sectors was statically non significant ($P > 0.05$).

DISCUSSION

This study reveals that raw milk, *Lben* and *Jben* that was consumed in the city of Fez in the period studied were contaminated with *L. monocytogenes* with an overall prevalence of 5.90%. Our results are comparable to other surveys conducted in other countries on raw milk and dairy products. Boubendir et al. (2011) reported 5.76% prevalence of *L. monocytogenes* in bovine raw milk produced in the North Eastern Algeria. Similar results found by Guerra et al. (2001), where the incidence of contamination was 5% in milk and dairy products sold in mainland Portugal. Gaya et al. (1998) reported also a low incidence of 3.6% of *L. monocytogenes* in raw milk produced in Spain. Whereas in China, the prevalence of *L. monocytogenes* in raw milk is very low (0.23 to 1.2%) (Ning et al., 2013). Common sources of *L. monocytogenes* in raw milk have been reported to be fecal (Husu, 2010) and environmental contamination during milking, storage and transport, infected cows in dairy farms and poor silage quality (Bemrah et al., 1998). *L. monocytogenes* have also been isolated from soil, decaying foliage, and bird and wildlife feces (Weis and Seeliger, 1975).

This study reveals a disparity of seasonal contamination with *L. monocytogenes* and more prevalent in one of the sectors (Table 2). These differences are probably due to seasonal and sectoral variations in the type and quality of feed given to cows. Silage was widely used as animal feed in certain zones around the city of Fez. Contamination of feeds is influenced by the conditions of preparation and storage practices.

Farmers in that area lack training in good agricultural practices for silage making, resulting in a product of poor quality (personal observation), may explain the high contamination of milk with *L. monocytogenes* in autumn and winter when silages are fed. Broseta et al. (2003) also reported that contamination of raw milk with *L. monocytogenes* is usually more common in the winter, most likely because silage feeding in many parts of the world is more common in that season. In addition, a large number of studies have indicated that clinical listeriosis in ruminants is often associated with feeding poor-quality silage (Boerlin et al., 2002).

The absence of *L. monocytogenes* in spring and low contamination in summer is most likely due to the pasture-base feeding during that period. Consistent with this hypothesis, Fenlon (1996) also found that it is uncommon for grazing animals to shed detectable levels of *L. monocytogenes*.

In addition to the initial charge of *L. monocytogenes* in raw milk used in the manufacture of traditional dairy products, the *Lben* and *Jben* can be contaminated during the manufacturing process in traditional dairies. It has been emphasized that the contamination of cheeses or dairy products with *Listeria* spp. is most likely due to post-process contamination from environmental sources and cross-contamination in the dairy plant and/or retail stores or inadequate processing (FAO/WHO, 2004) and colonization of *L. monocytogenes* in refrigerators in retail stores (Sergelidis et al., 1997). *L. monocytogenes* can survive a number of cheese-making processes and can remain viable in the final product for a considerable length period of time (Griffiths, 1989).

A lower prevalence of *L. monocytogenes* contamination in the *Lben* (5.20%) and *Jben* (4.16%) was noticed compared with raw milk (8.33%). This may be due to the acidic nature of *Lben* and *Jben* which have a low pH and high titratable acidity (Table 1), that can inhibit the growth of *L. monocytogenes*. The richness of Moroccans *Lben* and *Jben* by lactic acid bacteria (Ouadghiri, 2009) can also inhibit the growth of *L. monocytogenes*. In this context, Benkerroum et al. (2003) reported that bacteriocins produced by the lactic acid bacteria reduce counts of *L. monocytogenes* in cheese and yoghurt. Furthermore, several authors have confirmed that the growth or survival of *L. monocytogenes* in a food product depends on a variety of physico-chemical parameters, including pH, a_w and NaCl content (Conner et al., 1986; Learson et al., 1999).

Previous studies have specifically shown that *L. monocytogenes* does not grow at a pH below 5.3 when the a_w is lower than 0.93 (Tienungoon et al. 2000), or at a pH below 4.46 regardless of the a_w (Cole et al., 1990; Farber et al., 1996). In our study, the average physico-chemical parameters associated with *Lben* and *Jben* are as such that they should limit (if not prevent) the growth of *L. monocytogenes*.

In conclusion, raw milk and traditional dairy derivatives marketed in the north central region of Morocco has been contaminated with *L. monocytogenes*. The highest prevalence of contamination was recorded in autumn and winter and in an area where silage feeding is important.

The poor hygienic conditions during milking, transport, storage of milk and its use in the manufacture of *Lben* and *Jben* in traditional dairies, which do not respect the principles of food hygiene, can also be in favor of the contamination with *L. monocytogenes*. The results show that there is a risk of infection by strains of *L. monocytogenes* for consumers of raw milk and traditional dairy products marketed in the city of Fez.

So, we suggest that the general principles of food hygiene should still be enforced in order to minimize count of *L. monocytogenes* in milk and dairy products during the handling, storage and manufacturing in traditional dairies. Control of the feeding cattle and milk pasteurization can also limit the contamination with *L. monocytogenes*.

REFERENCES

Benkerroum N, Oubel H, Sandine WE (2003). Effect of nisin on yogurt starter and on growth and survival of Listeria monocytogenes during fermentation and storage of yogurt. Int. J. Food Saf. 1:1-5.

Bemrah N, Sanaa M, Cassin MH, Griffiths MW, Cerf O (1998). Quantitative risk assessment of human listeriosis from consumption of soft cheese made from raw milk. Prev. Vet Med. 37:129-145.

Boerlin P, Petzold FB, Jemni T (2002). Use of listeriolysin O and internalin A in a seroepidemiological study of listeriosis in Swiss dairy cows. J. Clin. Microbiol. 41:1055-1061.

Boubendir A, Hamidechi MA, Mostakim M, EL Abed S, Ibnsouda koraichi S (2011). Incidence de *Listeria spp.* et autres bactéries psychrotrophes dans le lait cru bovin dans le Nord Est Algérien. Revue Méd. Vêt. 162:265-269.

Boukadidda J, Sboui H, Monastiri H, Bibi M, Attia I, Snoussi N, Jeddi M (1994). La listeriose humaine en Tunisie: deux nouveaux cas chez le nouveau- ne. Med Mal Infect. 24:117-18.

Bou-M'handi N, El Marrakchi A (2003). Incidence de Listeria spp. en milieu marin au Maroc. Sciences des aliments. 23: 513-523.

Broseta S, Diot MA, Bastian S, Riviere J, Cerf O (2003). Estimation of low bacterial concentration: *Listeria monocytogenes* in raw milk. Int. J. Food Microbiol. 80:1-15.

Cohen N, Ennaji H, Bouchrif B, Hassar M, Karib H (2007). Comparative Study of Microbiological Quality of Raw Poultry Meat at Various Seasons and for Different Slaughtering Processes in Casablanca (Morocco). J. Appl. Poult. Res. 16:502-508.

Cole MB, Jones MV, Holyoak C (1990). The effect of pH, salt concentration and temperature on the survival and growth of *Listeria monocytogenes*. J. Appl. Bacteriol. 69:63-72.

Conner DE, Brackett RE, Beuchat LR (1986). Effect of temperature, sodium chloride, andpHon growth of *Listeria monocytogenes* in cabbage juice. Appl. Environ. Microbiol. 52:59-63.

Ennaji H, Timinouni M, Ennaji M, Hassar M, Cohen N (2008). Characterization and antibiotic susceptibility of *Listeria monocytogenes* isolated from poultry and red meat in Morocco. Infect. Drug Resist 1:45-50.

Farber JM, Cai Y, Ross WH (1996). Predictive modeling of the growth of *Listeria monocytogenes* in CO_2 environments. Int. J. Food Microbiol. 32:133-144.

Fenlon DR (1996). The incidence and level of *Listeria monocytogenes* contamination of food sources at primary production and initial processing. J. Appl. Bacteriol. 81:641-650.

Food and Agriculture Organization/World Health Organization (FAO/ WHO) (2004). Risk assessment of Listeria monocytogenes in readytoeat foods. Technical Reports. Microbiological Risk Assessment Series 5, Geneva.

Gaya P, Sanchez J, Medina M, Nuñez M (1998). Incidence of *Listeria monocytogenes* sand other *Listeria* species in raw milk produced in Spain. Food Microbiol.15:551-555.

Gillespie IA, Mook P, Little CL, Grant KA, Mclauchlin J (2010). Human Listeriosis in England, 2001-2007: association with neighbourhood deprivation. Eurosurveillance. 15 (27): pii=19609.

Griffiths MW (1989). Listeria monocytogenes: Its importance in dairy industry. J. Sci. Food Agric. 47:133-158.

Guerra MM, Mclauchlin JM, Bernardo FA (2001). *Listeria* in ready to eat unprocessed foods produced in Portugal. Food Microbiol. 18:423-429.

Husu JR (2010). Epidemiological Studies on the Occurrence of Listeria monocytogenes in the Feces of Dairy Cattle. Journal of Veterinary Medicine, Series B. 37: 276-282.

ISO (2004). Modification of the isolation media and the haemolysis test, and inclusion of precision data. ISO 11290-1:1996/Amd.1:2004. International Organization for Standardization. Geneva. Switzerland.

Kriem MR, El Marrakchi A, Hamama A (1998). Prevalence of *Listeria* spp. on a variety of meat products in Morocco. M.A.N. Microbiologie, Aliments, Nutr. 16:179-187.

Learson AE, Johnson EA, Nelson JH (1999). Survival of *Listeria monocytogenes* in commercial cheese brines. J. Dairy Sci. 82:1860-1868.

Ning N, Guo K, Cheng L, Xu L, Zhang C, Cui H, Cheng Y, Xu R, Liu W, Lv Q, Cao W *and* Zhang Y (2013). Pilot survey of raw whole milk in China for Listeria monocytogenes using PCR. Food Control. 31: 176-179.

NM 08.4.005 (2007). Laits et produits dérivés: Détermination de l'acidité titrable du lait. Service de Normalisation Industrielle Marocaine. SNIMA.

Ouadghiri M (2009). Biodiversité des bactéries lactiques dans le lait cru et ses dérivés « *Lben* » et « *Jben* » d'origine marocaine. THÈSE DE DOCTORAT N°: 2475. Université Mohammed V - Agdal, Faculté Des Sciences -Rabat. Maroc.

Ryser ET, Marth E (2007). Listeria, Listeriosis and Food Safety, 3rd edition, Taylor and Francis, Boca Raton, FL.

Tazi M (1981). Les listérioses humaines. Image de la forme maternofœtale. THESE. Rabat. Maroc.

Tienungoon S, Ratkowsky DA, McMeekin TA, Ross T (2000). Growth limits of *Listeria monocytogenes* as a function of temperature, pH, NaCl, and lactic acid. Appl. Environ. Microbiol. 66:4979-4987.

Sergelidis D, Abrahim A, Sarimvei A, Panoulis C, Karaioannoglou Pr, Genigeorgis C (1997). Temperature distribution and prevalence of Listeria spp. in domestic, retail and industrial refrigerators in Greece. Int. J. Food Microbiol. 34:171-177.

Weis J, Seeliger HP (1975). Incidence of *Listeria monocytogenes* in nature. Appl. Microbiol. 30: 29-32.

World Health Organization/Organisation Mondiale De La Santé (1988). Listerioses d'origine alimentaire. WHO/EHE/FOS/88.5, Geneve.

Changes in quality attributes of mixed roselle, water melon and grape juice using *Aframomum danielli* as a preservative

Ashaye O. A.[1] , Olanipekun O. T.[1], Ige E. A.[1], Farinde E. O.[1] and Akinkunmi S. O.[2]

[1]Institute of Agricultural Research and Training P.M.B 5029, Moor-Plantation Ibadan, Oyo State, Nigeria.
[2]Federal College of Agriculture, Ibadan, Oyo State, Nigeria.

Roselle, water melon and grape juice are excellent sources of vitamin C. The shelf life of most fresh fruit juices stored under room temperature is very short. The use of *Aframomum danielli* (*A. danielli*) as a preservative is not known. Roselle juices processed at 1:1 and 1:2 with or without *A. danielli* of 15% concentration were stored for two weeks at room temperature and evaluated for chemical properties and sensory attributes. Sensory evaluation of the juices were done by a 10 membered panel randomly selected from male and female adults. At zero day of storage, roselle grape juice of 1:2 ratio with *A. danielli* was significantly higher in total solids (0.637%) and ash content (0.072%). The vitamin C content of roselle-water melon juice of 1:2 ratio with *A. danielli* was higher at p<0.01. At one week of storage, there was a decrease in ash, specific gravity and vitamin C content of the juice. Roselle grape juice of 1:2 ratio with *A. danielli* was higher in ash (0.0675%). The vitamin C content of roselle-water melon of ratio 1:2 with *A. danielli* (17.125%) was significantly higher. At two weeks of storage, there was a drop in ash, specific gravity and vitamin C content of the juices. Higher sensory scores were given to roselle based sample with *A. danielli* at 0, one and two weeks of storage. Roselle grape juice of ratio 1:2 with *A. danielli* was rated high in general acceptability at zero and one week of storage. In summary roselle grape juice of ratio 1:2 with *A. danielli* had acceptable nutrient and sensory qualities at one week of storage. Beyond one week, the preservative potential of *Aframomum danielli* at 15% concentration was low.

Key words: Roselle juice, *A. danielli*, preservation, water melon juice, grape juice.

INTRODUCTION

Roselle (*Hibiscus sabdariffa* L.) is a member of the family Malvaceae to which okra, cotton and kenaf belong. The flowers of roselle are generally small. Both the leaves and the fleshy base of the flower (the calyx) are employed in the preparation of soups and sauces. The vegetable is a popular diet during the raining season. Roselle calyx is a cheap source of vegetable protein, fat and minerals. Roselle is also very rich in vitamins and minerals and an excellent source of anti oxidant which helps to reduce blood pressure (Peng-Kong et al., 2002, Fasoyiro et al., 2005). Regular consumption of roselle may reduce nutritional deficiency problems such as night blindness, scurvy and rickets (Ashaye and Adeleke, 2009, Babalola et al., 2001).

Roselle calyces are popularly used in Nigeria in the production of zobo drink. It is also used in the production of tea and jams. Water melon is a fruit rich in vitamin C and some essentials nutrient such as beta - carotene (Strandhagen et al., 2000), Grape contains polyphenols called flavonoids which lowers the risk of clogged arteries and it potentially lowers blood pressure. It is also rich in vitamin C and mineral such as calcium and potassium. *A. danielli* also known as alligator pepper can be used as flavouring in soups. *A. danielli* species has antibacterial effects and has been used in food preservation studies (Ashaye et al., 2006).

This research work on changes in quality attributes of mixed roselle, water melon and grape using using *A. danielli* as preservative is not well known. Hence this study aimed at assessing the chemical and sensory attributes of stored roselle - water melon and roselle grape juice using *A. danielli* as a preservative.

MATERIALS AND METHODS

Raw materials

Roselle calyces (*Hibiscus sabdariffa*) used for this research study were obtained from the experimental farm of Institute of Agricultural Research and Training, I.A.R.& T., Ibadan, while water melon and grape fruits were purchased from the local market.

Preparation of *Aframomum danielli* flour

Dry *A. danielli* pods was shelled and the seeds was grounded into powder and packed.

Preparation of roselle, water melon and grape juice

200 g of a dried roselle was weighed and washed in the water, it was poured inside a clean pot and three liters of water was added to it. It was boiled for ten minutes after; it was allowed to cool and sieved to obtain the juice.

Grape and water-melon juice were prepared by first peeling the grape and water-melon fruit, they were then chopped, crushed, blended and sieved. Roselle watermelon and roselle grape juices were prepared at ratios 1:1 and 1:2 with and without the addition of grounded seed flour of *Aframomum danielli* at 15% concentration.

Chemical analysis

Determination of titratable acidity (TTA)

1 ml of each sample was diluted to 100 ml, 10 ml aliquot of the diluents was pipette into a 100 ml volumetric flask and one drop of 1% phenolphthalein was added and mixed properly to give a pink colour. The mixture was titrated against 0.1N NaOH until the pink colour was discharged to a clear colourless solution at the equivalent point. The % TTA was calculated using the formula:

$$\%TTA = \frac{\text{Titre value} \times 0.1\text{N NaOH} \times \text{acid equivalent} \times \text{DF}}{\text{Volume of aliquot taken.}}$$

Where, DF = dilution factor; acid equivalent = respective acid to which the % TTA is expressed (A.O.A.C 1984).

Determination of specific gravity

10 ml of juice was measured using a 50 ml measuring cylinder into a previously weighed 10 ml beaker and weight W_1 was taken. 10 ml of water was also measured into the same 10 ml beaker of weighed W_0 and the overall weight W_2 taken on the analytical weighing balance. Specific gravity of the juice was obtained by using the formula:

$$\text{Specific gravity} = \frac{\text{Weight of 10 ml of juice } (W_1 - W_0)}{\text{Weight of equal volume of water } (W_2 - W_0)}$$

Determination of total solids

2 ml of the sample was weighed into a previously weighed crucible. The crucible plus sample taken was then transferred into the oven set at 100°C to dry to a constant weight for 24 h overnight. At the end of the 24 h, the crucible plus sample was removed from the oven for ten minutes and weighed.
If the weight of empty crucible plus sample is W_O, weight of crucible plus sample is W_1, weight of crucible plus oven dried sample is W_2.

$$\%\text{ Total solids} = \frac{W_2 - W_0}{W_1 - W_0} \times \frac{100}{1}$$

Determination of ash

20 ml of the sample were measured into a porcelain crucible. Then transferred into the muffle furnace at 550°C and left for about four hours until it has turned to white ash. The crucible and its content were cooled off to about 100°C in air, then room temperature in a desiccator and weighed. This was done in duplicate. The percentage ash was determined or calculated from the formula below:

$$\%\text{ Ash content} = \frac{\text{Weight of ash} \times 100}{\text{Original weight of sample}}$$

(A.O.A.C, 1989)

Determination of vitamin C

50 ml of the juice extract was pipette into a phosphoric acid as stabilizing agent was added and made up to the mark in water. 10ml of the solution was pipette into a 50 ml volume flask, 2.5 ml acetone was added and titrated with 5×10^{-5}g/l 2,6-dichlorophenol solution control, a faint pink colour which persist for 15 s was obtained (A.O.A.C, 1984).

Sensory evaluation

Sensory evaluation was carried out on the juice samples as described by Larmond (1977) using 10 panelists to assess the quality of the juices with regard to their colour, flavour, taste, texture and overall acceptability. The sensory scores were subjected to analysis of variance using completely randomized block design and their means were separated by Duncan multiple range test (Duncan 1955).

Table 1. Chemical Composition of Roselle based juice at zero day of storage

Sample	Titrable acidity (%)	Total solids (%)	Ash (%)	Vitamin c (mg/100 g)	Specific gravity
RJ	0.324^{b}	0.567^{g}	0.060^{f}	13.240^{j}	1.053^{b}
RW 1:1	0.314^{c}	0.538^{f}	0.050^{h}	17.060^{d}	1.062^{a}
RW 1:2	0.307^{d}	0.603^{d}	0.062^{e}	17.230^{b}	1.045^{c}
RG 1:1	0.303^{e}	0.603^{d}	0.060^{f}	14.086^{h}	1.015^{g}
RG 1:2	0.250^{h}	0.623^{b}	0.068^{b}	14.250^{f}	1.016^{f}
RJ AD	0.327^{a}	0.563^{h}	0.063^{de}	13.465^{i}	1.062^{a}
RW1:1 AD	0.284^{g}	0.694^{e}	0.057^{g}	17.140^{c}	1.053^{b}
RW1:2 AD	0.283^{g}	0.612^{c}	0.066^{g}	17.365^{a}	1.045^{c}
RG1:1 AD	0.297^{f}	0.614^{c}	0.064^{cd}	14.240^{g}	1.025^{e}
RG1:2 AD	0.220^{i}	0.637^{a}	0.072^{a}	14.365^{e}	1.027^{d}

Means in the same columns followed by the same letter are not significantly different from each other at P<0.01. RJ, Roselle juice (control); RW (1:1), roselle and water melon juice of ratio (1:1); RW (1:2), roselle and water melon juice of ratio (1:2); RG (1:1), roselle and grape juice of ratio (1:1); RG (1:2), roselle and grape juice of ratio (1:2); RJ AD, roselle juice preserved with *Aframomum danielli*; RW AD (1:1), roselle and water melon preserved with *Aframomum danielli* of ratio (1:1); RW AD (1:2), roselle and water melon preserved with *Aframomum danielli* of ratio (1:2); RG AD (1:1), roselle and grape juice preserved with *Aframomum danielli* of ratio (1:1); RG AD (1:2), roselle and grape juice preserved with *Aframomum danielli* of ratio (1:2).

Table 2. Chemical composition of roselle based juice at one week of storage.

Sample	Titrable acidity (%)	Total solids (%)	Ash (%)	Vitamin c (mg/100 g)	Specific gravity
RJ	0.352^{a}	0.560^{d}	0.056^{e}	12.575^{i}	0.999^{b}
RW 1:1	0.317^{c}	0.581^{f}	0.046^{g}	15.485^{d}	0.957^{e}
RW 1:2	0.290^{c}	0.592^{bc}	0.058^{d}	16.925^{b}	0.983^{d}
RG 1:1	0.307^{d}	0.596^{bc}	0.052^{f}	13.700^{h}	0.999^{b}
RG 1:2	0.254^{f}	0.617^{a}	0.063^{b}	13.915^{f}	0.999^{b}
RJ AD	0.331^{b}	0.557^{d}	0.060^{c}	12.125^{j}	0.991^{c}
RW1:1 AD	0.304^{d}	0.590^{bc}	0.053^{f}	16.855^{c}	1.00^{a}
RW1:2 AD	0.286^{c}	0.605^{cb}	0.062^{b}	17.125^{a}	0.991^{c}
RG1:1 AD	0.302^{d}	0.604^{ab}	0.058^{d}	13.850^{g}	0.999^{b}
RG1:2 AD	0.232^{g}	0.615^{a}	0.068^{a}	14.075^{e}	1.008^{a}

Means in the same columns followed by the same letter are not significantly different from each other at P<0.01. RJ, Roselle juice (control); RW (1:1), roselle and water melon juice of ratio (1:1); RW (1:2), roselle and water melon juice of ratio (1:2); RG (1:1), roselle and grape juice of ratio (1:1); RG (1:2), roselle and grape juice of ratio (1:2); RJ AD, roselle juice preserved with *Aframomum danielli*; RW AD (1:1), roselle and water melon preserved with *Aframomum danielli* of ratio (1:1); RW AD (1:2), roselle and water melon preserved with *Aframomum danielli* of ratio (1:2); RG AD (1:1), roselle and grape juice preserved with *Aframomum danielli* of ratio (1:1); RG AD (1:2), roselle and grape juice preserved with *Aframomum danielli* of ratio (1:2).

Statistical analysis

Data was subjected to analysis of variance and their means were separated by Duncan multiple range test Duncan 1955.

RESULTS AND DISCUSSION

Table 1 shows that roselle grape juice of ratio 1:2 with *A. danielle* at zero day of storage was significantly higher in total solids (0.639%) and ash content (0.072%). The higher content of total solids and ash may be due to presence of higher molar mass compounds such as pectin and starch. (Katia et al., 2013).

The vitamin C content of roselle water melon juice of 1:2 ratio with *A. danielle* was higher at p<0.01 when compared with others. This variation may be due to time of harvest, maturation degree and varietal differences (Katia et al., 2013). However, these values are within the range reported for most juices (Aina and Shodipe, 2006, Peng-Kong et al., 2002).

At one week of storage (Table 2), there was a decrease in ash, specific gravity and vitamin C contents of the

Table 3. Chemical composition of roselle based juice at two weeks of storage.

Sample	Titrable acidity (%)	Total solids (%)	Ash (%)	Vitamin c (mg/100 g)	Specific gravity
RJ	0.463^{e}	0.345^{d}	0.051^{ab}	12.265^{i}	0.989^{a}
RW 1:1	0.568^{d}	0.338^{d}	0.037^{c}	14.840^{d}	0.955^{a}
RW 1:2	0.578^{c}	0.306^{g}	0.047^{b}	16.220^{b}	0.987^{a}
RG 1:1	0.314^{h}	0.577^{c}	0.046^{c}	13.285^{h}	0.985^{a}
RG 1:2	0.317^{g}	0.590^{b}	0.056^{a}	13.740^{f}	0.978^{a}
RJ AD	0.457^{f}	0.326^{c}	0.053^{ab}	12.025^{j}	0.985^{a}
RW1:1 AD	0.584^{b}	0.321^{ef}	0.044^{bc}	16.100^{c}	0.958^{a}
RW1:2 AD	0.597^{a}	0.313^{fg}	0.056^{a}	16.655^{a}	0.988^{a}
RG1:1 AD	0.295^{j}	0.582^{bc}	0.050^{b}	13.565^{g}	0.987^{a}
RG1:2 AD	0.308^{i}	0.600^{a}	0.059^{a}	14.005^{e}	0.987^{a}

Means in the same columns followed by the same letter are not significantly different from each other at P<0.01. RJ, Roselle juice (control); RW (1:1), roselle and water melon juice of ratio (1:1); RW (1:2), roselle and water melon juice of ratio (1:2); RG (1:1), roselle and grape juice of ratio (1:1); RG (1:2), roselle and grape juice of ratio (1:2); RJ AD, roselle juice preserved with *Aframomum danielli*; RW AD (1:1), roselle and water melon preserved with *Aframomum danielli* of ratio (1:1); RW AD (1:2), roselle and water melon preserved with *Aframomum danielli* of ratio (1:2); RG AD (1:1), roselle and grape juice preserved with *Aframomum danielli* of ratio (1:1); RG AD (1:2), roselle and grape juice preserved with *Aframomum danielli* of ratio (1:2).

Table 4. Sensory evaluation of roselle based juice at zero day of storage.

Sample	Titrable acidity (%)	Total solids (%)	Ash (%)	Vitamin c (mg/100 g)	Specific gravity
RJ	6.400^{bc}	4.000^{a}	3.700^{b}	4.100^{d}	5.200^{c}
RW 1:1	5.500^{cd}	4.300^{b}	4.400^{b}	4.300^{d}	5.000^{c}
RW 1:2	4.600^{d}	3.900^{b}	4.100^{b}	3.800^{d}	4.400^{c}
RG 1:1	5.000^{cd}	4.300^{b}	4.200^{b}	5.000^{cd}	5.100^{c}
RG 1:2	4.500^{d}	4.300^{b}	4.300^{b}	5.200^{abc}	5.400^{bc}
RJ AD	8.000^{a}	7.000^{a}	7.100^{a}	6.800^{a}	7.400^{a}
RW1:1 AD	7.100^{ab}	7.100^{a}	6.900^{a}	6.700^{a}	7.000^{a}
RW1:2 AD	6.400^{bc}	6.800^{a}	6.900^{a}	6.500^{ab}	6.600^{ab}
RG1:1 AD	7.500^{ab}	7.200^{a}	7.300^{a}	6.100^{ab}	7.700^{a}
RG1:2 AD	7.100^{ab}	7.200^{a}	7.700^{a}	6.100^{ab}	7.200^{a}

Means in the same columns followed by the same letter are not significantly different from each other at P<0.01. RJ, Roselle juice (control); RW (1:1), roselle and water melon juice of ratio (1:1); RW (1:2), roselle and water melon juice of ratio (1:2); RG (1:1), roselle and grape juice of ratio (1:1); RG (1:2), roselle and grape juice of ratio (1:2); RJ AD, roselle juice preserved with *Aframomum danielli*; RW AD (1:1), roselle and water melon preserved with *Aframomum danielli* of ratio (1:1); RW AD (1:2), roselle and water melon preserved with *Aframomum danielli* of ratio (1:2); RG AD (1:1), roselle and grape juice preserved with *Aframomum danielli* of ratio (1:1); RG AD (1:2), roselle and grape juice preserved with *Aframomum danielli* of ratio (1:2).

juices with concomitant increase in their titratable acidity. This observation may be due to utilization of soluble solids by yeast cells during fermentation (Idolo et al., 2012; Omemu et al., 2006, Doughari et al., 2007; Braide et al., 2012).

Roselle-grape juice of 1:2 ratio with *A. danielle* was higher in ash (0.0675%). However there was no significant difference in the total solids of roselle -grape juice of ratio 1:2 with *A. danielli.* This finding may be due to antibacterial property of *A. danielli.*(Ashaye, 2006). The vitamin C content of Roselle -water melon of ratio 1:2 with *A. danielli* was significantly higher with a value of (17.125%). At two weeks of storage (Table 3), there was a further drop in ash, specific gravity and vitamin C contents of the juices. Also, titratable acidity content increased in all the juice samples. Increase in titratable acidity may be due to the increased activities of microorganisms resulting in production of organic acids (Ojokoh et al., 2002).

The ash contents of roselle -grape juice of 1:2 with *A. danielle* and roselle water melon juice of 1:2 ratio were not significantly different from each other at p<0.01. The vitamin C content of roselle -water melon of ratio 1:2 with *A. danielli* was the highest. At zero day of storage, roselle-grape juice of ratio 1:2 with *A. danielli* was also rated high in general acceptability (7.20) (Table 4).

Table 5. Sensory evaluation of roselle based juice at one week of storage.

Sample	Titrable acidity (%)	Total solids (%)	Ash (%)	Vitamin c (mg/100 g)	Specific gravity
RJ	6.900^{a}	4.300^{b}	3.800^{cd}	4.800^{b}	4.400^{c}
RW 1:1	5.800^{a}	4.800^{b}	3.700^{cd}	5.200^{b}	4.400^{bc}
RW 1:2	5.800^{a}	4.400^{b}	3.900^{bcd}	4.800^{b}	5.600^{ab}
RG 1:1	3.700^{b}	2.600^{c}	2.500^{de}	3.300^{c}	3.800^{c}
RG 1:2	3.300^{b}	2.100^{c}	2.000^{e}	2.500^{c}	2.400^{d}
RJ AD	7.100^{a}	5.400^{ab}	5.500^{ab}	6.100^{ab}	6.700^{a}
RW1:1 AD	6.800^{a}	5.400^{ab}	5.300^{abc}	6.000^{ab}	6.100^{a}
RW1:2 AD	6.000^{a}	5.800^{ab}	5.300^{abc}	5.600^{b}	6.400^{a}
RG1:1 AD	6.900^{a}	6.500^{a}	6.700^{a}	7.100^{a}	6.900^{a}
RG1:2 AD	6.000^{a}	5.700^{ab}	6.200^{a}	6.200^{ab}	5.900^{a}

Means in the same columns followed by the same letter are not significantly different from each other at P<0.01. RJ, Roselle juice (control); RW (1:1), roselle and water melon juice of ratio (1:1); RW (1:2), roselle and water melon juice of ratio (1:2); RG (1:1), roselle and grape juice of ratio (1:1); RG (1:2), roselle and grape juice of ratio (1:2); RJ AD, roselle juice preserved with *Aframomum danielli*; RW AD (1:1), roselle and water melon preserved with *Aframomum danielli* of ratio (1:1); RW AD (1:2), roselle and water melon preserved with *Aframomum danielli* of ratio (1:2); RG AD (1:1), roselle and grape juice preserved with *Aframomum danielli* of ratio (1:1); RG AD (1:2), roselle and grape juice preserved with *Aframomum danielli* of ratio (1:2).

Table 6. Sensory evaluation of roselle based juice at two weeks of storage.

Sample	Titrable acidity (%)	Total solids (%)	Ash (%)	Vitamin c (mg/100 g)	Specific gravity
RJ	4.900^{c}	3.600^{ab}	3.700^{a}	4.000^{c}	3.200^{cd}
RW 1:1	5.000^{bc}	2.700^{bcd}	2.200^{cd}	2.500^{bcd}	2.200^{d}
RW 1:2	5.000^{bc}	2.900^{abcd}	2.400^{bcd}	2.900^{abcd}	2.500^{cd}
RG 1:1	1.800^{e}	1.800^{d}	1.800^{d}	1.900^{d}	2.100^{d}
RG 1:2	1.500^{e}	2.100^{cd}	2.000^{cd}	1.800^{d}	2.700^{cd}
RJ AD	5.900^{abc}	4.200^{a}	3.600^{a}	3.700^{a}	3.700^{abc}
RW1:1 AD	6.000^{ab}	3.200^{abc}	3.400^{ab}	3.500^{abc}	4.800^{a}
RW1:2 AD	6.700^{a}	4.200^{a}	3.600^{a}	3.700^{ab}	4.600^{ab}
RG1:1 AD	3.200^{d}	2.900^{abcd}	2.600^{abcd}	2.500^{bcd}	3.500^{bc}
RG1:2 AD	3.600^{d}	3.000^{abcd}	3.000^{abc}	2.400^{cd}	3.100^{cd}

Means in the same columns followed by the same letter are not significantly different from each other at P<0.01. RJ, roselle juice (control); RW (1:1), roselle and water melon juice of ratio (1:1); RW (1:2), roselle and water melon juice of ratio (1:2); RG (1:1), roselle and grape juice of ratio (1:1); RG (1:2), roselle and grape juice of ratio (1:2); RJ AD, roselle juice preserved with *Aframomum danielli*; RW AD (1:1), roselle and water melon preserved with *Aframomum danielli* of ratio (1:1); RW AD (1:2), roselle and water melon preserved with *Aframomum danielli* of ratio (1:2); RG AD (1:1), roselle and grape juice preserved with *Aframomum danielli* of ratio (1:1); RG AD (1:2), roselle and grape juice preserved with *Aframomum danielli* of ratio (1:2).

However, higher sensory scores were given to roselle based sample with *A. danielli*. *A. danielli* is known to have a good flavor (Fasoyiro et al., 2005).

At one week of storage (table 5) roselle base juice with *A. danielle* was also rated highest in sensory parameters. This could be due to preservative potential of *A. danielle* (Ashaye et al., 2006). At two weeks (Table 6) acceptability decreased markedly with roselle based juice without *A. danielli* having the least. The colour of roselle-water melon of ratio 1:2 with *A. danielli* was significantly higher than the rest.

Conclusion

Roselle-grape juice of ratio 1:2 with *A. danielli* had acceptable nutrient and sensory qualities at one week of storage. Beyond one week, the preservative potential of *A. danielli* at 15% concentration was low.

REFERENCES

A.O.A.C (1989). Association of analytical chemists, Official methods of analysis (14th edition) Washington DC USA. A.O.A.C (1984).

Association of analytical chemists, Official methods of analysis (13th edition) Washington DC USA.
Aina JO, Shodipe AA (2006). Colour stability and vitamin C retention of roselle juice (Hibiscus Sabdariffa L) in different packaging materials. Nutr. Food Sci. 36:2:90-95.
Ashaye OA, Adeleke TO (2009). Quality attributes of stored Roselle Jam. Int. Food Res. J. 16:363-371.
Ashaye OA, Taiwo OO, Adegoke GO (2006): Effect of local preservative (*Aframomum danielli*) on the chemical and sensory properties of stored warakanshi Afr. J. Agric. Res. 1:10-16.
Babalola SO, Babalola AO, Aworh CO (2001). Compositional attribute of the calyces of Roselle (Hibiscus sabdariffa var sabdariffa L). J Food Technol. Afr. 6:4:133-134.
Braide W, Oranusi S, Peter-Ikechukwu AI (2012). Perspectives in the hurdle techniques in the preservation of a non alcoholic beverage, zobo. Afr. J. Food Sci. Technol. 3:2:46-52.
Doughari JH, Alabi G, Elmahmood AM (2007). Effect of some chemical preservatives on the shelf life of sobo drink. Afr. J. Microbiol. Res. 2:37-41.
Duncan DB (1955). Multiple range and Multiple F test. Biometrics 11:1:1-5.
Fasoyiro SB, Babalola SO, Owosinbo T (2005) chemical composition and sensory of quality of fruit flavoured roselle (Hibiscus sabdariffa) drinks. World J. Agric. Sci. 1:161-164.
Idolo I, Taiwo OO, Aina JO (2012). Production and quality attributes of vegetable wine from Hibiscus sabdariffa Lim. Afr. J. Food Sci. 6:7:212-215.
Katia R, Rodrigo RP, Silvia B, Claudimir AC, Jose CCP (2013). Effects of tangential microfiltration and pasteurization on the rheological, microbiological, physic-chemical and sensory characteristics of sugar cane juice. Int. J. Food Sci. Technol. 48:1-9.
Larmond E (1977). Laboratory method for sensory evaluation of foods Research Branch Canadian Department of Agriculture Publication. 1637:56-59
Ojokoh AO, Adetuyi FC, AkinyosoyeFA and Oyetayo VO (2002): Fermentation studies of Roselle (Hibiscus Sabdariffa) calyces neutralized with Trona. J. Food Technol. Afr. 7:3:75-78.
Omemu AM, Edema MO, Atayese AO, Obadina AO (2006). A survey of the microflora of H.ibiscus Sabdariffa (Roselle) and the resulting Zobo juice. Afr. J. Biotechnol. 5:3:254-259.
Peng-Kong W, Salmah Y, Ghazali YB, Che M (2002). Physico-chemical characteristics of roselle (L). Nutr. Food Sci. 32:2:68-73.
Strandhagen C, Hansson RO, Bosaeus I, Isakssun B., Eriksson H (2000). High fruit intake may reduce mortality among middle aged and elderly men. Eur. J. Clin. Nutr. 54:337-341.

17

Some quality parameters of crude palm oil from major markets of Douala, Cameroon

Ngando-Ebongue Georges Frank[1], Mpondo-Mpondo Emmanuel Albert[2] and Ewane Marcelle Astride[2]

[1]IRAD, Centre for Oil Palm Research of La Dibamba, P.O. Box 243, Douala, Cameroon.
[2]Department of Biochemistry, Faculty of Science, University of Douala, P.O. Box 24 157, Douala, Cameroon.

Palm oil is the major cooking oil in Cameroon, and crude palm oil (CPO) is an essential ingredient of many local recipes. However, the quality of this foodstuff is subject to doubt, considering the very often inadequate conditions of its extraction and storage in the fast growing small holder sector. In order to assess the quality of (CPO) available to Cameroonian consumers, 40 samples of this product were randomly collected in 10 major markets of the city of Douala and assayed through three physico-chemical parameters, namely oil acidity, peroxide value (PV) and moisture content. Results of this study showed that the quality of CPO samples was good regarding PV, as all values were below the 15 meqO_2/kg maximal limit. However, data from oil acidity determination were cause for concern, as 50% of the samples had values above the 5% acidity maximal limit for CPO. The worst was for moisture content which had all values above the 0.2% maximal limit. Considering that high moisture content inevitably enhances deterioration reactions involved in food spoilage, one can easily conclude that the quality of the CPO samples assessed was quite poor but above all could rapidly undergo further deterioration upon storage.

Key words: Oil palm, crude palm oil, food safety, quality, food deterioration.

INTRODUCTION

With an annual global production equating to about 39% of world production of vegetable oils (oil world, 2011), palm oil has outclassed soybean during the last decade to become the most important oil crop in the world. In Cameroon, palm oil accounts for about 90% of edible oil needs (Hirsch, 1999). Due to long term local eating habits, cheaper cost, and the presence of an important none industrial sector representing about 30% of total local production, palm oil is extensively used in its crude form (crude palm oil or CPO) for food purposes in Cameroon and also throughout Central and West African regions. This can be nutritionally beneficial, as CPO is a rich source of some essential nutrients such as carotenes, tocopherols and tocotrienols. Carotenes (mainly β-carotenes) are the most important vitamin A precursors in human nutrition, and deficiency of the latter can cause visual disorders. The β-carotene content in CPO is 15 fold higher than in carrot and 300 fold higher than in tomato, making CPO the richest source of this nutrient (Choo, 1994).

Tocopherols and tocotrienols are both precursors of vitamin E which plays important roles in the human reproductive system. The content of these two vitamins in palm oil is partially lost as a result of refining process. It has been reported that refined palm oil (RBD) retain approximately only 70% of the level of vitamin E in CPO, whereas vitamin A is almost entirely destroyed by the refining process (Sambanthamurthi et al., 2000). As a result, CPO has been widely used in many developing countries in supplementary feeding programmes (Latham, 1979; Tan et al., 1991; Qureshi et al., 1991; Gopalan et al., 1992; Seshadri, 1996). Many studies have also outlined the potential effect of tocotrienols on the resorption of tumour cells (Carrol et al., 1995; Guthrie et al., 1997a, b).

However, the consumption of CPO can also be detrimental to human beings, as CPO contains some components which are likely to enhance numerous reactions (hydrolysis, oxidation, etc.) involved in the degradation of this product. Moreover, these degradation reactions can also be initiated and/or accentuated by poor transportation and storage conditions of the product as it is generally the case among small holders. The most effective degradation process of CPO is acidification which was already mentioned by Desassis in 1957. Generally, fatty acids are present in oils as part of triacylglycerol molecules. The presence of free fatty acid molecules is an indication of the impairment of the quality of oils, as FFA are liberated from the triacylglycerol molecules under the action of enzymes from the acylglycerol hydrolase family, notably lipases and esterases. Acidification is generally assessed through determination of oil acidity or FFA content which is one of the most important criterion for determining the quality of virgin cooking oils.

Another degradation reaction of CPO regarding food safety is lipid peroxydation. This process involves unsaturated fatty acids moieties which undergo a chain reaction mechanism involving free radicals as intermediates and generating lipid peroxides as end products. The latter undergo additional chain cleavage at the level of the hydroperoxide group to form secondary oxidation products such as short chain aldehydes and products bearing ketone, epoxy or alcohol groups responsible for the rancid smell and taste of the oil. The determination of peroxide value (PV) gives an indication of the level of lipid peroxidation of cooking oils.

Alongside oil acidity and PV, moisture content is also one of the most important criterion for determining the quality of cooking oils and fats regarding food safety, as the presence of residual water is not a concern per see, but water is a catalyst of almost all chemical degradation reactions.

Based on the determination of these physicochemical parameters, previous studies by Ngando et al. (2011) showed that CPO from small holders' extraction sites was of lesser quality as compared to that from industrial oil mills regarding food safety. Considering the fact that about 30% of Cameroon's national production of CPO is provided by small holders, one can assume that the quality of this product which is freely available in local markets is subject to doubt.

The present study aimed to assess the quality (regarding food safety) of CPO available in the major markets of the city of Douala in the Republic of Cameroon. For this purpose, the three above-mentioned physicochemical parameters were chosen, namely acidity, moisture content and peroxide value. We decided to use them for our study as they were routinely used worldwide for the assessment of the quality of cooking oils.

MATERIALS AND METHODS

Collection of palm oil samples

CPO samples were collected in ten major markets of the city of Douala. Markets were chosen so as to provide equitable representation of all the geographical areas of the city. Market 10 is the main market of the city, whereas all the other could be ranged in the same category and be considered as of lesser importance. For each market, four samples were collected at seven days interval among vendors randomly chosen. The samples were collected in 200 ml screw capped glass flasks filled to the maximum, transported at room temperature and processed prior to laboratory analyses according to standard norms applicable in Cameroon (AFNOR, 1988). All samples were analyzed within 24 h after sampling. Additional information was also collected from the vendors regarding the origin of the CPO samples (industrial oil mill or from small holders). One industrial CPO sample collected at the oil mill was also used in this study as control.

Physiochemical analyses

For each sample, PV, free fatty acids (FFA) and moisture contents were assayed. Peroxide value was determined by titrating a chloroform/acetic acid/saturated KI solution of the oil with an aqueous solution of sodium thiosulfate using starch as colour indicator and expressed in Meq0_2/kg. For the FFA content, determination was performed by titrating the alcoholic solution of the oil with a 0.1 N solution of sodium hydroxide using phenolphthalein and alkaline blue as colour indicators. The FFA content was expressed as a percent of palmitic acid. Moisture content was determined through the gravimetric method by oven drying oil samples at 105°C to constant weight. All physicochemical analyses were performed according to AFNOR (1988) methods.

Statistical analysis

Each parameter was determined in triplicate for each sample, and average values for each market were expressed as mean ± SD. Kruskall-Wallis ANOVA were used for multiple comparisons. The Beauvais-Pearson test was used to assess the degree of linear dependence amongst the various parameters assessed.

RESULTS AND DISCUSSION

Result from Tables 1, 2 and 3 show an important variability for each of the three parameters assessed, with variation coefficients of 49, 59 and 72% for oil acidity, PV and moisture content, respectively. This variability was observed on samples collected within and amongst markets. Broadly, values ranged from 0.64 to 10.28% for oil acidity, 0.4 to 10.89 meq O_2/kg for PV and 0.2 to 3.6% for moisture content, respectively. No significant correlation was found amongst the three parameters assessed.

Table 1. Oil acidity measurements for the 41 samples of the study.

Market	M1	M2	M3	M4	M5	M6	M7	M8	M9	M10	Control (oil mill)
Sample 1	10.28	*5.14*	*3.22*	6.16	6.6	*6.85*	6.51	*0.64*	1.19	*1.45*	*0.55*
Sample 2	6.44	*3.58*	*7.35*	5.42	5.09	*7.27*	6.27	*5.00*	5.46	*6.54*	
Sample 3	*2.08*	2.21	1.28	*2.38*	*4.06*	2.76	*1.83*	1.06	*4.65*	4.98	
Sample 4	4.65	*5.36*	*5.8*	4.53	6.52	*5.59*	1.37	*6.01*	4.91	*2.63*	
Mean ± SD	5.86 ± 3.45	4.07 ± 1.47	4.41 ± 2.69	4.62 ± 1.64	5.57 ± 1.22	5.62 ± 2.03	4 ± 2.77	3.18 ± 2.72	4.05 ± 1.94	3.90 ± 2.29	0.55

Acidity was determined by titration with 0.1 N NaOH solution and expressed as percentage palmitic acid. Samples from industrial oil mills (according to information from vendors) are in italics. The control sample was collected directly at the mill.

Table 2. Peroxide value (PV) measurements for the 41 samples of the study.

Markets	M1	M2	M3	M4	M5	M6	M7	M8	M9	M10	Control (oil mill)
Sample 1	0.4	*1.31*	*3.18*	2.88	3.56	*2.63*	9.57	*2.99*	2.22	*9.76*	*1.11*
Sample 2	4.79	*4.20*	*5.61*	2.74	5.11	*4.47*	5.71	*0.92*	1.25	*1.57*	
Sample 3	*2.77*	6.5	5.99	*7.7*	*5.84*	3.42	*5.42*	3.37	*3.28*	6.31	
Sample 4	1.31	*5.22*	*5.26*	8.87	4.61	*5.75*	1.3	*3.45*	1.13	*10.89*	
Mean ± SD	2.32 ± 1.92	4.31 ± 2.21	5.01 ± 1.26	5.55 ± 3.2	4.78 ± 0.96	4.07 ± 1.35	5.50 ± 3.38	2.68 ± 1.19	1.97 ± 1	7.13 ± 4.19	1.11

PV was determined by titration with 0.002 N sodium thiosulfate solution and expressed in Meq 0_2/kg. Samples from industrial oil mills (according to information from vendors) are in italics. The control sample was collected directly at the mill.

Table 3. Moisture content determination for the 41 samples of the study.

Markets	M1	M2	M3	M4	M5	M6	M7	M8	M9	M10	Control (Oil mill)
Sample 1	0.26	*0.33*	*0.27*	0.40	0.86	*0.60*	0.20	*0.20*	1.46	*0.53*	*0.067*
Sample 2	0.80	*0.60*	*0.47*	0.73	0.20	*1.07*	0.27	*0.20*	0.33	*1.27*	
Sample 3	*0.73*	1.40	0.93	*1.27*	*0.73*	1.47	*1.53*	0.53	*1.33*	1.47	
Sample 4	0.27	*1.06*	*1.26*	1.40	0.73	*1.33*	0.20	*0.73*	0.20	*3.06*	
Mean ± SD	0.52 ± 0.29	0.85 ± 0.48	0.73 ± 0.45	0.95 ± 0.47	0.63 ± 0.29	1.12 ± 0.38	0.55 ± 0.65	0.42 ± 0.26	0.83 ± 0.66	1.58 ± 1.06	0.067

Moisture content was determined through the gravimetric method by oven drying oil samples at 105°C to constant weight and expressed as percentage. Samples from industrial oil mills (according to information from vendors) are in italics. The control sample was collected directly at the mill.

Oil acidity

Regarding oil acidity (Table 1), the highest single value (10.28%) was recorded on a sample from the market 1, as well as the highest average value (5.86%). On the other hand, the lowest single value (0.64%) was noticed in a sample from market 8 as well as the lowest average value (3.18%). From the 40 samples assessed, 20 had oil acidity values above the 5% maximal limit allowed for crude palm oil samples (Codex Alimentarius/FAO/WHO norms, 2011), but only 3 of the 10 markets assessed had mean values above that maximal limit. Markets 5 and 6 recorded the highest percent of poor quality samples (3 out of 4), thus appearing as the most unsafe regarding oil acidity. Meanwhile markets 8, 9 and 10 also appeared as the ones with the lowest percentage of poor quality samples (1 out of 4).

These results also show that values above the 5% norm were recorded in all the 10 markets assessed, and ANOVA tests showed no significant difference among mean values from different markets, thus confirming the overall poor quality of CPO sold in Douala markets regarding this parameter. FFAs are liberated from the triacylglycerol molecules under the action of enzymes from the acylglycerol hydrolase family, notably lipases and esterases.

Many studies have pointed out the presence of a very active endogenous lipase in the mesocarp of the fruit of the oil palm (Desassis, 1957; Abigor et al., 1985; Henderson and Osborne, 1990; Sambanthamurthi et al., 1995; Ngando et al., 2006). This lipase is activated in the fruit at maturity upon wounding and/or bruising and is responsible for the hydrolysis of triglycerides and the liberation of FFA. In order to limit the action of the lipase, fresh fruit bunches (FFB) must be processed rapidly after harvest. This is the case in industrial oil mills where the harvested FFB are generally steam sterilized rapidly or at the very worst less than two days after harvest so as to inactivate the lipase, thus limiting subsequent FFA accumulation in the resultant CPO. These facts are confirmed in our study by data of the control sample from industrial oil mill (SOCAPALM) whose FFA content is 0.55%, that is, far below the overall mean value of all the samples assessed (4.53%). These results are in accordance with those of Ngando et al. (2011) who already pointed out the poor quality of CPO produced by smallholders in Cameroon regarding oil acidity. Though CPO from industrial oil mills is of better quality, it is mostly dedicated to oil refineries and soap factories. As a result, most of the CPO sold in markets and dedicated to local consumption is produced by smallholders, as recent estimates consider the latter to provide about 30% of Cameroon national CPO production (Hoyle and Levang, 2012). Even in case CPO from industrial oil mills is destined to markets in Douala, poor packaging, transport and storage conditions of the product in force among smallholders negatively impact the quality and contribute to enhance acidity.

Peroxide value

As for PV (Table 2), the highest value (10.89 meq/kg) was recorded in a sample from market 10, whereas the lowest (0.4 meq/kg) belong to market 1. The highest average value was also for the market 10 (7.13 meq/kg), and the lowest for market 9 (1.97 meq/kg). The overall quality of the CPO samples assessed within this study was quite good regarding this parameter, as all the forty samples had PV below the 15 meq/kg maximal limit for cold pressed and virgin oils (Codex Alimentarius/FAO/WHO, 2011). The average mean value for the 40 samples of the study (4.34 meq/kg) was also below this 15 meq/kg maximal limit. ANOVA tests showed no significant difference among mean values from different markets. PV is used to assess the quality of cooking oils and fats through the measurement of the amount of lipid peroxides and hydroperoxides formed during the initial stages of oxidative degradation and thus, estimate to which extent spoilage of the oil has advanced. This parameter was chosen for our study as it is routinely used during food security controls to assess the quality of cooking oils. However, these peroxides are very unstable transitory products, and the determination of PV might not necessarily provide a correct estimate of the level of peroxidation, as it gives accurate information on the amount of peroxides and hydroperoxides but not on the secondary oxidation products.

Moreover, as all chemical processes involving chain reaction, peroxidation is a very dynamic process, and the amount of peroxides might change considerably within a short period of time in favourable conditions such as the presence of oxygen, sunlight, metallic ions or high moisture content. In this regard, the absence of samples with high PV in our study must be interpreted with care, as all the samples had high moisture content which is an enhancing factor for peroxidation. From another point of view, the overall low PV of CPO samples in our study is in the nature of things, as this oil is known to have a high oxidative stability. Thanks to its fatty acid composition with a balanced ratio of polyunsaturated/saturated fatty acids (51:49 w/w), CPO is less susceptible to oxidation and is widely used for frying of food (Matthaüs, 2007). Beside these visible harmful effects on the sensory quality of the oil, peroxidation also makes the oil dangerous for human health, as the free radicals generated by this process are proven to be carcinogenic (Rossel, 1999).

Moisture content

The highest value for moisture content (Table 3) was recorded in a sample from market 10 (3.06%), as well as the highest group average (1.58%). All the samples in our study had values above the 0.2% maximal limit for volatile matters at 105°C in oils and fats (Codex Alimentarius/FAO/WHO, 2011), and the overall average mean was up to 0.82%. ANOVA tests showed no significant difference among mean values from different markets. Water is an unusual component of oils and fats, as the two are non-miscible and the presence of water can be compatible only at very low proportions. However, even very low moisture contents can prove to be harmful to oils and fats products, as the presence of residual water is not a concern per see, but water is a catalyst of almost all chemical degradation reactions. In this regard, assaying the moisture content generally provides a good indication of the level of the other quality parameters and can also prove very helpful to forecast subsequent variation upon storage. As said earlier, the presence of high moisture content enhances oxidative degradation and thus PV. However, the harmful effect of high moisture content on CPO is more effective regarding oil acidity.

Though FFA accumulation in CPO is principally due to the action of the endogenous lipase of the mesocarp of the fruit of the oil palm, oil acidity can still be enhanced by other factors well after the lipase have been inactivated by the sterilization step during the extraction process. Thus, FFA can be formed in CPO through the action of contaminating lipases from microorganisms (Hiol et al., 1999; Houria et al., 2002; Tagoe et al., 2012). Moreover, in the presence of moisture content above the critical limit of 0.2%, FFA are also formed by a chemical process called autocatalytic hydrolysis where the FFA moieties initially present act as catalysts and highly enhance subsequent formation of other FFA (Loncin, 1952; 1956; 1965; Loncin and Jacobsberg, 1965; Chooi et al., 2006). Therefore, it is very important for CPO samples to maintain moisture content below this 0.2% critical limit, as the action of contaminating microbial lipases and autocatalytic hydrolysis is then very unlikely. In this regard, data from our study raised a real concern for CPO available in Douala markets regarding food safety, as all the samples assayed were above the 0.2% moisture content maximal limit. In such conditions, it is likely that these CPO samples already improper for consumption rapidly undergo additional degradation during storage.

Previous studies have demonstrated that oil acidification and peroxidation processes are significantly enhanced by high moisture content of CPO samples at the outset (Ngando et al., 2011). Analyses showed there were no differences amongst samples for the three parameters assessed regarding the size of the market or information provided by the vendor on the origin, thus strongly suggesting that the CPO (randomly) sampled in our study were probably all processed by smallholders, as the control sample from an industrial oil mill had moisture content far below the 0.2% maximal limit.

Data from our study were generally in accordance with those obtained by Aletor et al. (1990), Onwuka and Akaerue (2006) and Orji and Mbata (2008) in similar studies on Nigerian crude palm oil from small holders. For all the three parameters assessed, the overall important variability observed on data within and between markets is a clear indication of the volatility of quality indicators of CPO from small holders. This could be explained by the fact that the small holding sector is very fragmented, and CPO stockists generally have to collect from numerous suppliers in order to build up substantial stock.

Conclusion

Results from this study provided clear indication of the quality of CPO available in the markets of the city of Douala regarding food safety. It appeared that oil acidity was a real concern, as almost 50% of the samples assayed did not comply with recognized norms. On the other hand, all the samples assayed were in compliance with norms regarding PV, thus confirming the good oxidative stability of CPO. However, the most serious concern was raised by the assessment of moisture content, as all samples had values above the maximal limit and both oil acidity and PV are enhanced by high moisture content. These results clearly point out the overall poor quality of CPO available in Douala markets and the need to raise public awareness on the existence of norms related to food products as well as the necessity of CPO to comply with them. As CPO is a major component of local diets in Cameroon, and considering the increasing proportion of national production allocated to non-industrial oil extraction, it is obvious that despite its beneficial nutritional properties, this poor quality CPO may rapidly become a real concern regarding food safety.

REFERENCES

Abigor DR, Opoku RA, Opute FI, Osagie AU (1985). Partial purification and some properties of the lipase present in oil palm (*Elaeis guineensis*) mesocarp. J. Sci. Food Agric. 36:599-606.

AFNOR (1988). Recueil des normes françaises sur les corps gras, graines oléagineuses, produits dérivés, 4e édition. Association Française de Normalisation, Paris.

Aletor VA, Ikhena GA, Egharevba V (1990). The quality of some locally processed Nigerian palm oils: An estimation of some critical processing variables. Food Chem. 36:311-317.

Carrol KK, Guthrie N, Nesaretnam K, Gapor A, Chambers AF (1995). Anti-cancer properties of tocotrienols from palm oil. In nutrition, lipids, health and disease (Ong ASH, Niki E and Packer L, Eds). AOCS Press. pp.117-121.

Choo YM (1994). Palm Oil Carotenoids. Food Nutr. Bull. 15:130-137.

Chooi SY, Koh HF, Wren WG (2006). Autocatalytic hydrolysis and autoxidation of crude palm oil under various constant humidities. JOPR 18 278-287.

Codex Alimentarius Commission/FAO/WHO food standards (2011). Standard for named vegetable oils CODEX-STAN 210, Ed. FAO/WHO.

Desassis A (1957). Palm oil acidification. Oléagineux 12:525-534.

Gopalan C, Narasinga RBS, Seshadri S (1992). Special publication series No. 6. Nutrition Foundation of India. New Delhi.

Guthrie N, Gapor A, Chamber AF, Carroll KK (1997a). Inhibition of proliferation of estrogen receptor-negative MDA-MB-435 and -positive MCF-7 human breast cancer cells by palm oil tocotrienols and tamoxifen, alone and in combination. J. Nutr. 127:544s-548s.

Guthrie N, Gapor A, Chambers AF, Carroll KK (1997b). Inhibition of proliferation of oestrogen receptor-negative MDA-MB-435 and positive MCF-7 human breast cancer cells by palm oil tocotrienols and taximofen, alone and in combination. Asia Pac. J. Clin. Nutr. 6:41-45.

Henderson J, Osborne DJ (1991). Lipase activity in ripening and mature fruit of the oil palm. Stability *in vivo* and *in vitro*. Phytochemistry 30:1073-1078.

Hiol A, Comeau LC, Druet D, Jonzo MD, Rugani N, Sarda L (1999). Purification and characterization of an extracellular lipase from a thermophilic *Rhizopus oryzae* strain isolated from palm fruit. Enzyme Microb. Technol. 26:421-430.

Hirsch RD (1999). La Filière Huile de Palme au Cameroun dans une Perspective de Relance. Agence Française de Developpement, Paris (France), p. 79.

Houria A, Comeau L, Deyris V, Hiol A (2002). Isolation and characterization of an extra-cellular lipase from *Mucor* sp. Strain isolated from palm fruit. Enzyme Microb. Technol. 31:968-975.

Hoyle D, Levang P (2012). Oil palm development in Cameroon.

WWF/IRD/CIFOR report. p. 16.
Latham MC (1979). Human Nutrition In Tropical Africa. Rome. FAO. p. 194.
Loncin M (1952). L'hydrolyse spontanée des huiles glyceridiques et en particulier de l'huile de palme. Editions Couillet (Begium).
Loncin M (1956). Travaux récents dans le domaine des matières grasses. Rev. Franc. Des Corps Gras. 3:255-266.
Loncin M (1965). Influence de l'activité de l'eau sur les réactions chimiques et biochimiques. DECHEMA, Frankfurt-am-Main. monograph NR 926 – 992. pp. 193-205.
Loncin M, Jacobsberg B (1965). Research on palm oil in Belgium and the Congo. Conference: The Oil Palm, London. Trop. Prod. Inst. pp. 85-95.
Matthaüs B (2007). Use of palm oil for frying in comparison with other high-stability oils. Eur. J. Lipid Sci. Technol. 109:400-409.
Ngando Ebongue GF, Dhouib R, Carriere F, Amvam Zollo PH, Arondel V (2006). Assaying lipase activity from oil palm fruit (*Elaeis guineensis* Jacq.) mesocarp. Plant Physiol. Biochem. 44:611-617.
Ngando Ebongue GF, Mpondo Mpondo EA, Dikotto EEL, Koona P (2011). Assessment of the quality of crude palm oil from small holders in Cameroon. JSPPR 2:52-58.
Oil world (2012). Oil world 2011 annual Report. Mielke, Hamburg.
Onwuka GI, Akaerue BI (2006). Evaluation of the quality of palm oil produced by different methods of processing. Res. J. Biol. Sci. 1:16-19.
Orji MU, Mbata TI (2008). Effect of extraction methods on the quality and spoilage of Nigerian palm oil. AJBR 2:192-196.
Qureshi AA, Qureshi N, Hasler-Rapacz JO, Weber FE, Chaudhary V, Crenshaw TD, Gapor A, Ong ASH, Chong YH, Peterson D, Rapacz J (1991). Dietary tocotrienols reduce concentrations of plasma cholesterol, apolipoprotein B, thromboxane B2 and platelet factor 4 in pigs with inherited hyperlipidemias. Am. J. Clin. Nutr. 53:1042-1046.
Rossel JB (1999). Measurement of rancidity. In Rancidity in foods. Ed. By Allen JC and Hamilton RJ, UK. Aspen publishers. pp. 22-51.
Sambanthamurthi R, Oo KC, Parman SH (1995). Factors affecting lipase activity in *Elaeis guineensis* mesocarp. Plant Physiol. Biochem. 33:353-359.
Sambanthamurthi R, Sundram K, Tan Y (2000). Chemistry and biochemistry of palm oil. Prog. Lipid Res. 39:507-558.
Seshadri S (1996). In: Use of carotene rich foods to combat vitamin A defficiency in india. Special Report No.12. New Delhi. Nutrition Foundation of India.
Tagoe SMA, Dickinson MJ, Apetorgbor MM (2012). Factors influencing quality of palm oil produced at the cottage industry level in Ghana. Int. Food Res. J. 19:271-278.
Tan DTS, Khor HT, Low WHS, Ali A, Gapor A (1991). Effect of a palm-oil-vitamin E concentrate on the serum and lipoprotein lipids in humans. Am J. Clin. Nutr. 53:1027-1030.

Determination of heavy metals in exposed corned beef and chicken luncheon that sold in Sulaymaniah markets

Hozan Jalil Hamasalim and Hemin Nuradden Mohammed

Animal Production, Faculty of Agricultural Science, University of Sulaimani, Iraq.

In this study, the levels of heavy metals (Cr, Cd, Co, Pb, Cu, Ni, Fe, Zn, Mg and Mn) were determined in canned beef and four different chicken luncheon from Sulaymaniah markets using atomic absorption spectroscopy (AAS). Significant difference was observed in the chromium (Cr) content among the corned beef brands, while there was no-significant difference in Cr content in the all chicken luncheon trademarks. The results of cobalt (Co) for A, B, C and D were (0.03, 0.04, 0.04 and 0.08 µg/gm, respectively). Co concentration for Gadeer, F, G and H were 0.00, 0.03, 0.04 and 0.04 µg/gm, respectively. Lead (Pb) and copper (Cu) were significantly different among all trademarks. Manganese (Mn) and nickel (Ni) were not found in any of the corned beef and chicken luncheon trademarks. The iron (Fe) content for the corned beef trademarks was 1.43, 5.41, 3 and 4.4 µg/gm) for A, B, C and D respectively. The results also reveal that there was a significant difference in Fe content between chicken luncheon trademarks. B and C contained a high concentration of Zn while A and D had low concentration of Zn. Furthermore, the F trademark was significantly higher in Zn concentration than G trademark, which showed significantly lower concentration of Zn.

Key words: Canned beef, chicken luncheon and heavy metals determination.

INTRODUCTION

Applying food safety standards on a product is very important because it relates closely to human's health. Good food products have a high nutritional quality, as well as being free from physical, chemical and biological contaminations. The food industry development encourages food manufacturer's to produce more practical and durable products, but still must have high nutrition. For example, beef processing to produce meatballs, corned beef, beef burgers and sausages have the purpose to form more practical and durable products, as well as having high nutritional value. The presence of some metal pollutants in everywhere; especially, cadmium (Cd), lead (Pb), chromium (Cr), arsenic (As), nickel (Ni) and mercury (Hg); spread out into animal rations and food chains, thus it may increase the possibility of toxic effects in humans and animals (Farmer and Farmer, 2000; Javed et al., 2009).

Meat is a very rich and convenient source of nutrients, also to large extent microelements. The chemical composition of meat depends on both kind and degree of the feeding animal. The need for mineral compounds depends on the age, physiological state and feed intake as well as on living conditions (Baykov et al., 1996). Heavy metals from manmade pollution sources are continuously released into aquatic and terrestrial ecosystems and therefore the concern about the effect of anthropogenic pollution on the ecosystems is growing. Contamination with heavy metals is a serious threat

Table 1. Concentrations of heavy metals (µg/gm) in four trademarks of corned beef.

Trademark	Cr	Cd	Co	Pb	Cu	Ni	Fe	Zn	Mg	Mn
A	0.20 ± 0.100[b]	0.07 ± 0.004[b]	0.03 ± 0.005[b]	0.53 ± 0.020[d]	0.11 ± 0.005[d]	0.00 ± 0.00	1.43 ± 0.005[d]	2.43 ± 0.015[d]	2.51 ± 0.005[d]	0.00 ± 0.00
B	0.10 ± 0.000[b]	0.06 ± 0.001[b]	0.04 ± 0.020[b]	0.61 ± 0.020[c]	0.27 ± 0.020[b]	0.00 ± 0.00	5.41 ± 0.015[a]	3.81 ± 0.015[a]	4.32 ± 0.005[a]	0.00 ± 0.00
C	0.19 ± 0.095[b]	0.27 ± 0.060[a]	0.04 ± 0.010[b]	1.53 ± 0.035[b]	0.17 ± 0.010[c]	0.00 ± 0.00	3.00 ± 0.100[c]	3.16 ± 0.005[b]	3.80 ± 0.025[b]	0.00 ± 0.00
D	0.40 ± 0.100[a]	0.20 ± 0.050[a]	0.08 ± 0.030[a]	2.07 ± 0.060[a]	0.32 ± 0.025[a]	0.00 ± 0.00	4.40± 0.00[b]	2.48 ± 0.025[c]	3.61 ± 0.015[c]	0.00 ± 0.00

Means having the same letter in the same sections are not significantly different at $P \leq 0.05$ according to Duncan tests.

because of their toxicity, bioaccumulation and biomagnifications in the food chain (Demirezen and Uruc, 2006). The risk of heavy metal contamination in meat is of great concern for both food safety and human health because of the toxic nature of these metals at relatively minute concentrations (Santhi et al., 2008). Instances of heavy metal contamination in meat products during processing have been reported (Santhi et al., 2008 and Brito et al., 2005).

Heavy metal pollutants can contaminate the products during processing (through the raw material, spices, water and packaging) by inhalation of air and penetration through the skin's surface (Raikwar et al. 2008 and Santhi et al., 2008). In other cases, contaminated animal feed and rearing of livestock in proximity to polluted environment were reportedly responsible for heavy metal contamination in meat (Miranda et al., 2005 and Sabir et al., 2003).

The objective of this investigation was to find out whether the level of metallic elements in corned beef and chicken luncheon obtained from Sulaymaniah Market in Iraq for human consumption have exceeded the maximum permissible level specified by international standards.

MATERIALS AND METHODS

Sample collection

Samples included canned beef for four different trademarks (A, B, C and D) of Brazilian origin and Chicken luncheon for four different trademarks, E and F of Jordan origin while G and H of France origin, were collected from the Sulaymaniah Market in Sulaymaniah governorate, Iraq. The samples were collected beginning in May, 2012 and transported to the laboratory for analysis.

Ash content

Ash content was determined according to the method of AOAC (2000) by taking a specimen of known weight of flesh and placing in a muffle furnace at 550°C for 16 h.

Sample preparation

The ash was dissolved in 5 ml of HCl acid and 5 ml of distillated water and put on a hot plate for 5 min after which it was topped up to 25 ml with double distillated water.

Elemental analysis of samples

Cr, Cd, Co, Pb, Cu, Ni, Fe, Zn, Mg and Mn in corned beef and chicken luncheon samples were determined directly on each sample of the final solutions using atomic absorption spectroscopy (AAS).

Statistical analysis

All data were subjected to one-way analysis of variance (ANOVA) using XL Stat program for windows. Differences between the means were tested by Duncan tests. The level of significance was chosen at P<0.05 and the results are presented as mean (Steel et al., 1996).

RESULTS AND DISCUSSION

Tables 1 and 2 show the concentration of heavy metals in some trademarks of corned beef and chicken luncheon. The ranges of metal concentrations found in the canned corn beef samples were as follows: Cr: 0.10 to 0.40 µg/g ($\bar{x}$ = 0.22 µg/g); Cd: 0.06 to 0.27 µg/g ($\bar{x}$ = 0.60 µg/g); Co: 0.03 to 0.08 µg/g ($\bar{x}$ = 0.48 µg/g); Pb: 0.53 to 2.07 µg/g ($\bar{x}$ = 1.3 µg/g); Cu 0.11 to 0.32 µg/g ($\bar{x}$ = 0.21 µg/g); Fe: 1.43 to 5.41 µg/g ($\bar{x}$ = 3.42 µg/g); Zn: 2.43 to 3.81 µg/g ($\bar{x}$ = 3.12 µg/g) and Mg: 2.51 to 4.32 µg/g ($\bar{x}$ = 3.41 µg/g). The ranges of metal concentrations found in the canned Chicken luncheon samples were as follows: Cr: 0.10 µg/g; Cd: 0.06 to 0.10 µg/g ($\bar{x}$ = 0.08 µg/g); Co: 0.00 to 0.04 µg/g ($\bar{x}$ = 0.02 µg/g); Pb: 0.52 to 1.19 µg/g ($\bar{x}$ = 0.85 µg/g); Cu: 0.15 to 0.30 µg/g ($\bar{x}$ = 0.22 µg/g); Fe: 2.55 to 4.22 µg/g ($\bar{x}$ = 3.38 µg/g); Zn: 1.59 to 3.08 µg/g ($\bar{x}$ =2.33 µg/g) and Mg: 2.16 to 4.16 µg/g ($\bar{x}$ =3.16 µg/g).

The significant difference (p<0.05) in the chromium (Cr)

Table 2. Concentrations of heavy metals (μg/gm) in four trademarks of Chicken Luncheon.

Trademark	Cr	Cd	Co	Pb	Cu	Ni	Fe	Zn	Mg	Mn
E	0.10 ± 0.00^a	0.10 ± 0.00^a	0.00 ± 0.000^b	1.19 ± 0.010^a	0.24 ± 0.005^b	0.00 ± 0.00	3.14 ± 0.010^b	2.00 ± 0.005^c	2.16 ± 0.015^d	0.00 ± 0.00
F	0.10 ± 0.00^a	0.06 ± 0.003^c	0.03 ± 0.005^a	1.02 ± 0.005^a	0.30 ± 0.000^a	0.00 ± 0.00	4.22 ± 0.020^a	3.08 ± 0.010^a	4.16 ± 0.010^a	0.00 ± 0.00
G	0.10 ± 0.00^a	0.08 ± 0.001^b	0.04 ± 0.010^a	1.00 ± 0.300^a	0.15 ± 0.010^c	0.00 ± 0.00	2.55 ± 0.005^d	1.59 ± 0.000^d	3.06 ± 0.050^c	0.00 ± 0.00
H	0.10 ± 0.00^a	0.08 ± 0.002^b	0.04 ± 0.015^a	0.52 ± 0.005^b	0.15 ± 0.005^c	0.00 ± 0.00	2.67 ± 0.005^c	2.86 ± 0.005^b	3.93 ± 0.005^b	0.00 ± 0.00

Means having the same letter in the same sections are not significantly different at $P \leq 0.05$ according to Duncan test.

was recorded among the corned beef trademarks. The D trademark contained a high concentration of Cr, while the B trademark contained low concentration of Cr. There were no-significant differences in Cr content of all chicken luncheon trademarks.

Significant differences in cadmium (Cd) were shown among the corned beef trademarks, C trademark appeared to have higher content of Cd (0.27 μg/gm) but the B trademark had a lower content of Cd (0.06 μg/gm). The E trademark had the highest value (0.1 μg/gm) while the F trademark showed the lowest value (0.06 μg/gm). The distribution of cadmium into the food is often through water, air and soil. In humans and animals, cadmium contamination is usually caused by the food chain (Zhuang et al., 2009). Cadmium is absorbed around five percent and the remaining passes directly into the feces. Gastrointestinal absorption of cadmium in the digestive tract is influenced by several factors; one of them is low-calcium diet that causes higher absorption and deposition of cadmium into the intestinal mucosa, liver, and kidney (Kazantzis, 2004). The accumulation of cadmium in the liver and kidneys seems to be mainly dependent on the storage of cadmium in association with cadmium-binding protein, metallothionein (Ellah and Yahia, 2009). Cadmium may accumulate in the human body and may induce kidney dysfunction, skeletal damage and reproductive deficiencies.

The recommended consumer dietary exposure of Cd are 0.12 μg/kg body weight/day for adult, 0.315 μg/kg body weight/day for toddlers (1.5 to 4.8) and 0.22 μg/kg body weight/day for young people (FSIS, 2004). The testimonial consumer dietary exposures of Cr are 1.7 μg/kg body weight/day for toddlers and 1.15 μg/kg body weight/day for young people (FSIS, 2004).

The results of cobalt (Co) as shown in Table 1 for A, B, C and D were (0.03, 0.04, 0.04 and 0.08 μg/gm, respectively). Cobalt concentration for E, F, G and H were 0.00, 0.03, 0.04 and 0.04 μg/gm, respecttively. Significant differences were found between all trademarks ($P<0.05$). The results in Table 1 show significant differences in the lead (Pb) concentration among all trademarks. D trademark recorded higher Pb concentration (2.07 μg/gm) while A trademark contained (0.53 μg/gm) low levels of Pb. Lead concentrations for E, F, G and H were 1.19, 1.02, 1 and 0.52 μg/gm, respectively. All trademarks varied significantly ($P<0.05$). The concentrations of heavy metals Pb and Cd in the liver, kidneys and most of meat products have been described by researchers in other countries (Kirkpatrick and Coffin, 2006; Gonzaacutelez-Weller et al., 2006; Santhi, et al., 2008; Blanco-Penedo et al., 2009). The commendation consumer dietary exposure of Pb is 18 μg kg^{-1} for adult, 46 μg kg^{-1} for toddlers and 30 μg kg^{-1} for young people (FSIS, 2004). Lead would become toxic if it is consumed in the amount which exceeds the threshold and can cause acute and chronic poisoning symptoms (Jarup, 2003). Lead is known to induce reduced cognitive development and intellectual performance in children and increase blood pressure and cardiovascular disease in adults (Comission of the European, 2002). The source of Pb contamination of food livestock come from the air, water, food materials, cooking utensils and food packaging. The use of spices during processing of meatballs, corned beef, burger's beef and sausages is one of the sources of lead pollution. Plants absorb lead from the soil and the air around resulting in further accumulation (Landis and Ming-Ho, 2004). According to the recommendation of the Central Agency for Standardization and Quality Control of Iraq (1988), the Pb threshold is 1 mg / kg in luncheon meat.

There was significant difference in Copper (Cu) concentration which appeared in all trademarks; higher Cu concentration value was detected in D trademark but lower Cu concentration was detected in A. Table 2 shows the mean value for Cu concentration of chicken luncheon trademarks. Generally, Cu concentration were higher in F while E showed amid concentration, and G and H trademarks showed rather low concentrations of

Cu. Mariam et al. (2004) reported higher concentration of copper in mutton, beef and poultry meat from some region such as Lahore region. Copper can also cause public health hazards in high concentrations (Brito et al., 2005). According to the recommendation of the Central Agency for Standardization and Quality Control of Iraq (1988) Cu should not exceed 15 mg / kg in luncheon meat. The best consumer dietary exposure of Cu are 18, 46 and 30 μg/kg body weight/day for adult, toddlers and young people, respectively (FSIS, 2004).

The result does not show any trace of manganese (Mn) and nickel (Ni) in any of corned beef and chicken luncheon trademarks. Nickel (Ni) can cause respiratory problems and is a carcinogen (Agency for Toxic Substance and Disease Registry, 2004). The upper tolerable Ni intake for children (one to three years old) and adults (19 to 70 years old) is 7 and 40 mg/day, respectively. The highest mean level of Ni was found in turkey meat samples collected from Warri zone (20.78 $mg.kg^{-1}$). The levels of Ni found in these samples did not exceed the upper tolerable intake levels of this element.

Daily intake of small amounts of Mn is needed for growth and good health in humans, otherwise deficiency of Mn can cause nervous system problems (Demirezen and Uruç, 2006). Consumer dietary exposure of Mn are 67 μg/kg body weight/day for adult, 132 μg/kg body weight/day for toddlers (1.5 to 4.8) and 101 μg/kg body weight/day for young people (4 to 18) (FSIS, 2004).

The iron (Fe) concentration for the corned beef trademarks were (1.43, 5.41, 3 and 4.4 μg/gm) for A, B, C and D, respectively, significant differences were shown among all trademarks. The Fe concentrations of chicken luncheon trademarks are illustrated in Table 2. There was a significant Fe concentration ($P<0.05$) between chicken luncheon trademarks.

B and C contained high concentration of Zn (3.81and 3.16 μg/gm, respectively), while A and D had low concentration of Zn (2.43 and 2.48 μg/gm, respectively). Furthermore, the F trademark showed significantly higher Zn concentration than G trademark. Trace Zn can cause problems; however, too much Zn is harmful to human health (ATSDR, 2004). The recommended daily intakes of zinc are 15 mg for adult males and 12 mg for adult females. The highest average of zinc concentration (1.7 $\mu g\ g^{-1}$) was observed in "HAMOON" brand and the lowest average zinc concentration was observed in "DARYA" (0.41 $\mu g\ g^{-1}$). The United States Environmental Protection Agency and the European Commission (US-EPA and EC)(2001) have not considered any standards or limits for the zinc concentrations (Ashraf et al., 2006). According to the recommendation of the Central Agency for Standardization and Quality Control of Iraq (1988), Zn should not exceed 50 mg / kg in luncheon meat.

Conclusion

From the results of this study, the concentrations of all the heavy metals in the canned beef and chicken luncheon were found to be statistically significant. The levels of each heavy metals (Cr, Cd, Co, Pb, Cu, Fe, Zn and Mg) in the canned beef differed significantly; the heavy metal in the chicken luncheon (Cd, Co, Pb, Cu, Fe, Zn and Mg) also differed significantly among the different trademarks; while, the Mn and Ni were not found in all of each canned beef and Chicken luncheon.

REFERENCES

Agency for Toxic Substances and Disease Registry (ATSDR) (2004). Agency for Toxic Substances and Disease Registry, Division of Toxicology, Clifton Road, NE, Atlanta, GA. Retrieved from: http://www.atsdr.cdc.gov/toxprofiles/.

Ashraf W (2006). Levels of selected heavy metals in tuna fish. Arabian J. Sci. Eng. 31(1A):89.

Association of Official Analytical Chemists (AOAC) (2000). Meat and Meat products. In: Official Methods of Analysis. Association of Official Analytical Chemists Inc. Gaithersburg, U.S.A.

Baykov BD, Stoyanov MP, Gugova ML (1996). Cadmium and lead bioaccumulation in male chickens for high food concentration. Toxicol. Environ. Chem. 54:155-159.

Blanco-Penedo JL, Benedito RF, Shore M, Miranda M, García V, López-Alonso M (2009). Influence of farm type (organic, conventional and intensive) on toxic metal accumulation in calves in NW Spain. Agron. Res. 7:578-584.

Brito G, Díaz C, Galindo L, Hardisson A, Santiago D, Montelongo FG (2005). Levels of metals in canned meat products: Intermetallic correlations. Bull. Environ. Contam. Toxicol. 44(2):309-316.

Central Agency for Standardization and Quality Control (1988). Luncheon meat.UDC, 664-531. No. 1431.

Commission of the European communities (2001). Commission regulation (EC) No. 221/ 2002 of 6 February 2002 amending regulation (EC) No. 466/2002 setting maximum levels for certain contaminants in foodstuffs. Off. J. Eur. Commun. Brussels, 6 February 2002.

Demirezen D, Urue K (2006). Comparative study of trace elements in certain fish, meat and product. Meat Sci. 74(2):255-260.

Farmer AA, Farmer AM (2000). Concentrations of cadmium, lead and zinc in livestock feed and organs around a metal production center in eastern azakhstan. Sci. Total Environ. 257(1): 53-60.

Food Safety and Inspection Service (FSIS) (2004). 2000 Total Diet Study of 12 elements – aluminium, arsenic, cadmium, chromium, copper, lead, manganese, mercury, nickel, selenium, tin and zinc. Food Survey Information Sheets.48/04.www: http://www.food.gov.uk/science/surveillance/.

Gonzaacutelez-Weller D, Karlsson L, Caballero A, Hernaacutendez F, Gutieacuterrez A, Gonzaacutelez-Iglesias T, Marino M, Hardisson A (2006). Lead and cadmium in meat and meat products consumed by the population in Tenerife Island, Spain. J. Food Add. Contam. 23:757–763.

Jarup L (2003). Hazards of heavy metal contamination. Br. Med. Bull. 68:167-182.

Javed I, Jan I, Muhammad F, Zia-ur-Rahman, M Zargham, B Aslam and JI Sultan (2009). Heavy metal residues in the milk of cattle and goats during winter season. Bull. Environ. Contam. Toxicol. 82: 616-620.

Kazantzis G (2004). Cadmium, osteoporosis and calcium metabolism. J. Biometals 17(5):493-498.

Kirkpatrick DC, Coffin DE (2006). Cadmium, lead and mercury content of various cured meats. J. of Sci. of Food and Agric. 24(12):1595-1598.

Landis WG, Ming-Ho Yu (2004). Introduction to Environmental Toxicology; Impacts of Chemicals Upon Ecological System. Lewis Publishers, CRC Press, LLC. pp. 221-223.

Mariam I, Iqbal S, Nagra AS (2004).Distribution of some trace macro minerals in beef, mutton and poultry. Int. J. Agric. Biol. 5:816.

Miranda M, Lopez-Alonso M, Castillo C, Hernandez J, Prieto F, Benedito JL (2003). Some toxic elements in liver, kidney and meat

from calves slaughtered in Asturias (Northern Spain). Eur. Food Res. Technol. 216, 284-289.
Raikwar M, Kumar P, Singh M, Singh A (2008). Toxic effect of Heavy metals in livestock health. Vet. World 1(1): 28-30.
Sabir SM, SW Khan, I Hayat (2003). Effect of environmental pollution on quality of meat in district Bagh, Azad Kashmir. Pak. J. Nutr. 2(2): 98-101.
Santhi D, Balakrishnan V, Kalaikannan A, Radhakrishnan KT (2008). Presence of heavy metals in pork products in Chennai (India). Am. J. Food Technol. 3(3):192-199.
Steel RGD, Torrie JH, Dickey DA (1996). Principles and procedures of statistics. A biometrical approach. 3rd ed. McGraw Hill Book Company Inc, New York, USA. pp. 334-381.
Zhuang P, Zou H, Shu W (2009). Biotransfer of heavy metal in meat. J. Environ. Sci. 21(6):849-855.

Effect of boiling and frying on nutritional value and *in vitro* digestibility of rabbit meat

Yin Zhang, Xinhui Wang, Wei Wang and Jiaming Zhang

Key Laboratory of Meat Processing of Sichuan, Chengdu University, Chengdu, 610106, China.

Effect of boiling and frying on nutritional value and *in vitro* digestibility of rabbit meat were investigated. It was observed that boiling for 15 min increased the digestibility and nutritional value of rabbit meat, while boiling for 5 and 40 min led to their loss. Although, frying for 2 and 4 min produced an acceptable digestibility and nutritional value, the values were lower than those of the boiled rabbit meats. These results were supported by the analysis of amino acid and SDS-PAGE. Boiling or frying for longer time led to decrease in whiteness and protein content of rabbit meat. Micrographs showed that the meat myofibrils became clearer with the increase in boiling time, but a fuzzier surface appeared with the increase in frying time. From these results, it was concluded that a proper boiling or frying time is necessary to acquire a nutritious and high quality rabbit meat.

Key words: Rabbit meat, boiling, frying, amino acid, protein, essential amino acid index (EAAI).

INTRODUCTION

Rabbit meat has a high protein and low sodium contents (Combes, 2004). Its cholesterol and fat contents are much lower than chicken, turkey, beef and pork (Combes, 2004; Cavani et al., 2010). The unsaturated fatty acids content of rabbit meat is 63% (w/w) of its total fatty acids. The United States Department of Agriculture (USDA) has stated that the rabbit meat is the most nutritious meat available to man (Cavani et al., 2010). The rabbit meat was highly accepted for its appearance and color, juiciness and taste when boiled and fried as comparing to the Giant African Land Snail (*Archachatina marginata*) meat (Malik et al., 2011). All these features make rabbit meat a very healthy, nutritious and popular food suitable for general consumption and as a special diet including heart disease patients diet, low sodium diet, weight reduction diet, diets for the aged, etc (Dalle, 2002; Cavani and Petracci, 2004).

Cooking is necessary to prepare palatable and microbially safe foods including rabbit meat. Suitable cooking minimizes the nutrient loss and also improves the digestibility of food (Kimura and Itokawa, 1990). However, improper cooking will terribly harm the nutrients present in the food, and sometimes even produce toxic and hazardous substances. Investigations have demonstrated that there were positive associations between red processed meat and proximal colon cancer, pan-fried red meat and colorectal cancer (Miller et al., 2013), consumption of fried and/or well-done red meat was associated with increased risk of lung cancer in women (Sinha et al., 1998). Carcinogenic compounds formed by high-temperature cooking techniques like heterocyclic amines and polycyclic aromatic hydrocarbons were reported to contribute to the risk of developing colorectal tumors (Sinha et al., 1999; Wong et al., 2012). Meat cooked with high temperature techniques produced mutagenic compounds such as heterocyclic amines, which are strongly associated with the risk of colorectal adenomas (Sinha et al., 1999; Wong et al., 2012). The degree of browning that occurred during deep-frying of meat was related to the pathogenesis of breast cancer (Dai et al., 2002; Lyn-Cook et al.,

2011) and a high intake of processed meat (pork chops, sausages and hotdogs) was found to increase the risk of lung cancer in men (Tasevska et al., 2009). Fried meat was associated with a higher risk of cancer of oral cavity, pharynx and esophagus, roasting/grilling meat increased risk of prostate cancer (Di Maso et al., 2013).

Due to the health concerns caused by improper cooking of meat, cooking method has been paid special attention and many studies have been done to investigate the effects of heating on meat quality and its nutrients. It was found that the conventional and microwave heating of ground beef samples increased the nonheme iron concentrations (Schricker and Miller, 1983) and the fried ground beef patties had the highest flavor intensity (Berry and Leddy, 1984). The amount of heterocyclic amines increased with frying or boiling time increase during the production of braised chicken (Yao et al., 2013). Cooking had no significant effect on riboflavin content of dark meat, while frying significantly decreased the riboflavin content of light meat (Al-Khalifa and Dawood, 1993). The combination of roasting and microwave heating of hamburgers increased the lipid oxidation (Rodriguez-Estrada et al., 1997). Heat cooking considerably affected the proximate composition and mineral contents and baking and grilling were reported as the best cooking methods for rainbow trout (Gokoglu et al., 2004).

Effect of cooking on rabbit meat quality had been investigated in the past. Some of the investigations include the effect of electric oven cooking on oxidative stability of meat of rabbit fed with fodder containing α-tocopheryl acetate (Monahan et al., 1990). The effect of boiling (in a vacuum sealed polyethylene bag at 100±1°C for 8 min), frying (in sunflower oil at 175±5°C for 3 min), roasting (in an electric oven preheated to 200±10°C for 15 min) on pH, shear force, lightness, color, and water holding capacity of rabbit meat (Bosco et al., 2001), was also investigated. The effect of cooking (immersed in a constant temperature water-bath) temperature (50-90 °C) and time (10-120 min) on tenderness of rabbit muscle had been investigated (Combes et al., 2004). Recently, the effect of cooking methods and the related processes on nutrition and quality of chicken (Das et al., 2013), silver carp (Naseri et al., 2013), beef muscles, rainbow trout fillets (Choubert and Baccaunaud, 2010), and beef roast (Modzelewska-Kapituła et al., 2012; Walsh et al., 2010), were investigated. However, to our knowledge, the effect of cooking methods on the protein content and quality of protein present in the cooked rabbit meat have not been investigated or reported previously. Better knowledge of effect of cooking methods on the nutritional value of protein of rabbit meat will contribute to the development of nutrient rich rabbit meat products.

Heat cooking is being used as a general term for various processes including boiling, frying, roasting, baking, etc. Boiling and frying are most frequently used for cooking of meat (Fillion and Henry, 1998; Chen, 2006). However, the effect of doneness or cooking levels on protein qualities of rabbit meat was rarely investigated. This research is focused on investigating the effect of two commonly used heat cooking methods (boiling and frying) and different doneness levels on nutritional value of protein and *in vitro* digestibility of rabbit meat.

MATERIALS AND METHODS

Reagents

Amino acid standards used in this study were obtained from Sigma-Aldarich Corporation (St. Louis, MO). HPLC grade acetonitrile and methanol were used in this study. The trichloroacetic acid (TCA) was purchased from Jiangsu Yonghua Fine Chemical Co., Ltd (Shanghai, China). Sodium dodecyl sulfate (SDS), urea, Tris (hydroxymethyl) aminomethane (Tris), β-mercaptoethanol (β-ME) and other electrophoresis reagents were purchased from Shanghai Bio Life Science and Technology Co., Ltd (Shanghai, China). All chemicals used in this study were of analytical grade.

Meat samples

Totally, 140 numbers of 10 week old Sichuan Rex Rabbits, having a body weight of 2.35±0.1 kg were used in the experiments. The rabbits were slaughtered by the butchers of Hage Rabbit Industry Co. Ltd, in Sichuan Yilong County and brought to our laboratory within 2 h. The rabbit meat samples for cooking experiments (boiling and frying) were taken from hind legs. The meat was chopped into dice having uniform size of 20 ± 3 mm in both length and width, and 10 ± 3 mm in height. A total weight of 300 g hind leg meat was used for each treatment. Six treatments (boiling for 5, 15 and 40 min; frying for 2, 4 and 6 min) were performed for 9 times each.

Cooking methods

In order to investigate effect of heat cooking methods on nutritional value and *in vitro* digestion of rabbit meat, the diced meat was boiled and also fried using the procedure shown below. The boiling of the rabbit meat was done with six times volume of hot water in a temperature-controlled media MK2102 electromagnetic furnace (Media Group, China) at temperature of 95±3°C for a predetermined time of 5 (equals to medium), 15 (equals to well done) and 40 min (equals to very well done). The rabbit meat was fried with 1/3 times soybean oil in a temperature controlled Media MK2102 electromagnetic furnace (Media Group, China) set at 175±3°C for 2 (equals to medium), 3 (equals to well done) and 6 min (equals to very well done).

Determination of moisture and protein

Moisture contents of samples were analyzed according to the Association of Official Analytical Chemists method (AOAC 950.46, 1990). The crude proteins (N = 6.25) and dry matter contents were analyzed according to Association of Official Analytical Chemists (AOAC, 2006). All the tests were conducted in triplicate.

Color analysis

CIELAB color (L*, a* and b*) of cooked muscles were tested using a portable chroma meter CR-400 (Minolta Cameras, Osaka, Japan). The instrument was calibrated using the white plate

(Calibration Plate CR-A43, Minolta Cameras) at the beginning of each session. The chroma meter has a measuring area of 8 mm in diameter and 8 mm in thickness and illuminates the sample area with the diffused illumination from a pulsed xenon arc lamp. After choosing the Commission International illuminant (CIE) lighting conditions, the calibration channel selection was set at auto select mode. Reflectance measurements were obtained at a viewing angle of 0° and the spectral component was included. Color is defined by the three CIE coordinates, L* (lightness), a* ((+) redness/ (-) greenness) and b* ((+) yellowness/ (-) blueness). The whiteness (W) is calculated from the L*, a*, b* values using the equation (Lanier et al., 1991).

$$W = 100-((100-L^*)^2+a^{*2}+b^{*2})^{0.5}$$

The cooked samples were dried, milled and filled in a 40 × 25 mm weighing bottle. The powder in the weighing bottle was placed smoothly in the plates and the color was determined using the chroma meter. Six repeated determinations were done for each sample.

Analysis of the micro-structure of meat in optical microscopy

The surface micro-structure of cooked and uncooked rabbit meat was observed with a XSP-15 reading optical microscope (Jiangsu Optical Instrument Factory, Nanjing). In order to maintain the natural state, the meat samples were not treated with any reagents. The meat sample was fixed with the stage clips and was observed under diffused reflection light with a 10 times magnifying eyepiece lens and 16 times objective lens. The optical micro surface structure was obtained by an IXUS 60 6.0 MP Digital Camera.

Sequential *in vitro* protein digestion procedure

The *in vitro* digestibility of the proteins was evaluated by the method used by Fu et al. (2002), with some modification. Pepsin (Sigma, P7000, 1:10,000, 600-1000 units/mg) and trypsin (Genview, DH355-1, 1:250) were used for the *in vitro* digestion study. Five grams of meat sample from each treatment (boiled or fried meat were converted as the weight of uncooked) was chopped into pieces and dipped into 250 mL solution of hydrochloric acid (HCl) of pH 1.5. The dipped samples were pre-incubated in a water bath at 37°C for 3-5 min. Then, an amount of pepsin (20 mg of pepsin per mL of 0.1 M KH_2PO_4, pH 2, buffer) was added, and the ratio of enzyme to protein substrate was maintained at 1:100 (w/w). The pepsin and protein solutions were mixed well and incubated at 37°C. After the predetermined incubation periods: 0, 60, 120, 180, and 300 min, aliquots of the mixtures were taken, and the pH was adjusted to pH 7.0 by adding 1.0 mol/L NaOH to stop the enzymatic reactions. Additionally, the final pepsin-digested hydrolysates (pH 7.0) were further digested by the addition of trypsin (20 mg of trypsin per mL of 0.1 M Tris-HCl buffer, pH 7.0) at 37°C for 60, 120, 180, and 300 min. The amount of enzyme to initial protein substrate ratio was maintained at 1:20 (w/w).

Nitrogen release during digestion

The nitrogen release (%, w/w) during digestion was determined by the TCA-NSI method used by Iwami et al. (1986), with some modifications. 10 mL of the digested mixtures were mixed with 10 mL of 10% (w/v) TCA (and the final concentration of TCA was 5%, w/v). The mixtures were then centrifuged (20°C, 8000 g, 30 min) to obtain the precipitates. After washing with 10 mL of TCA (10%, w/v), the precipitates were obtained again by centrifugation at the same condition. The N content of the samples was determined by Kjeldahl method (N×6.25). The nitrogen release (%, w/w) during the digestion was calculated as:

$$\text{Nitrogen release} = (N_0-N_t)\times 100/N_{total}$$

Where, t is the digestion time (min), N_t (mg) is the TCA-insoluble N after digestion for t (min), N_0 (mg) is the TCA-insoluble N in the protein sample and N_{total} (mg) is the total N of protein sample. Triplicate determinations were done for each sample.

Amino acid analysis

The amino acid composition of the rabbit meat samples were determined using the method used by Bidlingmeyer et al. (1984). The samples were hydrolyzed with a 6 M HCl solution in a vacuum-sealed tube for 24 h at 110°C. The samples were subsequently centrifuged (1500 rpm for 5 min) and dried under vacuum for 1.5-2 h. The pH was adjusted by adding 20 µL of an ethanol : water : triethylamine (2:2:1) solution and the samples were dried for 1.5-2 h. The resulting sample was derivatised by adding 20 µL of ethanol : water : triethylamine : phenylisothiocyanate (7:1:1:1) derivatising solution which was then allowed to react at room temperature for 10 min prior to drying under vacuum (minimum of 3 h). The samples were re-suspended in 200 µL of Picotag sample diluent (Waters, Millford, MA, USA) and 8 µL sub-sample was injected for separation by HPLC under gradient conditions. Buffer A was a sodium acetate buffer (pH 6.4) containing 5000 ppm ethylenediaminetetraacetic acid (EDTA), 1:2000 triethylamine and 6% acetonitrile and buffer B consisted of 60% acetonitrile with 5000 ppm EDTA. A waters high performance liquid chromatography system (1525 HPLC with a binary gradient delivery, 717 auto-sampler and Injector, 1500 column heater, 2487 dual wavelength UV detector) and a Breeze data workstation (Waters, Millford, MA, USA) were used for the determination of amino acid composition. Triplicate determinations were done for each sample.

Nutrition analysis of amino acid

Biological values of cooked rabbit meat was analyzed by 5 parameters; they are total amino acid (TAA) content, essential amino acid ratio (EAAR), protein chemical score (PCS), essential amino acid index (EAAI) (FAO., 1985) and protein efficiency ratio (PER) (Alsmeyer et al., 1974). The specific calculation formula of those parameters as follow:

$$TAA = \sum_{i=1}^{n} a_i$$

$$EAAR = \sum_{j=1}^{m} b_j \Big/ \sum_{i=1}^{n} a_i$$

$$PCS = Min\{(aa/AA)_1, \cdots, (aa/AA)_k\}$$

$$EAAI = \sqrt[k]{(aa/AA)_1 \times \cdots \times (aa/AA)_k}$$

$$PER = -0.468 + 0.454 \times leucine - 0.105 \times tyrosine$$

where a - amino acid in meat sample; b - essential amino acid in meat sample; n - the number of amino acid; m - the number of essential amino acid; AA - the content of amino acid in FAO (1985)

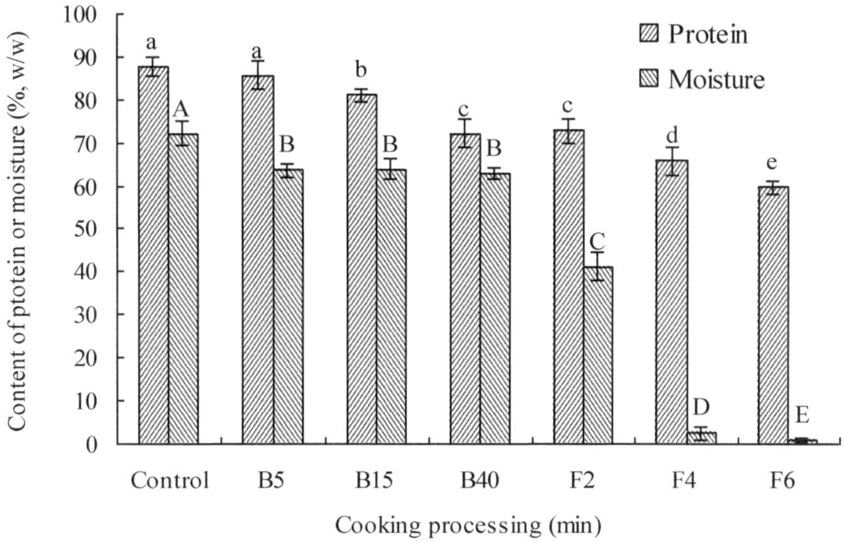

Figure 1. Effect of heat cooking on content of protein and water of rabbit meat. Protein content was calculated on dry basis; Means (n=3) without a common letter differ significantly (P<0.05); "Control" is uncooked rabbit meat, B5, B15 and B40 represents meats boiled for 5, 15, and 40 min, respectively; and F2, B4 and B6 represents meats fried for 2, 3 and 6 min, respectively.

suggested pattern of protein requirement; aa - the content of amino acid as compared to FAO (1985) suggested pattern of protein requirement and k- the number of amino acid type (FAO, 1985, Alsmeyer et al., 1974).

SDS–polyacrylamide gel electrophoresis (SDS-PAGE)

SDS-PAGE analysis was performed according to the method of Laemmli (1970). Two grams of cooked meat sample was added to 18 mL of SDS solution (5 g/100 mL). The mixture was then homogenized using a FJ-200 homogenizer (Shanghai Specimen and Model Factory, Shanghai, China) at a speed of 10,000 rpm for 1 min. The homogenate was incubated at 85°C in a temperature-controlled water bath for 1 h to dissolve total proteins. The sample was centrifuged at 12,000 g for 6 min. Solubilized samples were mixed at 1:1 (v/v) ratio with the sample buffer (0.5 mole/L Tris-HCl, pH 6.8, containing SDS [4 g/100 mL], glycerol [20 g/100 mL] and β-ME [10 g/100 mL]), and the mixture was boiled for 3 min. The samples (20 mg protein) were loaded into the polyacrylamide gel made with 10% running gel and 4% stacking gel, and were subjected to electrophoresis at a constant current of 15 mA. After separation, the proteins were stained with Coomassie Brilliant Blue R-250 (0.02 g/100 mL) in methanol (50 mL/100 mL) and acetic acid (7.5 mL/100 mL), and destained with methanol (50 ml/100 ml) and acetic acid (7.5 mL/100 mL), followed by methanol (5 mL/100 mL) and acetic acid (7.5 mL/100 mL). The electrophoresis gel was scanned with a Bio-Rad imaging scanning densitometer (Versa Dco 3000, Bio-Rad, Milan, Italy).

Analysis of electrophoresis gel

The scanned electrophoresis gel image (300 dpi) was subjected to the optical density (O.D) analysis. Optical density is the measure of the transmission of an optical medium for a given light. Higher O.D indicates lower transmittance and vice versa. The O.D of the electrophoresis gel image was determined by a Gel-Pro Analyzer 4.0 (Media Cybernetics, L. P., USA). The O.D is calculated using the equation given below:

$$OD(x, y) = \frac{-\log(INTENSITY(x, y) - BLACK)}{INCIDENT - BLACK}$$

Where: INTENSITY (x, y) is the intensity at pixel (x,y); BLACK is the intensity generated when no light goes through the material; INCIDENT is the intensity of the incident light.

Statistical analysis

The results were analyzed by ANOVA at a significance level of 5% (H0: P<0.05). The comparison of means was analyzed by Fisher's LSD tests using the SAS statistical package (Statistical Analysis Systems Institute (SAS), 2000).

RESULTS AND DISCUSSION

Effect of heat cooking on protein and moisture contents of rabbit meat

The protein and moisture contents of the raw (control) and the cooked rabbit meats are shown in Figure 1. Both boiling and frying reduced the moisture content of rabbit meat significantly (P<0.05) as compared to the control, but the rate of moisture reduction is significantly (P<0.05) higher for frying than boiling. Boiling of rabbit meat for 5 (B5), 15 (B15) and 40 min (B40) caused a moisture reduction of 12.03, 11.66 and 12.98%, respectively. The frying of rabbit meat for 2 (F2), 4 (F4) and 6 min (F6) resulted in the moisture reduction of 49.92, 96.56 and

Table 1. Effect of cooking on color of rabbit meat.

	Control	B5	B15	B40	F2	F4	F6
L*	64.82±0.07a	64.54±0.39a	63.31±1.46b	62.20±0.44c	47.38±0.25d	36.26±0.83e	21.50±0.14f
a*	-1.95±0.12d	-2.85±0.11e	-1.64±0.58d	-2.82±0.48e	4.40±0.15b	12.52±0.14a	3.49±0.19c
b*	25.48±0.34c	24.32±0.05 d	24.26±0.44d	26.04±0.52c	37.54±0.79a	28.18±0.35b	2.21±0.18e
W	56.52±0.26 a	56.90±0.33a	55.97±1.01a	54.01±0.30b	35.21±0.59c	29.19±0.73d	21.40±0.14e

Each data is a mean of three replicate; Means (n=3) without a common letter differ significantly (P<0.05); "Control" is uncooked rabbit meat, B5, B15 and B40 represents meats boiled for 5, 15 and 40 min, respectively; F2, F4 and F6 represents meat fried for 2, 4 and 6 min, respectively; L* (lightness), a* ((+)redness/(-)greenness) and b* ((+)yellowness/(-)blueness), W (whiteness).

98.88%, respectively. The above results showed that the boiling has less influence on moisture content of rabbit meat as compared to frying which led to severe moisture loss to the extent of 98.88% depending on the frying time. Similar results were reported for boiling and frying of rainbow trout meat by Gokoglu et al. (2004), who found that frying and boiling had considerable effect on the proximate composition and mineral contents of the cooked fish comparing with the raw meat.

All the heat cooking treatments performed in this study resulted in significant loss of protein content of rabbit meat as compared to the control except the boiling for 5 min (B5). The loss of protein content of rabbit meat after boiling for 5 (B5), 15 (B15) and 40 min (B40) were 2.51, 7.75 and 17.78%, respectively. These results are in agreement with the results obtained by Bosco et al. (2001) for the loss of crude protein of rabbit meat during boiling. The protein loss during frying treatments of 2 (F2), 4 (F4) and 6 min (F6) were 17.20, 25.17 and 32.06%, respectively. The loss of protein content increased with an increase in both boiling and frying times and the loss caused by frying is greater than that of boiling. Frying resulted in rabbit meat with a low protein content similar to the results obtained by Oz et al. (2007). New chemicals, such as heterocyclic amines and polycyclic aromatic hydrocarbons, were formed during the frying might be the reason of the low protein content of fried rabbit meat. Those chemicals have been confirmed to contribute to the risk of developing colorectal tumors (Sinha et al., 1999; Wong et al., 2012), cancer of oral cavity and pharynx and esophagus (Di Maso et al., 2013).

Water molecules are highly polar and attracted to the muscle proteins by ionizable basic and acidic groups such as arginine, histidine, lysine, glutamic acid, etc (Wierbicki and Deatherage, 1958). The meat protein was denatured by heating and a part of its ionized basic or acid groups were broken. This led to the release of the water molecules in meat, which resulted in low moisture content of cooked meats than that of the control. The moisture loss of fried meats was more than those of boiled ones. The possible reason for this might be that the high temperature frying (175±3°C) caused more and rapid denaturation of meat protein when compared with low temperature boiling, resulting in more loss of water molecules (Skog et al., 1995). Oil absorption during frying might be another reason for low moisture content of fried meats than that of the boiled meats.

Prolonged boiling caused hydrolyzes of part of connective tissue and other protein in meat (Mutilangi et al., 1996) and leaking out of sarcoplasmic proteins from the muscle fibers channel (Murphy and Marks, 2000). Hydrolysis of proteins and leaking out of sarcoplasmic proteins might be the reason for the higher loss of protein content from the meat boiled for 15, 40 min than the meat boiled for 5 min and the control. The increased loss of protein content with an increase in frying time was related to the loss of nitrogen in the form of volatile nitrogenous compounds during frying (Yu et al., 1993).

Effect of heat cooking on color of rabbit meat

The CIELAB color (L*, a*, b*) values obtained from the raw and cooked rabbit meat samples are shown in Table 1. The L* value of rabbit meat boiled for 5 min (B5) was similar to that of control, whereas the L* values of meat boiled for 15 (B15) and 40 min (B40) and fried for 2 (F2), 4 (F4) and 6 min (F6) were significantly (P<0.05) lower than that of control. The L* value decreased in the order of B15>B40>F2>F4>F6. The L* values showed that the lightness of the rabbit meat was less affected by boiling than the frying.

The a* values of meat obtained from frying treatments F2, F4 and F6 were significantly (P<0.05) higher than that of control, and the boiling treatments B5, B15 and B40 (Table 1). It was also found that a* of control, B5, B15 and B40 were negative, but the a* of F2, F4 and F6 were positive. The positive a* values indicated that the frying made the cooked meat to become red and boiling treatments maintained the greenness (-a*) of meat similar to that of control. There was no significant (P>0.05) difference observed in b* values for the boiling treatments B5, B15 and B40 as compared to control whereas the b* values of meat samples from frying treatments, F2, F4, F6 were found to be significantly (P<0.05) different from the control. The highest b* value was obtained for F2 and b* decreased with increase in frying time. The b* values indicated that the frying had a greater influence on b* of meat than that of the boiling. No significant (P>0.05) difference in the whiteness values was observed for meat samples of control and treatments B5 and B15, but the whiteness values of meat samples of B40, F2, F4 and F6 were found to be significantly (P<0.05) different than the control. The whiteness values are found to be in the ascending order of B40>F2>F4>F6. These results showed that either boiling for longer time or frying for even shorter time (2 min) caused the meat to become darker and the frying severely affected the whiteness of cooked meat. These results are consistent with the findings of Lien et al. (2002) for heat processing of pork patties.

Higher moisture content in the meat increased the lightness of meat gels (Park, 2006). The boiled meat samples had a higher moisture contents than the fried meats, which might be the reason for the higher L* values of boiled meats than those of the fried ones. Previous findings showed that the higher moisture content decreased yellowness (b*) (Park, 2006) and higher cooking temperature decreased the b* value of cooked meat (Brewer and Novakofski, 1999). These findings did not help much to explain the changes in b* values for the boiled or fried meat samples. The low a* values of meat samples of low temperature treatments can be corroborated to the prior findings that red color in cooked meat diminished due to a masking of the myoglobin by aggregation/co-precipitation of other myofibrillar and sarcoplasmic proteins at lower temperatures (Brewer and Novakofski, 1999). The darker color of fried meat was also observed by Sosa-Morales et al. (2006), who reported that frying resulted in a darker crust color of pork meat. Higher redness values in fried products could also be attributed to concentration of meat pigments as moisture content was decreased and the Maillard reaction happened during meat frying, which resulted in the formation of heterocyclic amines (Wong et al., 2012) and have been confirmed to contribute to the risk of developing colorectal tumors (Sinha et al., 1999), deep-frying of meat resulted in a browning color (Dai et al., 2002; Lyn-Cook et al., 2011). All these might be a possible reason why the a* of the fried meat samples had positive a* values and also higher than those of the boiled meats.

The equation used for calculating the whiteness values from the observed L*, a*, and b* values illustrated that L* influences the whiteness more than a* and b*. Therefore, the whiteness of meat samples changed similarly to the change in the lightness value.

Effect of heat cooking on surface feature of rabbit meat

In order to differentiate surface feature of cooked meat samples, the micro structure of raw rabbit meat (control) and the cooked rabbit meats were determined (Figure 2). When compared with the control, the myofilament of the boiled meats (B5, B15 and B40) were clearly seen with the increase in boiling time, but the myofilament of the fried meats (F2, F4, F6) became fuzzier with the increase in frying time, and a black coagulation on the surface was observed, especially on the meat fried for 6 min.

Boiling and frying caused moisture loss of meat samples (Figure 1). Water molecules were attracted to the muscle proteins (Wierbicki and Deatherage, 1958) and heating led to denaturation of meat protein as discussed earlier. The water molecules bounded at myofilament were transferred from myofibrils to the extracellular space (Bertram et al., 2002). This transformation of water within the meat myofibrils and the denaturation of meat protein caused the shrinkage of meat within the tissue matrix, and resulted in the myofibrils standing out (Murphy and Marks, 2000). Therefore, the myofibrils of the boiled meats were seen clearly with increase in boiling times. The surface of the fried meat samples looked fuzzier with the increase in frying time. The possible reason for this could be that the high temperature during frying caused severe denaturation of the myofibrillar protein and some of the frying oil is absorbed in fried meat replacing water (Fillion and Henry, 1998). Maillard reaction happened during frying and formation of dark color heterocyclic amines (Wong et al., 2012) is another possible reason causing the natural surface of the rabbit meat to look tougher and fuzzier.

Effect of heat cooking on *in vitro* digestibility of rabbit meat

The *in vitro* digestibility of cooked rabbit meats were evaluated by TCA-soluble nitrogen release during digestion of pepsin and trypsin in simulated gastric fluid (Figure 3). The nitrogen release rate for boiled (B5, B15 and B40) and fried (F2, F4, F6) meats in pepsin section showed a similar trend during 30 to 60 min, but the differences appeared after 60 min. The nitrogen release rates for B15, B40 was higher than that of B5, F2, F4 and F6, especially after 120 min. The nitrogen release rate for longer duration frying (F6) was the lowest and the nitrogen release rate for B5, F2 and F4 were in between. Similar trend was observed during the trypsin digestion, but the nitrogen release rate for B5, B15 and B40, was found to be higher than that of F2 and F4. The nitrogen release rate for 6 min (F6) frying had the lowest values.

The *in vitro* digestibility of boiled and fried rabbit meat samples implied that, the boiled meats were easy to be digested than the fried ones. The longer boiling was found to be beneficial to the digestibility of meat protein while longer frying made the meat protein hard to be digested. The possible reason is that the frying leads to formation of heterocyclic aromatic amines in protein-rich

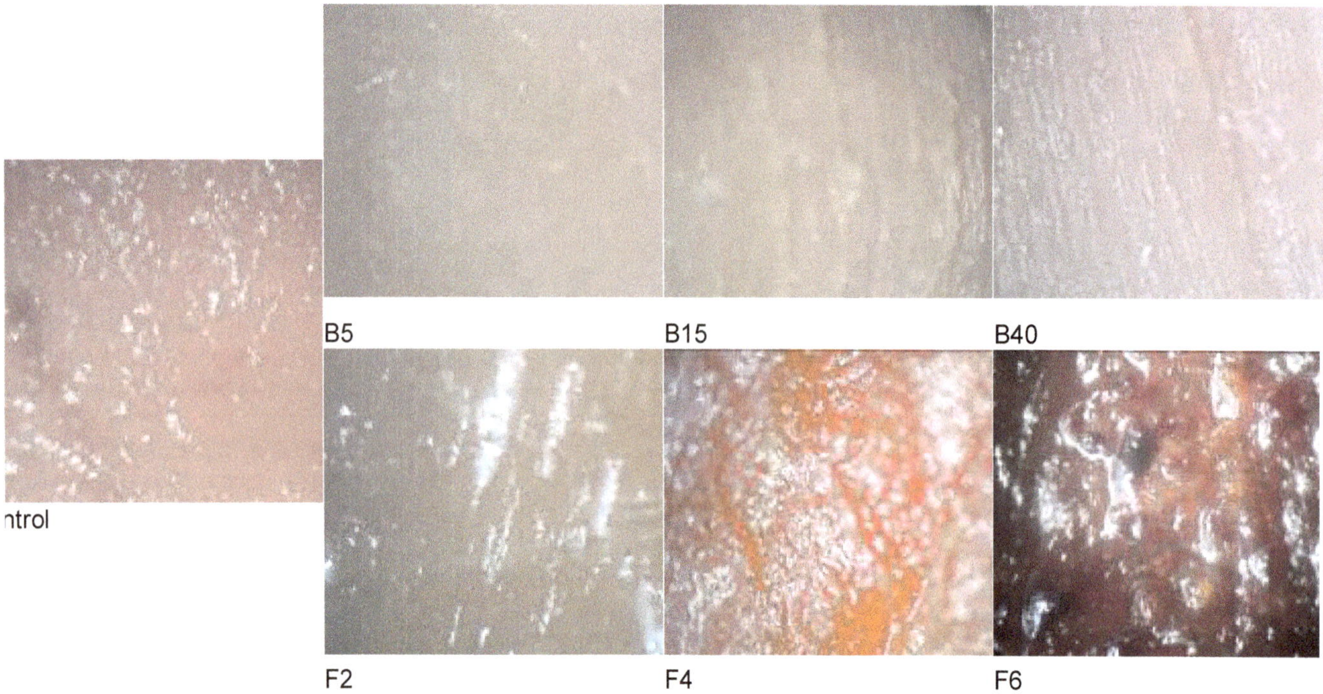

Figure 2. Effect of cooking on micro-surface of rabbit meat. "Control" is uncooked rabbit meat, B5, B15 and B40 represents meats boiled for 5, 15 and 40 min respectively; and F2, B4 and B6 represents meats fried for 2, 3 and 6 min, respectively.

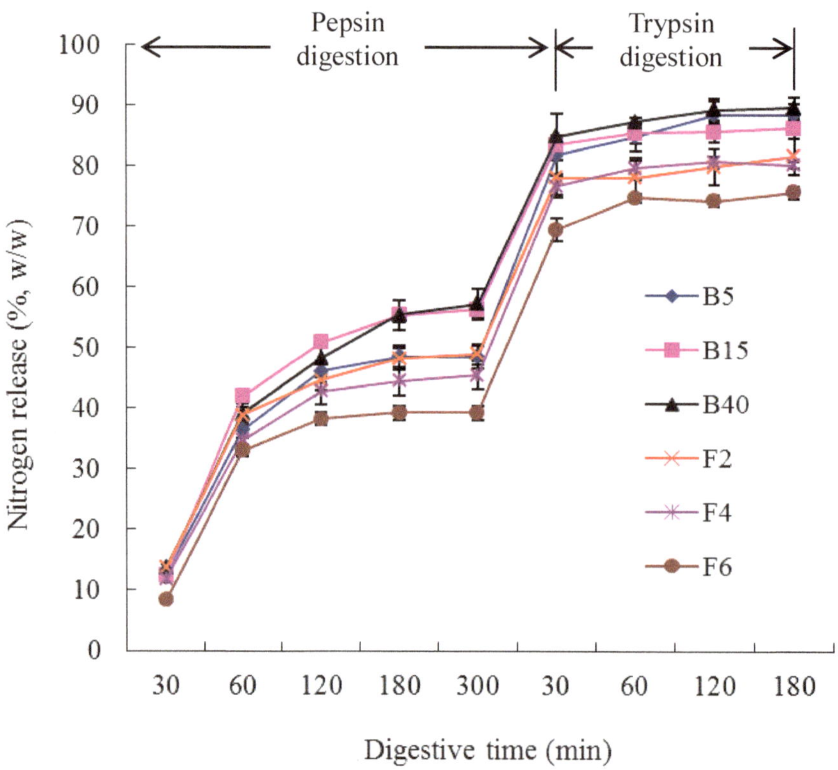

Figure 3. Effect of cooking on digestibility of rabbit meat *in vitro*. Means (n=3) without a common letter differ significantly (P<0.05); "Control" is uncooked rabbit meat, B5, B15 and B40 represents meats boiled for 5 min, 15 min, and 40 min respectively; and F2, B4 and B6 represents meats fried for 2, 3 and 6 min, respectively.

Table 2. Effect of cooking methods on amino acid of rabbit meat (mg/g protein).

Amino acid	B5	B15	B40	F2	F4	F6
Phenylalanine	9.012±0.018^{d}	11.414±0.020^{a}	10.427±0.131^{b}	9.096±0.006^{cd}	9.174±0.011^{c}	8.686±0.009^{e}
Alanine	18.922±0.067^{d}	21.145±0.034^{a}	21.227±0.094^{a}	18.157±0.075^{e}	19.698±0.094^{b}	19.300±0.057^{c}
Methionine	7.624±0.010^{c}	8.424±0.010^{b}	8.538±0.004^{a}	6.945±0.019^{e}	7.269±0.006^{d}	6.267±0.011^{f}
Proline	11.440±0.003^{d}	12.721±0.010^{a}	12.279±0.057^{c}	11.222±0.037^{e}	10.782±0.007^{f}	12.381±0.009^{b}
Glycine	14.736±0.160^{c}	16.257±0.086^{a}	16.106±0.075^{a}	15.311±0.094^{b}	15.323±0.037^{b}	14.640±0.057^{c}
Arginine	15.857±0.048^{c}	19.122±0.172^{a}	17.641±0.008^{b}	15.853±0.007^{c}	15.478±0.016^{d}	10.364±0.006^{e}
Lysine	20.829±0.058^{c}	25.043±0.064^{a}	22.817±0.075^{b}	20.066±0.031^{d}	16.667±0.008^{e}	12.453±0.004^{f}
Leucine	19.990±0.016^{c}	24.384±0.005^{a}	22.706±0.037^{b}	19.810±0.094^{d}	19.909±0.004^{c}	18.691±0.057^{e}
Tyrosine	8.554±0.005^{c}	10.584±0.052^{a}	9.632±0.094^{b}	8.550±0.057^{c}	8.277±0.037^{d}	7.877±0.024^{e}
Serine	10.913±0.144^{c}	12.267±0.052^{b}	12.478±0.094^{a}	10.278±0.094^{d}	9.184±0.057^{e}	3.318±0.112^{f}
Threonine	11.041±0.067^{c}	13.460±0.104^{a}	12.763±0.094^{b}	10.862±0.043^{d}	10.365±0.037^{e}	5.642±0.131^{f}
Glutamic acid	40.524±0.032^{d}	47.270±0.059^{a}	46.020±0.057^{b}	39.732±0.057^{e}	41.124±0.037^{c}	38.638±0.057^{f}
Aspartic acid	25.607±0.113^{b}	28.978±0.521^{a}	28.671±0.373^{a}	23.948±0.301^{c}	22.233±0.057^{d}	16.529±0.094^{e}
Isoleucine	8.521±0.006^{f}	11.630±0.031^{a}	9.732±0.013^{b}	9.097±0.075^{c}	8.836±0.004^{d}	8.686±0.006^{e}
Histidine	10.091±0.064^{e}	10.436±0.068^{d}	10.688±0.094^{d}	11.371±0.037^{c}	13.628±0.038^{a}	13.173±0.388^{b}
Valine	9.675±0.032^{e}	13.108±0.039^{a}	11.235±0.022^{b}	10.576±0.094^{c}	10.141±0.037^{d}	10.166±0.057^{d}
Cystine	0.609±0.010^{e}	0.582±0.012^{f}	1.567±0.008^{b}	0.674±0.009^{d}	1.197±0.011^{c}	2.126±0.008^{a}

Means (n=3) without a common letter differ significantly (P<0.05); B5, B15 and B40 represents meats boiled for 5, 15 and 40 min respectively; and F2, B4 and B6 represents meats fried for 2, 3 and 6 min, respectively.

foods such as meat and fish at temperatures mostly over 150°C (Oz et al., 2007), which are considered difficult to be decomposed and harmful as they are carcinogenic (Meng et al., 2009, Knize and Felton, 2005). Frying formed a hard dry layer on the surface of meat which prevented loss of internal nutrients of meat (Fillion and Henry, 1998) and also frying led to loss of meat protein (Yu et al., 1993). Harmful and hard to hydrolyzed chemicals, such as heterocyclic amines, formed during meat frying (Wong et al., 2012). All these facts could be cited as the possible reason for the slow nitrogen release rate of fried rabbit meats than those of the boiled meats.

Effect of heat cooking on amino acids content of rabbit meat

The mean amino acid contents of the boiled and fried rabbit meats are presented in Table 2 and the nutrition analysis of amino acids are presented in Table 3. All amino acids, except cystine, increased significantly (P<0.05) when rabbit meat was boiled for 15 min (B15) than boiled for 5 min (B5). The highest increase rate of 36.49% was observed for isoleucine, followed by valine with 35.48% increase. Meat boiled for 40 min also resulted in significant increase in the amino acid contents but at a lower rate when compared with 15 min boiling. These results are in acceptance with our findings on the digestibility of boiled meat which showed that the digestibility of B15 and B40 were easier than that of B5.

Most amino acids decreased significantly after frying for 4 and 6 min while frying for 2 min caused an insignificant (P>0.05) decrease in amino acids contents. Amino acids, which decreased during frying include proline, arginine, lysine, tyrosine, serine, threonine, aspartic acid, isoleucine and valine. Amino acids, lysine and serine had a higher decrease rate than other amino acids. Frying for 6 min caused a severe reduction in the amino acid content than the other two frying treatments, similar results were reported as four marine fishes commonly consumed in Nigeria (*Clupea harengus, Scomber scombrus, Trachurus trachurus* and *Urophycis tenuis*) were fried (Oluwaniyi et al., 2010). Though the boiling of meat for 15 and 40 min showed an increase in amino acid contents when compared with that of 5 min, meat boiled for 15 min contained more amino acids than the meat boiled for 40 min. Boiling resulted in denaturation of proteins in meat and the denatured meat proteins could be easily hydrolyzed (Adler-Nissen, 1976). When dissolved in hydrochloric acid for determination of amino acids, the meat boiled for longer time with high content of denatured proteins were easily hydrolyzed resulting in high amino acid contents than the meat boiled for shorter time. This is the reason for the increase in amino acid contents with the increase in boiling time. However, the prolonged boiling for 40 min resulted in more protein loss in meat and therefore the amino acid contents in meat boiled for 40 min were less than that of meat boiled for 15 min. This speculation agreed with that of Murphy et al. (2008), who reported that an endothermic transition at a temperature

Table 3. Effect of cooking methods on protein nutrition of rabbit meat.

Parameter	B5	B15	B40	F2	F4	F6
Total amino acids (mg/g, w/w)	243.9453±0.1403c	286.8250±0.8239a	274.5260±0.6699b	241.5478±0.1699d	239.2860±0.0980e	208.9370±0.0140f
Essential amino acid (% w/w)	35.5378±0.0621c	37.4667±0.0816a	35.7773±0.0281b	35.7906±0.0795b	34.4197±0.0289d	33.7858±0.1075e
*Chemical score	0.3043±0.0002e	0.4154±0.0011a	0.3476±0.0005b	0.3249±0.0027c	0.3156±0.0001d	0.2015±0.0047f
**Chemical score	0.6307±0.0040d	0.6523±0.0043c	0.6680±0.0059b	0.6998±0.0057a	0.6797±0.0003b	0.6269±0.0145d
*Essential amino acid index	0.3346±0.0005c	0.4041±0.0005a	0.3784±0.0009b	0.3354±0.0005c	0.3269±0.0003d	0.2871±0.0010e
**Essential amino acid index	0.6238±0.0008c	0.7535±0.0009a	0.7054±0.0017b	0.6253±0.0010c	0.6095±0.0006d	0.5353±0.0019e
Protein efficiency ratio	7.7091±0.0068 c	9.4911±0.0048a	8.8290±0.0227b	7.6279±0.0417d	7.7018±0.0022c	7.1908±0.0280e

Means (n=3) without a common letter differ significantly (P<0.05); B5, B15 and B40 represents meats boiled for 5, 15 and 40 min respectively; and F2, B4 and B6 represents meats fried for 2, 3 and 6 min respectively. *Based on amino acid requirements of school children (10-12 years); **Based on amino acid requirements of adults.

range of 59.6 to 68.4°C for collagen in chicken breast patties which resulted in an increase of soluble protein. The soluble protein dissolved in water caused decrease of protein content in chicken breast patties. Similar results were also obtained by Winegarden et al. (1952) for strips of collagenous material heated in distilled water.

The amino acid content of rabbit meat decreased with an increase in frying time. The possible reason is that frying led to formation of heterocyclic aromatic amines (Oz et al., 2007), and a hard and dry surface on the meat (Fillion and Henry, 1998), which made the meat proteins difficult to hydrolyze during amino acid analysis (Yu et al., 1993) and caused the meats fried for longer duration to have a low amino acid content. This result is consistent with that of longer frying time which resulted in low protein of cooked rabbit meat. Heterocyclic amines and polycyclic aromatic hydrocarbons formed during the frying cook, were found to contribute not only to the risk of developing colorectal tumors (Sinha et al., 1999), but also to lung adenocarcinoma (Butler et al., 2013).

The protein nutritional value of boiled and fried rabbit meats are shown in Table 3. The total and essential amino acid contents of boiled rabbit meats (B5, B15 and B40) are significantly (P<0.05) different from each other. The rabbit meat boiled for 15 min (B15) had the highest total and essential amino acid contents followed by meats boiled for 40 (B40) and 5 min (B5). The boiled rabbit meats (B5, B15 and B40) are also found to be significantly (P<0.05) different in the values of chemical score of school children (10-12 years) and adults, essential amino acid index of school children (10-12 years) and adults and the protein efficiency ratio. These results suggest that the boiling time is closely related to the nutritional facts of rabbit meat and there is an optimum time (15 min) to get maximum nutritional value in boiled meat. Boiling for a shorter or longer duration than the optimum time (15 min), resulted in decrease in the nutritional value.

When the protein nutritional value of fried meats were compared, it was found that the total amino acid contents in the meat fried for 4 min (F4) was significantly (P<0.05) lower than that for 2 min (F2), and the total amino acid content of F2 was significantly (P<0.05) lower than that for 6 min (F6). Similar significant (P<0.05) difference was discerned when the essential amino acid content, chemical score and essential amino acid index of F2, F4 and F6 were compared. Exceptionally, the protein efficiency ratio of F4 was found to be significantly (P<0.05) higher than that of F2, and the protein efficiency ratio of F6 was significantly (P<0.05) lower than that of both F2 and F4. The differences of protein nutritional value of F2, F4 and F6 implied that a proper and optimum frying time is essential to maintain the protein nutritional value of rabbit meat and frying for longer time resulted in a severe loss of protein nutritional value.

When comparing the protein nutritional value of boiled and fried rabbit meat, it can be found that the total amino acids, the essential amino acid index of both school children (10-12 years) and adults, and the protein efficiency ratio of boiled meats (B5, B15 and B40) are significantly (P<0.05) higher than that of fried meats (F2,F4 and F6). There is no similar significant (P>0.05) differences observed when the chemical score values of boiled and fried meats were compared. Except for chemical score of adults, all the protein nutritional value parameters of F6 is significantly (P<0.05) lower than B5, B15, B40, F2 and F4. This difference reiterated that frying rabbit meat for

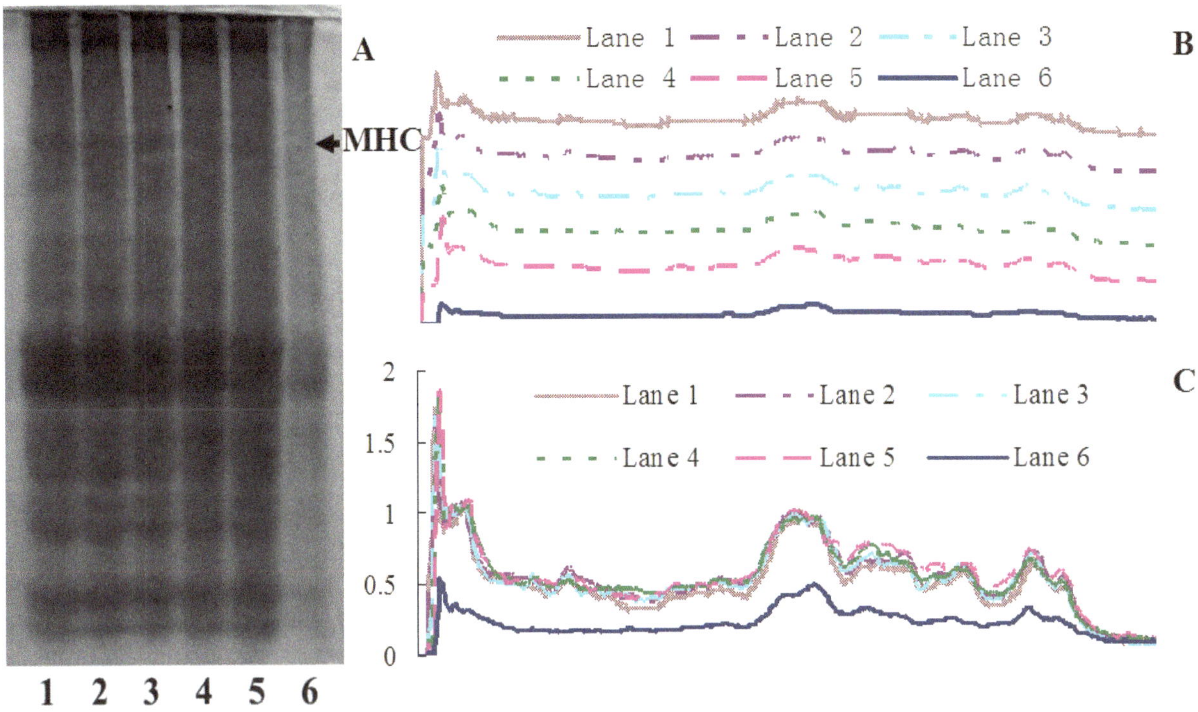

Figure 4. SDS-PAGE of the cooked rabbit meat. Lane 1, 2, 3 represents boiled 5, 15, 40 min, respectively; Lane 4, 5, 6 represents fried 2, 4 and 6 min, respectively; MHC is myosin heavy chains

longer time caused a severe nutrition loss. Beside the protein nutrition was decreased by frying, other investigations showed that frying process caused a significant alteration in the fatty acid profiles of hammour fish lipid and formation of harmful compounds in hammour fish fillets (Ganbi, 2011), heterocyclic aromatic amines were produced during pork frying, and their levels increased with increasing frying time and temperature (Zhang et al., 2013).

SDS-PAGE of cooked rabbit meat

In order to demonstrate the effect of boiling (B5, B15 and B40) and frying (F2, F4, F6) on protein pattern of rabbit muscle meat, sodium dodecyl sulfate polyacrylamide gel electrophoresis (SDS-PAGE) was performed. The results of SDS-PAGE analysis is shown in Figure 4A, along with the optical density (O.D) analysis of the gel (Figure 4B and C). From the Figure 4B and C, it was concluded that there is no difference between O.D values of B5, B15 and B40, but a notable difference is observed between the O.D curves of F2, F4 and F6. The O.D curve of F6 lies below the F2 and F4 (Figure 4 C) and also the O.D curve of F6 is different from all other heat treatments as shown in Figure 4 B.

The O.D values are proportional to the grey value of the bands in the electrophoresis image and a higher O.D value indicated larger grey value (Zhang et al., 2011). The difference of O.D in Figure 4B and C indicate that the protein pattern of the rabbit muscle meat was mildly affected by boiling for 5, 15 and 40 min and the protein pattern of rabbit muscle meat was moderately affected by frying for 2 and 4 min. But the protein pattern was seriously decreased by frying for 6 min. Boiling also resulted in the decrease of myosin heavy chains (MHC) (Figure 4A, B and C), which are polypeptides with a molecular mass of about 200 KDa (Sandri et al., 1999). This result agrees with the findings of Murphy and Marks (2000), who reported the decrease in the MHC of ground chicken breast after boiling.

Conclusions

The nutritional value of rabbit meat was significantly influenced when it was cooked by boiling or frying for different times. Boiling for 15 min enhanced the *in vitro* digestibility and nutritional value of rabbit meat, whereas boiling for 5 or 40 min led to their loss. Frying for 2 and 4 min helped to obtain an acceptable *in vitro* digestibility and nutritional value; even though the values were lower than that of boiled rabbit meat. The frying of rabbit meat for 6 min resulted in a severe impairment to the protein content, color, muscle surface structure, *in vitro* digestibility and protein nutrition. All these results strongly suggest that a proper boiling or frying duration is necessary to acquire a nutritious and high quality cooked rabbit meat.

ACKNOWLEDGEMENTS

The authors express their appreciation to the Department of Science and Technology of Sichuan Province (10ZC1292-4; 09ZC1719-4) and the Department of Science and Technology of China (2011GA810017) for the financial support.

REFERENCES

Adler-Nissen J (1976). Enzymic hydrolysis of proteins for increased solubility. J. Agric. Food Chem. 24:1090-1093.

Al-Khalifa AS, Dawood AA (1993). Effects of cooking methods on thiamin and riboflavin contents of chicken meat. Food Chem. 48:69-74.

Alsmeyer R, Cunningham A, Happich M (1974). Equations predict PER from amino acid analysis. Food Technol. 28:34-38.

AOAC: Official methods of analysis(2006). 18th edn. Gaithersburg Association of Official Analyical Chemists International.

AOAC 950.46 (1990). Official methods of analysis (15th ed.). Arlington, VA: AOAC International.

Berry B, Leddy K (1984). Effects of fat level and cooking method on sensory and textural properties of ground beef patties. J. Food Sci. 49:870-875.

Bertram HC, Purslow PP, Andersen HJ (2002). Relationship between meat structure, water mobility, and distribution: A low-field nuclear magnetic resonance study. J. Agric. Food Chem. 50:824-829.

Bidlingmeyer BA, Cohen SA, Tarvin TL (1984). Rapid analysis of amino acids using pre-column derivatization. J. Chromatogr. B Biomed. Sci. Appl. 336:93-104.

Bosco AD, Castellini C, Bernardini M (2001). Nutritional quality of rabbit meat as affected by cooking procedure and dietary vitamin E. J. Food Sci. 66:1047-1051.

Brewer MS, Novakofski J (1999). Cooking rate, pH and final endpoint temperature effects on color and cook loss of a lean ground beef model system. Meat Sci. 52:443-451.

Butler LM, Montague JA, Koh WP, Wang R, Mimi CY, Yuan JM (2013). Fried meat intake is a risk factor for lung adenocarcinoma in a prospective cohort of Chinese men and women in Singapore. Carcinogenesis 34:1794-1799.

Cavani C, Petracci M (2004). Rabbit meat processing and traceability. Proc. 8th World Rabbit Congr., Puebla, Mexico. 13:18-1336.

Cavani C, Petracci M, Trocino A, Xiccato G (2010). Advances in research on poultry and rabbit meat quality. Ital. J. Anim. Sci. 8:741-750.

Chen Y (2006). Cooking methods of Rabbit Meat. Chin. J. Rabbit Farming 1:34-36.

Choubert G, Baccaunaud M (2010). Effect of moist or dry heat cooking procedures on carotenoid retention and colour of fillets of rainbow trout (*Oncorhynchus mykiss*) fed astaxanthin or canthaxanthin. Food Chem. 119:265-269.

Combes S (2004). Valeur nutritionnelle de la viande de lapin. Productions animales 17:373-383.

Combes S, Lepetit J, Darche B, Lebas F (2004). Effect of cooking temperature and cooking time on Warner–Bratzler tenderness measurement and collagen content in rabbit meat. Meat Sci. 66:91-96.

Dai Q, Shu X, Jin F, Gao YT, Ruan ZX, Zheng W (2002). Consumption of animal foods, cooking methods, and risk of breast cancer. Cancer Epidemiol. Biomark. Prev. 11:801-808.

Dalle Zotte A (2002). Perception of rabbit meat quality and major factors influencing the rabbit carcass and meat quality. Livest. Prod. Sci. 75: 11-32.

Das R, Pawar DP, Modi VK (2013). Quality characteristics of battered and fried chicken: comparison of pressure frying and conventional frying. J. Food Sci. Technol. 50:284-292.

Di Maso M, Talamini R, Bosetti C, Montella M, Zucchetto A, Libra M, Negri E, Levi F, La Vecchia C, Franceschi S (2013). Red meat and cancer risk in a network of case-control studies focusing on cooking practices. Ann. Oncol. 24:3107-3112.

FAO (1985). Energy and protein requirements: Report of a joint FAO/WHO/UNU expert consultation, World Health Organization.

Fillion L, Henry C (1998). Nutrient losses and gains during frying: a review. Int. J. Food Sci. Nutr. 49:157-168.

Fu TJ, Abbott UR, Hatzos C (2002). Digestibility of food allergens and nonallergenic proteins in simulated gastric fluid and simulated intestinal fluid a comparative study. J. Agric. Food Chem. 50:7154-7160.

Ganbi HHA (2011). Alteration in Fatty Acid Profiles and Formation of Some Harmful Compounds in Hammour Fish Fillets and Frying Oil Medium Throughout Intermittent Deep-Fat Frying Process. World Appl. Sci. J. 12:536-544.

Gokoglu N, Yerlikaya P, Cengiz E (2004). Effects of cooking methods on the proximate composition and mineral contents of rainbow trout (*Oncorhynchus mykiss*). Food Chem. 84:19-22.

Iwami K, Sakakibara K, Ibuki F (1986). Involvement of post-digestion'hydrophobic'peptides in plasma cholesterol-lowering effect of dietary plant proteins. Agric. Boil. Chem. 50:1217-1223.

Kimura M, Itokawa Y (1990). Cooking losses of minerals in foods and its nutritional significance. J. Nutr. Sci. Vitaminol. (Tokyo) 36:S25-S32.

Knize MG, Felton JS (2005). Formation and human risk of carcinogenic heterocyclic amines formed from natural precursors in meat. Nutr. Rev. 63:158-165.

Laemmli UK (1970). Cleavage of structural proteins during the assembly of the head of bacteriophage T4. Nature 227:680-685.

Lanier TC, Hart K, Martin RE (1991). Manual of Standard Methods for Measuring and Specifying the Properties of Surimi. National Fisheries Institute, Arlington, VA.

Lyn-Cook LE, Tareke E, Word B, Starlard-Davenport A, Lyn-Cook BD, Hammons GJ (2011). Food contaminant acrylamide increases expression of Cox-2 and nitric oxide synthase in breast epithelial cells. Toxicol. Ind. Health 27:11-18.

Malik A, Aremu A, Bayode G, Ibrahim B (2011). A nutritional and organoleptic assessment of the meat of the giant African land snail (*Archachatina marginata Swaison*) compared to the meat of other livestock. Livest. Res. Rural Dev. 23:60-67.

Meng XX, Peng ZQ, Feng Y (2009). Effect Factors of Formation of Polycyclic Aromatic Hydrocarbon and Heterocyclic Amines in Fried Meat and Control Measures. Meat Res. 6:52-54.

Miller PE, Lazarus P, Lesko SM, Cross AJ, Sinha R, Laio J, Zhu J, Harper G, Muscat JE, Hartman TJ (2013). Meat-related compounds and colorectal cancer risk by anatomical subsite. Nutr. Cancer 65:202-226.

Modzelewska-Kapituła M, Dąbrowska E, Jankowska B, Kwiatkowska A, Cierach M (2012). The effect of muscle, cooking method and final internal temperature on quality parameters of beef roast. Meat Sci. 91:195-202.

Monahan F, Buckley D, Gray J, Morrissey P, Asghar A, Hanrahan T, Lynch P (1990). Effect of dietary vitamin E on the stability of raw and cooked pork. Meat Sci. 27:99-108.

Murphy R, Marks B (2000). Effect of meat temperature on proteins, texture, and cook loss for ground chicken breast patties. Poult. Sci. 79:99-104.

Murphy R, Marks B, Marcy J (2008). Apparent specific heat of chicken breast patties and their constituent proteins by differential scanning calorimetry. J. Food Sci. 63:88-91.

Mutilangi W, Panyam D, Kilara A (1996). Functional Properties of Hydrolysates from Proteolysis of Heat-denatured Whey Protein Isolate. J. Food Sci. 61:270-275.

Naseri M, Abedi E, Mohammadzadeh B, Afsharnaderi A (2013). Effect of frying in different culinary fats on the fatty acid composition of silver carp. Food Sci. Nutr. 1:292-297.

Oluwaniyi O, Dosumu O, Awolola G (2010). Effect of local processing methods (boiling, frying and roasting) on the amino acid composition of four marine fishes commonly consumed in Nigeria. Food Chem. 123:1000-1006.

Oz F, Kaban G, Kaya M (2007). Effects of cooking methods on the formation of heterocyclic aromatic amines of two different species trout. Food Chem. 104:67-72.

Park JAEW (2006). Surimi gel colors as affected by moisture content and physical conditions. J. Food Sci. 60:15-18.

Rodriguez-Estrada M, Penazzi G, Caboni M, Bertacco G, Lercker G (1997). Effect of different cooking methods on some lipid and protein components of hamburgers. Meat Sci. 45:365-375.

Sandri M, Rizzi C, Rossini K, Catani C, Cantini M, Spina M (1999). Purification of myosin heavy chain isoforms by electroendosmotic preparative gel electrophoresis: characterization of embryonic slow myosin heavy chain. BAM-PADOVA 9:71-78.

Schricker B, Miller D (1983). Effects of cooking and chemical treatment on heme and nonheme iron in meat. J. Food Sci. 48:1340-1343.

Sinha R, Chow WH, Kulldorff M, Denobile J, Butler J, Garcia-Closas M, Weil R, Hoover RN, Rothman N (1999). Well-done, grilled red meat increases the risk of colorectal adenomas. Cancer Res. 59:4320-4324.

Sinha R, Kulldorff M, Curtin J, Brown CC, Alavanja MCR, Swanson CA (1998). Fried, well-done red meat and riskof lung cancer in women (United States). Cancer Causes Control 9:621-630.

Skog K, Steineck G, Augustsson K, J Gerstad M (1995). Effect of cooking temperature on the formation of heterocyclic amines in fried meat products and pan residues. Carcinogenesis 16:861-867.

Sosa-Morales ME, Orzuna-Esp Ritu R, V Lez-Ruiz JF (2006). Mass, thermal and quality aspects of deep-fat frying of pork meat. J. Food Eng. 77:731-738.

Tasevska N, Sinha R, Kipnis V, Subar AF, Leitzmann MF, Hollenbeck AR, Caporaso NE, Schatzkin A, Cross AJ (2009). A prospective study of meat, cooking methods, meat mutagens, heme iron, and lung cancer risks. Am. J. Clin. Nutr. 89:1884-1894.

Walsh H, Martins S, O'Neill E, Kerry J, Kenny T, Ward P (2010). The effects of different cooking regimes on the cook yield and tenderness of non-injected and injection enhanced forequarter beef muscles. Meat Sci. 84:444-448.

Wierbicki E, Deatherage F (1958). Water content of meats, determination of water-holding capacity of fresh meats. J. Agric. Food Chem. 6:387-392.

Winegarden MW, Lowe B, Kastelic J, Kline EA, Plagge AR, Shearer PS (1952). Physical changes of connective tissues of beef during heatinga, b. J. Food Sci. 17:172-184.

Wong D, Cheng KW, Wang M (2012). Inhibition of heterocyclic amine formation by water-soluble vitamins in Maillard reaction model systems and beef patties. Food Chem. 133:760-766.

Yao Y, Peng Z, Shao B, Wan K, Wang F, Zhang Y, Li J, Hui T (2013). Effects of frying and boiling on the formation of heterocyclic amines in braised chicken. Poult. Sci. 92:3017-3025.

Yu TH, Wu CM, Ho CT (1993). Volatile compounds of deep-oil fried, microwave-heated and oven-baked garlic slices. J. Agric. Food Chem. 41:800-805.

Zhang Y, Yu C, Mei J, Wang S (2013). Formation and mitigation of heterocyclic aromatic amines in fried pork. Food Addit. Contam. Part A. 30:1501-1507.

Zhang Y, Zeng QX, Zhu ZW (2011). Effect of ultrasonic treatment on the activities of endogenous transglutaminase and proteinases in tilapia (*sarotherodon nilotica*) surimi during gel formation. J. Food Process Eng. 34:1695-1713.

Microbiological quality assessment of some National Agency for Food and Drug Administration and Control (NAFDAC) approved fruit juices sold in Ilorin metropolis

Olaniyi, O. O.

Department of Microbiology, Federal University of Technology, Akure, Nigeria.

This study was carried out to investigate on the microbiological qualities of some NAFDAC approved fruit juices sold in Ilorin metropolis. The brand of fruit juices assessed was coded as LL, LC and LB. The total viable counts, associated microorganisms and physicochemical properties of 10 randomly selected samples from each of the brand were determined. The highest mean bacterial load of 7.8 cfu/ml was found in fruit juice coded LL, while the lowest mean bacterial load of 3.5 cfu/ml was recorded for LB. The highest mean fungal load of 6.1 sfu/ml was observed with LL and the lowest mean load of 2.2 sfu/ml was found in LB. LL had no coliform growth, while LC and LB had the same mean load of 0.7 cfu/ml. The bacterial isolates identified from the samples were *Bacillus* sp., *Micrococcus* sp, *Proteus* sp, *Escherichia coli* and *Lactobacillus* sp, while the fungal isolates were *Aspergillus nidulans*, *Aspergillus niger*, *Cladosporum* sp. and *Schzosaccharomyces pombe.* The pH values for LL, LC and LB ranged from 3.78 to 3.83, 2.40 to 2.43 and 2.39 to 2.41, respectively. LC had the highest mean total titratable acidity which ranged from 0.14 to 0.66, followed by LB ranging from 0.10 to 0.16, while LL had the lowest values ranged from 0.07 to 0.13. *Bacillus* sp. is the most frequently encountered in all the samples with a percentage occurrence of 46.67%, while *Lactobacillus* sp., *Micrococcus* sp. and *Escherichia coli* had the lowest percentage occurrence of 36.67% each. In case of fungi, *Cladosporum* sp had the high percentage occurrence of 43.33%, while *A. nidulans* had the least percentage occurrence of 23.33% of all the samples examined.

Key words: Fruit juice, microbial quality, microbial load, microbial contamination.

INTRODUCTION

Traditionally, fruit juices are considered susceptible to spoilage only by yeast, mycelial fungi and lactic acid bacteria (Chang and Kang, 2004). Foods with a pH lower than 4 are considered as high in acid and are generally regarded as not being susceptible to spoilage by a variety of microorganisms (Jay, 1998). The low pH is considered sufficient to prevent the growth of almost all bacteria spore formers. Spores of *Clostridium botulinum* cannot germinate or produce the lethal *botulinum* toxin in an environment with a pH below 4.6 (Chang and Kang, 2004). The microorganisms present in fruit juice often originate from the natural flora of the raw materials used for the preparation and those introduced during the course of the processing (Splittstoesser et al., 1994). The number and types of organisms are determined by the properties of the food product and activity of the organisms in the product. In some cases, the micro-organisms have no discernable deleterious effects and the food is consumed without harm (Yeh et al., 2004). In other cases, however, the presence of microorganisms have manifested in form of spoilage, food-borne illness and fermentation (Yeh et al., 2004). During the heat treatment of foods, pathogens and most non spore forming microorganisms are killed, but a heat process

Table 1. Physical characteristics of samples.

Sample	Condition of the pack	Visual observation of the sample
LL	Intact plastic bottle	Normal colour of pale orange and no foreign particles
LC	Intact plastic bottle	Normal colour of caramel and no foreign particles
LB	Intact paper pack	Normal colour of pale orange and no foreign particles

LL, A brand of fruit juice; LC, a brand of fruit juice; LB, a brand of fruit juice.

Table 2a. Total bacterial and coliform counts (cfu/ml) and fungal counts (sfu/ml) of the fruit juice LL.

Brand/sample	TBC	TFC	TCC
LL1	17	5	ND
LL2	ND	8	ND
LL3	ND	9	ND
LL4	13	6	ND
LL5	16	1	ND
LL6	18	7	ND
LL7	6	6	ND
LL8	8	4	ND
LL9	ND	6	ND
LL10	ND	9	ND
Mean	7.8	6.1	0

TBC, Total bacterial count; TFC, total fungal count; LL, a brand of fruit juice; ND, not detectable.

sufficient to destroy all the microbial spores will have a detrimental effect on the organoleptic quality of the product (Walls and Chuyata, 2000). The purpose of this study is to evaluate the microbiological quality of some brands of fruit juice sold in Ilorin metropolis, Nigeria.

The presence of 'registration number' from 'regulatory agencies' may not necessarily give assurance of quality, as these can be faked.

MATERIALS AND METHODS

Brands of fruit juice used

The fruit Juices employed in this study were manufactured by companies in Nigeria and were purchased randomly at different locations in Ilorin metropolis. A total of 3 brands (LL, LC and LB) were used in the study and each brand contains 10 samples of different batches.

Physical observation of the samples

The packed juices were examined for leakage or bulged ends and changes in the colorations of the liquid content.

Determination of pH of the samples

The pH of the various samples was immediately determined using sterile probes of the pH meter (Corning 35).

Determination of total titratable acidity (TTA)

This was determined according to the method of Akharayi and Omoya (2005). Twenty-five millilitres of the samples with three drops of 1% phenolphthalein as an indicator was titrated against 0.1 M NaOH. The end point was recorded when pick colour was noted. The TTA was then calculated as lactic acid thus: volume of 0.1 M NaOH × 100 over volume of weight of sample used in titration.

Microbiological analysis

The experiment was carried out without further diluting of the fruit juice. This was done to determine the presence of viable micro-organisms in the fruit juices at the concentration at which they were consumed. Microbiological analysis included identification and enumeration of potential pathogens which was carried out according to standard procedures (Buchanan and Gibbons, 1974). The total colony count was done by pour plate method using nutrient agar for bacteria and potato dextrose agar for fungi (Lateef, 2004; Lateef et al., 2005). The presence of faecal coliforms was determined by using eosine methylene blue agar (Titarmare et al., 2009). All inoculated at the requisite time and temperature. The bacterial isolates were then identified following standard microbiological procedures as described by Buchanan and Gibbons (1974) and Cheesbrough (2002), while fungal isolates were identified based on the taxonomic schemes and descriptions by Ainsworth et al. (1973) and Mislivec et al. (1992).

RESULTS AND DISCUSSION

Physical characteristics of samples from the 3 brands were examined for leakage or bulged ends and changes in the coloration of the content (Table 1). The samples from the brand of fruit juice coded as LB were intact in paper pack, while LL and LC were intact in plastic bottles. The visual observation of samples from all the brands (LL, LC and LB) retained their normal colour without any suspended foreign particles. The physical characteristics of all the samples are in line with good manufacturing practices as stated in fruit juice and nectar regulations Act 2005. Fruit juices are well appreciated by consumers because of their taste, nutritional value and availability at the right time. Inspite of these potential benefits offered, concerns over their safety and quality have been raised. Result of this study (Table 2a, b and c) indicated the presence of bacteria, fungi and coliform in some of the samples selected randomly from the brands. At the concentration at which they were consumed, samples from

Table 2b. Total bacterial and coliform counts (cfu/ml) and fungal counts (sfu/ml) of the fruit juice LC.

Brand/sample	TBC	TFC	TCC
LC1	3	8	1
LC2	5	2	ND
LC3	ND	5	2
LC4	13	5	1
LC5	ND	2	ND
LC6	2	ND	ND
LC7	16	3	2
LC8	ND	7	ND
LC9	4	6	1
LC10	ND	ND	ND
Mean	4.3	3.8	0.7

TBL, Total bacterial count; TFC, total fungal count; LC, A brand of fruit juice; ND, not detectable.

Table 2c. Total bacterial and coliform counts (cfu/ml) and fungal counts (sfu/ml) of the fruit juice LB.

Brands/sample	TBC	TFC	TCC
LB1	20	3	1
LB2	ND	5	ND
LB3	ND	1	2
LB4	ND	4	1
LB5	1	4	ND
LB6	ND	1	ND
LB7	ND	3	2
LB8	ND	ND	ND
LB9	12	ND	1
LB10	2	1	ND
Mean	3.5	2.2	0.7

TBL, Total bacterial counts; TFC, total fungal counts; LB, a brand of fruit juice; ND, not detectable.

Table 3a. pH and total titratable acidity determination of the fruit juice LL.

Brand/sample	pH	TTA (ml)
LL1	3.80	0.07
LL2	3.81	0.13
LL3	3.79	0.08
LL4	3.80	0.07
LL5	3.83	0.07
LL6	3.80	0.10
LL7	3.78	0.08
LL8	3.78	0.11
LL9	3.79	0.13
LL10	3.80	0.13
Mean	3.80	0.10

LL , A brand of fruit juice; pH, hydrogen ion concentration; TTA, total titratable acidity .

Table 3b. pH and total titratable acidity determination of the fruit juice LC.

Brand/sample	pH	TTA (ml)
LC1	2.40	0.16
LC2	2.42	0.14
LC3	2.41	0.16
LC4	2.42	0.21
LC5	2.43	0.16
LC6	2.42	0.17
LC7	2.40	0.17
LC8	2.43	0.16
LC9	2.41	0.16
LC10	2.42	0.17
Mean	2.42	0.17

LC, A brand of fruit juice; pH, hydrogen ion concentration; TTA, total titratable acidity.

from LL recorded the highest mean total bacterial count of 7.8 cfu/ml followed by LC (4.3 cfu/ml), while the least mean value of 3.5 cfu/ml was recorded for LB. The mean total fungal counts for the samples from LL, LC and LB were 6.1, 3.8 and 2.2 sfu/ml, respectively. The samples from LC and LC indicated the presence of coliform, while LL had no growth after 24 h of incubation. The values obtained in this study for total bacterial and fungal counts fall below the range of 10^2 to 10^5 cfu/ml reported for microbial populations in fruit juices (Hatcher et al., 1992). However, the presence of bacteria and fungi could be an indication of spoilage. Also, low or high bacterial and fungal counts may be indicative of improper hygiene and may perhaps be a result of poor quality fruit being used (Lateef et al., 2004).

According to the microbiological guidelines as stated by Hatcher et al. (1992), still and carbonated beverages must be free of coliforms. The presence of coliform is not only an indication of poor hygienic quality of this juices but also places consumers at high risk of contacting food borne infections. The mean pH and titratable acidity (TTA) are shown in Table 3a, b and c. The results of pH determination showed that the samples from all the brands were acidic in nature. Samples collected from brand LB had the lowest mean pH of 2.40, while the highest mean pH of 3.80 was observed with samples from LL. The acidic nature of juices probably favours the growth of yeasts. In this study, a strain of

Table 3c. pH and total titratable acidity determination of the fruit juice LB.

Brand/samples	pH	TTA (ml)
LB1	2.40	0.10
LB2	2.39	0.16
LB3	2.39	0.13
LB4	2.39	0.10
LB5	2.41	0.12
LB6	2.40	0.16
LB7	2.41	0.10
LB8	2.40	0.13
LB9	2.40	0.13
LB10	2.41	0.10
Mean	2.40	0.12

LB = A brand of fruit juice, pH = Hydrogen ion concentration, TTA = Total titratable acidity.

Table 4. Percentage occurrence of bacterial and fungal isolates from the samples.

Bacterial isolates	% occurrence	Fungal isolates	% occurrence
Bacillus sp.	46.67	*Aspergillus niger*	40.00
Proteus sp.	40.00	*A. nidulans*	23.33
Lactobacillus sp.	36.67	*Schizosaccharomyces pombe*	33.33
Micrococcus sp.	36.67	*Cladosporium* sp.	43.33
Escherichia coli	36.67		

Schizosaccharomyces was isolated from the fruit juices. The isolation of these yeasts from fruit juice has been previously reported (Deak and Beuchat, 1993; Lateef et al., 2004). The pH of fruit juices is usually too low for the growth of pathogenic bacteria (Lateef et al., 2004; Jackson et al., 2010), but the incidence of such bacteria in the juices used in this study might not be unconnected with the nature of the juices. All the juices are ready-to-serve fruit juices and the water activity (aw) values are sufficiently high to allow microbial growth (Lateef et al., 2004). There was a slight variation in the TTA of the samples examined and this could be attributed to the differences in chemical constituents.

Microbial isolates and their percentage occurrence from the samples are shown in Table 4. The microbial isolates consisted of five bacteria and four fungi. The bacterial isolates were *Bacillus* sp, *Proteus* sp, *Lactobacillus* sp, *Micrococcus* sp and *Escherichia coli*, while the fungal isolates included *A. niger*, *A. nidulans*, *Schizosaccharomyces pombe* and *Cladosporium* sp. Frazier (1984) observed that the natural microflora plus contaminating microorganisms from the soil constitute the organisms found on the surface of healthy fruits. The presence of notable bacterial pathogens such as *Bacillus* sp, *Proteus* sp, *Micrococcus* sp and *E. coli* in fruit juices is considered a safety concern. *B. cereus*, an aerobic spore former is associated with food borne illnesses (Lateef et al., 2004), while *B. substilis* has been associated with food borne diseases. Some strains of *E. coli* synthesize heat stable enterotoxins and are responsible for diarrhoeal disease in humans and domestic animals. There is no justification for processed ready to eat food being contaminated with these organisms, and their presence even in small numbers results in such foods being of unacceptable quality or potentially hazardous (PHLS, 2000). The processing units of the juices are likely primary causes of bacterial and fungal load. The maintenance of proper hygienic conditions and use of good quality fruit and water will certainly improve the microbiological quality of these juices, and make them acceptable to quality conscious markets both locally and abroad. In this line, the establishment of quality control unit/laboratory becomes imperative to detect contamination of either the raw materials or the products early enough.

In addition, operators in these sectors should utilize the technical assistance of NAFDAC towards attaining acceptable quality standard.

Conclusion

The study indicates that all brands of fruit juices examined are contaminated which might be due to poor unhygienic conditions related to washing of utensils, use of contaminated water and ice, poor personal and domestic hygiene. The practice of consuming fruit and vegetable juices cannot be stopped on unhygienic grounds or prohibited from selling such items, since it is a source of their livelihood. However, government health agencies must adopt measures to educate the producers on food safety and hygienic practices. Regular monitoring of the quality of fruit juices for human consumption must also be enforced.

REFERENCES

Ainsworth GC, Sparrow FK, Sussman AS (1973). The Fungi Vol.1 VA:A Taxonomic Review with keys; Ascomycetes and Fungi imperfecti, London: Academic Press. pp. 13-67.

Akharayi FC, Omoya FO (2005). Physicochemical determination and sensory evaluation of wine produced from selected tropical fruits. Biosci. Biotechnol. Res. 3:21-28.

Buchanan RE, Gibbons NE (1974). Bergey's Manual of Determinative Bacteriology, Baltimore. Williams and Wilkins Co. 8th edn. pp. 34-89.

Chang S, Kang D (2004). *Alicyclobacillus* spp. in the fruit juice Industry: History, Characteristics and Current Isolation/Detection Procedures. Crit. Rev. Microbiol. 30: 55-74.

Cheesbrough M (2002). Biochemical Tests to Identify Bacteria. In: Laboratory Practice in Tropical Countries, Cheesbrough M (eds). Cambridge edn. pp. 63-70.

Deak T, Beuchat LR (1993). Yeasts associated with fruit juice concentrates. J. Food Prot. 56:777-782.

Frazier WC (1984). Wine and wine yeasts. In Food Microbiology 2nd Edition, Mc Grawhill New York. pp. 418.

Gilbert RJ, de Louvois J, Donovan T, Little C, Nye K, Ribeiro CD, Richards J, Roberts D, Bolton FJ (2000). Guidelines for the microbiological quality of some ready-to-eat foods sampled at the point of sale. PHLS Advisory Committee for Food and Dairy Products. Commun. Dis. Public Health 3(3):163-167.

Hatcher WS, Weihe JL, Splittstoesser DF, Hill EC, Parish ME (1992). Fruit Beverages. In: Compendium of methods for the microbiological examination of foods. Vanderzant C, Splittstoesser D.F (eds). American Public Health Association, Washington, DC.

Jackson C, Ibezim E, Akeem A, Udofia M, Odo H (2010). Determination of bioload of commercially available brands of fruit juices in Uyo, Nigeria. Res. Pharm. Biotechnol. 2(4): 51-53.

Jay JM (1998). Intrinsic and Extrinsic Parameters of Foods that affect Microbial Growth. In: Modern Food Microbiology, 5th ed. Chapman and Hah., New York. pp. 354-355.

Lateef A (2004). The Microbiology of a Pharmaceutical effluent and its Public health implications. World J. Microbiol. Biotechnol. 20: 167-171.

Lateef A, Oloke JK, Gueguim Kana EB (2004). Antimicrobial resistance of bacterial strains isolated from orange juice products. Afr. J. Biotechnol. 3(6): 334-338.

Lateef A, Oloke JK, Gueguim Kana EB (2005). The Prevalence of bacterial resistance in clinical, food, water and some environmental samples in Southwest Nigeria. Environ. Monit. Assess. 100: 59-69.

Mislivec PB, Beuchat LR, Cousin MA (1992). Yeasts and molds In: Compendium of methods for the microbiolocal examination of foods. Vanderzant C, Splittstoesser DF (eds). American Public Health Association, Washington, DC.

PHLS Advisory Committee for Food and Dairy products (2000).Guidelines for the microbiological quality of some ready –to-eat foods sampled at the point of sale. Comm. Disease and Public Health. 3:163-167.

Splittstoesser DF, Churey JJ, Lee CY (1994). Growth characteristics of aciduric spore-forming bacilli isolated from fruit juices. J. Food Prot. 57: 1080-1083.

Titarmare A, Dabholka P, Godbole S (2009). Bacteriological Analysis of Street Vended Fresh Fruit and Vegetable Juices in Nagpur City, India. Internet J. Food Safe. (11):1-3.

Walls I, Chuyate R (2000). Isolation of *Alicyclobacillus acidoterrstris* from fruits juices. J. AOAC Int. 83: 1115-1120.

Yeh JY, Ellis H, Chen J (2004). Influence of calcium lactate on the fate of spoilage and pathogenic microorganisms in orange juice. J. Food Prot. 67:1429-1433.

Safety of bread for human consumption in an urban community in Southwestern Nigeria

Olusegun T Afolabi[1], Olufemi O Aluko[1], Olaniran O[2], Olusola Ajao[1], Bolade K Ojumu[1] and Oluwatoyosi Olawande[3]

[1]Department of Community Health, Obafemi Awolowo University, Ile-Ife, Osun State, Nigeria.
[2]Department of Microbiology and Parasitology, Obafemi Awolowo University, Ile-Ife, Osun State, Nigeria.
[3]Institute of Human Virology, P. O. Box 9396, Garki, Abuja, Nigeria.

Bread is an important staple food that does not require further processing before consumption. Despite the awareness created and efforts of government to ensure that quality bread are available for consumption of all, there are several routes through which bread is being contaminated. The aim of the study was to assess the hygiene involved from the point of production to sale of bread and document pathogens isolated from bread. This was a descriptive cross-sectional study with 10 out of the 30 registered bakeries spread across wards randomly selected and only six consenting to participate (60% response rate), and a total of 10 representative bread sellers were purposively selected with respect to their spread in Ile Ife. Observational checklist was used to assess environmental and personal hygiene while microbiological specimens taken from bread were assessed for pathogenic contamination. Only a third of bakeries assessed had good hygiene while hygiene of bread sellers was suboptimal. Bread samples from bakeries and bread sellers yielded *S. aureus* in 33 and 90%, respectively. Several points of contamination were identified and poor bread handling practices documented. Existing laws should be enforced and bread sellers educated on proper handling of bread to prevent outbreaks of food borne illnesses.

Key words: Bread, food safety, hygiene, food-borne.

INTRODUCTION

Globally, access to safe food is a basic human right. It is so important that this year the World Health Organization has made food safety the theme of the World Health Day 2015.The food sector is broad and diverse, with different regulations guiding practices of stakeholders from micro to macro sector players. Due to rapid rural urban migration, several urban dwellers depended on ready to eat food to satisfy their food requirements. In this regard, ready to eat food (RTE) refers to food that could be eaten as purchased and does not require further significant processing other than reheating or completion of a cooking process (Singh et al., 2014).

One of the readily available RTE food is bread, which is a staple food that do not require further processing before consumption. It is produced in various forms and eaten in homes, restaurant and hotels in Nigeria (Emeje et al., 2010). In addition, the consumption of bread cut across socio-cultural and religious barriers and is a food of choice for both rich and poor in Nigeria (NAFDAC, 2010). The dough of bread is usually baked but in some eateries is steamed, fried or baked on an un-oiled platform. Bread is prized for its taste, aroma, quality, appearance and texture and, retaining its freshness is important to keeping it appetizing and appealing to consumers. Bread is made from low protein flour made from wheat or cassava and contains several ingredients that improve its quality. Some of the ingredients added to bread are table salt, sugar, flavour and at least an efficient oxidising additive to assist in the raising process and to produce a texture in the finished product that is appealing to the consumers (Emeje et al., 2010).

In Nigeria, commercial bread making is regulated by guidelines, through the National Agency for Food and Drug Administration and Control (NAFDAC), through the 4-point safety standards that bread bakers should comply with in Nigeria, as follows:

1. Consistent' production of flour fortified with Vitamins A&C and other " safe enhancers in approved quantities to prevent bakers from adding dangerous chemicals like potassium bromate as flour/bread improvers,
2. Compliance with set standards of Good Manufacturing Practice (GMP) and Hazard Analysis Critical Control Points (HACCP) guidelines and requirements of the Agency,
3. Stop the use of dangerous ingredients which are not on the list of substances generally regarded as "safe" and
4. Bakers should stop the habit of distributing loaves of bread to consumers without proper packaging and in unhygienic conditions (NAFDAC, 2010).

Nigeria has several food safety legislations, food is prepared in diverse socio and environmental settings and frequently contaminated with naturally occurring pathogenic microorganisms (Abdalla et al., 2008). In addition, concern on safety of bread have increased following the indiscriminate use of potassium bromate while the storage, handling process, the condition of the bread processing environment and health of the workers in the cottage industries and street hawkers are important considerations in ensuring the safety of consumers (Isong et al., 2013). Moreover, despite the awareness created and efforts to ensure that quality bread is available for citizens, there are several routes through which bread could be contaminated in the processing chain, especially during packaging, respectively at the factory and by vendors in containers to reduce dessication (HPA, 2009). Bread contamination and growth of pathogens alter its quality and a potential source of infection to consumers since improper handling and poor personal hygiene of food, and in particular during packaging is implicated in most food-borne illnesses through cross contamination and poor personal hygiene of bread handlers (Ehavald, 2009).

When food handlers do not practice safe personal hygiene, they may become vehicle for transmission of pathogens, through hands, mouth, skin, among others (HPA, 2009). Therefore, the aim of the study was to assess the hygiene involved from the point of production to sale of bread and document pathogens isolated from bread.

MATERIALS AND METHODS

The study was a descriptive, cross sectional design. The study was carried out in Ile-Ife, an urban city in Nigeria. The city is inhabited by 167,204 people, comprising 88,403 male and 78,801 females (NPC, 2009). Ile Ife has 11 wards and thirty registered bakeries.

Ten (10) out of the thirty registered bakeries spread across wards were randomly selected with only six consenting to participate (60% response rate) and all bakers on duty in each of the bakeries, totaling 40 were assessed for the study. Also, a total of 10 representative bread sellers were purposively selected with respect to their spread in Ile Ife from the following locations Mayfair, Campus gate, Sabo, General phase 1, Owode, Ilode, Arubiidi, Iyekere, Idi-omo and road 7, respectively.

Ten (10) different brands of bread produced by the bakeries were purchased from 10 market vendors at Ile Ife after being supplied and packaged. Observational checklist was used to assess hygiene in bakery houses and bread sellers while interviewer-administered questionnaires was employed to assess personal hygiene of bakery workers and swabs were taken from freshly baked breads, cutting tables, baking pans and mixing table while loaves of bread were collected as samples from the bakeries for laboratory analysis.

Microbiological assessment of bread samples

Swab samples obtained using sterile materials were collected from various baking surfaces, cultured into MacConkey and Chocolate agar and incubated at 37°C and 20% Carbon dioxide gas for 24 h. Also, 20 g of various bread samples were weighed and inoculated into sterile peptone (broth) and incubated for 24 h after which the growths were examined and isolates were identified by colony characteristics, Gram stain, catalase and coagulase tests as described by Monica Cheesbrough (1984).

Hygiene assessment of bakery houses in the study area

In assessing the level of hygiene in bakeries, a score of one (1) was assigned to correct hygienic practice and zero (0) for each wrong practice in each bakery which yielded a maximum score of 13. A composite score was graded as follows; less than 50% (<7), between 50-70% (7-9) and 70% (>9) and above signifies poor, fair and good level of hygiene, respectively in each of the six bakeries assessed.

Data analysis

Data were analysed using SPSS version 16 and presented using tables and charts as appropriate.

Informed consent

The informed consent of the bakery owners, workers and

Table 1. Socio-demographic characteristics of Respondents.

Variable	Characteristics	Bakery workers		Bread vendors	
		Frequency (N=40)	Percentage (%)	Frequency (N=10)	Percentage (%)
Age (years)	18-24	9	22.5	2	20
	25-31	17	42.5	5	40
	>31	14	35.0	3	30
Gender	Male	28	70	0	0
	Female	12	30	10	100
Educational status	Primary	9	22.5	3	30
	Secondary	27	67.5	7	70
	Tertiary	4	10.0	0	0

Table 2. Environmental hygiene characteristics of selected bakeries in Ile-Ife.

Variable	A	B	C	D	E	F
Is the bakery close to a residential area	1	1	1	1	1	1
Is the bakery close to any refuse dumping ground	1	0	0	0	0	0
Are there any unsanitary drainages around the bakery	1	1	1	0	0	1
Does the bakery have a proper refuse disposal system	0	1	0	0	1	0
Does the bakery have adequate space for storage of raw materials	0	1	1	1	1	0
Is the bakery surrounding bushy	0	0	0	0	0	0
Does the bakery have a toilet	0	1	1	1	1	1
If yes is the toilet separated from the processing area	0	1	1	1	1	1
Does the bakery have adequate supply of potable water	0	1	0	1	1	0
Are the doors and windows of the storage room protected against insects/pests	0	1	0	0	1	0
Does the bakery have handwashing facilities	0	0	0	0	1	0
Are the equipment cleaned regularly	0	0	1	1	1	0
Is the bakery premises cleaned regularly	0	1	0	1	1	1
Total score (%)	3 (27)	9 (70)	6 (46)	7 (55)	10 (75)	5 (40)

Key: 0=No; 1=Yes.

commercial bread vendors were obtained while ethical approval was given by the ethics committee of the Institute of Public Health.

RESULTS

The socio demographic information of 40 bakery workers and 10 randomly selected bread vendors that participated in the study is presented in Table 1. The result revealed that 65% of the bakery workers age ranges between 18-31 years with 24 years as the median age. 70% of them were males and majority of the bakers (77.5%) completed at least secondary schools education whilst, 70% of the bread vendors had ages ranging between 18 and 31 years with a median age of 26 years. Interestingly, all the bread sellers were females and most of them (70%) had educational attainment up to secondary school level.

The level of hygiene in the bakeries assessed was suboptimal with only two of the six facilities assessed having good level of hygiene with 1 having fair hygiene while three had poor hygiene (Table 2).

Though the use of head gear by bakery workers is mandatory, the study shows that 6% of the respondents had never use head gear while only 30% always wear it and only 10% use face mask. About two-thirds (65%) of bakery workers wear aprons regularly while only 45% wear hand gloves regularly. Hand washing before

Table 3. Personal Hygiene of Bakery workers and commercial bread sellers.

Hygiene practice	Bakery workers			Commercial bread sellers	
	Always N (%)	Sometimes N (%)	Never N (%)	Yes N (%)	No N (%)
Use of head gear	12(30)	4 (10)	24 (60)	10 (100)	0 (0)
Use of face mask	4 (10)	0 (0)	36 (90)	-	-
Use of foam to clean bread	-	-	-	10 (100)	0 (0)
Use of apron	26 (65)	8 (20)	6 (15)	2 (20)	8 (80)
Washing of hands	38 (95)	0(0)	2 (5)	4 (40)	6 (60)
Use of hand gloves	18 (45)	4 (10)	18 (45)	-	-
Adequate covering of bread	-	-	-	10 (100)	0 (0)
Cutting of finger nails	38 (95)	2 (5)	0(0)	10 (100)	0 (0)

Table 4. Microbiological isolates from bakery surfaces and bread samples.

Surface	A	B	C	D	E	F
Finishing table	+§	-	+§	-	-	-
Baking pan	-	-	-	-	-	+§
Cutting table	+§	+§	-	-	+§	+§
Mixing table	-	+§	-	-	-	-
Bread sample	+*	+§	+§	+§	+§	+*

* -pathogenic organism (*Staphylococcus aureus*)
§-nonpathogenic organism (aerobic spores).

processing bread was almost universally reported by bakery workers.

Among the commercial bread sellers, use of head gear was universally practiced. Only 60% reported washing their hands before handling or packaging bread with use of apron practiced by 20% of bread sellers. The study further revealed that all bread sellers use" foam" to clean bread before packaging (Table 3).

The study further assessed the microbial contamination of various surfaces in use in bakeries, namely the finishing table, baking pan, cutting table and mixing table of the six bakeries assessed. The analysis of the swab samples collected from each of the surfaces revealed that the cutting tables in most of the bakeries (66.7%) developed growth of non-pathogenic aerobic spores of bacilli while a significant 17.6% of the swab samples from bakeries finishing table, mixing table and baking pan respectively revealed growth of non-pathogenic aerobic microbes. A third of bread samples collected from the bakeries grew pathogenic organism (*S. aureus*) (Table 4).

The widespread handling of bread from point of transportation to point of sale is depicted in Figures 1 to 3. Figure 1 showed the supplier handling bread with bare hands while Figure 2 showed the bread sellers handling bread with bare hands and "cleaning" the bread with foam and Figure 3 shows the bread uncovered whilst

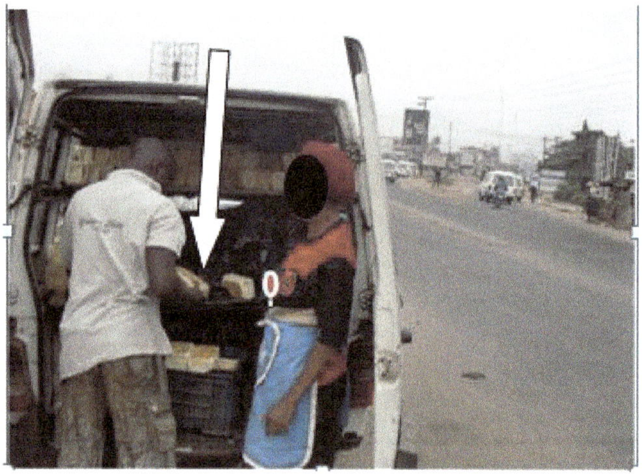

Figure 1. Handling practice of bread supplier and seller.

Figure 2. Bread seller using foam to clean bread.

the seller was "cleaning" the bread.

Similarly, the study further revealed that most bread

Figure 3. Bred seller cleaning bread with foam while bread left exposed.

Table 5. Microbiological analysis of bread samples from commercial bread vendors.

Bakery	Gram reaction	Isolated Micro-organism
C1	Gram +ve Cocci	*Staphyloccus spp.*
D1	Gram +ve Bacilli	*Aerobic spore(NP)*
E.	Gram +ve Cocci	*Staphyloccus aureus*
F	Gram +ve Cocci	*Staphyloccus aureus*
A1	Gram +ve Cocci	*Staphyloccus aureus*
B1	Gram +ve Cocci	*Staphyloccus spp.*
A2	Gram +ve Cocci	*Staphyloccus aureus*
B2	Gram +ve Cocci	*Staphyloccus aureus.*
C2	Gram +ve Cocci	*Staphyloccus aureus*
D2	Gram +ve Cocci	*Staphyloccus aureus.*

samples collected from vendors (80%) were contaminated by pathogenic *S. aureas* while 20% showed growth of non-pathogenic aerobic spore (Table 5).

DISCUSSION

Living organisms in this environment are always looking for food to stay alive, therefore food safety is very important to ensure wholesomeness of food for human consumption. Studies have shown that poor handling practice tend to cause food borne illness (Clayton et al., 2002). Bread being a meal that is usually eaten without further processing makes it a good source of food borne illness if improperly handled.

This study reveals that over 60% of bread sellers do not wash their hand before packaging bread, which is contrary to ideal hygiene practice as stated by Ehavald (2009). This finding is much higher than reported findings by Ifeadike et al. (2014), Isara et al (2013), Altekruse et al. (1996), Yang et al. (1998) and Shiferaw et al. (2000). This finding may be attributed to the fact that respondents may not necessarily refer to bread as raw food and thus little risk of contamination. Placing this finding in perspective of the frequent handling of bread by sellers it is imperative to note that whilst only a third of the bread sampled from the bakery were contaminated by *S. aureus*, almost all the bread sampled from the bread sellers were contaminated with *S. aureus*. This buttresses the assertion from studies on Hazard analysis and critical control points (HACCP) that shows that most foods are contaminated along the processing chain (Daniyan and Nwokwu, 2011, Adesetan et al., 2013).

The unwholesome practice of cleaning bread with foam has never been reported in literature and is a major mechanism for contamination of bread as it provides ample opportunity to expose the bread to handling with hands that are rarely ever washed before handling bread; the hands that are used to receive money in between transactions which has been reported in a study by Nurudeen et al. (2014).

Food hygiene is the set of basic principles employed in the systematic control of the environmental conditions during production, packaging, delivery/transportation, storage, processing, preparation, selling and serving of food in such a manner as to ensure that food is safe to consume and is of good keeping quality. Hygiene of the bread sellers was assessed and in this study it was discovered that use of head gear among bread sellers is universally practiced perhaps due to the fact that they were majorly female and it is culturally and religiously inappropriate for most women to leave their head uncovered.

This finding is however at variance with less than half (47%) reported by Nurudeen et al. (2014). Use of apron was reported by none of the bread sellers and by 65% of bakery workers which is similar to reports from study by Nurudeen et al. (2014).

The bread sellers often left their bread exposed granting access to flies which are common in a humid, hot environment. These flies have been shown by various studies to transmit food borne pathogens which can cause food borne illnesses such as Cholera, Campylobacteriosis, *E. coli* gastroenteritis, Salmonellosis, Shigellosis, Typhoid and paratyphoid fevers, Brucellosis, Amoebiasis, and Poliomyelitis (De Jesus et al., 2004; Keiding 1986; WHO, 1997).

Hygiene of bakeries assessed was poor as only a third of bakeries assessed had good hygiene which is similar to findings by Huq et al. (2013) in Bangladesh. This seems to be a common occurrence in developing countries where quality of bread for human consumption is doubtful as premised by the fact that not much is published in literature from developed countries.

Among the bakery workers, 45% wears hands gloves regularly while 15% uses it occasionally and 45% do not see any reason why they should use it. This is also similar to findings by Chukuezi (2010) where 47% of bakers uses bare hands while working in bakeries, against the standard practice but much higher than the 16.7% reported by Huq et al. (2013). However, the use of gloves should protect the hands of workers during packaging and sealing the nylon and is expected to minimize direct hands contact with finished bread. At Ile Ife, most of the bakeries utilize manual processes of feed, to slicing and packaging. Therefore, non-use of hand gloves compromises bread quality safety and health of people who eat them since bread post production is sold to the residents and travelers having stopovers at Ile Ife.

Conclusions

The hygiene conditions in bakeries, handling of bakeries processes by bakers and attitude of bread vendors is suboptimal and predisposes bread to contamination by pathogenic and non-pathogenic microorganisms. Therefore, regulatory agencies, States ministries of health and environmental health units of local government areas should ensure compliance and adherence of bakeries and bread sellers to the regulations and public health ordinance guiding the approval and monitoring of bakeries as a regulated premise.

There is need for re-orientation of bakeries on the basic ideal practice which has to be enforced to sound the seriousness of government to entrench sound bakery operations and practice in Nigeria. Therefore, all bakery workers should receive training in food hygiene and handling. This should be supported by country-wide health education, training and registration of formal bread vendors to promote compliance with best practices globally.

Conflict of interests

The authors did not declare any conflict of interest.

REFERENCES

Abdalla MA, Sihanm ES, Alian YYHA, Amel OB (2008). Food safety knowledge and practice of street food-vendors in Khartoum city. Sud. J. Vet. Sci. Anim. Husb. 47:126-136.

Adesetan TO, Ilusanya OAF, Sobowale AA, Jumani UP (2013) Bacteria commonly associated with bakery equipments in selected areas around Olabisi Onabanjo University environ, Ago-Iwoye, Nigeria. Adv. Environ. Biol. 7(1):177-188.

Altekruse SF, Street DA, Fein SB, Levy AS (1996). Consumer knowledge of food-borne microbial hazards and food-handling practices. J. Food Prot. 59:287-294.

Cheesbrough M (1984) Medical laboratory manual for tropical Countries Vol 2 Microbiology. Cambridge, Kentshire.Tropical Health Technology/ Butterworth Heinemann.

Chukuezi CO (2010) Food safety and hygiene practices of street food vendors in Owerri, Nigeria. Stud. Sociol. Sci. 1(1):50-57

Clayton DA, Griffith DJ, Price P, Peters AC(2002). Food handlers' beliefs and self-reported practices. Int. J. Environ. Health Res. 12:25-39.

Daniyan SY, Nwokwu OE (2011) Enumeration of microorganisms associated with the different stages of bread production in FUTMIN bakery, Nigeria. Int. Res. J. Pharm. 2(7):88-91.

De Jesus AJ, Olsen AR, Bryce JR, Whiting RC (2004) Quantitative contamination and transfer of Escherichia coli from foods by houseflies, Musca domestica L. Int. J. Food Microbiol. 93(2):259-62.

Ehavald H (2009). Food safety risk management in bakeries. Presentation at the 3rd SAFOODNET seminar St Olav's Hotel, Tallinn, Estonia; May 4-6, 2009.

Emeje MO, Ofoefule SI, Nnaji AC, Ofoefule AU, Brown SA (2010). Assessment of bread safety in Nigeria: Quantitative determination of potassium bromated and lead. Afr. J. Food Sci. 4(6):394-397.

Health Protection Agency (HPA) (2009) Guidelines for assessing microbiological safety of ready- to- eat foods. Available from www.gov.uk/government/publication/ready-to-eat-foods-microbiological-safety-assessment-guideline [accessed January 15 2015].

Huq AKO, Uddin JM, Haque KMF, Roy P, Hossain MB (2013). Health, hygiene practices and safety measures of selected baking factories in Tangail region, Bangladesh. Int. J. Curr. Microbiol. Appl. Sci. 2(10):68-75.

Ifeadike CO, Ironkwe OC, Adogu PO, Nnebue CC (2014) Assessment of the food hygiene practices of food handlers in the Federal Capital Territory of Nigeria. Trop. J. Med. Res. 17:10-5.

Isara AR, Aigbokhaode AQ, Onwuso OV, Onyeulo EC, Orumwense SO (2013). Food hygiene and safety practices of food service staff in University of Benin Teaching Hospital, Benin city, Nigeria. J. Med. Biomed. Res 12:2.

Isong NB, Akpan MM, Udota HI, Barber L (2013). Antibacterial and microbial assessment of breads in Abak local government area, Nigeria, West Africa. J. Microbiol. Biotechnol. Res. 3 (3):155-159.

Keiding J (1986). The housefly biology and control. Training and information guide (advanced level). Geneva, WHO (WHO/VBC/86.937) accessed 14th January 2015.

NAFDAC (2010). NAFDAC Okays New Safety Standard for Bread Producers. Archived in "This Day live" (27thNovember, 2010). http://www.thisdaylive.com/articles/nafdac-okays-new-safety-standard-for-bread-producers/73242/ accessed on_15th June, 2014

Nurudeen AA, Lawal AO, Ajayi SA (2014). A survey of hygiene and sanitary practices of street food vendors in the Central State of Northern Nigeria. J. Public Health Epidemiol. 6(5): 174-181.

Singh S, Rana T, Kaur N, Parihar R, Kumar R, Kumar S, Kumar R (2014). Studies on Microbiological Quality of Ready to Eat Food from Solan City of Himachal Pradesh. Int. J. Univ. Pharm. BioSci 3(3):181-189.

WHO (1997) Vector Contol: methods for use by individuals and communities. Available from www.who.int>resources>vectoritoxii [accessed January 5 2015].

Yang S, Left MG, Mctague D, Horvath KA, Jackson-Thompson J, Murayi T, Boeselager GK, Melnik TA, Gildemaster MC, Ridings DL, Altekruse SF, Angulo FJ (1998). Multistate surveillance for food-handling, preparation, and consumption behaviors associated with food borne diseases: 1995 and 1996. MMWR Morb. Mortal Wkly. Rep. 47:S33-57.

Assessment of bacterial and fungal spoilage of some Nigerian fermented and unfermented foods

Adebayo CO[1], Aderiye BI[2] and Akpor OB[3]

[1]Department of Science Technology, Federal Polytechnic, Ado-Ekiti, Ekiti State, Nigeria.
[2]Department of Microbiology, Ekiti State University, Ado-Ekiti, Ekiti State, Nigeria.
[3]Department of Biological Sciences, Landmark University, Omu-Aran, Kwara State, Nigeria.

The study was aimed at evaluating the microbial spoilage of selected Nigerian fermented and unfermented foods. A total of four fermented and unfermented food samples were used for this investigation. Microbial and sensory evaluations of the food products during storage were carried out using standard procedures. During storage, the bacteria counts were observed to range from 2.2×10^5 to 4.8×10^5 CFU/mL and from 2.5×10^3 to 5.0×10^4 CFU/mL, for the unfermented and fermented food products, respectively. Similarly, the fungal counts ranged from 1.8×10^3 to 2.9×10^3 CFU/mL and from 0 to 5.70×10^3 CFU/mL for the unfermented and fermented food samples, respectively. *Klebsiella aerogenes*, *Lactobacillus plantarum*, *Leuconostoc* sp., *Micrococcus varians*, *Proteus mirabilis*, *Streptococcus faecalis*, *Staphylococcus epidermidis*, *Aspergillus niger*, *Aspergillus flavus*, *Cladosporium herbarum*, *Geotrichum candidum*, *Mucor mucedo*, *Neurospora sitophilia* and *Penicillium* sp. bacteria and fungi that were recovered from the food samples during storage. The sensory evaluation of the food products showed the fermented ones being more acceptable to panelists than the unfermented ones. This could indicate that the palatability of the fermented food samples only experienced slight or no changes during storage, when compared to the unfermented ones.

Key words: Fermented food, unfermented food, microbial spoilage.

INTRODUCTION

It is reported that fermented foods constitute about 25% of the foods consumed worldwide. Fermented foods have many advantageous attributes, which include improved nutritional value and safety against pathogens over non-fermented foods. Generally, these foods are normally considered to be safe against foodborne diseases because of the advantage of low pH, which is due to the presence of organic acids produced during fermentation. In Africa, because a majority of the fermented foods are produced at household levels, the issue of hygiene is a major concern (All and Dardir, 2009; Gadaga et al., 2004).

Both fermented and unfermented food products are known to be susceptible to spoilage during storage. Their spoilage during storage is attributed to the presence of microorganisms and extracellular enzymes produced, which breakdown the food product into new substances resulting into changes in their organoleptic properties

(Fadahunsi et al., 2013). Chance inoculation, activities of spoilage organisms and the humid condition in the Tropics are some of the factors responsible for spoilage of these foods during storage. While starter cultures have been used to ensure uniformity in the composition of some fermented foods, their reduced shelf life still remains a major problem worldwide. The ingestion of products contaminated with these microorganisms could be a potential health threat hence there is the need to control microbial contamination of fermented foods and feeds.

Filamentous moulds and yeast are common spoilage organisms of food products as fermented milk products, cheese, bread, stored crops and feed such as hay and silage (Filtenborg et al., 1996). *Penicillia* and *Aspergilli* species have been reported as spoilage organisms during storage of a wide range of foods where they may produce a number of mycotoxins (Samson et al., 2002). Several authors (Pitt and Hocking, 1999; Magnusson and Schnurer, 2001) have also implicated yeasts such as *Candida parapsilosis, Rhodotorula mucilaginosa, Kluyveromyces marxians and Derbaromyces hansenii* as common spoilage organisms of yoghurt and other fermented food products.

Apart from food spoilage, health hazards from fungal mycotoxins had been well documented. For instance, the large consumption of grains infected with sclerotia of *Claviceps pupurea* caused several thousand deaths in the 10th and 11th century in central Europe (Pohland, 1993). It is estimated that between 5 and 10% of the world's food products is lost due to fungal deterioration (Pitt and Hocking, 1999). The growth of these undesirable fungal species in foods has been a worldwide concern over the years because of its effects on man's health and consequent economic loss due to food storage. During food spoilage, which is a metabolic process, the spoilt food becomes undesirable and unacceptable for human consumption due to changes in sensory characteristics.

Fermented foods such as gari, ogi, fufu and mawe are found all over the world. Fermentation has offered a means of food security in Nigeria; where over 52% of respondents in a previous study in 2003 indicated that about 25% of their monthly income is spent on fermented food products (Aderiye and Laleye, 2003). This paper therefore examines and discusses the desirability of consuming part of spoiled fermented food products as reported by Aderiye and Laleye (2003). It is also aimed at evaluating the types and extent of microbial spoilage of some selected Nigerian fermented and unfermented foods.

MATERIALS AND METHODS

A total of eight different food samples, divided into unfermented (pounded yam, yam, rice and beans) and fermented ('Eba', 'Eko', 'Fufu' and 'Kati') food products were used in this study. The raw food products were purchased from the local market in Ado-Ekiti, Ekiti State in Nigeria. To ensure uniformity and prevent any pre-storage contamination, the purchased raw food products were prepared locally in the laboratory using aseptic techniques. Each of the prepared food samples was transferred into and wrapped with sterile polythene bags (*ca* 0.2 mm thick), while some were left unwrapped. Food samples from 'Eko' and 'Kati' were wrapped in the leaves of *Thaumatoccus daniella*. This was to ensure that the samples were in the same state as they were normally packaged in the locality. Detail description of the food products and their preparation have been described elsewhere (Aderiye and Lalaye, 2003). All the food products were stored under hygienic conditions at refrigeration and ambient temperatures.

Microbiological analyses were carried out using conventional microbiological procedures. The total microbial count on the food sample was determined by the pour plate method using standard microbiological techniques. Nutrient agar and Malt extract agar were used to culture bacteria and fungi, respectively. All bacterial and fungal counts were expressed as colony forming units per milliliter (CFU/mL) and propagules and cells/mL respectively. In all cases, each analysis was carried out in triplicates. The isolation and identification of the bacteria and fungi were carried out as described elsewhere (Barnett and Hunter, 1992; Burgess and Sum, 1994; Holt et al., 1994). The occurrence of each isolate in the food samples was determined as a percentage of the total number of food samples and recorded as the frequency of occurrence. The frequency of occurrence of isolates in the food samples was calculated using the following formula:

$$\%\ \text{frequency of occuerence} = \frac{\text{No of microbial occurrence in the food samples}}{\text{Total number of food samples}}$$

Sensory properties such as appearance, colour, odour, taste, texture and general acceptability were also used in an 11-point 'Hedonic' scale to test the acceptability and rejection of the food samples. The detection of spoilage was done through physical examination of the stored food samples. The wrapped food samples were observed daily for changes in appearance, texture, aroma and taste. The day of onset of each defect or sign of spoilage and the types were recorded.

Each of the experiments was carried out in triplicates and conducted twice. All the statistical analyses were carried out using the SPSS computer statistical software. The comparison of means was done using the One-Way Analysis of Variance Test.

RESULTS

The microbial quality of the food samples is shown in Figures 1 and 2. When the food samples were examined within one hour of processing, no microbial contamination was recorded. As shown in Figure 1, in all the food samples, bacterial colonization was observed within 24 h after production. The bacterial counts in the food samples were observed to increase with increase in storage time. At the end of 96 h storage period, the bacterial counts from the unfermented food samples ranged from 2.2×10^5 to 4.8×10^5 CFU/mL, for yam and pounded yam, respectively. In the fermented food samples, the bacterial count was observed to range from 2.5×10^3 to 5.0×10^4 CFU/mL, for 'Kati' and 'Eba', respectively (Figure 1). In all cases, the bacterial counts in the unfermented food samples were significantly higher than those in the fermented food samples ($P \leq 0.05$). Even after 96 h storage, 'Eba' still

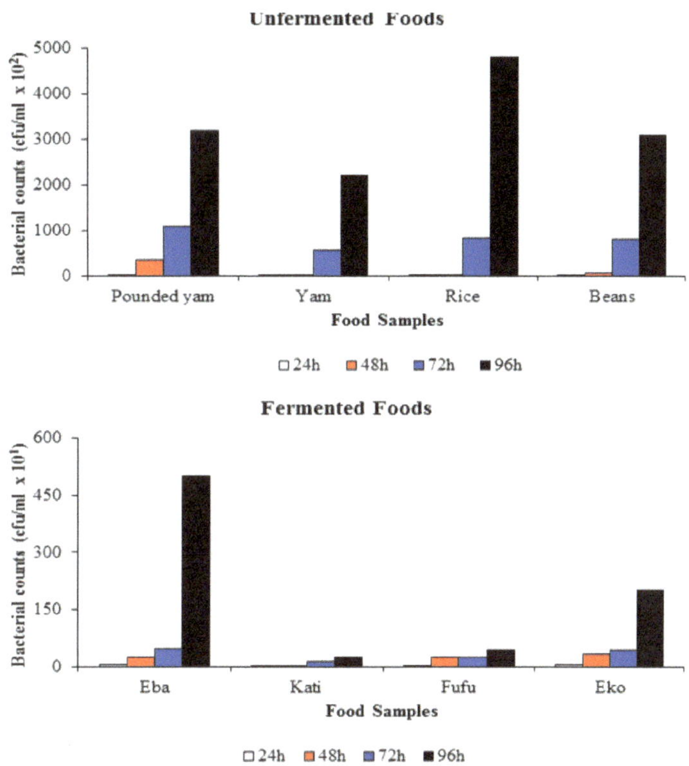

Figure 1. Average bacterial counts of the food samples during the period of storage.

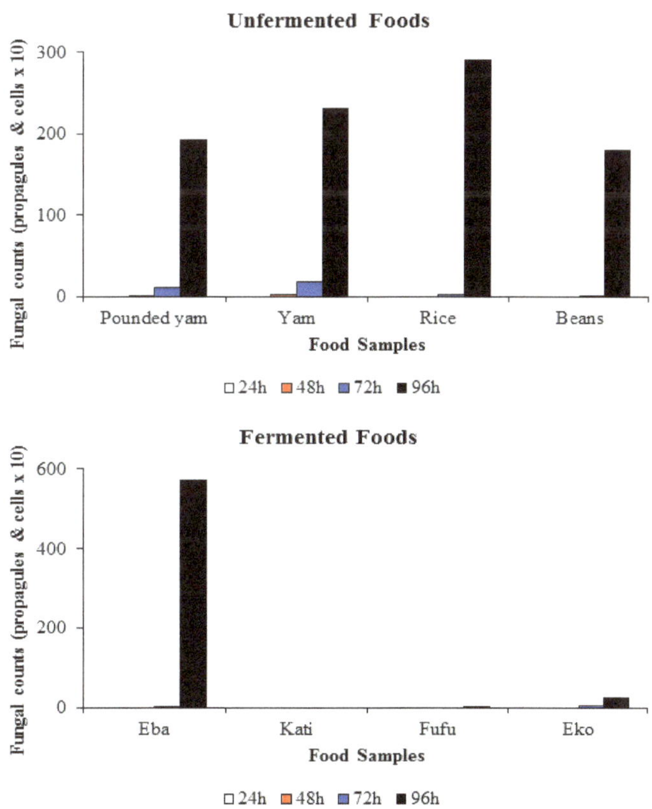

Figure 2. Average fungal counts of the food samples during the period of storage.

Table 1. Occurrence and distribution of bacteria in the food samples.

Food sample	*Kleb.*	*Lact.*	*Leuc.*	*Micr.*	*Prot.*	*Strep.*	*Staph.*
Pounded yam	+	-	-	+	+	+	+
White rice	+	-	-	+	+	-	+
Beans	-	-	-	-	+	-	+
Yam	-	-	-	-	-	+	+
Eba	-	-	-	-	+	+	+
Kati	-	+	+	-	-	-	+
Fufu	-	+	-	-	-	-	+
Eko	-	-	-	+	-	-	+

'+' and '-' mean present and absent respectively. *Kleb.= Klebsiella aerogenes, Lact.= Lactobacillus plantarum, Leuc.= Leuconostoc* sp., *Micr.= Micrococcus varians, Prot.= Proteus mirabilis, Strep.= Streptococcus faecali, Staph.= Staphyloccus epidermidis.*

Table 2. Occurrence and distribution of fungi in the food samples.

Food sample	*A. fla*	*A. nig*	*Clad*	*Geot*	*Muco*	*Neur*	*Peni*
Pounded yam	-	-	-	-	-	+	+
White rice	-	-	-	+	-	+	-
Beans	+	-	-	+	-	-	-
Yam	-	+	+	-	-	-	-
Eba	-	+	-	-	+	-	+
Kati	-	-	-	-	-	-	-
Fufu	+	-	-	-	-	-	-
Eko	+	-	-	+	-	+	+

'+' and '-' mean present and absent, respectively. *A. fla= Aspergillus flavus, A. nig= Aspergillus niger, Clad= Cladosporium herbarum, Geot= Geotrichum candidum, Muco= Mucor mucedo, Neur= Neurospora sitophilia, Peni= Penicillium* sp.

had a load (5×10^4 CFU/mL) far lower than the pounded yam stored for 24 h (2.2×10^5 CFU/mL).

As shown in Figure 2, fungi colonization in the unfermented and fermented food samples was observed to commence within 48 and 72 h, respectively. At the end of 96 h storage, the fungal counts were observed to range from 1.8×10^3 propagules/ cells and 2.9×10^3 propagules /cells in the unfermented food samples. In the fermented food samples, fungal count was observed to range from 0 to 5.70×10^3 propagules/cells, respectively. In all the fermented foods, there were no traces of fungal growth for 48 h. Throughout the 96 h storage period, no fungi colonization was observed in 'kati' (Figure 2). As was observed with the bacterial counts, the fungal counts in the unfermented food samples were significantly higher than counts in the fermented food samples (P≤0.05).

During the storage period, a total of eight bacterial strains were isolated from the food samples. The isolated bacterial strains include *Klebsiella aerogenes*, *Lactobacillus plantarum*, *Leuconostoc* sp., *Micrococcus varians*, *Proteus mirabilis*, *Streptococcus faecalis* and *Staphylococcus epidermidis* (Table 1). In the case of fungi, eight strains were isolated from the food samples. The strains were *Aspergillus niger*, *Aspergillus flavus*, *Cladosporium herbarum*, *Geotrichum candidum*, *Mucor mucedo*, *Neurospora sitophilia* and *Penicillium* sp. (Table 2).

With respect to the frequency of occurrence of microbial isolates from the food samples, a rank of the bacteria revealed that *Leuconostoc* sp (13%) and *Staphylococcus epidermidis* (45%) were the least and the most occurring bacteria in the food samples, respectively. In the case of fungi, *Aspergillus niger* and *Penicillium* sp. were observed to be the least (40%) and the most occurring fungi in the food samples, respectively (Figure 3).

With respect to the organoleptic attributes of the food samples during storage, although there was a significant difference between the different food samples, a general trend was that the fermented food samples were more acceptable to the panellists than the unfermented food samples. Only a few number of respondents accepted

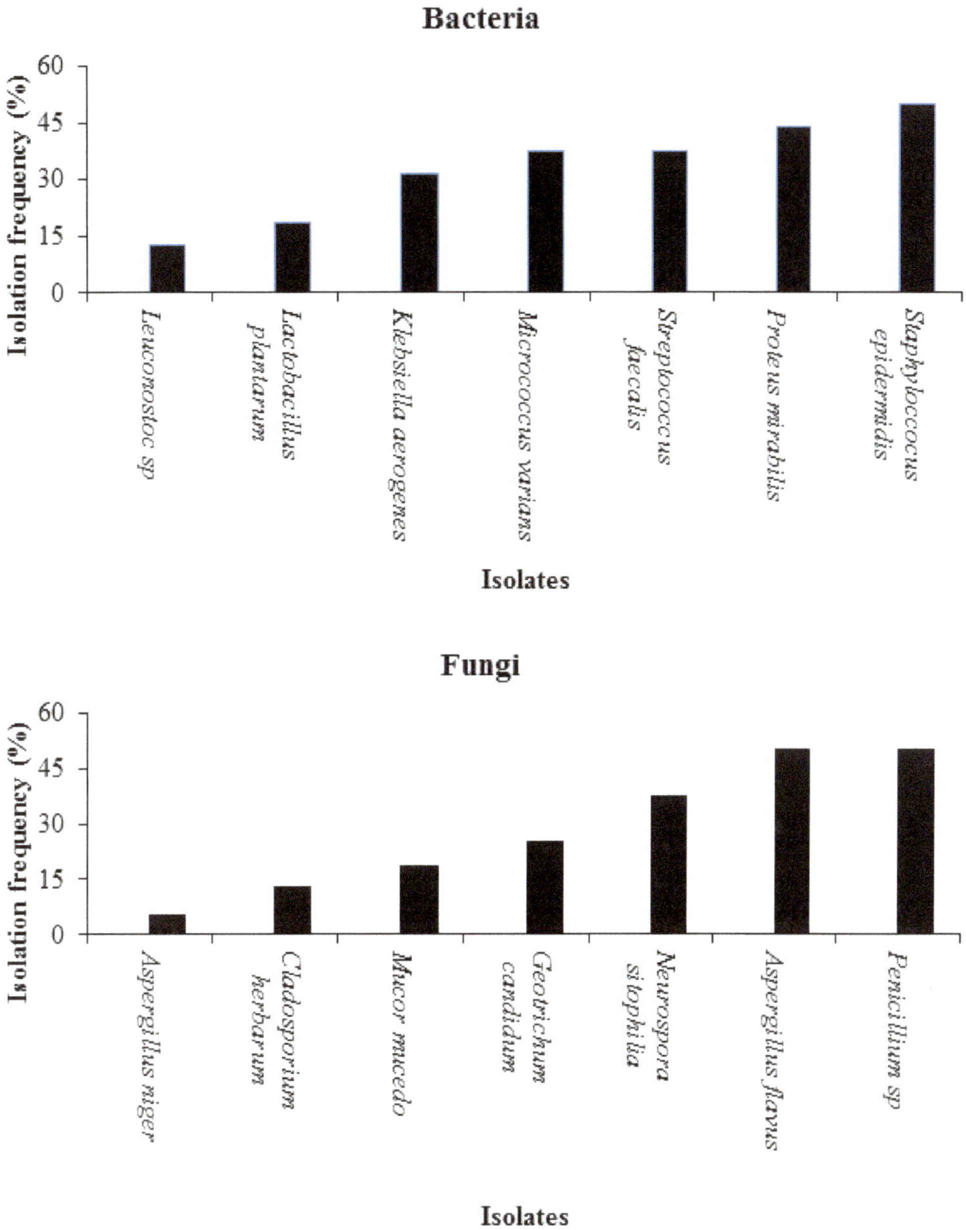

Figure 3. Rank of the frequency of occurrence of the microbial isolates in the food samples.

beans stored for 96 h while 'Kati' was the most acceptable of the food samples. A daily decrease in the means of the attribute was observed in all the unfermented food samples while the arithmetic difference in the means of the fermented food samples was observed to be negligible (Table 3).

A comparison of storage at 25 and 4°C revealed that off-colour, mouldiness and objectionable taste were the commonest indicators of spoilage in all the food samples. Stringiness and sliminess were observed to be limited to the unfermented food samples. The results revealed that all the food samples were affected by one or more fungal species as shown by mouldiness (Table 4). The foods stored at refrigeration temperature had longer acceptable days (5 to 76 days) and a longer shelf life (4 to 56 days) than those stored at room temperature, which were 2 to 22 days and 2 to 17 days for acceptable days and longer shelf life, respectively).

DISCUSSION

A general observation in this study was the presence of higher microbial load during storage in the unfermented food samples than in the fermented. The higher storability and lower spoilage rate of the fermented food samples are therefore not surprising as fermented foods had been

Table 3. General Sensory evaluation of the stored food samples.

Food sample	Fresh	24 h	48 h	72 h	96 h
Pounded yam	9.8^{a}	5.6^{c}	4.7^{c}	2.6^{d}	1.5^{d}
Yam	9.6^{a}	6.0^{c}	5.4^{b}	4.6^{b}	2.4^{c}
Rice	10.0^{a}	6.5^{b}	5.1^{c}	3.5^{c}	1.98^{c}
Beans	9.6^{a}	6.1^{c}	4.8^{c}	2.5^{d}	1.2^{d}
Eba	9.9^{a}	8.6^{a}	8.5^{a}	8.0^{a}	7.0^{b}
Kati	9.9^{a}	8.5^{a}	8.5^{a}	8.2^{a}	8.0^{a}
Fufu	9.8^{a}	8.3^{a}	8.0^{a}	7.7^{a}	7.5^{b}
Eko	10.0^{a}	8.2^{a}	8.2^{a}	7.8^{a}	7.5^{a}

Values in the same column not followed by the same superscript are significantly different ($p \leq 0.05$).

Table 4. Observed days of spoilage symptoms of the food samples stored at 25 and 4°C.

Parameter	Pounded yam	Yam	Rice	Beans	*'Eba'*	*'Kati'*	*'Fufu'*	*'Eko'*
Appearance								
Off-colour	3 (4)	2 (3)	2 (4)	4 (6)	5 (17)	17 (56)	9 (35)	5 (15)
Mouldiness	4 (7)	3 (6)	5 (8)	5 (9)	7 (20)	20 (64)	12 50)	8 (21)
Texture								
Stringiness	3 (-)	2 (-)	2 (5)	- (-)	- (-)	- (-)	- (-)	- (-)
Sliminess	- (-)	1 (4)	2 (4)	1 (4)	- (-)	- (-)	- (-)	5 (-)
Watery	3 (-)	- (-)	2 (5)	3 (-)	- (-)	- (-)	- (-)	7 (-)
Softness	2 (-)	3 (-)	2 (4)	- (-)	6 (20)	- (-)	9 (-)	6 (-)
Separated	2 (3)	- (-)	- (-)	- (-)	- (-)	- (-)	- (-)	7 (18)
Aroma								
Off-odour	2 (4)	2 (4)	2 (4)	2 (5)	7(22)	19 (74)	12(50)	8 (17)
Putrid	- (-)	- (-)	2 (3)	2 (5)	- (-)	- (-)	- (-)	- (-)
Taste								
Objectionable	3 (4)	2 (4)	1 (5)	2 (6)	5 (21)	22 (72)	10(62)	9 (19)
Sourness	2 (3)	- (-)	- (-)	- (-)	- (-)	18 (70)	- (-)	9 (5)
Rejected	2 (5)	2 (5)	3 (7)	2 (7)	8 (22)	22 (76)	15(52)	11(24)

Values in parenthesis represent storage at 4°C while all other values indicate at 25°C. '-' indicate not detected. *'Kati' and 'Eko'* were wrapped in leaves during the storage period.

reported to have a higher shelf life (Steinkraus, 1997; Ogunbanwo et al., 2004). This could be attributed to the release of antimicrobial metabolites by the fermenting organisms into these food substrates (Adebayo and Aderiye, 2010). The antimicrobial compounds include bacteriocin, diacetyl, hydrogen-peroxide and organic acid (Adebayo and Aderiye, 2007).

The differences in the microbial contamination of pounded yam and yam both from the same raw material could be attributed to the higher level of moisture content in the pounded yam. Frazier and Westhoff (2007) had reported that high water activity favoured the growth of contaminating microorganisms. The higher storability of the refrigerated food samples (4°C) could be attributed to the fact that the spoilage organisms are mesophiles hence the low temperature (4°C) inhibited their metabolic activities and thus led to limited growth, inhibition or death in causes of some organisms (Bracket, 1997). Also, the highest shelf life (76 days) recorded by 'Kati' may be due to its low pH, low moisture and high fibre contents. Such food tends to be more stable microbiologically (Frazier and Westhoff, 2007). One of the findings of this study, that mouldiness is a perceived indicator of deterioration in the selected Nigerian foods, predicates the need to evolve means to prevent fungal spoilage of fermented Nigerian foods.

A total of seven fungal strains belonging to six genera were found to be responsible for the spoilage of the food samples used in this study. The most predominant genera were *Aspergillus* and *Penicillium* which constituted more than 66% of the total fungal isolates. Previous workers have reported the predominance of these organisms in other food systems like mawe (Hounhouighan et al., 1993); 'burukutu' (Sanni et al., 1999) and dried sausages (Mataragas et al., 2002).

The prevalence of *Aspergillus* and *Penicillium* species could be due to their sporulating ability; hence they easily contaminate the environment (Frisvad and Samson, 2007). Food samples from 'Eko' recorded the highest number of fungal contamination (66.66%). This could be attributed to its having the highest moisture content (55.04%). Foods with high moisture content generally encourage fungal growth because spore germination and hyphal growth require moisture (Odigie, 2000). Another reason could be due to the wrapping of 'Eko' with leaves of *Thaumatococcus daniella*, which could be a potential source of contamination. According to Fadahunsi and co-workers (2013), the occurrence of the isolates could be attributed to the biodegrading potential of the microflora to convert diverse substrates by secretion of extracellular enzymes.

In the present study, the spoilage fungi isolated from the food samples are common microorganisms associated with the environment. Contamination of the food substrates by these fungi was expected as the ambient temperature of storage and pH (4.1-4.6) of the fermented food samples favour fungal proliferation (Brock and Madigan, 2003). Many of the isolated fungi are of clinical importance because of their potentials in causing human diseases. For instance, strains of *Aspergillus* and *Penicillium* spp. are known to produce mycotoxins which can result in mycotoxicoses (Smith et al. 1995). The toxins can damage liver, causing cirrhosis and can also induce tumor. Hence the assumption of some consumer of the fermented foods in southwest Nigeria that these mould - infested foods are only unaesthetic but not dangerous to health should be discouraged, as such foods could be potential sources of mycotoxicosis.

The 'Eko' had the highest occurrence and distribution of genera of fungal contaminants (5) while 'Fufu' had the least (2). The differences in the occurrence of these fungi in the food samples may be attributed to the differences in their nutrient status. For instance, Adebayo and Aderiye (2009) reported that 'Eko' had higher metabolisable sugar (1.63%) and moisture (55.04%) content, than 'Fufu' (0.048 and 42.15%) thus predisposing the former to a higher level of fungal attack. The delay in the incidence of the spoilage organisms until after 72 h could be attributed to inadequate moisture level required for spore germination prior to 72 h of storage or to the antimicrobial effect of some compounds liberated by LAB into the food substrates during fermentation (Ryhz et al., 2008; Adebayo and Aderiye, 2010).

Foods generally become more susceptible to pathogens as storage progressed due to the absorption of moisture from air and thus make the tissues more accessible to hydrolytic enzymes of the attacking pathogens. It is indicated that although microbial fermentation has played a vital role in food processing in Nigeria, because many of the indigenous fermented foods are processed through spontaneous and uncontrolled natural fermentation, despite its advantages, the accompanying microflora are indicated to be causes of spoilage and in some case toxicity (Aderiye and Adebayo, 1999).

Infestation by spoilage fungi caused some physical defects like off-colour, loss of firmness, loss of aroma, and softness on the food samples. Frazier and Westhoff (2007) attributed the change in texture to the activity of the spoilage organisms that encouraged the extraction of mineral nutrient from the food substrates. Loss of aroma was however attributed to the fermentation of the non-proteinaceous constituents of the food substrates. Fungal activity has been reported to cause a rapid deterioration of quality of food substrates (Teniola and Odunfa, 2002; Samson, 2007; Adebayo and Aderiye, 2009).

Conflict of Interests

The author(s) have not declared any conflict of interests.

Conclusion

This study which was aimed at evaluating the microbial spoilage of selected fermented and unfermented foods in Nigeria have revealed that during storage, both fermented and unfermented foods are susceptible to spoilage, although the onset of spoilage is faster in the unfermented one. A variety of bacteria and fungi species were also observed to be responsible for spoilage of the food samples during storage. With storage at room temperature, the onset of spoilage was remarkably faster in the fermented food samples than the fermented ones. In all, a sensory evaluation showed that the fermented foods were more acceptable to panelists than the unfermented ones after storage. This could indicate that the palatability of the fermented food samples only experienced slight or no changes during storage, when compared to the unfermented ones.

REFERENCES

Adebayo CO, Aderiye BI (2007). Ecology and antibacterial potential of lactic acid bacteria associated with fermented cereals and cassava. Res. J. Microbiol. 2 (5): 426-435.

Adebayo CO, Aderiye BI (2009). Nutrient depletion of some stored Nigerian fermented foods. J. Appl. Environ. Sci. 5 (1): 5-10.

Adebayo CO, Aderiye BI (2010). Antifungal activity of bacetriocvins of lactic acid bacteria from some Nigerian fermented foods. Res. J. Microbiol. 5 (11): 1070-1082.

Aderiye BI, Adebayo CO (1999). Fermentation of yam II: effect of the use of starter cultures on the physic-chemical properties of fermented yam and its products. J. Technoscience 3: 1-5

Aderiye BI, Laleye SA (2003). Relevance of fermented food products in southwest Nigeria. Plant Foods Hum. Nutr. 58: 1-6.

All AAAA, Dardir HA (2009). Hygienic quality of local traditional fermented skimmed Milk. World J. Dairy Food Sci. 4 (2): 205-209

Brock TD, Madigan MT (2003). *The Biology of Microorganisms.* 8th Edition Prentice Hall International, London. p. 840.

Fadahunsi IF, Ogunbanwo ST, Fawole AO (2013). Microbiological and nutritional assessment of burukutu and pito (indigenously fermented alcoholic beverages in West Africa) during storage. Nat. Sci. 11 (4): 98-103

Filtenborg O, Frisvad JC, Thrane U (1996). Moulds in food spoilage. Int. J. Food Microbiol. 2 (4): 1-5.

Frazier WC, Westhoff DC (2007). Food Microbiology, 4th Edn., Tata Mcgraw-Hill Book Co. Ltd. New Delhi. p. 538.

Frisvad JC, Samson RA (2007). Identification of *Aspergillus* and *Penicillium*. In: Food Mycology; Emerging mold problems and spoilage in food and beverages (Samson R.A., Dijksterhuis J., Houbraken J., Rico E., Johnson S., eds). Proceedings of Food Mycology, pp. 124–127.

Gadaga TH, Nyanga LK, Mutukumira AN (2004). The occurrence, growth and control of pathogens in African fermented foods. Afr. J. Food Agric. Nutr. Dev. 4 (1): 5358-537

Holt JG, Krieg NR, Sneath PHA, Staley JT, Williams ST (1994). Bergey's Manual of Determinative Bacteriology. 9th Edition, Williams and Wilkins, USA.

Magnusson J, Schnurer J (2001). *Lactobacillus coryneformis* subsp. *Coryneformis* strain SI3 produces a broad spectrum proteinaceous antifungal compound. Environ. Microbiol. 67(1):1-5

Mataragas M, Melaxopoulous J, Drosinos EA (2002). Characterization of two bacteriocins produced by *Leuconostoc mesenteroides* L124 and *Lactobacillus curvatus* L442, isolated from dry fermented sausages. World J. Microbiol. Biotechnol. 18: 847-856.

Odigie EE (2000). Effect of some environmental factors on the growth and sporulation of *Curvularia clavata.* Niger. J. Microbiol. 14 (1): 83 – 89.

Ogunbanwo ST, Sanni AI, Onilude AA (2004). Effect of bacteriocinogenic *Lactobacillus* spp. on the shelf life of fufu, a traditional fermented cassava product. World J. Microbiol. Biotechnol. 20:57-63.

Pitt JI, Hocking AD (1999). Fungi and Food Spoilage. Chapman and Hall, New York. p. 110.

Samson RA (2007). Overview of current spoilage problems in food and beverages. In: Food Mycology: Emerging mold problems and spoilage in food and beverages. In: Samson RA; Dijksterhuis J; Houbraken J; Rico E and Johnson S. eds). Proceedings of Food Mycology, June 7–8, Key West, Florida, U.S.A.: 1–3

Samson RA, Hoekstra ES, Frisvad JC, Filtenborg O (2002). Introduction to food and air-borne fungi. 6th edn. Central bureau Voor Schimmelcultures. Utrecht, The Netherlands. p. 210.

Smith JE, Solomon G, Lewis C, Anderson JG (1995). Role of mycotoxins in humans and animal nutrition and health. Nat. Toxins 3: 187 – 192.

Steinkraus KH (1997). Classification of fermented foods: worldwide review of household fermentation techniques. Food Control 8 (5/6): 311 – 317.

Teniola OD, Odunfa SA (2002). Microbial assessment and quality evaluation of 'ogi' during spoilage. World J. Microbial Biotechnol. 18: 731 – 737.

23

Nutritional suitability of bred sorghum (*Sorghum bicolor*) accessions from East Africa

Kiprotich Felix[1] , Mwendia M. Charles[1], Cheruiyot K. Erick[3] and Wachira N. Francis[2]

[1]Biochemistry and Molecular Biology Department, Egerton University, P.O. Box 563-20115, Egerton, Kenya.
[2]Association for Strengthening Agricultural Research in Eastern and Central Africa (ASARECA) P.O. Box 765, Entebbe, Uganda.
[3]Department of Crops, Horticulture and Soil, Egerton University, P.O. Box 536-20115, Egerton, Kenya.

In many semi-arid and tropical areas of the world especially in sub Saharan Africa, sorghum is a staple food grain and has great potential to be used in various industries. Thirty sorghum highland and open pollinated varieties were analysed for their biochemical and physiological characteristic to determine their industrial suitability and breeding impacts. The results show that sorghum varieties have the capability to be used for different industries and can be good alternatives to other cereal varieties. Majority of the varieties like *Ainamoi #1, Siaya # 24-2, Kipkelion # 2, Kipkelion # 1, Nyangezi, Uasin Gishu #1* and *Uasin Gishu #2* with high starch and amylopectin contents also recorded high tannin contents and vice versa. Hybrids are bred to give low tannins, but unfortunately this also affects their starch amounts and in the long run, decreases the suitability of sorghum for industrial and domestic use. In addition, there was a significant correlation between yielding ability and plant height.

Key words: Breeding, industrial suitability, biochemical characteristics.

INTRODUCTION

Sorghum, *Sorghum bicolour,* is the fifth most important cereal crop after rice, wheat, maize and barley (Smith and Frederiksen, 2000; FAO, 2005), and it contributes significantly to the protein and energy requirements of millions of people, especially the poor in Africa and Asia (Elkhier and Hamid, 2008). Sorghum has the potential to drive economic development in Africa. In developing countries, the commercial processing of these locally grown sorghum grains into value-added food and beverage products is an important driver for economic development (Taylor, 2004). Subsistence farmers in Africa cultivate sorghum widely as a staple food for home consumption. In particular, sorghum is mainly important for people in sub-Saharan Africa as a food crop because it is adapted to a wide range of ecological conditions and can tolerate adverse conditions such as hot, dry, wet and water logged conditions. It can also adapt to poor fertility and high salinity soils (FAO, 2012).

Sorghum grain is an important staple food in developing countries of the semi-arid tropics and is also used as an animal feed in both developed and developing countries. The use of sorghum not only provides farmers with incomes after marketing their products but also saves foreign exchange, which would otherwise be required to import cereals. More than 35% of sorghum is grown primarily for human consumption, while the rest is used for feeding animals, brewing alcohol among other uses (Amir et al., 2009). Cereal grains like sorghum can constitute major energy sources and starch as raw materials for several end uses such as in baking, brewing, poultry and livestock industries. Despite this, maize, wheat and barley still remains the primary energy source in these industries. Ajaja et al. (2002) stated that maize is still the main energy source in compounded diets and constitutes about 50% of poultry ration. Like other cereals, starch is the principal storage form of carbohydrate and sorghum has an average starch content of 69.5% (Jambunathan and Subramanian, 1987). High fiber content and poor digestibility of nutrients are characteristic features of sorghum grains, which severely influences its consumer acceptability. Proteins form the second major component of sorghum grains. The protein content of sorghum is affected by both genetic and environmental factors (Arun et al., 2009).

Phenolic compounds in cereal grains encompass a diverse group of secondary plant metabolites. They can be conveniently divided broadly into phenolic acids, flavanols, polymeric flavanols and condensed tannins. Tannins bind proteins, carbohydrates and minerals, thereby affecting the nutritional and functional value of the bound constituents. Sorghum varieties rich in tannins are recommended for obese individuals and diabetic patients. Animal studies observed a 50% weight loss when fed with sorghum containing high tannin levels (Ambula et al., 2001). This is because they have a longer emptying time in the stomach (Awika and Rooney, 2004). The low digestibility of high tannin sorghums is due to the inhibition of hydrolytic enzymes and their potent antioxidant activities (Dicko et al., 2005) may be interesting from a nutritional standpoint for obese persons.

Anthocyanins are nowadays regarded as an important nutraceuticals mainly due to their possible antioxidant effects. They have a potential therapeutic role related to some cardiovascular diseases, cancer treatment, inhibition of certain types of virus including human immunodeficiency virus type 1 (HIV-1) and improvement of visual acuity (Stintzing et al., 2002; Sandvik, 2004; Talavera et al., 2006; Beattie et al., 2005). Mild levels of reactive oxygen species (ROS) in food have been shown to induce proliferation of cancer cells (Arora et al., 1999; Del Bello et al., 1999). Therefore, foods rich in antioxidant phytochemicals are important for the prevention of diseases related to oxidative stress such as heart disease and cancer.

Sorghum hybrids have contributed significantly to increased grain and forage yields in several countries. A large number of hybrids have been developed and released for commercial cultivation in East Africa. Achievements in sorghum breeding in Africa have mainly been in the development and release of improved varieties based on higher grain yield and resistance to diseases, insect pests and Striga (Obilana, 2004). Little focus has been put on improving the nutritional aspects of cereal grains.

In plant breeding traits such as yielding ability and quality should be the fundamental objectives. However before a new cultivar is released to farmers, it is important that its nutritional quality is properly evaluated for their biochemical characteristics. Therefore, to screen varieties for consumer acceptability, simplified laboratory tests, applicable to a very large number of samples are required. The objective of this study was to determine biochemical and physiological characteristics of sorghum grain that lead to food and nutritional security.

MATERIALS AND METHODS

Sample preparation

Thirty (30) sorghum materials used in the study were obtained from the East African region. The mid lowland sorghum were grown in Kampi Ya Moto (035°. 56' E and 00°. 05' S) at an altitude of 1660 m a s l while the highland sorghum were grown at Egerton University at an altitude of 2,250 m a s l . Both sorghum materials were sown in a randomized complete blocking design and replicated 3 times during the April – August season 2013. The grain from two middle rows in each experimental unit was harvested, dried, threshed and used for subsequent laboratory and industrial tests (Table 1).

Determination of protein content

Total nitrogen and protein was determined using Kjeldahl method (AOAC, 1999). Sorghum grain was finely milled after which 0.1 g was weighed and transferred into a digestion tube. Selenium catalyst mixture weighing 1 g was mixed with the sample and 5 ml of sulphuric acid (96%) was added into the tube. The tubes were then heated cautiously in the digestion apparatus, at the fume cupboard until the digest was clear. The sample was transferred to a 100 ml volumetric flask, and distilled water was added into 100 ml graduated flask upto the mark. Boric acid indicator solution of 5 ml was then transferred to 100 ml conical flask containing 5 drops of mixed indicator and was placed under the condenser of the distillation apparatus. 10 ml of the clear supernatant was then transferred into the apparatus; 10 ml of 46% sodium hydroxide was added and then rinsed again with distilled water. Colour changed from pink to green when the first distillation drops reached the boric acid indicator solution. A total of 150 ml of the distillate was collected and was titrated with 0.0174 N sulphuric acids. Total nitrogen (N) was then determined as follows:

$$N\,(\%) = \left\{\frac{a \times N \times M_w \times 100}{b \times c}\right\} \times 100\%$$

Where a = ml of sulphuric acid used for titration of the sample; N = normality of sulphuric acid (0.0174); a = titer volume (10 ml); Mw = molecular weight of N_2 (0.014); c = ml digest taken for distillation (10 ml); b = g sample taken for analysis (0.1 g); % crude protein = 6.25 × % N.

Table 1. A list of the selected sorghum varieties used in the study.

Code	Variety	Code	Variety	Code	Variety
1	Ainamoi #1	11	IS 25547	21	ICSA 276 X ICSR 162
2	Siaya # 24-2	12	SDSH 90003	22	1S 25546
3	Kipkelion # 2	13	Uasin Gishu #1	23	IESH 22012
4	Nyiragikori	14	1S 25561	24	ICSA 12 X WAHI
5	Kipkelion # 1	15	Uasin Gishu #2	25	SDSA 29 X KARI MTAMA 1
6	MB 27	16	IESH 22006	26	IESV 92033 SH
7	Kabamba	17	Gadam Hamam	27	IESV 91104 DL
8	Nyangezi	18	ICSA 276 X ICSR 38	28	ICSA 371 X ICSR 108
9	BM 32	19	Kisanana	29	Busia # 21
10	1S 11162	20	Siaya # 2-3	30	IESV 92043 DL

Determination of starch content

Percent starch content was estimated by the Anthrone method (Hodge and Hofreiter, 1962). A powdered sample (0.25 g) was homogenized in hot 80% ethanol to remove sugars. The residue was then centrifuged and retained. The residue was dried well over a water bath. To the residue, 5.0 ml of distilled water and 6.5 ml of 52% perchloric acid was added, and then extracted at 0°C for 20 min. The supernatants were centrifuged, pooled and made up to 100 ml. Of this supernatant, 0.1 ml was pipetted out and topped to the 1 ml mark with distilled water. The standards were prepared by taking 0.2, 0.4, 0.6, 0.8 and 1 ml of the working standard and the volume made up to 1 ml in each tube with water. To this, 4 ml of anthrone reagent was then added to each tube and sample heated for eight minutes in a boiling water bath. Sample was cooled rapidly and the intensity of green to dark green colour was read using a spectrophotometer at 630 nm. The glucose content in the sample was determined using the standard calibration graph, and then the value was multiplied by a factor of 0.9 to arrive at the starch content.

Determination of amylose content

Amylose was determined using the Mc Cready et al., (1950) method where a sample (0.1 g) of the powdered flour was weighed, and 1 ml of distilled ethanol added followed by 10 ml of 1 N NaOH. The sample was heated for 10 min in a boiling water bath. The volume was made up to 100 ml. To a 2.5 ml extract, 20 ml of distilled water was added followed by three drops of 0.1% phenolphthalein. Dropwise HCl 0.1 N was then added until the pink colour disappeared. To this solution, 1 ml iodine reagent was added till the volume was 50 ml and the colour read at 590 nm using a spectrophotometer. Standard amylose solution 0.2, 0.4, 0.6, 0.8 and 1 ml was taken and the colour developed as in the case of the test samples. The amount of amylose present in the sample was calculated using the standard graph. The blank was obtained by diluting 1 ml of iodine reagent to 50 ml with distilled water. Amylose content was obtained thus:

$$\text{Amylose (\%)} = \left(\frac{x}{2.5} \right) \times 100 \text{ mg amylose}$$

Where, x is the absorbance obtained. The amylopectin content was obtained thus: Starch (%) - amylose (%).

Determination of tannin content

A sample of (0.5 g) of the powdered flour was weighed and transferred to a 250 ml conical flask, and then 75 ml of water added. The flask was heated gently and boiled for 30 min, then centrifuged at 2000 rpm for 20 min. The supernatant was collected in a 100 ml volumetric flask. A measure of 1 ml of the sample extract was transferred to a 100 ml volumetric flask containing 75 ml water.

5 ml of folin regent, 10 ml of 35% sodium carbonate solution were added, and then diluted to 100 ml with water. The sample was shaken and the absorbance read at 700 nm after 30 min. A graph was prepared using 0 - 100 mg tannic acid, where 1 ml contained 100 mg tannic acid. The tannin content of the sample was calculated as tannic acid equivalent from the standard curve. Tannins content was determined by the Folin-Denis method (Schanderl, 1970).

Determination of height and yielding ability

The height of the mature sorghum varieties was determined in centimeters by getting the average of three plants in each plot of each variety in every treatment, then the average of each plot in the three treatments.

The yielding ability of sorghum varieties was determined in grams by getting the average yielding ability of three plants in each plot of each variety in every treatment, then the average yield of each plot in the three treatments.

Data analysis

Data obtained from this study was statistically analyzed using one way analysis of variance (ANOVA) with JMP 10.0.0 Software, 2012 SAS Institute Inc. Turkey's test was used to find the difference among the means. Pearson's correlation analysis was also carried out to determine the relationship between biochemical parameters at $p \leq 0.05$ and 0.001

RESULTS

The starch content of dry seed weight of the genotypes ranged from 25.5 to 81.1% with a mean of 52.2% while amylose and amylopectin ranged from was 12.3 to 26.7%

and 6.6 to 59.8% with means of 12.3 and 6.7%, respectively. Proteins ranged from 3.1 to 18.1% with a mean of 9.8%. The tannin content ranged from 2.5 to 100.0 mg/100 g with a mean of 35.5 mg/100 g. Yield values were 8570 to 863.6 kg/ha with a mean of 3879.0 kg/ha while height ranged from 341.5 to 97.0 cm with a mean of 198.4 cm.

There was a statistically significant correlation between starch and amylopectin (0.97, $p \leq 0.001$) while starch and protein had a negative significant correlation (-0.55, $p \leq 0.001$). The correlation between starch and tannins was 0.49, $p \leq 0.05$, whilst starch and height had a correlation of 0.47, $p \leq 0.05$. Other significant correlation includes amylopectin and proteins (- 0.58, $p \leq 0.001$), tannins and height (0.47, $p \leq 0.05$); 0.52, $p \leq 0.05$. Protein and tannin contents had a negative correlation of – 0.52, $p \leq 0.05$ whereas protein and height had a correlation of –0.46, $p \leq 0.05$. The yielding ability and height has a positive significant correlation of 0.64, ≤ 0.001.

Hierarchical clustering (method = ward)

Cluster 4 varieties had high starch and high tannins contents and also had high amylopectin as compared to their respective amylose amounts (Figure 2). They also had medium height, yield and good protein amounts. On the other hand, cluster 3 varieties had medium starch, height and good protein amounts. They exhibited high amylopectin as compared to their respective amylose high yields and tannins.

Cluster 2 varieties had high amylopectin as compared to their respective amylose amounts, medium starch, height and average protein contents lower than cluster 1 and 2. They possess the highest yields and height as compared to the rest of the clusters. Cluster 1 varieties had the lowest starch, tannins and height as compared to other clusters. Most varieties with lower amylopectin to amylose ratio are in this cluster. They registered highest proteins and are mostly the hybrids while the yields of the varieties varied. These clusters are useful when identifying the varieties that are favourable for different end uses.

DISCUSSION

The varieties used in this study were both hybrids and open pollinated and they portrayed a wide range of nutritive contents (Table 2). Starch is the main component of sorghum grains (BSTID-NRC, 1996) as evident, thus this makes it favorable to be explored for different industries. This study has demonstrated that sorghum varieties possess biochemical characteristics that can match up to other cereals thus should be considered as alternative to these industries. Sorghum with starch as high as 70% *Ainamoi, kipkelion, Siaya* 24-2 (Figure 1 and Figure 2, Cluster 4) is a reliable source of energy that can be explored by many industries. In terms of nutritive value, cost and availability sorghum grains are probably the next best alternative to maize in poultry feed as reported by Lepleaideur (2004). As per this study, most of the varieties with good starch contents are open pollinated while the hybrids showed moderate to low levels of starch (Table 2). Normally, hybrids are improved varieties thus expected to have good starch amounts but this was not the case as shown in Table 2. More needs to be done to improve the starch contents of these hybrids if they are to be accepted in these industries.

Apart from the high starch content, the ratio of amylose to amylopectin is vital for the baking and brewing industries because it affects the quality of the end products. Generally, the amylopectin are higher than their respective amylose contents in sorghum. This study demonstrated some exceptions (Table 2 and Figure 2, Cluster 1) like *IESV 92033 SH, SDSA 29 X KARI MTAMA 1, ICSR 276 X ICSR 38* and *Kisanana*. These exceptions are favourable for the baking industry because high amylose gives stabilization to bread quality as stated by Taylor et al. (2006). Sorghum alone cannot be used for baking due to its lack of gluten, but addition of 20 to 50% sorghum flour to wheat flour results to good quality bread (Hugo et al., 2003). Sorghum is a good candidate for the brewing industry and has been adopted by the East African industry to some extent. This is because of lack of specific varieties that are tailored for this industry. The amylose to amylopectin ratio in this study was 1:2, and this is lower than the 1:3 ratio reported by Taylor et al. (2006). On the other hand, varieties favourable for brewing ought to have more amylopectin than amylose contents because Sharma et al. (2008) reported that high amylose affects starch digestibility. As per this study, most varieties suitable for brewing were mostly the hybrids.

The protein contents of sorghum varieties ranged from 3 - 18% and this was similar to 6 – 18% reported by Lasztity (1999). Varieties *ICSA 12 X WAHI* and *SDSA 29 X KARI MTAMA 1* had notably high protein contents (Table 2). These varieties would be good for animal and human foods, thus boost food security. The wide variety of sorghum protein gives it an advantage to be utilized for different uses depending on the protein requirements of that industry. The brewing and baking industry needs varieties with moderate proteins while animal and human feeds needs maximum proteins, and sorghum offer a wide choice. All the hybrids recorded high protein value unlike the open pollinated varieties. This can be attributed to sorghum breeding efforts as originally sorghum was more subsistence than industrial. Sorghum protein is gluten free and this is safe to be used by celiac patients (Ciacci et al., 2007) and high proteins means high food quality. This gives a good alternative to wheat flour due to their neutral flavor and the hybrids with white pericarp (Normell et al., 2010).

Table 2. ANOVA output of biochemical and physiological traits of sorghum accessions.

Genotype	Starch (%)	Amylose (%)	Amylopectin (%)	Protein (%)	Tannins mg/100 (g)	Yielding ability (kg/ha)	Height (cm)
Ainamoi #1	81.1^{a}	22.2^{abc}	58.9^{a}	7.8^{efgh}	79.8^{b}	3362.2^{cdef}	241.5^{de}
Siaya # 24-2	76.4^{ab}	19.4^{bcde}	56.9^{ab}	9.5^{def}	41.7^{g}	3245.5^{cdef}	125.0^{jk}
Kipkelion # 2	76.0^{ab}	16.2^{cdef}	59.8^{a}	7.4^{fgh}	58.7^{f}	3617.9^{bcdef}	209.2^{efg}
Nyiragikori	75.2^{ab}	18.2^{cdef}	56.9^{ab}	8.2^{efgh}	15.5^{kl}	3014.5^{def}	260.4^{cd}
Kipkelion # 1	71.6^{abc}	19.1^{bcdef}	52.5^{abc}	5.2^{hi}	59.9^{ef}	2263.1^{ef}	233.5^{de}
MB 27	68.9^{abcd}	14.0^{def}	54.9^{abc}	3.1^{i}	18.9^{kl}	6975.4^{abcd}	218.0^{def}
Kabamba	68.9^{abcd}	12.3^{f}	56.5^{ab}	6.9^{fgh}	100.0^{a}	1591.3^{f}	226.2^{de}
Nyangezi	64.4^{bdce}	17.1^{cdef}	47.2^{abcde}	8.7^{defg}	69.1^{cd}	3045.5^{def}	290.1^{bc}
BM 32	62.7^{bcdef}	18.1^{cdef}	44.5^{abcdef}	8.1^{efgh}	33.1^{h}	7460.9^{abc}	306.7^{abc}
1S 11162	61.7^{bcdefg}	13.4^{ef}	48.2^{abcd}	8.6^{defg}	35.1^{h}	7909.5^{ab}	341.5^{a}
IS 25547	57.4^{cdefgh}	16.1^{cdef}	41.3^{bcdefg}	5.9^{ghi}	42.3^{g}	6459.3^{abcde}	317.7^{ab}
SDSH 90003 (H)	57.0^{cdefgh}	18.6^{cdef}	38.3^{cdefgh}	7.9^{efgh}	24.4^{ij}	4777.3^{abcdef}	124.8^{jk}
Uasin Gishu #1	55.2^{defghi}	21.5^{abc}	33.6^{defghi}	5.8^{hi}	65.1^{de}	3253.8^{cdef}	173.1^{fghi}
1S 25561	54.5^{defghij}	15.4^{cdef}	39.0^{cdefgh}	11.8^{cd}	44.6^{g}	4972.6^{abcdef}	305.8^{abc}
Uasin Gishu #2	51.6^{efghijk}	18.3^{cdef}	33.2^{defghi}	8.7^{defg}	73.7^{g}	3382.1^{cdef}	204.3^{efg}
IESH 22006 (H)	48.7^{fghijk}	18.4^{cdef}	30.2^{efghi}	13.4^{bc}	9.4^{mno}	5020.7^{abcdef}	139.0^{hijk}
Gadam Hamam	47.4^{ghijk}	16.5^{cdef}	30.9^{efghi}	10.9^{cde}	26.2^{i}	863.6^{f}	97.0^{k}
ICSA 276 X ICSR 38 (H)	44.6^{hijkl}	25.5^{ab}	19.0^{ijkl}	15.6^{ab}	13.8^{lmn}	2872.3^{def}	131.2^{hijk}
Kisanana	44.5^{hijkl}	26.7^{a}	17.7^{ijkl}	7.2^{fgh}	68.3^{d}	3186.2^{cdef}	155.6^{hij}
Siaya # 2-3	43.4^{hijkl}	13.8^{def}	29.6^{fghi}	5.1^{hi}	19.5^{jk}	4151.4^{bcdef}	203.8^{efg}
ICSA 276 X ICSR 162 (H)	43.4^{hijkl}	18.5^{cdef}	24.8^{ghijk}	11.8^{cd}	6.3^{op}	4323.2^{abcdef}	176.6^{fgh}
1S 25546	41.9^{ijklm}	14.2^{def}	27.6^{fghij}	7.1^{fgh}	36.0^{h}	8570.1^{a}	300.7^{abc}
IESH 22012 (H)	41.6^{ijklm}	15.7^{cdef}	25.9^{ghijk}	16.8^{a}	9.7^{mno}	2738.1^{def}	127.4^{ijk}
ICSA 12 X WAHI (H)	40.1^{jklmn}	17.6^{cdef}	22.4^{hijkl}	18.1^{a}	7.7^{op}	1245.4^{f}	134.7^{hijk}
SDSA 29 X KARI MTAMA 1(H)	38.4^{klmn}	21.7^{abc}	16.6^{ijkl}	18.0^{a}	2.5^{p}	3888.2^{bcdef}	173.6^{fghi}
IESV 92033 SH	36.8^{klmn}	18.0^{cdef}	18.7^{ijkl}	9.1d^{efg}	8.6^{no}	3448.2^{cdef}	163.4^{ghij}
IESV 91104 DL	32.1^{lmn}	13.9^{def}	18.2^{ijkl}	10.8^{cde}	9.1^{no}	2095.3^{ef}	154.0^{hij}
ICSA 371 X ICSR 108 (H)	27.0^{mn}	20.4^{abcd}	6.6^{l}	16.0^{ab}	4.5^{op}	2915.8^{def}	147.2^{hij}
Busia # 21	26.9^{mn}	17.6^{cdef}	9.3^{jkl}	11.8^{cd}	66.2^{d}	2869.8^{def}	121.0^{jk}
IESV 92043 DL	25.5^{n}	13.3^{ef}	12.1^{jkl}	10.8^{cde}	14.6^{klm}	2849.5^{def}	146.5^{hij}
Standard Error (SE)	2.69	1.22	3.08	0.58	0.96	785.5	8.42

Means with the same letter are not significantly different, H– hybrids.

Sorghum grains are characterized by their high content of condensed tannins which limits their suitability for several end users. Tannins are considered undesirable due to their capacity to bind proteins making them less digestive, and also producing astringent taste (Ambula et al., 2003). Sorghum genotypes with low tannin contents have been produced due to plant breeding efforts and this is evident in the hybrids varieties *ICSA 12 X WAHI, SDSA 29 X KARI MTAMA 1, ICSV 92033 SH* and *IESV 91104 DL* (Table 2 and Figure 2, Cluster 1). The open pollinated varieties on the other hand had considerably high tannin contents (Figure 2, Cluster 4). Low tannin content sorghum especially the hybrids are favorable for poultry feeding due to their proteins and carbohydrate contents. This is because high tannic acid varieties are not good for monogastrics feed as reported by Smithhard (2002). He also observed that when monogastrics were fed with high tannic sorghum based diet, there was a general poor performance due to the binding effect of tannins to digestive enzymes. However, when used in conjunction with maize, these effects were minimized. Thus moderate to low tannin sorghum can be used to supplement the poultry industry in an effort to reduce overdependence on maize.

Legumes like lucerne and clovers contain saponins and on fermentation release vast quantities of soluble proteins this is because these legumes lack tannins and this result to bloat in livestock (Jansman, 1993). According to Jones et al. (1973), this can be minimized by using high tannin sorghum that have been shown to posses the ability to control bloat. It has also been reported that after the digestion of tannin foods, the

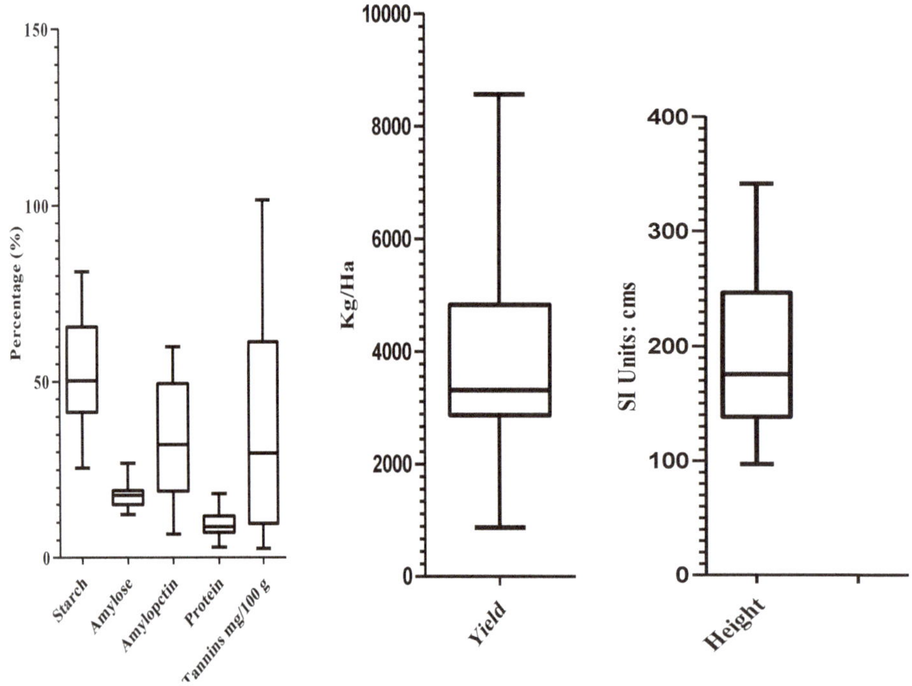

Figure 1. Biochemical and physiological traits of 30 sorghum varieties showing their ranges and means.

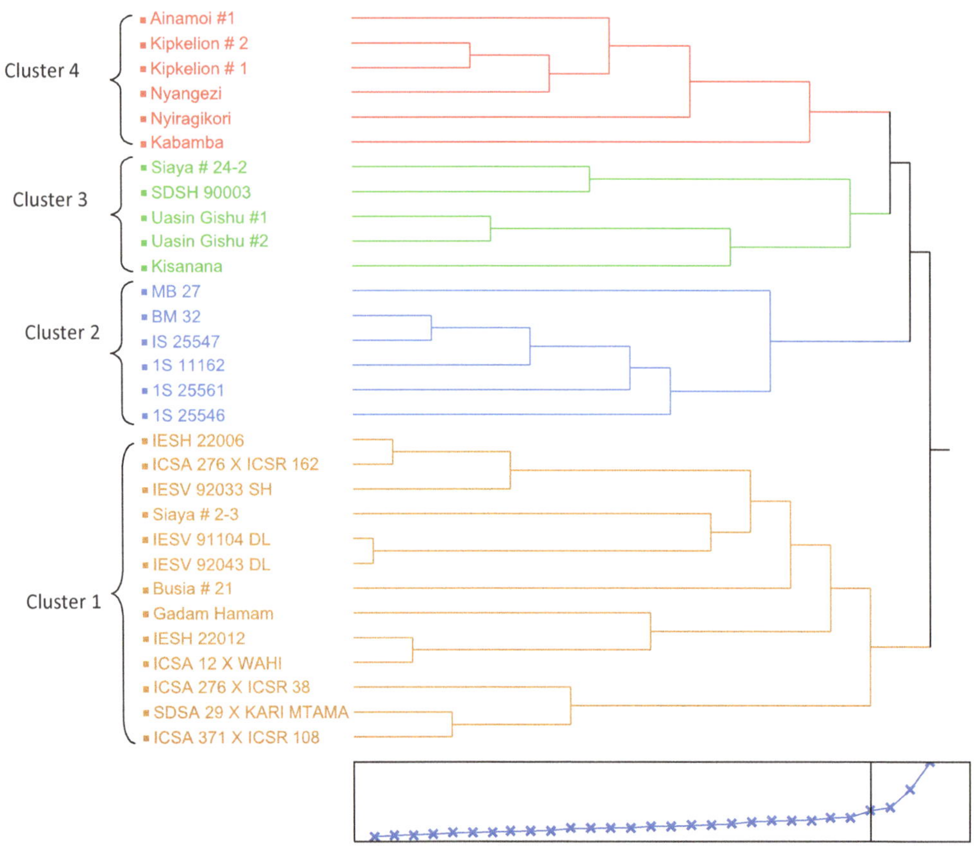

Figure 2. Dendrogram of 30 sorghum varieties shown by ward cluster analysis based on biochemical and morphological data set.

Table 3. Pearson's correlation coefficient results for 30 sorghum accessions.

N = 30;	Starch	Amylose	Amylopectin	Protein	Tannins	Yield	Height
Starch	1.00	0.003 ns	0.97***	- 0.55***	0.49**	0.16 ns	0.47**
Amylose		1.00	- 0.21ns	0.22 ns	0.05 ns	-02.3 ns	- 0.31
Amylopectin			1.00	- 0.58***	0.47**	0.20 ns	0.52**
Protein				1.00	- 0.52**	- 0.34 ns	- 0.46**
Tannins					1.00	- 0.10 ns	0.31ns
Yield						1.00	0.64***
Height							1.00

N = 30; Significant at**- P ≤ 0.05, ***-0.001; NS – non significant at P ≤ 0.05 and 0.001.

tannin-protein complex formed reacts with gut wall proteins and stimulates growth hormones that in turn increases lipid turnover and nitrogen retention (Muir et al., 1983). The significance is that it protects plant protein degradation in the rumen and this is the advantage of sorghum grain in animal feed. Despite sorghum being undesirable in many industries, animal feed industry can take advantage of the availability of sorghum tannins. The varieties that can be explored as per this study include *Ainamoi, Kipkelion # 1, Nyiragikori, Kipkelion # 2, Kabamba* and *Nyaangezi* (Figure 2, Cluster 4).

Sorghum varieties rich in tannins can be recommended for obese individuals and diabetic patients. According to Ambula et al. (2001) animals fed with sorghum containing high level tannins demonstrated 50% weight loss. This was attributed to their long emptying time of the stomach and their low digestibility through the inhibition of hydrolytic enzymes together with their antioxidant activities (Dicko et al., 2005).

In agronomical point of view, tannins are desirable due to their protection against bird, insect and disease damage (Waniska et al., 2001). According to Bullard and Elvis (1980) the astringent taste of tannins is due to tannin-saliva protein complex in the mucous epithelium of the oral cavity thus decreasing its palatability. This largely contributes among other factors to the high yield in tannin sorghum (Table 2).

Yielding ability is a vital parameter when selecting varieties for different end uses because high yielding varieties are economical both to farmers and the industry. The varieties in this study demonstrated a wide range in yielding ability as shown in Figure 1. This study demonstrated yielding ability as high as 8570 kg/ha with some varieties such as *1S 25546* and 7909.5 kg/ha as in *1S 11162*. Industries that focus greatly on yielding ability can adopt these particular varieties. This is vital because due to climate change wheat and maize production have been greatly affected but sorghum is known to be resilient to harsh climatic conditions and give good yield (Dicko et al., 2006). Plant height is also an important parameter especially for large scale farmers because short varieties reduce the cost of production due to the possibility of mechanization. As shown (Table 2), the hybrid varieties were shorter than the open pollinated varieties with less than 176 cm in length while the open pollinated had height as high as 317 cm *1S 25546* and 341 cm *1S 11162* (Figure 2, Cluster 2). These are the same varieties that registered high yields. Thus, it is important to note that yielding ability positively, significantly correlated with plant height in this study. The consequence is that in the event of trying to reduce height via breeding, might lead to reduction of yielding ability.

Correlation analysis

Most of the varieties with high starch and amylopectin contents also recorded high tannin contents with a few exceptions (Table 2). These correlations suggest a genetic association between the parameters both in the hybrids and the open pollinated varieties. The positive significant correlation between tannins and starch of these varieties is a bad characteristic for breeding since starch is desirable in most end uses unlike tannins (Table 2). On the other hand, the negative correlation between tannins and proteins is a good selection indicator for varieties in sorghum breeding (Table 3). The protein was negatively correlated with starch and amylopectin content which is expected, as protein and starch make the major component in grains. This is evident in Table 2 where *SDSA 29 X KARI MTAMA 1* and *ICSA 12 X WAHI* are hybrids with high protein and low starch and tannin amounts. On the other hand, *kabamba, ainamoi # 1, kipkelion* are open pollinated varieties with high tannins and starch but low protein contents (Table 2). Hybrids are bred to give low tannins but unfortunately this also affects their starch content and in the long run decrease the suitability of sorghum for some industrial uses. There was a significant correlation between yielding ability and plant height, as similar to results of Alhassan et al. (2008). This demonstrates the correlation between plant height and yielding ability and it might be attributed to genotypes with good plant height being able to receive more sunlight for photosynthesis, which translates to more photosynthates resulting to increased yield. Plant height also showed a significant positive correlation with starch and amylopectin contents. This is due to increased photosynthesis due to height which positively affects the yields.

Cluster analysis

Cluster 4 varieties had high starch and high tannins contents and also had high amylopectin as compared to their respective amylose content (Figure 2). They also had medium height, yielding ability and good protein content. On the other hand, cluster 3 varieties have medium starch, height and good protein content. They exhibited high amylopectin as compared to their respective amylose, high yields and tannins. Cluster 2 varieties had high amylopectin as compared to their respective amylose amounts, medium starch, height and average protein contents lower than cluster 1 and 2. They possess the highest yields and height as compared to the rest of the clusters. Cluster 1 varieties had the lowest starch, tannins and height as compared to other clusters. Most varieties with lower amylopectin to amylose ratio are in this cluster. They registered highest proteins and are mostly the hybrids while the yields of the varieties varied. These clusters are useful when identifying the varieties that are favourable for different end uses.

Conclusion

Success of any food grain or its product depends on acceptance by the consumers. In the process of achieving food production targets, agricultural scientists concentrate their efforts on developing high-yielding varieties giving little importance to other quality characteristics. From the study, we find that proteins and tannins were in high contents among the hybrids and the open pollinated varieties (OPVs) respectively. Most of the hybrids had greater amylose than amylopectin and the vice versa was true for the OPVs. It is clear that breeding programmes that aim to reduce tannin contents also result to lower starch as evident in the hybrid varieties. Plant breeders should put into consideration not only agronomical characteristics but also biochemical characteristics. There is need for further research through biotechnological techniques to identify genes responsible for specific biochemical traits if sorghum and other cereals are to be completely adopted by the huge African market.

Conflict of interests

The authors did not declare any conflict of interest.

REFERENCES

Ajaja K, Agbede JO, Aletor VA (2002). Replacement value of sorghum dust for maize in diets for broiler chicks. Proc 27th Ann. Conf. Nig. Soc. for Anim. Prod. March 17-21, Federal University of Technology, Akure. pp. 109-112.

Alhassan U, Yeye MY, Aba DA, Alabi SO 2008. Correlation and path coefficient analyses for agronomic and malting quality traits in some sorghum (Sorghum bicolor (L.) Moench) genotypes. J. Food Agric. Environ. 6 (3&4):285-288.

Ambula MK, Oduh GW, Tuitoek JK (2003). Effects of high tannins sorghum and bentonite on the performance of laying hens. Trop. Anim. Health Prod. J. 35:285-292.

Ambula MK, Oguho GW, Tuitoek JK (2001). Effects of sorghum tannins,a tannin binder (polyvinylpyrrolidone) and sorghum inclusion level on the performance of broiler chicks. Asian-Austr. J. Anim. Sci. 14:1276-1281.

Amir MA, Muralikrishna G, El Tinay AH, Mustafa Al (2009). Characterisation of Tannins and Study of in vivo Protein Digestibility and Mineral Profile o Sudanese and Indian Sorghum Cultivers. Pak. J. Nutr. 8(4):469-476.

AOAC (1999) Official Methods of Analysis Method 988.05. Chapter. 4, AOAC International Press, Gaithersburg. p 13.

Arora KP, Lucchesi PA, Wurster RD (1999). Proliferation of cultured human astrocytoma cells in response to an oxidant and antioxidant. J. Neurooncol. 44:213-221.

Arun G, Kulamarva A, Venkatesh R, Sosle A, Vijaya Raghavan GS (2009). Nutritional and Rheological Properties of Sorghum. Int. J. Food. Prop. 12 (1): 55-69

Awika JM, Rooney LW (2004) Sorghum phytochemicals and their potential aspects on human health. Phytochemistry 65: 1199 – 1221.

Beattie J, Crozier A, Duthie GG (2005). Potential health benefits of berries. Curr. Nutr. Food Sci. 1: 71- 86.

BSTID-NRC (Board on Science and Technology for International Development-National Research Council) (1996). Lost crops of Africa. Academic Press, Washington DC, pp.127-213.

Bullard RW, Elias DJ (1980). Sorghum polyphenols and bird resistance. In Polyphenols in Cereals and Legumes, [Ed J. Hulse,]. Ottawa: International Development Research Centre. p. 4349.

Ciacci C, Maiuri L, Caporaso N, Bucci C, Del Giudice L, Massardo DR, Pontieri P, Di Fonzo N, Bean SR, Ioerger B, Londei M (2007). Celiac disease: *In vitro* and *in vivo* safety and palatability of wheat-free sorghum food products. J. Clin. Nutr. 26(6):799-805.

Del Bello B, Paolicchi A, Comporti M, Pompella A, Maellaro E (1999). Hydrogen peroxide produced during gamma-glutamyl transpeptidase activity is involved in prevention of apoptosis and maintenance of proliferation in U937 cells. FASEB J. 13:69-79.

Dicko MH, Gruppen H, Traore AS, Van Berkel WJH, Voragen AGJ (2005). Evaluation of the effect of germination on phenolic compounds and antioxidant activities in sorghum varieties. J. Agric. Food. Chem. 53:2581-2588.

Dicko MH, Gruppen H, Traoré AS, Voragen AGJ, Van Berkel WJH (2006). Sorghum grain as human food in Africa: relevance of content of starch and amylase activities. Afr. J. Biotechnol. 5: 384 - 395.

Elkhier MKS, Hamid AO (2008). Effect of malting on the chemical constituents, anti-nutrition factors, and ash composition of two sorghum cultivars (Feterita and Tabat). J. Agric. Biol. Sci. 4(5):500-504.

Food and Agricultural Organization (FAO) (2005). FAOSTAT agricultural database. Available on line http://faostat.fao.org/faostat/.

Food and Agriculture Organization (FAO) (2012). Sorghum bicolor (L.) Moench. In: Grassland Species Profiles Database [online]. www.fao.org/ag/agp/agpc/doc/gbase/ data/pf000319.h tm (accessed 12 March 2015).

Hodge JE, Hofreiter BT (1962). In: Methods in Carbohydrate Chemistry, (Editors. Whistler, R.L. and Be Miller, J.N.), Academic Press, New York.

Hugo LF, Rooney LW, Taylor JRN (2003). Fermented sorghum as a functional ingredient in composite breads. Cereal Chem. 80: 495-499.

Jambunathan R, Subramanian V (1987). Grain quality and utilization of sorghum and pearl millet. In: Biotechnology in tropical crop improvement. Proceedings of the International Biotechnology Workshop, ICRISAT. Patancheru, India.

Jansman AJM (1993). Tannins in feedstuffs for simple- stomached animals. Nutr. Res. Rev. 6:20Y-236.

Jones WT, Anderson L B, Ross MD (1973). Bloat in cattle. Detection of protein precipitants (flavolans) in legumes. N. Z. J. Agric. Res. 16:441-446.

Lasztity R (1999) Cereal chemistry. Akademiai Kiado Budapest, Hungary.

Lepleaideur M (2004). Poultry farming - A disease called competition. In: SPORE Magazine. 114:4-5.

Mc Cready RM, Guggolz J, Siliviera V, Owens HS (1950). Starch. Anal. Chem. 22: 11- 56.

Muir LA, Wien S, Duquette PF, Rickes EL, Cordes EH (1983). Effect of exogenous growth hormone and diethylstilbestrol on growth and carcase composition of growing lambs. J. Anim. Sci. 56:1315-1323.

Normell JM, Sajid A, Scott RB (2010). Sorghum Proteins: The Concentration, Isolation, Modification, and Food Applications of Kafirins. J. Food. Sci. 75:5

Obilana T (2004). In Sorghum genetic enhancement- Research process, dissemination and impacts (Bantilan, M.C.S., Deb UK, Gowda CLL).

Sandvik L (2004). Anthocyanins useful for the treatment of diabetes, cardiovascular disorders and to lower the risk of adverse effects of hormone replacement therapy. PCT Int. Appl. p. 28.

Schanderl SH (1970). In: Method in Food Analysis. Academic Press, New York. p. 709.

Sharma A, Yadav BS, Ritika BS, (2008) Resistant starch: physiological roles and food applications. Food Rev. Int. 24:193-234.

Smith CW, Frederiksen, RA (2000). Sorghum: Origin, history, technology and production, John Wiley and Sons Inc., New York. NY 824, p. 668.

Smithhard R (2002). Secondary Plant Metabolites in Poultry Nutrition. In: Poultry Feedstuff: Supply, Composition and Nutritive Value. Mc Nab JM, Boorman KN (eds.) CAB International, Wallingford, UK, p. 254.

Stintzing FC, Stintzing AS, Carle R, Frei B, Wrolstad RE (2002). Colorand Antioxidant Properties of Cyanidin-Based Anthocyanin Pigments. J. Agric. Food Chem. 50: 6172 - 6181.

Talavera S, Felgines C, Texier O, Besson C, Mazur A, Lamaison JL, Remesy C (2006). Bioavailability of a bilberry anthocyanin extract and its impact on plasma antioxidant capacity in rats. J. Sci. Food Agric. 86:90-97.

Taylor JRN (2004). Grain production and consumption: Africa. In: Wrigley C, Corke H, Walker CE. (Eds.), Encyclopedia of Grain Science. Elsevier, London. pp. 70-78.

Taylor JRN, Schober TJ, Bean SR (2006). Novel food and non-food uses for sorghum and millets. J. Cereal Sci. 44:252-271.

Waniska RD, Venkatesha RT, Chandrashekar A, Krishnaveni S, Bejosano FP, Jeoung J, Jayara J, Muthukrishnan, S, Liang GH (2001). Antifungal proteins and other mechanisms in the control of sorghum stalk rot and grain mold. J. Agric. Food. Chem. 49:4732-4742.

Microbiological quality of food sold by street vendors in Kisangani, Democratic Republic of Congo

L. K. Makelele[1] , Z. A. Kazadi[1], R. W. Oleko[1], R. Foma[2,] Rosette Kabwang a Mpalang[3], Koto - te - Nyiwa Ngbolua[4] and Bongo Ngiala Gédeon[4,5]

[1]Department of Biotechnology, Faculty of Sciences, University of Kisangani, PO. Box 2012, Kisangani, Democratic Republic of Congo.
[2]Department of Food Chemistry, Agronomy Institute Faculty, P.O. Box 1232, Yangambi, Democratic Republic of Congo.
[3]Laboratory of Expertise, Hygiene and Technology of Foods of Animal Origin, Faculty of Veterinary Medicine, University of Lubumbashi, P.O. Box 1825, Lubumbashi, Democratic Republic of Congo.
[4]Department of Biology, Faculty of Sciences, University of Kinshasa, P.O. Box 190, Kinshasa, Democratic Republic of Congo.
[5]Sokoine University of Agriculture, Faculty of Veterinary Medicine, P.O. Box 3000, Chuo Kikuu, Morogoro, United Republic of Tanzania.

Food sold by roadside vendors was compared with French Standards (AFNOR, 1996) in order to determine the microbiological quality of cooked meals. Forty-two samples of fresh and smoked fish and bushmeat were collected between March and May 2013 in Kisangani (The Democratic Republic of Congo), and analysed. Analysis of variance (ANOVA) and post-hoc Tukey tests were used to analyse the level of contamination according to the category of cooked food. Results were significant at the 0.05 threshold. For all three categories of dishes, the average bacterial counts (total aerobic plate count) were above the critical threshold: bushmeat (6.70 ± 0.15 log cfu g^{-1}), smoked fish (6.44 ± 0.09 log cfu g^{-1}) and fresh fish (5.97 ± 0.33 log cfu g^{-1}). The difference in levels of contamination between groups was statistically significant ($p < 0.05$, ANOVA test). Bushmeat was the most contaminated category ($p < 0.05$, Tukey test). Most of the 42 samples were of unsatisfactory microbiological quality: 38 (90.5%) due to total aerobic plate count; 24 (57.1%) to *Salmonella* sp. and 21 (50%) to *Staphylococcus aureus*. The application of hygienic practices during the preparation and sale of street food could reduce the microbial risk. Such training is highly recommended for roadside food vendors.

Key words: Microbiological quality, street food, food contamination, bacteriological count, Kisangani, Democratic Republic of Congo.

INTRODUCTION

Consumption of food prepared and sold by street vendors is growing rapidly in developing countries. Street food, particularly beverages and cooked snacks or meals, are sold in public places such as roadsides, markets and similar locations (Muyanja et al., 2011). They represent an important part of the daily diet for millions of low- and middle-income consumers in urban areas (FAO, 2003).

Street food plays an important socio-economic role: it provides a regular source of income for millions of low- or unskilled men and women in developing countries (FAO, 2010). In Indonesia, street food also contributes to local economic growth. This informal activity generates an annual estimated income of US$ 1 million in taxes (FAO, 2010; Cohen, 1986).

In the Democratic Republic of Congo (DRC), meals sold by the street vendors (called *Malewa* in Lingala, a local dialect) are very popular and are primarily consumed in Kisangani. The menu usually consists of a meal cooked in sauce, either based on beans (*madesu* in Lingala), meat (smoked or fresh) or fish (fresh or smoked). These stews are often accompanied by cassava-based starches (*fufu*), mashed plantains (*lituma* in Swahili) or rice (*loso* in Lingala). These street meals are sold in small restaurants next to public places (markets, schools, hospitals), along main roads or on the ground in the street.

Salmonella sp. and *Staphylococcus aureus* are the most common foodborne pathogens and are responsible for food poisoning and food-related infections (Akbar and Anal, 2013). Manguiat and Fang (2013) showed that contamination of street food in the Philippines was mainly due to *S. aureus*, *Salmonella* sp. and *Vibrio cholerae*, while *Salmonella* sp. was isolated in 15% of samples from grilled pork and chicken.

According to a study carried out in Egypt by Moustafa El-Shenawy et al. (2011), 24% of street food was infected with *Listeria* sp. This bacterium is responsible for listeriosis and is manifested by septicaemia, meningitis and intrauterine infections, leading to spontaneous abortion in pregnant women. In Nigeria (Omemu and Aderoju, 2008), Kenya (Muinde and Kuria, 2005) and South Africa (Holy et al., 2006) researchers have shown that the overall hygiene of pre-prepared street food is poor.

One strategy to reduce microbial risk in the consumption of street food is the World Health Organisation resolution AFR/RC53/R5/2003 (WHO, 2003) that recommends capacity strengthening of health authorities in signatory countries to control the hygienic quality of street food, which must comply with international standards.

However, relation to hygiene at the locations where food is sold (rubbish in the streets, blocked drains, the hygiene of cooked street food in Kisangani (D. R. Congo) remains questionable and addition to problems asked) and quality checked by Health Inspectors (and other health services) are often irregular.

Although, the phenomenon of the sale of street food is a major public health problem (Rane, 2011), unlike other African countries (Muyanja et al., 2011; Barro et al., 2006; Diouf, 1992) there are very few international studies of the situation in the DRC. A rare example is a study by the Food and Agriculture Association (FAO, 1996) in Kinshasa, which identified the following variables: consumer typology, the role of female sellers, daily revenue earned through the sale of street food and the support of United Nations in this sector. However, there was no evaluation of the hygiene of cooked street food.

The objective of this study was therefore to evaluate the microbiological quality of meals cooked in sauce (bushmeat, smoked and fresh fish) sold in the surroundings of the central market in Kisangani, DRC. Total aerobic plate count at 30°C (as indicator of hygienic quality of cooked meals), *S. aureus* and *Salmonella* sp. (as pathogens) were investigated. The contamination level of the different categories of meals was also assessed in order to identify the dishes with high microbial risk.

MATERIALS AND METHODS

Sample collection

Cooked meals sold by street vendors were collected by a simple random sampling method from around the central market in Kisangani (DRC). The sampling period (three months) spanned from March to May 2013. A total of 42 sites (restaurants) were investigated. The number of consumers per day ($n \geq 30$) was a determining factor in the choice of sampling sites. Thus, the samples (n=42) were randomly collected at noon for microbiological analysis. Samples of meals cooked in sauce consisted of bushmeat (n=16); smoked fish (n=18) and fresh fish (n=8). The samples (weighing approximately 100 g) were collected in sterile borosilicate glass bottles which were then placed in a cooler. Each sample was given an alpha-numeric code that represented the sample number and the meal category. Samples were immediately brought to the laboratory for microbiological analyses.

Microbiological analysis

Microbiological analyses were performed according to the protocol described by Manguiat and Fang (2013). A 25 g sample was diluted in 225 ml of peptone water (Merck KGaA, Germany). The mixture was macerated in a sterile Stomacher bag for two minutes. The supernatant representing the stock solution was then diluted up to 10^{-6} using sterile physiological water and the corresponding plate agar media were seeded. Counts (colony-forming units per gram) were expressed as a logarithm (log cfu g^{-1}).

Total aerobic plate count at 30°C

Total aerobic plate count (TAPC) was determined by the pour plate method using Nutrient Agar (Liofilchem, Italy). Then, one millilitre of the 10^{-3} dilution was introduced aseptically into the Petri dish and was incubated at 30°C for 72 h. After incubation, the colonies were counted. The number of colony-forming units (cfu) obtained was multiplied by the reciprocal of the dilution.

Staphylococcus aureus

S. aureus plate count was determined by the pour plate method using Mannitol Salt Agar (Biomérieux, France). After thoroughly

Table 1. Hygienic quality of street food cooked in sauce (Central Market, Kisangani, the DRC).

Hygienic quality	Critical threshold AFNOR (1996) $log_{10}cfu\ g^{-1}$	Fish				Meat		Total (n = 42)	
		Fresh (n = 8)		Smoked (n = 18)		Bushmeat (n = 16)			
		n	%	n	%	n	%	n	%
TAPC at 30°C									
Satisfactory	<5.5	1	12.5	0	0	0	0	1	2.4
Acceptable	5.5–6.0	3	37.5	0	0	0	0	3	7.1
Unacceptable	> 6.0	4	50	18	100	16	100	38	90.5
***Salmonella* sp.**									
Satisfactory	0	2	25	9	50	7	43.75	18	42.9
Unacceptable	> 0	6	75	9	50	9	56.25	24	57.1
***S.aureus* (coagulase+)**									
Satisfactory	< 2	1	12.5	8	44.4	0	0	9	21.4
Acceptable	2–3	7	87.5	4	22.2	1	6.25	12	28.6
Unacceptable	> 3	0	0	6	33.3	15	93.75	21	50

TAPC = Total aerobic plate count at 30°C; *S. aureus* coagulase + = *S. aureus* coagulase-positive; $log_{10}cfu\ g^{-1}$ = colony-forming units per gram of food (logarithm).

mixing the food sample (25 g) in peptone water (225 mL) in 1: 10 dilution, the inoculum (1 ml) was introduced aseptically into the Petri dish. Incubation lasted for 24 h at 37°C. Mannitol positive colonies, surrounded by a yellow halo were counted. Coagulase tests on *S. aureus* strains were carried out using conventional techniques (AFNOR, 2004).

Salmonella sp.

The sample was enriched by incubating the parent suspension for 18-24 h at 37°C. Then, one millilitre of the (enriched) parent sample was added to 9 ml of Selenite broth (Merck KGaA, Germany). This was incubated at 37°C for 24 h. Positive samples (cloudy broth) were cultivated in Salmonella Shigella Agar (Biomérieux, France). The sample taken from the positive Selenite broth was cultured by the streaking method on the Salmonella Shigella agar medium, after which it was incubated at 37°C for 24 h. Translucent *Salmonella* sp. colonies with a black centre appeared and were counted.

Interpretation and statistical analysis

Logarithmic bacterial counts ($log_{10}cfu\ g^{-1}$) were compared against French standards in order to determine the hygienic quality of the cooked dishes (Table 1). An analysis of variance (Anova) compared the level of contamination between categories of dishes. In cases where a significant difference was found between cooked food categories, the Tukey test (post-Anova test) was used to determine the most contaminated food categories (Walpole et al., 2002). Microsoft Excel (2007 edition) and R (version 2.15) software was used for statistical analyses. The *p*-value was considered to be significant at the 0.05 threshold.

RESULTS

The hygiene of dishes cooked in sauce was evaluated (satisfactory, acceptable and unacceptable) according to French standards (AFNOR, 1996) by an assessment of the TAPC, *S. aureus* and *Salmonella* sp. (Table 1).

Food quality: TAP count at 30°C

Out of the total of 42 meals analysed, 38 (90.5%) were considered to be of unsatisfactory quality following the TAPC bacterial analysis (Table 1). The difference in level of contamination was statistically significant according to the meal category ($p<0.05$, ANOVA test) (Table 2). Bushmeat was the most contaminated ($p<0.05$, Tukey's post-ANOVA test) (Table 2).

Food quality: *Salmonella* sp.

Out of a total of 42 samples, 24 (57.1%) were non-compliant with French standards (Table 1) following the bacterial analysis. Fresh fish samples (n=8) were the most contaminated category of food with *Salmonella* sp. (2.4 ± 1.5 log cfu g^{-1}, Table 3). A comparison of bacterial counts of *Salmonella* sp. in the three categories of dishes showed no statistically significant difference ($p> 0.05$, ANOVA test, Table 3).

Food quality: *S. aureus*

Out of the 42 dishes street analysed, 21 (50%) were declared unsatisfactory with respect to French standards for cooked foods (Table 1) and unfit for consumption

Table 2. TAPC bacterial count (log cfu g^{-1}) according to meal category.

Meal category	TAPC at 30°C(log cfu g^{-1})		Anova			Tukey
	X̅	SD	df	f	p^1	p^2
Fresh fish (n = 8)	5.96	0.33	2	45.03	<0.05*	<0.05*
Smoked fish (n = 18)	6.44	0.09				
Bushmeat (n = 16)	6.71	0.15				
Average	6.37	0.19				

X̅ = mean bacterial count; SD = standard deviation; n = number of dishes analysed; * = significant at $p = 0.05$ (difference in contamination levels between categories); $p^1 = p$ calculated using the ANOVA Test; $p^2 = p$ calculated using Tukey's post-ANOVA test; f = F-test of the equality of two variances.

Table 3. *Salmonella* sp. count (log cfu g^{-1}) according to meal category.

Meal category	*Salmonella* sp. (log cfu g^{-1})		Anova test		
	X̅	SD	df	f	p
Fresh fish (n = 8)	2.4	1.5	2	0.915	> 0.05
Smoked fish (n = 18)	1.5	1.5			
Bushmeat (n = 16)	1.6	1.5			
Average	1.8	1.5			

X̅ = mean bacterial count; SD = standard deviation; df = degrees of freedom; f = F-test of the equality of two variances; *p* = non-significant.

Table 4. *S. aureus* count (log cfu g^{-1}) according to meal category.

Meal category	*S. aureus* (log cfu g^{-1})		Anova			Tukey
	X̅	SD	df	f	p^1	p^2
Fresh fish (n = 8)	0.4	1.0	2	13.92	<0.05*	<0.05*
Smoked fish (n = 18)	1.7	1.5				
Bushmeat (n = 16)	3.3	0.8				
Average	1.8	1.1				

X̅= mean bacterial count; SD = standard deviation; df = degree of freedom; * = value is significant (difference in contamination between cooked food categories); p^1 = p-value calculated from the ANOVA test; p^2 = p-value calculated from Tukey's post-ANOVA test.

following the analysis of *S. aureus*. A comparison of *S. aureus* bacterial counts in the three categories of meals was statistically significant ($p<0.05$, ANOVA test, Table 4). Bushmeat was the category most affected by *S. aureus*, demonstrated by the assessment of the bacterial count (3.3 ± 0.8 log cfu g^{-1}, Table 4) using the post-Anova Tukey Test ($p< 0.05$).

DISCUSSION

The study demonstrated that the popularly sold street foods (cooked meals) were been contaminated by TAPC at 30°C, a microbiological indicator of hygien, demonstrated that the popularly sold street food (cooked meals) were contamined. Food pathogens (*S.aureus and Salmonella sp.)* were also found. This situation puts the health of the population at risk. According to Rane (2011) and Ghosh et al. (2007), street foods are perceived to be a major public health risk due to lack of basic infrastructure and services, difficulty in controlling the large numbers of street food vending operations because of their diversity, mobility and temporary nature.

Food quality: TAPC at 30°C

AFNOR Standards (AFNOR, 1996) are used in the DRC

for microbiological food evaluation and provide the criteria for the microbiological evaluation of samples. Studies carried out in the Philippines on grilled pork (Manguiat and Fang, 2013), in South Korea on grilled meat (Cho et al., 2011) and in Nigeria on roasted chicken (Ologhobo et al., 2010) showed an average TAPC of ≤ 6.0 log cfug^{-1}. This count is similar to that observed in our study (6.37±0.19 log cfu g^{-1}, Table 2) and suggests non-compliance with good hygiene practices during the handling, cooking and storage of street food in Kisangani, DRC.

Food quality: *Salmonella* sp.

The prevalence of *Salmonella* sp. (57.1%, Table 1) in Kisangani was very high as compared to that of, for example, in Mexico (5%) (Estrada-Garcia et al., 2004). According to Rane (2011), the contamination of street food due to *Salmonella* sp. can be explained by the use of dirty dishwater (from dirty dishes) or lack of good hygiene practices of vendors when handling street food.

Food quality: *S. aureus*

In Mexico, Diaz-Lopez et al. (2011) detected *S. aureus* in 4 out of 43 (9.3%) street dishes. According to Guven et al. (2010) and Harakeh et al. (2005) meals prepared in the street provide a suitable culture medium for the emergence of *S. aureus* strains that are resistant to multiple antibiotics. These antibiotic-resistant strains are thus transmitted to humans through eating contaminated street food.

Conclusions

This study on the hygienic quality of cooked meals sold in the main public places in Kisangani (DRC) showed that these foods are mostly unfit for consumption and present a significant risk of food poisoning to consumers.

With the proliferation of street restaurants in Kisangani, this situation puts the health of the population at risk. This should be a wake-up call for those responsible for health services to carry out regular quality checks and ensure compliance with best hygiene practices for the safety of cooked food sold in the streets of Kisangani.

The application of these measures could reduce morbidity from diarrheal diseases related to contaminated cooked meals sold by street vendors. A training programme in good hygiene practices (food handling and sale) is highly recommended for these vendors.

Conflict of interests

The authors did not declare any conflict of interest.

ACKNOWLEDGEMENTS

This paper was translated (French into English) by Miss Elaine Seery (see http :// www. elaineseery.wordpress.com/).

REFERENCES

AFNOR (*Association Française de Normalisation*) (1996). Analyse Microbiologique : Contrôle de la qualité des produits alimentaires (Microbiological analysis : Food Quality Control). In: *Méthodes Horizontales*, Volume 1. AFNOR, Paris

AFNOR (*Association Française de Normalisation*) (2004). Method for the enumeration of Coagulase-positive *Staphylococci* (*Staphylococcus aureus* and other species) NF V08 - 057-2, Part 2: Technique using Rabit Plasma Fibirnogen Agar Medium. Paris.

Akbar A, Anal AK (2013). Prevalence and antibiogram study of *Salmonella* and *Staphylococcus aureus* in poultry meat. Asian Pac. J. Trop. Biomed. *3*(2):163-168.

Barro N, Bello AR, Savadogo A, Ouattara CAT, Iiboudob AJ, Traoré AS (2006). Hygienic status assessment of dish washing waters, utensils, hands and pieces of money from street food processing sites in Ouagadougou (Burkina Faso). Am. J. Biotechnol. 5 (11):1107-1112.

Cho JI, Cheung C-Y, Lee S-M, Ko S-I, Kim K-H, Hwang, I-S, Kim S-H, Cho S.-Y, Lim C-J, Lee K-H, Kim K-S, Ha S-D (2011). Assessment of microbial levels of street-vended foods in Korea. J. Food Saf. (31): 41-47.

Cohen M (1986). The influence of the street food trade on women and child health. In D.B. Jelliffe& E.F.P, Jelliffe (eds). Advances in international maternal and child health. Vol. 6, Oxford, Clarendon Press.

Diaz-Lopez A, Cantu-Ramirez RC, Garza-Gonzales E, Ruitz-Tolentino L, Tellez-Luis S.J, Riviera G, Bocanegra – Garcia V (2011). Prevalence of foodborne pathogens in grilled chicken from street vendors and retail outlets in Reynosa, Tamaulipas, Mexico. J. Food Prot. 74: 1320–1323.

Diouf F (1992). Contribution à l'étude de la qualité hygiénique des aliments vendus sur la voie publique (AVP) dans la région de Dakar (Sénégal) (Study on the hygienic quality of foodsold on the main road in Dakar, Senegal). Doctoral thesis,Université Cheik Anta Diop, E.I.S.M.V., Dakar, Senegal.

Estrada-Garcia T, Lopez-Sancedo C, Zamarripa-Ayala B, Thompson MR, Gutierrez L (2004). Prevalence of *Escherichia coli* and *Salmonella spp*. in street vended food of open markets (tianguis) and general hygienic and trading practices in Mexico City. Epidemiol Infect.132:1181-1184.

FAO (1996).Streetfoods. Availablefrom ftp :// ftp.fao.org/docrep/fao/W3699T/W3699t01.pdf. Accessed on december 5, 2014

FAO (2003). The State of Food Insecurity in the World. FAO. Available from ftp://ftp.fao.org/docrep/fao/006/j0083e/j0083e00.pdf Accessed on November 18, 2014.

FAO (2010). The State of Food Insecurity in the World. FAO. Available from http://www.fao.org/docrep/013/i1683e/i1683e.pdf (Accessed on November 26, 2014.)

Ghosh M, Wahi S, Kumar M, Ganguli A (2007). Prevalence of enterotoxigenic *Staphylococcus aureus* and *Shigella* spp. in some raw street vended Indian foods. Int. J. Environ. Health Res.17(2):151-157.

Guven K, Mutlu MB, Gulbandilar A, Cakir P (2010). Occurrence and characterization of *Staphylococcus aureus* isolated from meat and dairy product consumed in Turkey. J. Food Saf. 30:196–212.

Harakeh S, Yassine H, Gharios M, Barbour E, Hajjar S, El-Fadel M, Toufeili I, Tannous R (2005). Isolation, molecular characterization and antimicrobial resistance of *Salmonella* and *Escherichia coli* isolates from meat-based fast food in Lebanon. Sci. Total Environ. 341:33-44.

Holy AV, Makhoane FM (2006). Improving street food vending in South Africa: Achievements and lessons learned. Int. J. Food Microbiol.

111(2):89-92
Manguiat LS, Fang TJ (2013). Microbiological quality of chicken- and pork-based street-vended foods from Taichung, Taiwan and Laguna, Philippines. Food Microbiol. 36(1):57-62.
Moustafa El-Shenawy, Mohamed El-Shenawy, Jordi Mañes, Jose M Soriano (2011). *Listeria* spp. in street-vended ready-to-eat foods. Interdiscip. Perspect. Infect. Dis. 2011: 968031
Muinde OK, Kuria E (2005). Hygienic and sanitary practices of vendors of street foods in Nairobi, Kenya. Afr. J. Food Agric. Nutr. Dev. 5(1):1-14.
Muyanja C, Nayiga L, Brenda N, Nasinyama G (2011). Practices, knowledge and risk factors of street food vendors in Uganda. Food Control 22:1551-1558.
Ologhobo AD , Omojola AB, Ofongo ST, Moifroi S, Jibir M (2010). Safety of Street vended meat products – chicken and beef suya. Afr. J. Biotechnol. 9 (26):4091-4095.
Omemu AM, Aderoju ST (2008). Food safety knowledge and practices of street food vendors in the city of Abeokuta, Nigeria. Food Control 19:396-402
Rane S (2011). Street Vended Food in Developing World: Hazard Analyses. Indian J. Microbiol. 51(1):100-106.
Walpole RE, Myers RH, Myers SL, Keying Ye (2002). Probability & statistics for Engineer & Scientists, 7thedition, Prentice Hall, New Jersey, 730 p
WHO (2003). Résolutions : Salubrité des aliments et santé : Analyse de la situation et perspectives. http://whqlibdoc.who.int/publications/2003/9241562374.pdf . Accessed on November 26, 2014.

Nutritional potential of yam chips (*Dioscorea cayenensis* and *Dioscorea rotundata* Poir) obtained using two methods of production in Togo

B. Djeri[1,2], Tchobo P. F.[2], Y. Adjrah, D. S. Karou[1], Y. Ameyapoh[1], Soumanou M. M.[2] and C. Souza[1]

[1]Laboratoire de Microbiologie et de Contrôle de qualité des Denrées Alimentaires, Ecole Supérieure des Techniques Biologiques et Alimentaires (ESTBA), Université de Lomé BP 1515, Lomé Togo.
[2]Unité de Recherche en Génie enzymatique et Alimentaire de l'Ecole Polytechnique d'Abomey Calavi URGEA/ EPAC Université d'Abomey Calavi, Benin.

Yam is one of the most staple foods in West African countries and provides an important part of the energetic for people. This study examines chemical composition of *Dioscorea cayenensis* and *Dioscorea rotundata* Poir species dried yam chips obtained using two methods of production in Togo. Three local varieties of yams (*Koukou, Kéki* and *Laboco)* were processed in chips and were sun dried (at 28-30°C) or oven-dried (at 50°C). Nutritive components (carbohydrates, protein fats, mineral salt, vitamin C and anti-nutritional factors) of yam chips were assessed and compared with those of fresh yam. Sugar is a major component of yam chips followed by proteins, vitamin, fats, mineral salt and anti-nutritionals factors. For the same variety of yam, the nutritional quality depends on the method of production followed and the drying methods.

Key words: Yam chips, *Dioscorea cayenensis, Dioscorea rotundata*, nutritional potential.

INTRODUCTION

Yam serves as an important source of carbohydrate and a major source of income in countries where they are cultivated. In 2007, 96% of the worldwide production of yam (52 million tons) was from Africa while 94% of the yam was from West Africa with Nigeria alone producing 71% (http://www.iita.org/yam). Yams are usually processed into dry-yam tubers/slices and flour in West African countries such as Ghana, Benin and Nigeria (Bricas et al., 1997).

Yam tubers are consumed in forms of chunks, flour, chips, fufu and slices, which are obtained from any of the processes of boiling, frying, drying, fermentation, milling, pounding, roasting and steaming (Iwuoha, 2004). The genus *Dioscorea* contains a wide range of yam species used as food. There are many varieties of yam species widespread throughout the humid tropics, the most economically important species which are grown are white yam (*Dioscorea rotundata)*, yellow yam (*Dioscorea cayensis)*, water yam (*Dioscorea alata),* Chinese yam (*Dioscorea esculenta)* aerial *yam (Dioscorea bulbifera)* and trifolate yam (*Dioscorea dumentorum*) (Ike and Inoni, 2006). White yam (*D. rotundata)* originated from Africa and is the most widely grown and preferred yam species. The tuber is roughly cylindrical in shape; the skin is smooth and brown and has a white firm flesh. A large number of white yams exist with difference in the production and post-harvest characteristics. Yellow yam (*D. cayensis*) derives its common name from its yellow flesh, which is caused by the presence of carotenoids. It is also native to West Africa, the yellow yam has a longer period of vegetation and a shorter dormancy than white yam. In the past, yellow yam and white yam were considered as two separate species but most taxonomists now regard them as the same species, there are over 200 cultivated varieties between them. The *"Kokoro"* variety is important in making dried yam chips. *D. rotundata* which is referred to as dried yam contains a large amount of energy, starch, iron, the smallest amount of dietary fiber, Sn, Ca, P, Mg, Ca and Mn. The protein content is low.

Yam is a good root crop with high energy, acceptable protein and iron content but lowest in Ca and Zn. There is a high concentration of protein and minerals in the peels, which forms about 19% of the tubers. Dried yam is lower than water yam in ash content but high in soluble carbohydrate. Yam also provides protein three times more superior than the one of cassava and sweet potato. Apart from food, yams are also sources of pharmaceutical compounds like saponins and sapogenins, which are precursors of cortisone used medically in the treatment of arthritis and some allergies (Ezeocha and Ojimelukwe, 2012).

The perishability nature of yam due to its high moisture content suggests the need to process it into less perishable products such as yam chip through drying process (Karim et al., 2013). Up till today, this age-old traditional method is still being used for the processing of yam, to dry yam (yam chips) and yam flour. The quality of the chips and flour varies from processor to processor and from location to location (Ojokoh and Gabriel, 2010; Adewale et al., 2014). In Togo, this practice follows two methods: one method is according to Bassar region, the northern of part Togo and Ghana. The second method is according to Est-Mono region, Benin and Nigeria. The objective of the present study was to explore the effects of the two processing methods used in Togo on yam chips qualities, including proximate composition and anti-nutritionals factors content.

MATERIALS AND METHODS

Sampling

Yam tubers of varieties *"Koukou, Laboko, kéki"* of *Dioscorea cayenensis and Dioscorea rotundata* where collected in April 2008 at Est–Mono and Bassar and dried to yam chips.

Processing of yam to dry-yam

Yam tubers were processed in to chips since May 2008 in the Microbiology Laboratory of Food stuff and Quality Control in the University of Lomé. Two methods described during the survey with farmers were adopted. Three varieties of yam were used (*Laboco, Kéki* and *Koukou*).

Method A

This process was used in Bassar region and in northern Togo. Yam tubers were peeled and cut into slices (0.5 to 1 cm of diameter and 5 to 10 cm of length). After washing, slices were divided into two lots. The first one was sun-dried at 28-32°C and the second was oven-dried at 50°C.
Samples process in method A (method of Bassar region):

KoBSRS: Chips of *"koukou"* variety processed following method of Bassar with sun drying
KoBSRE: Chips of *"koukou"* variety processed following method of Bassar with oven drying
KeBSRS: Chips of *"kéki"* variety processed following method of Bassar with sun drying
KeBSRE: Chips of *"kéki"* variety processed following method of Bassar with oven drying
LaBSRS : Chips of *"Laboco"* variety processed following method of Bassar with sun drying
LaBSRE: Chips of *"Laboco"* variety processed following method of Bassar with oven drying.

Method B

This process was generally used in the Est-Mono region. Yam tubers were first peeled and cut into slices (0.5 to 1 cm of diameter and 5 to 10 cm of length) and cooked at 60-70°C. Cooked slices were divided into two lots. The first was oven dried at 50°C and the second were sun-dried at the ambient air (28-32°C).
Samples process in method B (method of Est-Mono region):

KoEMNS: Chips of *"koukou"* variety processed following method of Est-Mono with sun drying
KoEMNE: Chips of *"koukou"* variety processed following method of Est-Mono with oven drying
KeEMNS: Chips of *"kéki"* variety processed following method of Est-Mono with sun drying
KeEMNE: Chips of *"kéki"* variety processed following method of Est-Mono with oven drying
LaEMNS: Chips of *"Laboco"* variety processed following method of Est-Mono with sun drying
LaEMNE: Chips of *"Laboco"* variety processed following method of Est-Mono with oven drying.

Carbohydrate content determination

Total sugar

One gram of yam flour was dissolved in 10 mL of dimethylsulfoxyde (DMSO: (C_2H_6OS) 25% v/v). After 15 min of incubation in bain-marie (90-100°C), 0.1 mL of the mixture was diluted into 9.9 mL of distilled water, 0.5 mL of the last mixture were added to 0.5 mL of phenol (5%). And then, 2 mL of H_2SO_4 (75%) were added. The absorbance was read at 492 nm (Fox and Robyt, 1991).

Starch

0.1 g of yam chip flour was dissolved in 5 ml of KOH 1 N. After homogenization, the unit was neutralized by 5 ml of HCl 1 N. The mixture thus obtained was allowed to boil for 15 min in a water-bath. After filtration, volume was adjusted to 10 mL. 0.05 mL were diluted to 5 mL with 4.85 mL of distilled water and 0.1 mL of reagent (I_2/KI) and incubated for 10 min. The absorbance of samples was carried out at 580 and 720 nm (Jarvis and Walker, 1993).

Fats content determination

Lipid content was determined using a Soxhlet apparatus and hexane reagent. 10 g of yam chip flour were weighed and were putted in a balloon. Vacuum was given before introducing 175 mL of hexane. The weight of lipids extracted was obtained by difference between the final weight P1 after evaporation of solvent and drying of the balloon and the initial weight of the empty balloon (PO).

Proteins content determination

The protein content was determined by the Kjeldahl method (AOAC, 2000) according to the following protocol:

A mixture of two catalyst tablets, 12 mL of concentrated sulfuric acid, pumice stone and 1 g of yam chips was placed in a fume hood digestion unit at 420°C for 3 h and a clear liquid was obtained. This product was diluted with 50 mL of distilled water and 75 ml of sodium hydroxyde solution 38% and then distilled in 25 mL of another mixture consisting of 4% boric acid, 25 mL (w/v) of 1-methyl red (w/v) of 1% bromocresol green (w/v) and sodium hydroxide 4% (W/V). The product obtained was titrated with 0.1 N hydrochloric acid solution until the color changes from blue to pale pink. The protein content wass expressed as a percentage:

$$\text{Proteins (\%)} = \frac{1.401 \times 6.25 \times (Ve-Vb) \times T}{\text{Weight of sample in grams}}$$

T = Concentration of HCl solution; Ve = volume of hydrochloric acid used to titrate the sample (containing chips); Vb = volume of hydrochloric acid used to titrate the blank containing no chips to titrate the blank; 6.25 = the nitrogen conversion factor protein; 1.401 = constant.

Mineral content determination

Ten grams of yam chips were incinerated in a furnace (Prolabo Volca V50) at 550°C. After incineration, ash was dissolved into 100 mL of distilled water and shaked for 30 min. The solution obtained was filtered on Whatman paper No. 1. Mineral content was obtained using ionic chromatography and molecular spectrophotometry of absorption.

Ascorbic acid content determination

The ascorbic acid content of samples was determined by the volumetric method using Tillmans reagent (Sawadogo, 1993). 2.5 g of yam chip flour were dissolved in 50 mL mixture of acid solution (metaphosphoric 5%/acid acetic 10%). After 1 h of agitation, the mixture obtained was filtered. The proportioning of the vitamin in the material was carried out by using 2,6-Dichloro indol phenol (2,6-DIP) solution (0.5 mg/mL). Then, 25 mL of the filtered solution were titrated by the 2,6-DIP until the red color at the point of equivalence persists for 5 s (dryness). The quantity of vitamin was given using the reference established by the proportioning of 10 mL of pure ascorbic acid.

Determination of anti-nutritional factors (ANF)

Total oxalate

The total oxalate was determined by the method of Day and Underwood (1986). 1 g of yam chip flour was dissolved in 75 mL of sulfuric acid 15 N. The mixture was homogenized for one hour and filtered on Whatman No.1 paper. 25 mL of the filtrated solution were titrated with 0.1 N of permanganate of potassium.

Phytic acid

Phytic acid was determined according to the method of Reddy and Love (1999). This method consists of adding 4 g of flour of yam ships to 100 mL of hydrochloric acid 2% under magnetic agitation for 5 h. After filtration, 25 mL were added to 5 mL of ammonium thiocyanate 0.3%. The mixture was titrated with ferric chloride until the yellow color brown was obtained, persisting for 5 min.

Tannins

Tannins are proportioned by the method of Trease and Evans (1978). 1 mL of methanolic extract was treated with 5 mL reagent of Folin-Dennis in basic medium. The absorbance of the mixture was read at 760 nm. The content of tannins was given using a curve standard built starting from a range of concentration of gallic acid.

Saponins

Saponins was carried out according to the method of Birk et al. (1968) modified by Hudson and El Difrawi (1979). According to this method, 1 g of flour of sample was dissolved in 20 ml of the ethanol 20% (ethanol-water) under the magnetic agitation for 12 h at 55°C. The solution was filtered on Whatman No.1 and adjusted with 40 ml then 20 mL of the filtered solution was added to diethyl ether followed by a vigorous manual agitation. The aqueous phase was adjusted to pH 4.5 with a hydrochloric acid solution 0.2 N. 60 mL of the solution of N butanol were added to this phase and the unit was washed with 10 mL of NaCl 5%. The organic phase was then evaporated to have the content of saponins. The saponin was determined by the difference between the weight of the container after evaporation and the tare weight of the same container.

Statistical analysis

All analyses were carried out in triplicates and the data were subjected to analysis of Epi Info 3.5.1 for windows which was the statistical software used.

RESULTS AND DISCUSSION

Proximate composition of yam chip

The proximate compositions of yam chips manufactured by Bassar and Est-Mono processing, sun and oven dried are presented in Table 1 for carbohydrate, fat, proteins, vitamin C content, in Table 2 for anti-nutritional factors content and Table 3 for mineral content

Carbohydrate content of the chips of *"Koukou"* variety resulting from method of Bassar was higher to a significant degree ($P<0.05$) than those produced with

Table 1. Proximate composition of yam chips product from *"koukou"," kéki"* and *"Laboco"* (g/100 g).

Yam varieties	Samples processed	Protein (g/100)	Carbohydrate (g/100)	Fat	Vitamin C (mg/100)
"Koukou"	KoBSRS	3.47±0.2	62.61±0.05	0.11±0.02	1.19±0.39
	KoBSRE	5.71±1.23	86.57±2.72	0.14±0.02	0.71±0.07
	KoEMNS	4.05±0.07	66.40±0.07	0.19±0.01	0.84±0.21
	KoEMNE	4.34±0.45	67.79±0.04	0.1±0.09	0.73±0.07
"Kéki"	KeBSRS	4.68±0.06	44.99±0.08	0.11±0.01	0.94±0.01
	KeBSRE	4.89±0.55	55.05±0.04	0.1±0.02	0.83±0.07
	KeEMNS	3.74±0.13	37.75±0.11	0.1±0.01	0.65±0.05
	KeEMNE	4.68±0.07	43.45±0.11	0.11±0.02	0.61±0.10
"Laboco"	LaBSRS	5.58±0.16	44.21±0.24	0.09±0.05	0.05±0.00
	LaBSRE	5.43±0.23	42.40±0.06	0.13±0.02	0.2±0.02
	LaEMNS	03.6±0.28	48.85±0.11	0.13±0.01	0,3±0.01
	LaEMNE	3.92±0.52	46.62±0.09	0.13±0.06	0,1±0.03

Table 2. Anti-nutritionals factors content of yam chips (mg/100).

Yam varieties	Treatment	Oxalate	Phytic	Saponin	Tannins
"Koukou"	KoBSRS	3,3±0.01	0.15±0.01	0.7±0.02	0.38±0.01
	KoBSRE	1.4±0.02	0.16±0.03	0.2±0.01	0.32±0.03
	KoEMNS	1.65±0.05	0.26±0.01	0.57±0.01	0.31±0.01
	KoEMNE	0.82±0.01	0.19±0.05	0.77±0.1	0.3±0.06
"Kéki"	KeBSRS	0.93±0.05	0.22±0.08	0.52±0.02	0.21±0.01
	KeBSRE	0.88±0.03	0.24±0.01	0.44±0.02	0.08±0.03
	KeEMNS	0.88±0.01	0.13±0.05	0.31±0.01	0.02±0.01
	KeEMNE	0.77±0.03	0.13±0.01	1.03±0.06	0.87±0.02
"Laboco"	LaBSRS	0.79±0.05	0.53±0.01	0.8±0.02	0.13±0.03
	LaBSRE	0.56±0.02	0.22±0.03	0.42±0.01	0.12±0.03
	LaEMNS	0.66±0.07	0.2±0.05	0.46±0.01	0.13±0.03
	LaEME	0.65±0.05	0.41±0.01	0.72±0.02	0.13±0.01

the method of Est-Mono, Moreover, oven drying results to less loss in carbohydrate than sun drying (Table 1).

With regards to *"keki"* variety, the chips resulting from method of Bassar associated with oven drying causes significant degree less loss in carbohydrate than those obtained from method of Est-Mono (Table 1)

For the *"Laboco"* variety, the chips resulting from method of Est- Mono associated with oven drying had less loss in carbohydrate content than the other (Table 1).

It emerges from these results that the carbohydrate content of the chips were influenced by varietal factor and transformation. In view of these results, to avoid lost of carbohydrate during the transformation of yam into chips, it is proper to choose Bassar method for *koukou* and *"Kéki"* yam varieties and Est-Mono method for *"Laboco"* variety.

In general, it is desirable to use the oven drying because it is done quickly without allowing microorganisms to grow and alter nutrients. These results are in agreement with earlier reports of Okigbo and Nwakammah (2005) and Adewale et al. (2014). However, carbohydrate content were contrary to those obtained by Ojokoh and Gabriel (2010) and Adewale et al. (2014) with regards to *"Laboco"* and *"kéki"* varieties.

Proteins content

According to proteins content, the results indicate that:

Table 3. Mineral content of yam chips product from *''koukou''*, *'' keki''* and *''Laboco''* (mg/100 g dry basis).

Yam variety	Treatment	Fe	PO_4	Na	K	Mg	Ca	Cu	Zn
''Koukou''	KoBSRS	0.37±0.01	1859.53±4.62	105.7± 2.41	7845.2±51.03	145.82±0.97	16.69±0.11	13.53±0.02	15.48±00
	KoBSRE	0.34±0.1	2695.4±59.22	38.85±1.56	8840.8±41.238	142.1±2.49	13.6±0.29	4.2±1.69	13.19±2.19
	KoEMNS	0.5±0.01	2661.16±75.5	180.32±9.63	7314.37±10.54	15.88±1.42	20.09±0.18	12.21± 2.46	90.14±4.93
	KoEMNE	0.3±0.10	2024±1.16	440.29±8.78	7763.85±29.04	139.98±0.53	17.71±0.29	3.76±0.01	5.64±0.25
''Kéki''	KeBSRS	0.37±0.01	65.11±2.40	61±0.91	7679.46±27,48	315.66±0,36	23.57±0.43	6.82±3.21	28.02±7.45
	KeBSRE	0.23±0.01	36.63±0.25	26.83±3.46	6050.42±13.13	857.15±1.02	11.70±0.08	2.53±0.01	17.71±3.57
	KeEMNS	0.43±0.01	306.41±2.48	232.29±4.01	7222.56±72.96	131.09±1.45	17.34±0.07	4.81±2.26	4.81±2.26
	KeEMNE	0.32±0.01	32.98±0.33	560.05±4.01	6519.37±47.63	123.01±1.68	17,47±0.01	7.44±00	16.74±2.74
''Laboco''	LaBSRS	0.56±01	1678±0.81	56.28±1.24	6008.14±31.04	272.31±1.61	17.97±0.18	2.53±00	57.51±9.03
	LaBSRE	0.36±01	1834.71±1.22	136.71±2.11	7266.33±15.83	246.97±5.30	17.34±0.07	4.26±00	30.9±7.28
	LaEMNS	0.58±0.,03	1494.67±3.19	67.67±0.06	5488.95±35.21	260.75±13.68	25.1±1.13	6.01±2.84	24.06±2.25
	LaEMNE	0.47±0.01	1415.33±2.56	107.06±0.90	6683.2±39.22	427.16±2.67	30.55±0.58	9.3±2.25	22.9±3.23

1. Regarding "*koukou*" variety, the method of Bassar combined with oven drying would preserve the content of proteins better with significant degree (P<0.05) than the method of Est-Mono.
2. For *''kéki''* variety, method of Bassar combined with oven drying caused no significant degree less loss of proteins. With *''kéki''* variety, no influence of method production and drying method on the protein content was seen.
3. *''Laboco''* variety gave the same observations like *''koukou''* variety.

The protein contents obtained were contrary to those obtained by Oyeyiola et al. (2014), Adewale et al. (2014) and Ojokoh and Gabriel (2010) who obtained lower protein content of yam chips.

Fats content

The results reveal that the content of fat is very low and which is about 0.1, varietal influence and transformation process are not significant (Table 1). This content were in agreement with earlier reports (Oyeyiola et al., 2014) but were contrary to those reported by Adewale et al. (2014) and Ojokoh and Gabriel (2010) who obtained higher content of fat.

Vitamin C content

The results indicated that vitamin C content varies from 0.71 to 1.19 mg for *''koukou''* chips, 0.6 to 0.94 mg for chips of *''kéki''* and 0.05 to 0.3 mg for *''Laboco''* chips. Process and variety did not influence significantly vitamin C content (Table 1). Vitamin C content was contrary to those obtained by Gbolagade et al. (2011) who reported higher content of ascorbic acid (4.44 -6.46 mg/100 g).

Anti-nutritionals factors content of yam chips

The edible, matured yam does not contain any anti-nutritional factors however, bitter components tends to accumulate in immature tuber tissues of *D. rotundata* and *D. cayenesis*. They may be attributed to polyphenols or tannin-like compounds. Some workers (Asuzu and Undie, 1986; Okeola and Machuka, 2001; Ajibade et al., 2005) have identified the presence of some anti-nutritional factors (ANF) such as alkaloids, flavonoids, saponins, lectin, trypsin inhibitors, phytate and oxalate in the African yam. List of the anti-nutritional factors in the African yam bean includes trypsin inhibitor, haemagglutinating, tannic acid (tannins), phytic acid, oxalate (Apata and Ologhobo, 1997). In addition to the above list of ANF are α-galactosides (stachyose) and lectin (Oboh et al., 1998).

For *''Koukou''* variety, the oxalates deteriorate more with method A with sun drying, whereas the content of phytate does not vary to a significant degree according to the yam chips methods of production used (P > 0.05). The method of Est-Mono (MN) associated with oven drying and those of Bassar (BSR) with sun drying caused more loss of total phenols. The content of tannins and saponins varies less with the method of Est-Mono associated with sun drying.

As compared to *"keki"* variety, the oxalates content decreases with method of Est-Mono associated with sun and oven drying (50°C) and the drying oven with 50°C. The phytate content remains constant. Total phenols and saponins content deteriorated by the bleaching with oven drying. The content of tannins varies less with the method of Bassar (BSR) associated with sun and oven drying, whereas the method of Est-Mono (EMN) with sun and oven drying caused more deterioration in yam ships

For *"Laboco"* variety, the phytate and oxalate content of yam ships decreases less with the method of Bassar (BSR) associated with sun drying, whereas it decreases much more with method of Est-Mono (EMN) with sun and oven drying (50°C). The tannins and saponins contents remain constant with the other experimental conditions. However, these variations observed between the various treatments are not significant ($P > 0.05$).

The anti-nutritionals factors are substances of reserves of plants which complex some nutrients like rock salt, proteins and reduce their biodisponibility during digestion. It is the case of oxalates, the phytate, saponins, tannins and total phenols, etc. These results of content of anti-nutritional factors were lower than those reported by other authors on the yam and these studies relate to only the fresh tubers. Thus, Jau-Tien et al. (2009) found saponin values in the yam from 247 to 619 µg/g. FAO (1991) found 637 mg/100 g for the phytate.

Rock salt content of yam ships

The samples of yam ships contain a broad rock salt range the cations (Mg, Na, K and Ca), anions (Cl, SO_4, PO_4, Br, NO_2 and NO_3) and oligo elements (Zn, Cu and Fe). Only the principal ones are illustrated here.

For *"Koukou"* variety, the most representative oligo elements were zinc with contents between 5.64 and more than 90 mg per 100 g of dry matter, followed by copper with 3.76 to 13.53 mg. The content of iron is very weak and varies from 0.3 to 0.5 mg per 100 g.

Among the anions, phosphorus is most significant with contents from 1859 to 2695 mg per 100 g whereas among the cations, potassium is most significant with values ranging from 7314 to 8840 mg per 100g. According to *"Kéki"* variety, potassium and phosphorus are the principal biogenic salts with contents respectively from 6050 to 7679 mg and 1625 to 2634 mg/100 g. Oligo significant elements were Zn with content varying from 8 to 28 mg/100 g followed by copper whose values were from 2.53 to 7.44 mg/100 g. Concerning *"Laboco"* variety, potassium and phosphorus are the principal biogenic salts with contents respectively from 5488 to 7266 mg and 1415 to 1834 mg/100 g. The Mg, Na and Ca were slightly represented in all the samples of the variety. Oligo significant elements were Zn whose content varies from 22.9 to 5.51 mg /100 g followed by the copper whose values went from 2.53 to 9.3 mg/100g. Iron is the oligo element most slightly represented in the studied yam chips.

This study revealed that yam chips contain majorly potassium and phosphorus. FAO (1970) showed mineral salt dominating in yam is potassium thus, yam could cover a substantial part of the requirements of manganese and phosphorus for the adults. Our results were contrary to those obtained by Agbor-Egbé and Streche (1995) which found 65-125 mg/100 g of phosphorus; 8-36 mg for calcium; 590-1740 mg for potassium; 19-72 mg for magnesium.

Conclusion

The proximate compositions of yam chips for each yam variety, for drying and processing methods showed that these products have an important nutritional potential. Outside carbohydrates, protein, fat, vitamin C, minerals, the chips also contains anti-nutritional factors. There are no significant effects of processing ($P>0.05$) on lipid, vitamin C, mineral and anti-nutritional factors. In contrast, yam variety, drying and processing methods have significant effects on carbohydrates and proteins content. This can be exploited to provide food for diabetics, as an ideal food source of carbohydrates without affecting their disease.

Conflict of interests

The authors did not declare any conflict of interest.

ACKNOWLEDGEMENTS

The financial support from the Service de Coopération et d'Action Culturelle (SCAC) of France Embassy and the collaboration of URGEA Laboratory (Unité de Recherche en Génie enzymatique et Alimentaire de l'Ecole Polytechnique d'Abomey Calavi URGEA/EPAC Université d'Abomey Calavi) of the Republic of Benin are gratefully acknowledged.

REFERENCES

Adewale OO, Bukunola OB and Ifeoluwa O (2014). Changes in nutritional composition, functional, and sensory properties of yam flour as a result of pre-soaking. Food Sci. Nutr. 2(6):676-68.

Agbor-Egbé T, Streche E (1995). Evaluation of the Chemical Composition of Cameroonian Yam Germplasm. J. Food Comp. Anal. 8:274-283.

Ajibade SR, Balogun MO, Afolabi OO, Ajomale KO, Fasoyiro SB (2005). Genetic variation in nutritive and anti nutritive content of African yam bean. Trop. Sci. 5:144-148.

AOAC (2000). Official methods of analysis Association of Official Chemists. EUA Analysis, 15th ed., Virginia Assoc. Official Analytical Chemists, USA.

Apata DF, Ologhobo, AD (1997). Trypsin inhibitor and other Anti nutritional factors in tropical legume seeds. Trop. Sci. 37:52-59.

Asuzu IU, Undie A (1986). Some observations on the toxic effects of the seed extract of *Sphenostylis stenocarpa* (Hochst ex A. Rich.) Harms. on intestinal muscle. Qual. Plant Foods Hum. Nutr. 36:3-9.

Birk Y, Bondi A, Gestetner B, Ishaya IA (1968). Thermostable hemolytic factor, in soybeans. Natural 197:1089-1090.

Bricas N, Vernier P, Ategbo E, Hounhouigan J, Mitchikpe E, N'kpenu KE et al. (1997). Le développement de la filière cossettes d'igname en Afrique de l'Ouest. Les Cahiers de la Recherche-Développement 44:100-114.

Day RA, Underwood LA (1986). Quantitative analysis. 5th ed. New Delhi, India: Prentice-hall public. 701 p.

Ezeocha VC, Ojimelukwe P (2012). The impact of cooking on the proximate composition and anti-nutritional factors of water yam (*Dioscorea alata*). J. Stored Prod. Postharvest Res. 3 (13):172-176.

FAO (1970). Amino acid content of food and biological data on proteins nutritional studies, 24, 285 p.

FAO (1991). Composition en éléments nutritifs des racines et tubercules. In Racines, tubercules, plantains, bananes dans la nutrition humaine. Catalogage avant publication de la bibiothèque David Lubin FAO, Rome, Italie. ISBN 92-5-202862-5 200.

Fox JD, Robyt JF (1991). Miniaturization of three carbohydrate analyses using a microsample plate reader. Anal. Biochem. 195: 93-96.

Gbolagade J, Ibironke A, Yetunde O (2011). Nutritional compositions, fungi and aflatoxins detection in stored 'gbodo' (fermented Dioscorea rotundata) and 'elubo ogede' (fermented Musa parasidiaca) from South western Nigeria. Afr. J. Food Sci. 5(2):105-110.

Hudson BJF, El-Difrawi EA (1979). The sapogenins of the Seeds of Four Lupin Species. J. Plant Foods 3:181-186.

Ike PC, Inoni OE (2006). Determinants of yam production and economic efficiency among small-holder farmers in south-eastern Nigeria. J. Cent. Eur. Agric. 7(2):337-342.

Iwuoha CI (2004). Comparative evaluation of physicochemical qualities of flours from steam-processed yam tubers. J. Food Chem. 85:541-551.

Jarvis CE, Walker JRL (1993). Simultaneous, rapid, spectrophotometric determination of total starch, amylose and amylopectin. J. Sci. Food Agric. 63:53-57.

Jau-Tien L, Su-Lin C, Shih-Chuan L, Deng-Jye Y (2009). Effect of Harvest Time on Saponins in Yam (*Dioscorea pseudojaponica* Yamamoto). J. Food Drug Anal. 17(2): 116-122.

Karim OR, Kayode RMO, Oyeyinka SA, Oyeyinka AT (2013). Proximate, Mineral and Sensory Qualities of 'Amala' Prepared from Yam flour fortified with Moringa leaf powder. Food Sci. Qual. Manag. 12:10-22.

Oboh HA, Muzquiz M, Burbano C, Cuadrado C, Pedrosa MM, Ayet G, Osagie AU (1998). Antinutritional constituents of six underutilized legumes grown in Nigeria. J. Chromatogr. A. 823(1-2):307-12.

Ojokoh O, Gabriel RAO (2010). A comparative study on the storage of yam chips (gbodo) and yam flour (elubo). Afr. J. Biotechnol. 9(21):3175-3177.

Okeola OG, Machuka J (2001). Biological effects of African yam bean lectin on *Clavigralla tomentosicollis* (*Hemiptera: coreidae*). J. Econ. Entomol. 94:28-34.

Okigbo R, Nwakammah P (2005). Biodegradation of white yam (*Dioscorea rotundata* poir) and water yam (*Dioscorea alata* L.) slices dried under different conditions. Sci. Technol. J. (3): 577-586.

Oyeyiola GP, Oyeniyi OR, Arekemase MO, Ahmed RN (2014). A comparative study on the microbiological and nutritional properties of stored chips and flours. World J. Biol. Res. 6(2):1-6.

Reddy MB, Love M (1999). The impact of food processing on the nutritional quality of vitamins and minerals. Adv. Exp. Med. Biol. 459: 99-106.

Sawadogo PH (1993). Valorisation technologique de la variété Amelie de mangue au Burkina Faso. Thèse de doctorat de 3e cycle. Shalini, Ph.D,

Trease GE, Avens WC (1978). A text book of pharmacognosy. 11: Bailliere-Tindall, London. 622P.

Food expenditure and household welfare in Ghana

Samuel A. Donkoh[1], Hamdiyah Alhassan[1] and Paul K. Nkegbe[2]

[1]Department of Agricultural and Resource Economics, University for Development Studies, Tamale, Ghana.
[2]Department of Economics and Entrepreneurship Development, University for Development Studies, Ghana.

The main objective of this study was to investigate the determinants of household food expenditure and its effects on welfare. As a result of potential simultaneity between food expenditure and welfare, a simultaneous equations model was estimated using the two-stage least squares method. The findings confirm the theoretical and empirical evidences that households reduce the percentage share of their food expenditure as they become richer. Also, increases in the food budget share lead to a reduction in welfare. Different households which spent greater percentages of their incomes on food were as follows: female headed households; households headed by the aged; households whose heads had little or no formal education; households whose heads were married; smaller households; rural households; households in the forest and savannah belts; and households living farther from the nation's capital. Also, welfare was greater for the following households: female headed households; households headed by the aged, households whose heads had formal education, smaller households, households who owned assets; households living in the urban centres, as well as those living closer to the nation's capital. Households that must be targeted for support include male-headed households, households headed by the relatively young, larger households, rural households and households farther from the nation's capital, including those in the savannah belt.

Key words: Food expenditure, Ghana, households, two stage least squares, welfare.

INTRODUCTION

The Food and Agriculture Organization (FAO, 2011) noted that even though world food supply is enough to feed the population, there are profound disparities across countries, towns and households, in terms of access to food. For instance, in 2010, while the average caloric intake per person per day in least developed and developing countries were 2,120 and 2,640 kcal, respectively, that of developed countries was 3,430 kcal. The FAO (2012) stressed that despite the considerable efforts taken to curb global hunger, 925 million people were undernourished in 2010, while the number of obese people rose to 1.5 billion in 2008. The World Health Organization (WHO, 2011) noted that these discrepancies call for much concern.

The irony is that while the rich spend more on food than the poor in absolute terms, the latter allocate high proportions of their income to food consumption with smaller portions left for the consumption of high quality non-food items (Engel, 1857) necessary for a higher standard of living. The over consumption of the rich and the under nourishment of the poor have significant health, economic and environmental implications (UNEP, 2012).

The main objective of this study was to investigate the socioeconomic determinants of food consumption and the effects on the living standards of Ghanaian households. Specifically, we sought to investigate the extent to which the Engel's law applies to the consumption behaviours of Ghanaian households.

Economic performance of the Ghanaian economy

Over the past two decades, the Ghanaian economy is viewed as a model of development. The country has experienced strong and sustained economic growth in a relatively stable and democratic political environment. Ghana became the first African country to achieve the first Millennium Development Goal (MDG) by halving the poverty of the 1990s even before the targeted year of 2015 (UNDP, 2008). Between 2000 and 2010, Ghana's growth rate at 4.9% was higher than the sub-Saharan African (SSA) average, and more than twice the growth rate of a decade earlier. Similarly, Ghana was more effective in translating the economic growth into poverty reduction than her counterparts in the developing world. However, the Ghana Statistical Service (GSS, 2007) reported that even though poverty has reduced since the 1990s, inequality has not changed. Poverty is the main cause of food insecurity (WFP, 2012). In its comprehensive food security and vulnerability analysis, WFP (2012, p.17) found that "poorer households have lower levels of education, spend a larger share of their limited means on purchasing food, have smaller harvests, and are more often buying their staple foods when the market prices are highest when compared with wealthier households".

Household expenditure in Ghana in 2006/2007

In 2006/2007, the average annual household expenditure in Ghana was GH¢1,918.00 whilst the mean annual per capita consumption expenditure was GH¢644.00.

Food expenditure accounted for 40% of total household expenditure, while the imputed value of own-produced food consumed by households represents a further 10.5%. At the time of the survey, Ghanaian households were spending on an average, about GH¢2,680 million per annum on food (including non-alcoholic beverages). This represented about a third of the total expenditure while non-food expenditure represented about 70% of the total household expenditure.

Food accounted for about half of the total expenditure of households in the highest quintile and also formed about 60% of the expenditure of households in the lowest quintile. In the localities, households in urban centres spent about 44% on food (actual and imputed), while households in the rural areas spent more than 60% on food. In terms of regions, household expenditure represented more than 70% of the budget in Volta, followed by the Northern Region (65.2%). The figure for the Greater Accra was 40%.

MATERIALS AND METHODS

Theoretical framework: Consumption theory

The theory of consumption is central to the model of Keynes' General Theory, which is often considered to be the origin of macroeconomics. Keynes specified a simple linear consumption function in which consumption is a positive function of disposable income. Though consumption depends on disposable income, there is a part of consumption which does not depend on disposable income and this is called autonomous consumption. The portion of the consumption function which depends on disposable income is called induced consumption and the marginal propensity to consume shows how much consumption will change when income changes. Three points can be deduced from the consumption function stipulated by Keynes. First, the value of marginal propensity to consume is constant and less than one, thus consumer increases their consumption as their income increases, but not as much as the increase in their income. Secondly, the average propensity to consume which is the ratio of total consumption to total income falls as the level of income increases. For Keynes, individuals consider saving as a luxury and this explains why the rich save a higher proportion of their income than the poor. The third deduction from Keynes' consumption postulated that current consumption depends only on current income. Consumption models built on the initial work by Keynes are as follows: Irving Fisher's Inter-temporal choice model; The Life Cycle Hypothesis (LCH) developed by Modigliani and Brunberg (1954); and the Permanent Income Hypothesis (PIH) by Friedman (1957).

The Engel curve

The Engel curve describes the relationship between household expenditure and income. Engel (1857) initiated the studies on household food expenditure survey and found that food expenditure was an increasing function of income and family size, but the food budget shares declined with income. This finding led to the formulation of the Engel's law which states that "the poorer a family is, the larger the budget share it spends on nourishment" (Engel, 1857, pp. 28-29). Thus, the Engel's law can be used to evaluate the general welfare of households and in particular the conditions of poor households.

According to Engel (1857), food expenditure is an essential expenditure which dominates low income household expenditure patterns; a fall in households' income thus, tends to crowd out expenditure on other non-essential goods. However, when households' incomes increase, a smaller percentage of it is spent on food while a large portion goes into non-food items. Similarly, the finding by Engel (1857) shows that the proportion of income allocated to food is directly related to household size, where larger households spend a higher share of their income on food than smaller households.

Analytical approach

Some food expenditure studies (Yemer, 2011) estimate a Tobit model with the explanation that a significant number of households record zero values. Thus, in the case where there are no zero values the ordinary least squares method (OLS) is appropriate. Also, most of these studies (Ayo et al., 2012; Begum et al., 2010; Umeh and Asogwa, 2012) estimate a single equation with income as one of the main explanatory variables. The problem with the single equation estimation is that income is assumed to be exogenous, yet income is not truly exogenous; while income determines food expenditure, it is determined itself by other variables including food expenditure. This means that estimating a single equation and making income exogenous results in simultaneous bias (Koutsioyannis, 1977; Gujarati, 2004). The right model should be a simultaneous equation involving two equations, each for food expenditure and income. The most appropriate estimator for a simultaneous equation model involving two equations is that of the two stage least squares (2SLS). Lastly, even though in practice, income is equivalent to welfare as defined by the Ghana Statistical Service (2007), in principle, welfare is broader than income because the former includes other indicators of wellbeing in addition to income. Besides, people tend to underestimate their incomes and so the use of expenditure as a measure of welfare as used in the GLSS is preferred.

Simultaneous equations: Two stage least square (2SLS)

Gujarati (2004) shows that when we estimate a single equation by OLS, but the equation has one or more explanatory variable(s) that is/are endogenous, it results in simultaneous equation bias. The right approach is to develop a simultaneous equation system and estimate it by a two stage or three stage least squares depending on the number of endogenous variables. Given the following structural equations:

$$y_1 = b_{12}y_2 + \gamma_{11}x_1 + \gamma_{12}x_2 + \cdots + \gamma_{1i}x_i + u_1 \quad (1)$$

$$y_2 = b_{22}y_1 + \gamma_{21}x_1 + \gamma_{22}x_2 + \cdots + \gamma_{2i}x_i + u_2 \quad (2)$$

where y_1 and y_2 are endogenous variables; x_i are predetermined variables; b_{12} and b_{22} are coefficients of the endogenous variables and γ_{ji}'s are coefficients of predetermined variables; u_1 and u_2 are the random terms with zero mean, constant variance and zero covariance, but non-zero covariance between the y_i and the u_i.

The reduced form of the structural model is obtained by solving the structural equations simultaneously as follows:

Substituting Equation 2 into 1 we obtain:

$$y_1 = b_{12}(b_{22}y_1 + \gamma_{21}x_1 + \gamma_{22}x_2 + \cdots + \gamma_{2i}x_i + u_2) + \gamma_{11}x_1 + \gamma_{12}x_2 + \cdots + \gamma_{1i}x_i + u_1$$

$$y_1 = \frac{1}{(1-b_{12}b_{22})}[b_{12}(\gamma_{21}x_1 + \gamma_{22}x_2 + \cdots + \gamma_{2i}x_i + u_2) + \gamma_{11}x_1 + \gamma_{12}x_2 + \cdots + \gamma_{1i}x_i + u_1] \quad (3)$$

Thus;

$$y_1 = \pi_{11}x_1 + \pi_{12}x_2 + \pi_{13}x_3 + \cdots + \pi_{1i}x_i + v_1$$

where

$$\pi_{11} = \frac{b_{12}\gamma_{21}+\gamma_{11}}{((1-b_{12}b_{22}))}; \pi_{12} = \frac{b_{12}\gamma_{22}+\gamma_{12}}{((1-b_{12}b_{22}))}; \pi_{1i} = \frac{b_{12}\gamma_{2i}+\gamma_{1i}}{((1-b_{12}b_{22}))} \text{ and } v_1 = \frac{b_{12}u_1}{(1-b_{12}b_{22})}$$

Similarly, substituting Equation 1 into 2 we obtain:

$$y_2 = b_{22}(b_{12}y_2 + \gamma_{11}x_1 + \gamma_{12}x_2 + \cdots + \gamma_{1i}x_i + u_1) + \gamma_{21}x_1 + \gamma_{22}x_2 + \cdots + \gamma_{2i}x_i + u_2$$

$$y_2 = \frac{1}{(1-b_{12}b_{22})}[b_{22}(\gamma_{11}x_1 + \gamma_{12}x_2 + \cdots + \gamma_{1i}x_i + u_1) + \gamma_{21}x_1 + \gamma_{22}x_2 + \cdots + \gamma_{2i}x_i + u_2]$$

The reduced form is represented as;

$$y_2 = \pi_{21}x_1 + \pi_{22}x_2 + \pi_{23}x_3 + \cdots + \pi_{2i}x_i + v_2$$

where

$$\pi_{21} = \frac{b_{22}\gamma_{11}+\gamma_{21}}{((1-b_{12}b_{22}))}; \quad \pi_{12} = \frac{b_{22}\gamma_{12}+\gamma_{22}}{((1-b_{12}b_{22}))}; \quad \ldots; \pi_{1i} = \frac{b_{22}\gamma_{1i}+\gamma_{2i}}{((1-b_{12}b_{22}))} \text{ and } v_2 = \frac{b_{22}u_2}{(1-b_{12}b_{22})}$$

Thus, the reduced forms of the structural model for the two endogenous variables are:

$$y_1 = \pi_{11}x_1 + \pi_{12}x_2 + \pi_{13}x_3 + \cdots + \pi_{1i}x_i + v_1 \quad (4)$$

$$y_2 = \pi_{21}x_1 + \pi_{22}x_2 + \pi_{23}x_3 + \cdots + \pi_{2i}x_i + v_2 \quad (5)$$

It can be observed that y_i are correlated with u_i because from the reduced-form y_i are correlated with v_i. Hence we cannot obtain consistent estimates of the coefficients of the structural equations if we estimate them by OLS. The equations need to be estimated by the two-stage-least squares (2SLS).

Using the two stage least squares method to estimate the simultaneous equations system, we need to first apply the ordinary least squares to the reduced form Equations (4) and (5) to obtain estimates of the $\hat{\pi}$'s. Using the reduced form coefficients we obtain a set of computed values for the endogenous variables $\hat{y}_1$ and $\hat{y}_2$.

In the second stage the estimated endogenous variables $\hat{y}_1$ and $\hat{y}_2$ are substituted into the structural Equations (1) and (2) to obtain the transformed equation as:

$$y_1 = b_{12}\hat{y}_2 + \gamma_{11}x_1 + \gamma_{12}x_2 + \cdots + \gamma_{1i}x_i + u_1^* \quad (6)$$

$$y_2 = b_{22}\hat{y}_1 + \gamma_{21}x_1 + \gamma_{22}x_2 + \cdots + \gamma_{2i}x_i + u_2^* \quad (7)$$

where

$$u_1^* = u_1 + b_{22}v_2; \; u_2^* = u_2 + b_{12}v_1$$

Solving the transformed structural Equations (6) and (7) by using the ordinary least squares we obtained the 2SLS estimates of the structural parameters. However, to be able to estimate the model by 2SLS it must satisfy the order and rank conditions. These are well spelt out in Koutsoyiannis (1977) and Gujarati (2004).

Table 1. Summary definition of variables.

Variable	Description
Sex of household head	Dummy variable; 1 if head is male and 0 if female
Age of household head	No of years(in logarithm)
Age squared	No of years squared (in logarithm)
Education of household head	No of years of formal education
Marriage of household head	Dummy; 1 if head is married, 0 otherwise
Household size	No of members in the household(in logarithm)
Land	Dummy; 1 if household own land; 0 if otherwise
vehicle	Dummy; 1 if household own a commercial vehicle, 0 if otherwise
Durable assets	Total value in millions of Cedis of household durable assets(in logarithms)
Locality	Dummy; 1 if household lives in urban center and 0 if in rural area
Coastal zone (Coastal)	0 if household lives in forest zone and 1 if otherwise
Savannah zone (Savannah)	0 if household lives in forest zone and 1 if otherwise
Regional distance	Distance in kilometres from Accra (the national capital) to the capital of the region in which a household lives
Welfare	Household total nominal expenditure divided by the product of Accra price index and the national equivalence scale
Food Expenditure	Percentage of household expenditure on food
Poverty status	Categorical; 0 if welfare is below the lower poverty line; 1 if welfare is between the lower and upper poverty lines; 2 if welfare is above the upper poverty line

10,000 (old) Cedis = (New) Gh ¢ 1; Exchange rate at the time was $1= Gh ¢1.

Table 2. Descriptive statistics of continuous variables used in the model.

Variable	No.	Minimum	Maximum	Mean	Std. deviation
Age	3941	15	99	45.53	15.820
Household size	3941	1	27	4.18	2.823
Educational level	3941	0	16	6.18	5.326
Durables	3941	0	127000000	971743.59	4421766.889
Food expenditure	3941	00.369117	95.79412	56.9049777	16.06479846
Welfare	3941	75020.39	81700000.00	2209776.84	2499793.64

Empirical model

The empirical model for this present study consists of two main equations, namely, the food expenditure and welfare equations as specified below. Note that they satisfy both the order and rank conditions.

$$Pfexp = \alpha_0 + \alpha_1 Sexhead + \alpha_2 Agehead + \alpha_3 Agesqd + \alpha_4 Educhead + \alpha_5 Marhead +$$

$$\alpha_6 HHsize + \alpha_7 Locality + \alpha_8 Coastal + \alpha_9 Savanna + \alpha_{10} Region + \alpha_{11} Welfare + u_1$$

(Food Expenditure)

$$Welfare = \beta_0 + \beta_1 Sexhead + \beta_2 Agehead + \beta_3 Agesqd + \beta_4 Educhead + \beta_5 HHsize +$$

$$\beta_6 Land + \beta_7 Vehicle + \beta_8 Durables + \beta_9 Locality + \beta_{10} Region + \beta_{11} Pfexp + u_2$$

(Welfare)

The variables are defined in Table 1.

Data and descriptive statistics of continuous variables

The study uses data from the fifth round of the Ghana Living Standards Survey (GSS, 2008). From Table 2, the average age of a household head was 45 years while the average size of a household was 4. This is exactly the national average. Generally, rural households are larger than urban households. In terms of ecological zone, rural savannah recorded the highest average household size (5.4) in 2007, followed by rural forest (4.1), and the rural coastal (GSS, 2008). GSS (2008) noted that a combination of

Table 3. Frequency Distribution of Percentage of food expenditure.

Class interval	Frequency	Percentage
0-19	56	1.4
20-39	517	13.1
40-59	1549	39.3
60-79	1551	39.4
80-99	268	6.8
Total	3941	100.0

Table 4. Two stage and OLS estimation results of food expenditure equation.

Variable	2SLS		OLS	
	Coefficient	Standard Error	Coefficient	Standard Error
Constant	88.0407	3.1915	75.3627	3.1457
Sexhead	-1.4569**	0.5270	1.5166***	1.4993
Agehead	-0.6062***	0.0773	-0.5299***	0.0789
Agesqd	6.2095***	0.7410	5.6218***	0.7561
Educhead	-0.4701***	0.0542	-0.8568***	0.0487
Marriage Head	1.0464***	0.0030	1.0464***	0.0030
Household size	-2.0748***	0.1511	-0.5128***	0.1124
Locality	-5.7584***	0.6043	-11.4464***	0.4826
Coastal	-10.1243***	0.6127	-8.8675***	0.6202
Savannah	18.7627***	1.2422	14.8633***	1.2414
Region	0.0689***	0.0129	0.0463***	0.0131
Welfare	-4.5086***	0.3413	0.1944***	0.1420

Adjusted R-Squared = 0.99.

factors determines the household size in Ghana. The main ones are a desire, especially on the part of traditional families, to have large families. In the case of urban families, it is the extended family system, which compels them to take care of dependants, other than their immediate family members.

The average number of years that a household head spent in school was 6. This means that on average household heads completed primary six which takes a minimum of six years. The national statistics (as of 2008) was as follows: about 31% of all adults had never been to school; less than 17.1% attended school but did not obtain any qualification; 39% had MSLCE/BECE/VSCE certificate as their highest qualification, while a small percentage of 13.6 had secondary or higher qualification (GSS, 2008). The average wealth of durable assets (such as television set, bicycles and sewing machines) was GH¢997.00 an equivalent of about $500.00 today. Also the average value of welfare was GH¢220.97 an equivalent of about $110.00 in current terms. Lastly, the mean percentage of income spent on food was 65%. However, from Table 3, the majority of the respondents (78.7%) spent between 40 and 79% of their incomes on food.

Estimation results of the 2SLS and OLS compared

From Tables 4 and 5 we notice some differences in the 2SLS and the OLS estimation results. For instance, in Table 4 while the welfare variable is negative in the 2SLS it is positive in the OLS. Also, while sexhead has a negative coefficient in the 2SLS, it is positive in the OLS. Similarly, in Table 5, region has a negative coefficient in the 2SLS but is positive in the OLS. Obviously, the true relationship between food expenditure and welfare would have eluded us if we had used OLS estimator instead of the 2SLS.

The determinants of food expenditure

All the variables that were suspected to influence household food expenditure were significant (Table 4), most of them maintaining their expected signs. The Adjusted R-squared of 0.99 shows that the explanatory variables were able to explain 99% of the variation in food expenditure. Thus, the model was good. Also, the negative significant coefficient of the welfare variable means that an increase in household welfare leads to a reduction in the percentage of household budget on food expenditure. This is consistent with the Engel theory (1885) that households reduce their budget share of food as they become richer. At this stage, it is important a distinction is drawn between the findings of studies that used absolute food expenditure values (Begum et al., 2010; Yimer, 2011; Akpan et al., 2013) and those which used the budget share of food expenditure like our present study (Umeh and Asogwa, 2012).

Generally, in the former studies, income had a positive effect on the level of consumption, though in some instances, the relationship was negative. For instance, in Yimer (2011) study in Ethiopa, while income had a positive effect on the consumption of teff, it had a negative effect on maize consumption. Similarly, in Akpan et al. (2013) study, while workers' salaries had positive effects on food consumption in Southern Nigeria, other sources of income such as farm and non-farm income as well as income from other family

Table 5. Two stage and OLS estimation results of welfare equation.

Variable	2SLS		OLS	
	Coefficient	Standard error	Coefficient	Standard error
Constant	9.7659	0.5883	3.6695	0.3407
Sexhead	-0.3458***	0.0512	-0.5681***	0.0490
Agehead	-0.0806***	0.0078	-0.0306***	0.0069
Agesqd	0.7835***	0.0753	0.2698***	0.0645
Educhead	0.0246***	0.0068	0.0786***	0.0054
Household size	-0.3784***	0.0114	-0.3505***	0.0114
Vehicle ownership	0.3148***	0.0444	0.3609***	0.0451
Land ownership	0.0169***	0.0019	0.0126***	0.0019
Durable assets	0.0002	0.0003	0.0029***	0.0017
Locality	0.7420***	0.0604	1.2299***	0.0474
Region	-0.0041***	0.0011	0.0007***	0.0011
Food Expenditure	-0.0632***	0.0051	-0.0025***	0.0016

Adjusted R-Squared = 0.99.

members had negative effects on food consumption. However, in the study by Umeh and Asogwa (2012) where the dependent variable was food consumption budget, the variable was consistently negatively related to household income, consistent with our study and for that matter the Engel theory.

Other variables like education and household size appeared to go the same direction as income depending on whether the study used absolute or percentage share of food expenditure (Yimer, 2011; Meng et al., 2012; and Akpan et al., 2013). To sum up, the findings in Table 4 indicate that households which spent greater percentages of their incomes on food were as follows: female headed households; households headed by the aged, households whose heads had little or no formal education; households whose heads were married; smaller households; rural households; households in the forest belt as opposed to those in the coastal zone; households in the savannah belt; households living farther from the nation's capital; and poorer households. As indicated earlier most of these findings are consistent with that of the above mentioned studies.

The determinants of welfare

The welfare model was also good in explaining the variations in household welfare, considering the 99% Adjusted R-Squared value reported in Table 5. It can also be observed that all the variables were significant, except durable assets. The negative sign of the food expenditure variable shows that as the percentage of a household income spent on food increased, the welfare of that household reduced. In general, in this study, welfare was greater for the following households: female headed households; households headed by the aged, households whose heads had formal education, smaller households, households who owned assets such as land and commercial vehicles; households living in the urban centres, as well as those living closer to the nation's capital. These findings are also consistent with that of similar studies (Gibson and Rozelle, 2003; Datt and Jollife, 2005; Coulombe, 2008).

RESULTS AND DISCUSSION

As noted earlier, many theoretical and empirical studies have established a negative relationship between food expenditure and income, implying that the higher the income of a household the lower the percentage of the income spent on food, and the lower the income the higher the proportion allocated to food consumption. In this present study, the negative relationship between households' welfare and their food budget share implies that poorer households spend a greater percentage of their incomes on food than richer households. In line with Engel (1885), Umeh and Asogwa (2012) noted that poorer households spend large percentage of their incomes on necessities, including food. However, as their incomes increase they divert more of their incomes to buying higher quality goods and services, thereby reducing the proportion that goes into food. It should be noted that the emphasis is on the proportion of income that goes into food and not the absolute income. Obviously, the absolute income spent on food by a rich household is greater than that of a poorer household, *ceteris paribus*, but in terms of the percentage of income spent on food, that of the latter is likely to be higher. This is validated further by Figure 1a and b. It can be observed that in terms of the percentage of food expenditure, poorer households recorded higher, but in terms of food expenditure in absolute terms they recorded lower than their richer counterparts. The implications of this for research is what we mentioned earlier that a distinction should be drawn between studies that use the absolute food expenditure value and those that use the budget share of food expenditure as the dependent variable. While the former generally establish a positive causal relationship between income and food expenditure, the latter, like our present study establish a negative relationship.

The negative sign of food expenditure in the welfare equation also confirms the fact that higher food expenditure makes households poorer *ceteris paribus*. This is understandable, considering the fact that a

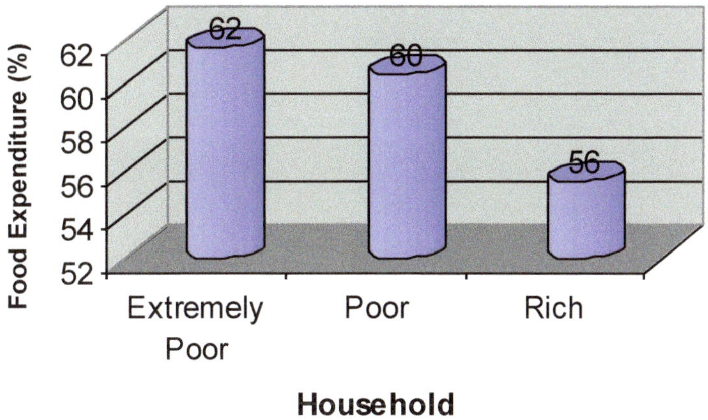

Figure 1a. Poverty status of households and percentage food expenditure.

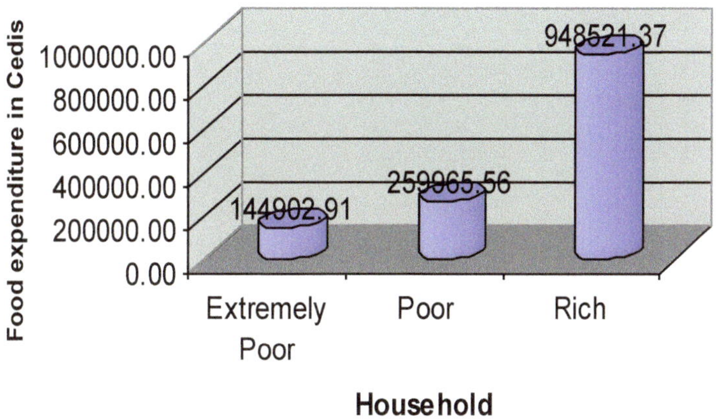

Figure 1b. Poverty status of households and food expenditure.

household spending a greater percentage of its income on food means that it has a smaller percentage left for other goods and services which are vital for its welfare. In Umeh and Asogwa (2012) study, while per capita income negatively influenced the share of food expenditure, therelationship between per capita income and the share of non-food expenditure was positive, confirming the fact that as a household's income rises it increases its share of expenditure on housing, clothing, education and health, among others, to improve upon its standard of living.

A close study of the results from the two equations reveals two main categories of households; (i) households who spent a greater percentage of their income on food but were richer (female headed households, households headed by the old and smaller households) and (ii) households who were poorer and also spent a smaller percentage of their income on food (male-headed households, households headed by the relatively young, larger households, rural households and households living farther away from the nation's capital). The latter group must be targeted for support. Other households that need support are households living in the savannah zone as well as households that had no assets such as land and commercial vehicles.

Female-headed households emerging as richer than male-headed households were contrary to our *a priori* expectations. We also observe in Figure 2a and b that in terms of both the food budget shares and the absolute food expenditure, they recorded higher figures than their male counterparts. Thus, the finding of the present study does not support the issue of feminization of poverty as found by Rodriguez (2000). Following the Beijing Conference on gender inequality a lot of support from both governmental and non-governmental organizations has been given to women in Ghana. Perhaps, this explains why female-headed households are now doing better in terms of welfare, than their male-headed counterparts in Ghana. While the support for women and for that matter, female-headed households should continue, their male counterparts should also be supported since by the findings of this study, they are not

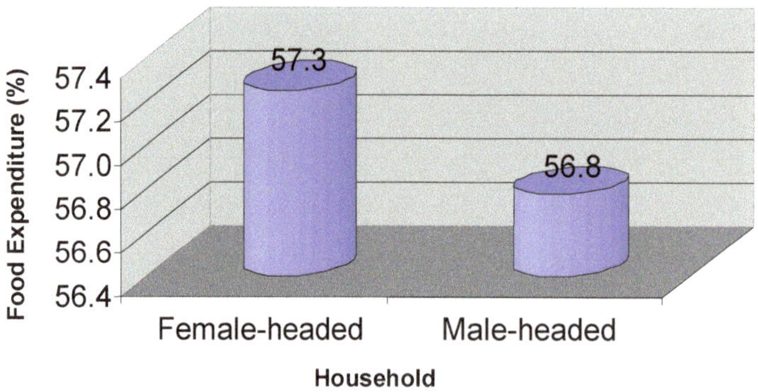

Figure 2a. Sex of Household Head and Percentage of Food Expenditure

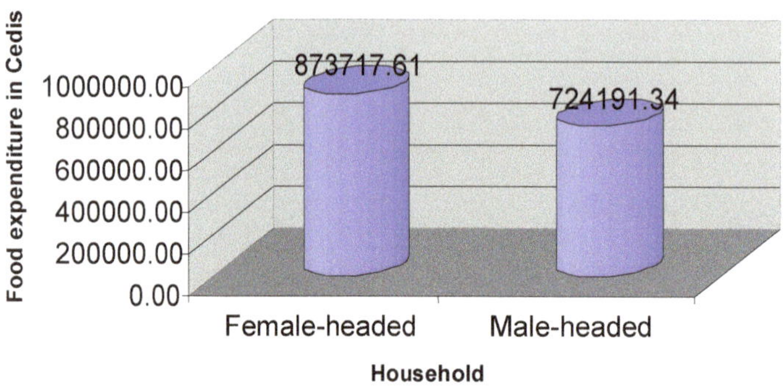

Figure 2b. Sex of Household Head and Food Expenditure.

only poorer but have smaller budget share of food expenditure.

An equally important group that needs support, as per these findings, is large households. As argued by Umeh and Asogwa (2012), the budget share of food expenditure for large households is smaller because they are likely to spend more on non-food items such as education and health. Recall that in the welfare model, larger households were poorer than smaller households, which means that the non-food expenditure here is not a sign of affluence but a necessity. Thus, by virtue of the household size being large, they are forced to cut down the budget share of food expenditure in order to take their children to school or seek medical attention. This is different from an affluent household spending more on luxurious or high quality products because they have a significant increase in their income. For instance, from Figures 3a, b, 4a and b we notice that even though the budget share of food expenditure for households headed by illiterates and households in the rural areas were higher, in terms of the absolute food expenditure, their figures were lower than households headed by literates as well as those in the urban centres, respectively.

It is also important that the gap, in terms of economic resources and opportunities, is closed between the nation's capital and those farther away. This calls for equitable distribution of the national cake and the implementation to the letter of the decentralization policy that is being pursued. Again, Figures 5a, b, 6a and b reveal that while the savannah zone and the three northern regions in general, recorded high percentages of food expenditure, the absolute food expenditure figures for these regions were lower than their counterparts in the south. For a long time, the economic disparity between the south and the north of the country has caught the attention of many analysts and policy makers. Consistently, the Ghana Living Standards Survey data have established that in addition to the central region, the three northern regions are the poorest, of the ten regions of Ghana (GSS, 2007). Food insecurity is associated with poverty or lack of wealth (WFP, 2012). According to WFP(2012), the Upper East has the highest proportion of households who are food insecure (28%), followed by Upper West (16%) and the Northern region (10%).

Conclusion

The main objective of this study was to investigate the determinants of household food expenditure and its effects on welfare. Specifically, the study sought to find out the extent to which household budget share of food

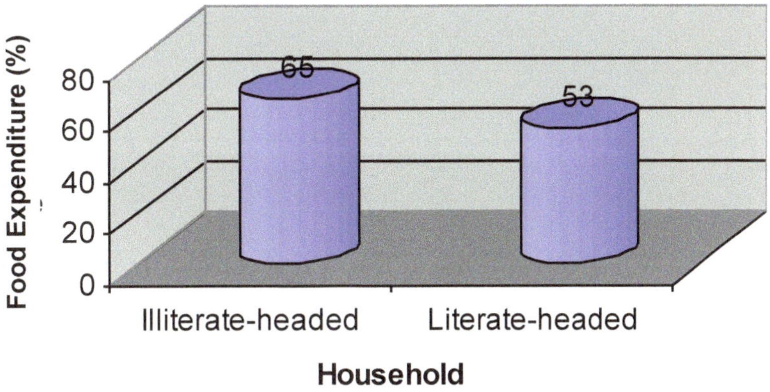

Figure 3a. Household heads' educational background and percentage of food expenditure.

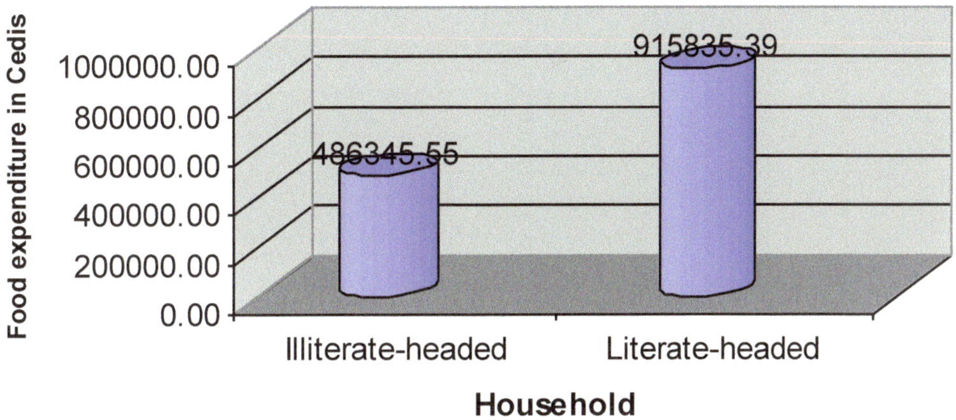

Figure 3b. Household heads' educational background and food expenditure.

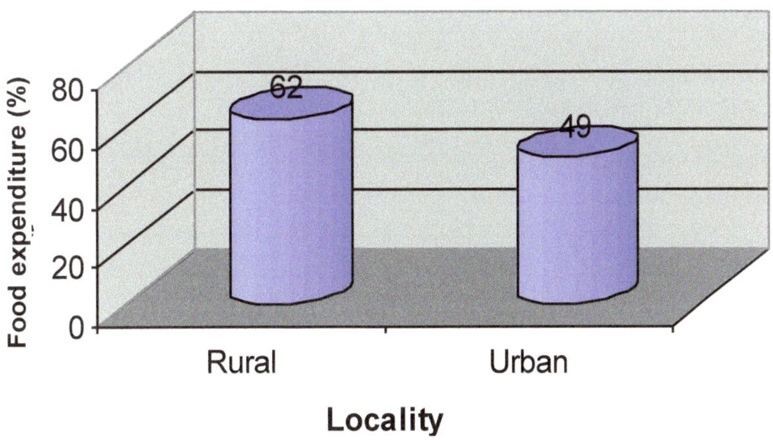

Figure 4a. Locality and percentage of food expenditure.

expenditure (percentage of income spent on food) determined their welfare and *vice versa*. As a result of the simultaneity between food expenditure and welfare, we estimated a simultaneous equations model by the two-stage least squares method. The negative and significant coefficients of the food expenditure and welfare variables

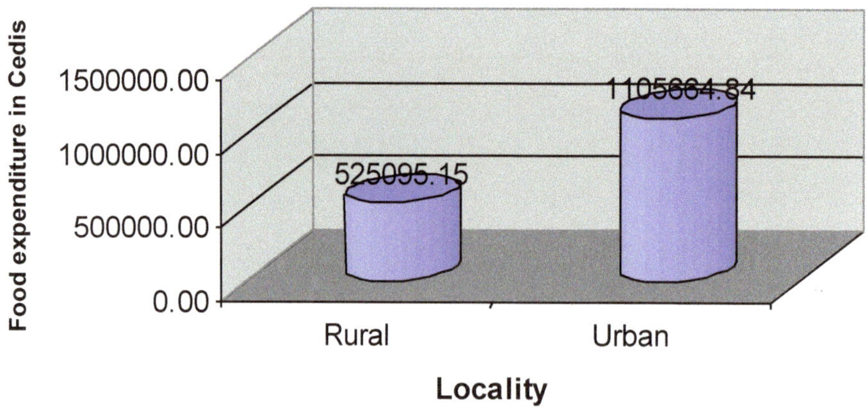

Figure 4b. Locality and Food Expenditure.

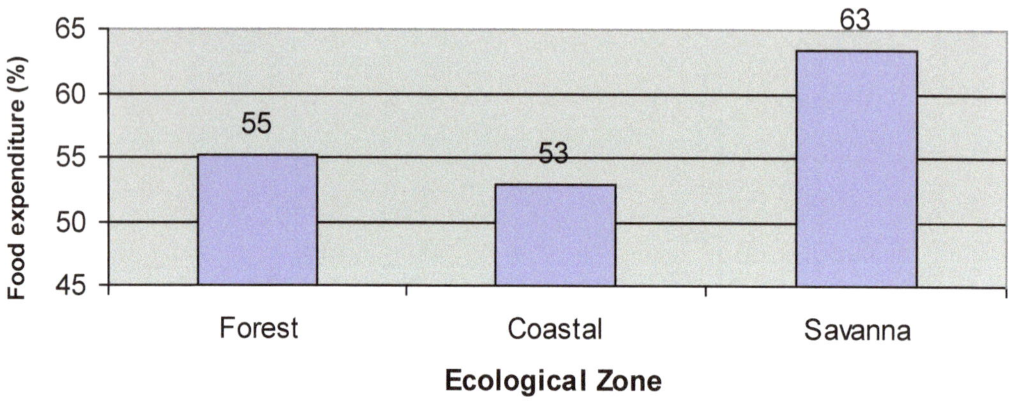

Figure 5a. Ecological zone and percentage of food expenditure.

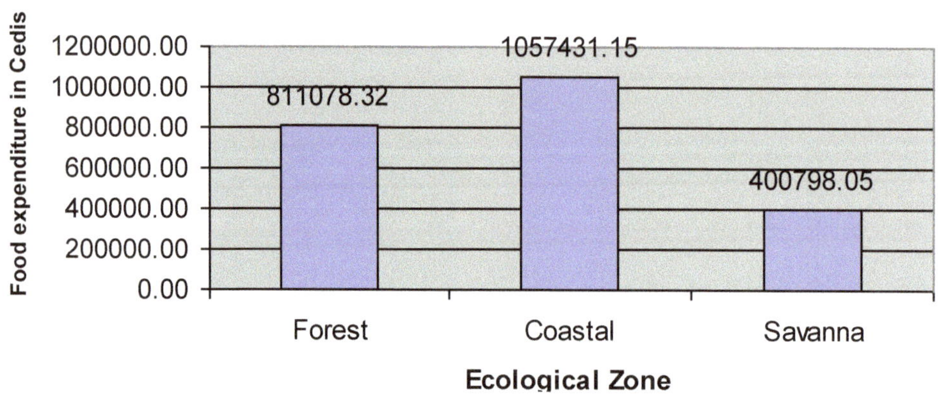

Figure 5b. Ecological zone and food expenditure.

confirm the theoretical and empirical evidence that households reduce the percentages of their incomes allocated to food expenditure as they become rich.

The relatively rich households were as follows: female headed households; households headed by the old; households headed by literates; smaller households; households who owned land, commercial vehicles and other durable assets; and households in the urban centres or closer to the nation's capital. Generally, apart from female-headed households and households headed

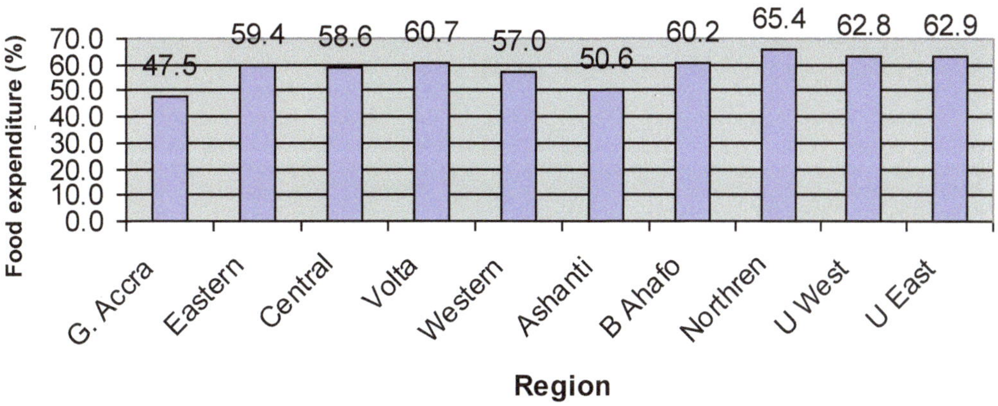

Figure 6a. Region and percentage of food expenditure.

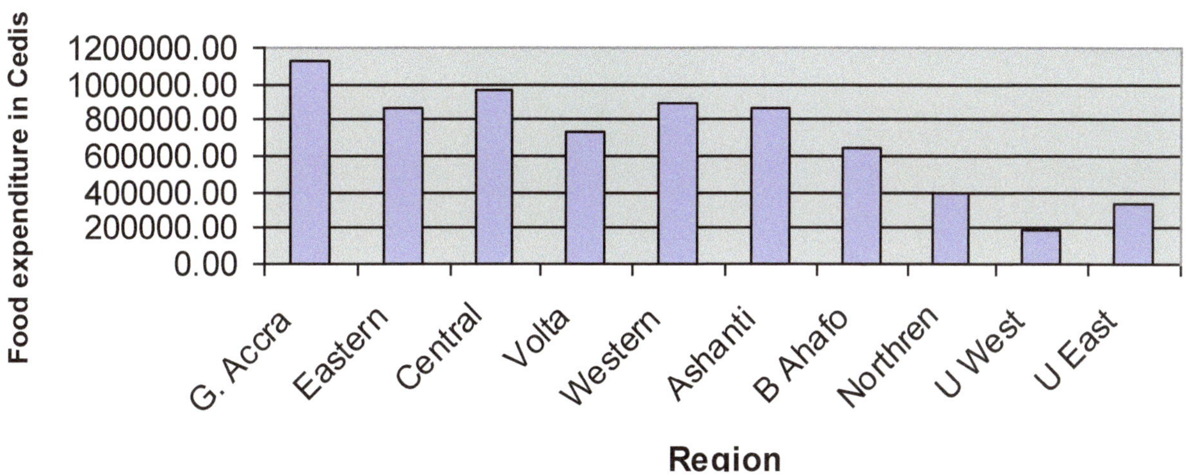

Figure 6b. Region and food expenditure.

by the old, these households spent smaller percentages of their incomes on food, but their absolute food expenditure figures exceeded their relatively poor counterparts such as male-headed households, households headed by the relatively young, larger households, rural households and households farther from the nation's capital, including those in the savannah belt. This latter group must have priority in terms of support.

The contribution of this study to existing literature is three-fold; first while many studies used absolute food expenditure as the dependent variable, we have used the budget share of food expenditure (percentage of household income spent on food). Second, in place of income, we have used welfare, which in principle is broader in scope and much more realistic than income which respondents tend to underestimate. Finally, instead of a single equation, we have estimated a simultaneous equation system; because food expenditure and welfare are both endogenous.

Conflict of Interests

The author(s) have not declared any conflict of interests.

REFERENCES

Akpan SB, Ini-mfon VP, Udoka SJ, Okon UE (2013). Determinants of Food Consumption Expenditure among Agro-Based Firms Workers in Southern Nigeria: Simultaneous Equation Approach. Asian J. Agric. Ext. Econ. Sociol. 2(1):69-84.

Ayo SA, Bonabana-Wabbi J, Sserunkuuma D (2012).Determinants of Fast Food Consumption in Kampala, Uganda. Afr. J. Food Agric. Nutr. Dev. 12(5):6567-6581.

Begum S, Khan M, Farooq M, Begum N, Shah IU (2010).Socio-Economic Factors Affecting Food Consumption Pattern in Rural Area of District Nowshera, Pakistan. Sarhad J. Agric. 26(4):649-653.

Datt G, Jolliffe D (2005). Poverty in Egypt: Modelling and Policy Simulations. Econ. Dev. Cult. Change 53(2):327-346.

FAO (2011). FAO Food Outlook: Global Market Analysis (June 2011). Food and Agriculture Organization of the United Nations, Rome. Accessed on 26th October, 2013 at http://www.fao.org/docrep/014/al978e/al978e00.pdf

FAO (2012). Focus: The Developing World's New Burden: Obesity. Food and Agriculture Organization of the United Nations, Rome.

Accessed on 26th October, 2013 at http://www.fao.org/FOCUS/E/obesity/obes1.htm.

Friedman M (1957).A Theory of Consumption Function. Princeton, NJ: Princeton University Press.

GSS (2007). Pattern and Poverty Trends in Ghana: 1991-2006. Ghana Statistical Service, Accra, Ghana.

GSS (2008).Ghana Living Standards Survey Report of the Fifth Round (GLSS 5).Ghana Statistical Service, Accra, Ghana.

Gujarati D (2004). Basic Econometrics (4th edition).The McGraw-Hill Companies.

Koutsoyiannis A (1977). Theory of Econometrics (2nd edition). Palgrave MacMillan.

Meng T, Wojciech JF, Kolavalii S, Ibrahim M (2012). Food Expenditures in Rural Households in the Northern Region of Ghana. Selected paper prepared for presentation at the Southern Agricultural Economics Association annual meeting, Birmingham, AL.

Modigliani, F. and Brumberg, R., (1954). Utility Analysis and the Consumption Function: An Interpretation of Cross-Section Data in Post-Keynesian Economics. Ed. K. Kithira. New Brunswick, NJ: Rutgers University Press.

Rodriguez JG (2000). The Determinants of Poverty in Mexico. Unpublished PhD Thesis, University of Missouri.

Umeh JC, Asogwa BC (2012). Determinants of Farm Household Food Expenditure: Implications for Food Security in Rural Nigeria International conference on Ecology, Agriculture and Chemical Engineering (ICEACS 2012) Dec. 18-19, 2012 Puce (Thailand).

UNEP (2012).The Critical Role of Global Food Consumption Patterns in Achieving Sustainable Food Systems and Food Source. United Nations Environment Programme Discussion Paper.

WFP (2012).Comprehensive Food Security and Vulnerability Analysis, Ghana 2012. Focus on northern Ghana. World Food Programme.

Yimer S (2011). Determinants of Food Consumption Expenditure in Ethiopia. Int. J. Econ. Res. 2(5):151-165.

WHO (2011). Obesity and Overweight Factsheet: 2011, World Health Organization, Geneva. Accessed on 26th October, 2013 at http://www.who.int/mediacentre/factsheets/fs311/en/index.html.

Nutritional contents of *Balanites aegyptiaca* and its contribution to human diet

Debela Hunde Feyssa[1], T. J. Njoka[2], Zemede Asfaw[3] and M. M. Nyangito[2]

[1]Jimma University, College of Agriculture and Veterinary Medicine, Jimma, Ethiopia.
[2]Nairobi University, College of Agriculture and Veterinary Sciences, Faculty of Agriculture, Dépatement of LARMAT, Nairobi, Kenya.
[3]Addis Ababa University, Collège of Natural Science, Department of Biology and Biodiversity Management, Addis Abeba, Ethiopia.

Wild edible plants are used as food and energy sources. However, their uses are not as the potential inherent in the resources. *Balanites aegyptiaca* is a multipurpose species in semiarid areas including Ethiopia. Hence, quantitative nutrition study in specific habitat is essential for sustainable use of the species. Composite fruits sample of *B. aegyptiaca* was collected from six areas of east Shewa, Ethiopia for nutrient analysis following standard laboratory procedure. The results reveal that the fruits are rich in P, Ca, Fe, Zn, Cu, Na, K, Mg and Mn. The nutritional content varied ($P<0.05$) across land uses. Mean calculated energy value of lipids ranged from 0.09-027 kcal 4.2-7.68 for *B. aegyptiaca* and the total energy from carbohydrate was 342.2-354.24 kcal. Therefore, the fruit of *B. aegyptiaca* is promising in terms of nutrient content to human's diet diversification. It is a valuable species particularly during dry season for coping and adapting to climate variability/change. In spite of the promising potential, the nutritional contribution of this species to the people's diet remained underutilized. Therefore, the utilization of *B. aegyptiaca* is justified to be considered for integration in dryland agrobiodiversity systems and nutrition research to enhance the contribution to the diet of people and enhance its sustainable utilization.

Key words: Food security, nutrition, diet.

INTRODUCTION

Consumption of wild edible plants (WEPs) was practiced for ages. Nebel et al. (2006) explained that there has been renewed interest worldwide in consumption of wild food plants. Despite the increased reliance by agricultural societies' on conventional crop plants, the tradition of eating wild plants has not completely disappeared; their nutritional role and health benefits are being reported in many studies (Pardo-de-Santayana et al., 2007). Wild food plants are used as a source of food energy, sources of vitamins and minerals. They were important as dietary supplements, providing trace elements, vitamins and minerals (Gillman, 2008, Debela et al., 2011).

In spite of many studies on WEPs focused on the

functions within ecology and ecosystems there are few research works that compared the food plants of various communities (Ladio et al., 2006). Some research works have compared medicinal floras and other useful plants (Pieroni and Quave, 2005). Such comparative studies therefore contribute to the understanding of why edible species are widely consumed or given little attention and provide insights into food and nutrition security potential of different communities.

Food and nutritional security are key issues for human wellbeing while WEPs are underutilized in many countries including in Ethiopia. Fentahun and Hager (2009) have reported that wild edible fruits contribute to nutrition and health security of rural people as they contain proteins, vitamins and minerals. In the first phase of this work which is already published in Debela et al. (2011) and others, the proximate (carbohydrate, lipids and total minerals) were addressed. However, the information on the fruits nutritional quality of the species is not exhaustively documented. The specific minerals and vitamins contents were not determined in this study. Therefore, these studies curve this research gap through further analysis of the nutrient composition macro and micronutrients of *Balanites aegyptiaca*. It also indicates future line of research on the contribution of *B. aegyptiaca* to nutrition and food security of the semiarid people. Therefore, this study was aimed at analysing the nutritional content of *B. aegyptiaca* and implications to food and nutrition of people in semiarid areas.

MATERIALS AND METHODS

The study area

The study was conducted in 'Fantalle' and 'Boosat' districts, in East Shewa Zone of Oromia National Regional State, Ethiopia. East Shewa is located between 7°12'-9°14'N latitudes and 38°57'-39°32'E longitudes in the northern part of the Great East African Rift Valley.

Collection and preparation of fruit sample for laboratory analysis

Frequency distribution and density of *B. aegyptiaca* in the local vegetation was systematically determined by total count of shrubs and trees from 20 x 20 m of 66 plots (33 from each district) laid at 200 m interval on the 5 km line transects.

Before undertaking laboratory nutrient analysis, fruit samples of *B. aegyptiaca* were identified through focus group discussion (FGD), interview and field observations as described in Martin (1995) and Cotton (1996) by repeated field visits between 2009 and 2010. Composite fruit samples of the species were collected in a sample bag from Fantalle (Galcha, Qobo and Dheebiti villages) and Boosat (TriiBiretti, DigaluTiyo and Xadacha villages) districts for nutrient analysis.

Minerals, vitamins A and C and tannin analysis

The mineral elements comprising sodium, calcium, potassium, magnesium, iron, zinc and phosphorus were determined according to the method of Shahidi et al. (1999), Sundriyal and Sundriyal (2004), AOAC (1990) and Nahapetian and Bassiri (1975). Two gram of each of the processed samples was weighed and subjected to dry ashing in a well-cleaned porcelain crucible at 550°C in a muffle furnace.

The resultant ash was dissolved in 5.0 ml of $HNO_3/HCL/H_2O$ (1:2:3) and heated gently on a hot plate until brown fumes disappeared. To the remaining material in each crucible, 5 ml of de-ionized water was added and heated until a colourless solution was obtained. The mineral solution in each crucible was transferred into a 100 ml volumetric flask by filtration through Whatman No.42 filter paper and the volume was made to the mark with de-ionized water. This solution was used for elemental analysis by atomic absorption. A 10 cm long cell was used and concentration of each element in the sample was calculated on percentage of dry matter, i.e. mg/100g sample. Phosphorus content of the digest was determined colorimetrically according to the method described by Nahapetian and Bassiri (1975).

For tannin determination samples were dried at a maximum of 60°C immediately after collection to minimize any chemical changes and extracted with 50%v/v acetone. Determination was by a modification of the vanillin method of Ranganna (1977) and Broadhurst and Jones (1978), which utilizes the formation of coloured complexes between vanillin and condensed tannins and Catechin was used for the standard, and results were expressed as catechin-equivalents.

Determination of vitamin A was carried out by spectrophotometer following Davies (1976) and (AOAC, 1990). Ascorbic acid (vitamin C) was determined by redox titration using iodate (I_3) following (Pearson, 1976) and Helmenstine (2007).

Determination of percentage titrable acidity and total soluble sugar (TSS)

Titrable acidity was determined by titration using 0.1 N NaOH. Titration was done by adding 0.1 N NaOH from a burette to the Erlenmeyer flask containing 10 ml of fruit solution prepared in measuring while swirling until the solution just starts to change colour to pink/purple (slightly pink). The amount of NaOH added to the flask was recorded immediately at this end point. Multiplied the volume of sodium hydroxide (volume of titre) added by acid conversion factor of the fruit acid to get the value of the acid (in grams per 100 ml); that is, TSS of fruit indicator used is 2-3 drops of phenolphthalein indicators from burette filled, along with phenolphthalein indicators (AOAC, 1990, OECD, 2005). Soluble solids (TSS) content was determined directly in the fruit juice, using a portable/handle held refractometer (Pocket PAL-1, UK) at room temperature in degree Brix (°Brix) following (Khemiss et al., 2005; Franco et al., 2009). Degrees Brix (°Bx) is a unit representative of the sugar content of an aqueous solution.

Determination of energy value

Energy of the WEPs fruits was calculated in kilocalories (kcal) multiplying by energy factor composition (4, 4 and 9) of percentage proteins, fats and carbohydrates respectively as used in FAO (1968, FAO, 2011), USDA (1999) and Asibey-Berko and Tayie (1999). The conversion factors are for physiological energy, which is the energy value remaining after losses due to digestion and metabolism and deducted from gross energy (USDA, 1999) where one kcal equals 4.184 kJ. Organic carbon (O.C) in the fruit was obtained by subtracting total ash from 100 (Adams et al., 1951).

Data analysis

Statistical analysis for nutritional content of *B. aegyptiaca* was done through analysis of variance and means were separated by LSD at P< 0.05 using GenStat (VSN International, 2011). Organic carbon (O.C) in the fruit of threatened WEP was calculated using formula % O.C = (%VS/1.8) x100, where, %VS = (100-%Ash) following Adams et al. (1951). TSS obtained directly from Refractometer was expressed in °Brix. Titrable acidity was calculated by:

$$TA\,(\%) = \frac{mL\ of\ NaOH\ (Titre) \times 0.1N\ NaOH \times acid\ meq.factor}{mL\ of\ juice\ titrated} \times 100$$

and was expressed as g malic acid 100 g-1 fresh weight (F.W.) (AOAC, 1990; Franco et al., 2009).

RESULTS AND DISCUSSION

Abundance and distribution of *B. aegyptiaca*

The present study indicated that, the average % relative frequency is 24.24 and Av% relative density was 0.96 in settled farmers land use and av % frequency was 33.33 and Av % relative density was 2.27 in transhumance land use for *B. aegyptiaca*. Direct repeated field visit and observation in the study areas revealed that the natural forest is more degraded in settled farmers area due to charcoal and firewood production, expansion of agriculture and settlements in settled farmers areas. Commercial fuel wood (charcoal and firewood) production was observed. *B. aegyptiaca* has declined in natural forest due to overharvesting. Cutting dawn local vegetation for building of houses for the increased settlement in the area has affected *B. aegyptiaca*. As reported by informants during field observations, overgrazing/browsing by camels and goats is a serious threat to the vegetation of the area including *B. aegyptiaca*.

Local consumption of WEPs as food was not regarded as a threat to the survival of WEP species unless when demand becomes higher than sustainable harvest in the future. Gamado-Dalle et al. (2005) and Asfaw (2009) reported that local people maintain species which are useful to their livelihood at farm boarders, live fences and sacred areas.

Nutritional potential of *B. aegyptiaca*

The results indicate the existence of variations of nutrient and energy from *B. aegyptiaca* from different study sites. This can be attributed to environmental factors such as soil, temperature and rainfall. In spite of the variations within and between land uses, the result indicated that *B. aegyptiaca* has significant amount of nutrients and energy to supplement household nutrition and a good income source if properly valued.

Besides direct nutritional contributions through carbohydrates, lipids, proteins, minerals and energy; the diversity of wild fruits by itself is a source of variety and taste in the local meals of rural communities including Ethiopia. Earlier studies made by FAO (1995) on non-timber forest products focusing on nutrition, and in India on nutrient composition of specific plants (Parvathi and Kumar, 2002) have reached similar conclusion. In east Shewa, the nutritional contribution of *B. aegyptiaca* and value are hardly recognized in the formal production system as it is overshadowed by the values of conventional food crops and charismatic species such as *Eragrostis tef, Triticum aestivum, Sorghum bicolor, Zea mays* and others.

Comparison of the results of the present study with the nutrient content of some cultivated fruits noted by Srivastava and Kumar (1998) revealed that they are superior in protein content to banana (1.2 and 0.3%), guava (0.9 and 0.3%) respectively; mango (0.6 and 0.4%, respectively) and papaya (0.6 and 0.1%) respectively.

Test also contributes to the nutritional quality of a fruit. In this regard, the relatively high tannin content of *B. aegyptiaca* may be a point of argument to fully recommend the fruit for common use for human diet. Chapagain and Wiesman (2007) also reported nine compounds from *B. aegyptiaca* of which six of them were saponins with molecular masses of 1196, 1064, 1210, 1224, 1078 and 1046 Da.

The same report revealed that the compound of mass 1210 Da being the main saponin (ca. 36%). Saponins with masses of 1224 and 1046 Da have not been previously reported in *B. aegyptiaca*. Chapagain and Wiesman (2007) also reported that in all saponins, diosgenin was found to be the sole aglycone. Though people use *B. aegyptiaca* fruits, the controversial issues of saponins and consequent bitter test and the impact consumption pattern and nutritional quality is question not addressed by the present study.

Interaction effect of fruit of *B. aegypiatca* and land use on mineral and tannin contents

The mineral and tannin contents of *B. aegyptiaca* significantly varied across land uses (P<0.05) (Table 1). Phosphorus content is highest in *B. aegyptiaca* for sampled collected from transhumance land use. Sodium content was highest in *B. aegyptiaca* followed for samples collected from settled farming area (Table 1).

Interaction effect of land use on Vitamin A and C contents of *B. aegyptiaca* species

Vitamin C content of *B. aegypiatca* was 50.30 and 50.90 on mg/ 100 g dry matter basis for SF and TH land uses respectively and not significantly varied (P>0.05) across land uses. Vitamin A content (Beta carotine REs, Retinol Equivalents) for *B. aegypiatca* was 233.70 and 266.33 for SF and TH land uses respectively which significantly varied between land uses (P<0.05) with higher mean value from sample collected from transhumance land use systems.

Table 1. Interaction effects of fruit tree species with land use system on mineral and tannin contents of *B. aegyptiaca* in mg/100 g dry matter basis.

Nutrient contents across land uses	Nutrient and tannin contents									
	P	Ca	Fe	Zn	Cu	Na	K	Mg	Mn	CT
SF* *B. aegyptiaca*	102.625	74.550	39.700	0.5500	0.0950	22.100	1990.30	87.070	0.8200	1219.88
TH* *B. aegyptiaca*	103.55	76.650	40.360	0.3150	0.1002	20.100	1992.50	87.635	0.7850	1222.10
LSD (5%)	0.1806	0.7824	0.1594	0.8667	0.03726	0.1816	1.589	0.6627	0.05051	2.07
s.e.d	0.0783	0.3393	0.0691	0.03758	0.00808	0.0787	0.689	0.2874	0.02190	0.898
Sig*	0.014	<0.001	<0.001	<0.001	<0.001	<0.001	0.002	0.005	<0.001	0.008

SF=settled farmers; TH=transhumant land uses; Means with the same letter are not significantly different ,P=Phpsphorus; Ca=Calcium; Fe=Iron,Zn=Zink,Cu=Copper; Na=Sodium; K=potassium; Mg=magnesium; Mn=manganese; CT=condensed tannin,*= significant at 0.05 level, s.e.d = standard error of the differences.

Percentage titrable acidity and total soluble solids in composite fruits pulp samples

Analysis of variance between locations revealed that in % TTA (1.37) and TSS (4.35) and TSS/ %TTA is not significant (P>0. 05). However, the mean values %TTA and TSS °Brix related that the fruits *B. aegyptiaca* have relatively high characteristics of sourness combined with a high degree of sweetness taste. By extrapolation, the total soluble solid was lower. The TSS/%TTA ratio (3.18) also indicated the fruit can be consumed by humans.

Comparison of energy content of *B. aegyptiaca* fruits

The result of the present study indicated that mean calculated energy value of lipids ranged from *B. aegyptiaca* from 0.09-027 kcal to 4.2-7.68 for *B. aegyptiaca*. Energy for total carbohydrate was 342.2-354.24 kcal for *B. Aegyptiaca* (FAO, 1968; 2011).

Conclusions

The fruit of *B. aegyptiaca* is promising in terms of nutrient content to humans and livestock. The results of the nutrient analyses showed that B. aegyptiaca fruits can be important indigenous sources of nutrients to supplement other major food sources. This is a valuable plant particularly during dry season for coping and adapting to climate variability/change. Though it has good potential, the nutritional contribution of this species to the human diets is not sufficiently utilized. On top of this, there is a need of more research on the identification of the bitter components and their effects on human nutritional utilization and quality. Hence, *B. aegyptiaca* need be considered for integration in dryland production system such as agrobiodiversity systems to improve the livelihoods of people and promote its sustainable utilization by improving nutritional quality through more research.

Conflict of interests

The authors did not declare any conflict of interest.

ACKNOWLEDGEMENTS

We thank RUFORUM-Rockefeller foundation, IDRC for funding the study, and the Association of African Universities for covering part of the research work. We also thank the local people and district management of east Shewa, Ethiopia for sharing valuable information and facilitating the field work. Also, the University of Nairobi for hosting the principal researcher and administrative support by Jimma University, College of Agriculture and Veterinary Medicine for permitting the use of the laboratory and ICT facilities and field vehicle is thankfully appreciated.

REFERENCES

Adams RC, MacLean FS, Dixon JK, Bennett FM, Martin GI, Lough RC (1951). The utilization of organic wastes in N.Z.: Second interim report of the inter-departmental committee. New Zealand Eng. 15:396-424.

AOAC (1990). Official methods of analysis, 14th Edition, Association of Official Analytical Chemists, Washington DC.

Asfaw Z (2009). The Future of Wild Food Plants in Southern Ethiopia Ecosystem Conservation Coupled With Management of the Roles of Key Social Groups. In: International Symposium on Underutilized Plants for Food Security, Nutrition, Income and Sustainable Development International Society for Horticultural Science-ISHS, January 2009, pp. 86-87.

Asibey-Berko E, Tayie FAK (1999). Proximate analysis of some underutilized Ghanaian Vegetables. Ghana J. Sci. 39: 91-92.

Broadhurst RB, Jones WT (1978). Analysis of condensed tannins using acidified vanillin. J. Sci. Food Agric. 29:788 -794.

Chapagain BP, Wiesman Z (2007). Determination of saponins in the kernel cake of Balanites aegyptiaca. Phytochem. Anal. 18(4): 354–362.

Cotton CM (1996). Ethnobotany: Principles and applications. Chichester, New York John Wiley and Sons Ltd.

Davies BH (197). Analytical Methods: Carotenoids. In: Chemistry and Biochemistry of plants pigments. Godwin, T.W. (ed). Academic press, London, Vol. 4: 125-127.

Debela HF, Njoka JT, Zemede A, Nyangito MM (2011). Wild edible fruits of importance for human nutrition in semiarid parts of East Shewa Zone, Ethiopia: Associated Indigenous Knowledge and Implications to Food Security. Pak. J. Nutr. 10(1):40-50.

FAO (1968). Food composition table for Use in Africa. Energy Factor, p. 226.

FAO (1995). Non-wood forest products in nutrition. In: Non-Wood Forest Products for Sustainable Forestry, Yogyakarta, Indonesia, 17-27 January, 1995. Non-Wood Forest Products, Vol. 3. FAO, Food and Nutrition Division, Rome.

FAO (2011). Calculation of the energy content of foods - energy conversion factors. http://www.fao.org.(Accessed March, 2011).

Fentahun MT, Hager H (2009). exploiting locally available resources for food and nutritional security enhancement: wild fruits diversity, potential and state of exploitation in the Amahara Region of Ethiopia. J. Food Sec. 1:207-219.

Franco MV, Fanco-Mora O, Lopez-Sandoval JA, Perez-Lopez DJ, Balbuena-Melgarejo A (2009). Characterization of wild plum (*Ximenia americana* L. var. *americana*; Olacaceae) fruit growing at Tepexi de Rodrı´guez, Puebla, Mexico. Genet. Resour. Crop Evol. 56:719-727.

Gemedo-Dalle T, Maass BL, Isselstein J (2005). Plant biodiversity and ethnobotany of Borana pastoralists in Southern Oromia, Ethiopia. Econ. Bot. 59(1):43-65.

Gillman S (2008). The Cattail - Best of the Wild Edible Plants. Free Online Articles Directory. http://www.articlesbase.com/camping-articles/the-cattail-best-of-the-wild-edible-plants-495731.html

Helmenstine AM (2007). Vitamin C Determination by Iodine Titration. doi:10.1016/j.foodchem.2007.02.037.

Khemiss F, Ghoul-Mazgar S, Moshtaghie AA, Saidane D (2005). Study of the effect of aqueous extract of *Grewia tenax* fruit on iron absorption by everted gut sac. J. Ethnopharmacol. 103:90-98.

Ladio A, Lozada M, Weigandt M (2006). Comparison of traditional wild plant knowledge between aboriginal communities inhabiting arid and forest environments in Patagonia, Argentina. J. Arid Environ. 2007:695-715.

Martin GJ (1995). Ethnobotany: amethods manual. London, UK: Chapman and Hall.

Nahapetian A, Bassir A (1975). Changes in concentration and interrelationships of phytate, P. Mg, Cu, Zn in wheat during maturation. J. Agric. Food Chem. 32:1179-1182.

Nebel S, Pieroni A, Heinrich M (2006). Ta chòrta: Wild Edible Greens Used in the Graecanic area in Calabria, southern Italy. Appetite 47(3):333-342.

OECD (2005). Guidance on Objective Tests to Determine Quality of Fruits and Vegetables and Dry and Dried Produce. Organization for Economic Co-Operation and Development. Http://Www.Ble.De/Nn_680006/Shareddocs/

Pardo-de-Santayana M, Javier Tardío J, Blanco E, Carvalho AM, Juan José Lastra JJ, Miguel ES, Morales R (2007). Traditional knowledge of wild edible plants used in the northwest of the Iberian Peninsula (Spain and Portugal): a comparative study. J. Ethnobiol. Ethnomed. 3:27.

Parvathi S, Kumar VJF (2002). Studies on chemical composition and utilization of the wild edible vegetable athalakkai (*Momordica tuberosa*). Plant Foods Hum. Nutr. 57 (3-4):215-222.

Pearson D (1976). The Chemical analysis of foods, Seventh ed.London and NewYork, Pp, 575.

Ranganna S (1977). Manual of analysis of fruit and vegetables products. New Delhi: Tata McGraw-Hill Publishing Company Limited, p. 634.

Shahidi F, Chavan UD, Bal AK, Mckenzie DB (1999). Chemical Composition of Beach Pea (*Lathyrus Maritimus* L.) Plant parts. Food Chem. 64: 39-44.

Sundriyal M, Sundriyal RC (2004).Wild Edible Plants of the Sikkim Himalaya: Values of Selected Species. Econ. Bot. 58(2):286-299.

USDA (1999). US Department of Agriculture, Agricultural Research Service, USDA Nutrient Data Laboratory. USDA Nutrient Database for Standard Reference, Release 13.

VSN International (2011). GenStat Discovery Edition 3, software for bioscience. VSN International Ltd. http://www.vsni.co.uk/software/genstat-discovery/

Mycoflora of fresh shrimps (*Penaeus aztecus*) from different markets in Port Harcourt, Nigeria

Obire, O[1], Okigbo[2], R. N. and Minimah, T. R.[1]

[1]Department of Applied and Environmental Biology, University of Science and Technology, P.M.B. 5080, Port Harcourt, Nigeria.
[2]Department of Botany, Nnamdi Azikiwe University, Awka, Nigeria.

Fungi were isolated from fresh brown shrimps (*Penaeus aztecus*) purchased from three different markets in Port Harcourt using standard mycological methods. The total counts of heterotrophic fungi range from 2.0 x 10^4 spore forming units per gram (sfu/g) to 7.1 x 10^4 sfu/g, while the total counts for pathogenic fungi range from 1.7 x 10^4 sfu/g to 7.1 x 10^4 sfu/g. The heterotrophic fungi and their percentage occurrence in the head, exoskeleton and telson/uropod (H/E/T) before deterioration and about deterioration were: *Aspergillus clavatus* (20%), *Aspergillus flavus* (20%), *Penicillium* sp. (20%), *Rhizopus* sp. (20%), *Rhodotorula* sp. (20%), *Mucor hiemalis* (14.3%), *Penicilllium* sp. (14.3%), *Rhizopus* sp. (14.3%), *Rhizopus stolonifer* (14.3%), yeast sp. (14.3%). While the pathogenic fungal occurrence before deterioration and about deterioration in the H/E/T were: *A. clavatus* (11.1%), *A. flavus* (22.2%), *Penicillium* sp. (33.3%), *Rhizopus oryzae* (11.1%), *Rhodotorula* sp. (11.1%), yeast sp. (11.1%), and *A. clavatus* (12.5%), *A. flavus* (12.5%), *M. hiemalis* (12.5%), *Penicillium* sp. (12.5%), *Rhodotorula* sp. (12.5%) and yeast sp. (37.5%). The presence of these fungi in the shrimps is attributed to contamination from the environment and from shrimp handlers (mongers). Also, some of these fungi are normal flora of the shrimp which unfortunately happens to be opportunistic pathogens or pathogens of humans. The maintenance of high personal and environmental hygiene as well as proper heating and cooking will improve fresh shrimp quality and prevent food-borne diseases.

Key words: Shrimps, contamination, pathogenic fungi, food-borne diseases.

INTRODUCTION

Brown shrimp is a small free-swimming edible crustacean with 10 legs. It is basically marine commonly found in estuaries and along coastal waters. However, its primary habitat is muddy bottom areas from inter tidal zone to approximately 110 m with greatest density occurring at depth between 27 to 55 m (Williams, 1984).

Brown shrimps are scavengers generally termed bottom-feeding opportunistic omnivores, which means they feed on most organic materials- dead plant or animal matter that they encounter at the bottom. They consume some algal species (that is, filamentous green algae, benthic diatoms, plant detritus), invertebrates such as copepods, mollusks, and annelids (Dall, 1968; Odum and Herald, 1972). Shrimps between the sizes of 65 to 100 mm in length became predatory, feeding on annelid worms, amphipods, zooplankton larvae, and nematodes (Jones, 1973).

These edible crustaceans are composed of 41% (w/w)

high value protein and omega-3 fatty acids which have anti-inflammatory effects and are able to prevent the formation of blood clots). They are also important sources of vitamin D and B12 needed to keep low levels of homocysteine - a molecule that can directly damage blood vessel walls and is considered a significant risk factor for cardiovascular diseases.

Shrimps, apart from being of great nutritional value to the consumer; also serves as an ideal culture medium for microbial growth, which connotes spoilage. Despite all the benefits associated with shrimps, one thing still remains clear and yet unsolved; shrimps are highly prone or susceptible to rapid deterioration immediately after harvest.

Microorganisms are undoubtedly responsible for spoilage of shrimps and these microorganisms include bacteria and fungi, both the heterotrophic and pathogenic forms. One important characteristic of microbes is that, they grow rapidly and increases in number (Harrigan and McCance, 1990). Molds which belongs to fungi, grows rapidly on food (shrimp) when held under moist conditions. These organisms take advantage of the moist condition available and through the aid of enzymes which they possess, they weaken and penetrate the protective outer layer of the shrimp and cause spoilage (Donald et al., 1999).

It is almost always possible to detect a range of human pathogens on any shrimp that has not received any microbiocidal treatment (Okigbo and Ajalie, 2005). Some of these pathogens may constitute part of the normal flora on the shrimp or be present as a result of unavoidable contamination (ICMSF, 1996). Also, a normal microflora of humans can cause spoilage of shrimp by way of contamination through handlers, as they touch the shrimp with bare hands in the bid of harvesting and selling the shrimps. Also shrimp spoilage based on microbial origin, originates mainly from the shrimp's environment (which is aquatic) and in this environment, contamination may occur as a result of the presence or introduction of microorganisms through faecal means into the water bodies.

Fresh brown shrimps sold in the common markets in Port Harcourt do not undergo any form of microbiocidal treatment, and they are not iced before sale. The shrimp mongers also handle them with bare hands during transactions. Some shrimp consumers cook the shrimps whole (that is, including the head, exoskeleton and telson/uropods); others remove the head, exoskeleton and telson/uropods of the shrimps before cooking (only flesh) while some consumers after removing the head, exoskeleton and telson/uropods, grind them and add the paste to the soup or sauce.

There is therefore the need to ascertain the microorganisms (fungi) associated with these various parts of shrimps and the health hazards associated with the types of microorganisms. The knowledge of the types of microorganisms involved in their spoilage will also help to control shrimp spoilage.

MATERIALS AND METHODS

Sampling sites and collection of fresh shrimp samples

Fresh shrimp samples were purchased from shrimp mongers in Creek Road Market, Town Market, and Mile One Market all in Port Harcourt Local Government Area of Rivers State. These markets were visited at about 7:30am on the days of collection of samples.

Disposable gloves were worn, before collection of the fresh shrimps from the tray into a sterile transparent plastic container containing ice. The plastic container was covered with its lid and transported immediately to the laboratory for analysis.

Preparation of media and preparation of diluent

Potato dextrose agar (PDA) and Sabouraud dextrose agar (SDA) media were used for isolation of fungi. Heterotrophic fungi were cultured on PDA and pathogenic fungi on SDA medium. Serial dilution plate method was adopted for isolation of fungi. Exactly 0.85 g of sodium chloride (NaCl) was weighed and dissolved in 100 ml of sterile distilled water; 9 ml of the normal saline was transferred into sterile lost tubes and autoclaved at 121°C for 15 min at 15 pounds/pressure and allowed to cool. The diluent was used to carry out serial dilution of the fresh shrimp samples. Normal saline was used as a diluent because it helps to reactivate stressed microorganisms.

Cultivation and enumeration and isolation of fungi in the fresh shrimp

The fresh shrimp samples were divided into two portions. One portion comprised the head, exoskeleton, and telson/uropod (HET) while the other portion was the flesh (F). The samples were analyzed separately to have a thorough sampling study of the presence of fungi in the samples purchased from the three (3) different markets.

Each portion (sample) was separately ground in a sterile ceramic mortar and pestle. One gram (1.0 g) of each portion was transferred into 9 ml of the normal saline and thoroughly shaken to make a 10^{-1} dilution. This dilution was further diluted serially up to the 10^{-3} dilution. This was done so as to obtain discrete colonies or colony forming units (CFU) when plated out on the different media for fungi.

From the 10^{-2} and 10^{-3} dilution, an aliquot (0.1 ml) was collected aseptically onto separate freshly prepared PDA and SDA plates to which 0.2 ml of 0.5% ampicilin was added to inhibit the growth of bacteria and allowing the growth of fungi (Harrigan and McCance, 1990). A sterile glass spreader was used to spread the inoculum evenly on the whole surface of the media. The cultured plates were also prepared in duplicates and inoculated plates were incubated at ambient temperature for three to five days. Colonies or spore forming units which developed on the PDA and SDA plates after five days were counted and the average count for duplicate cultures were recorded as total viable heterotrophic fungi and total viable pathogenic fungi in the sample, respectively. The colour and colonial morphologies or characteristics were also recorded.

Discrete colonies were subcultured onto freshly prepared PDA and SDA plates and incubated at 28°C for five to seven days to further purify the fungal isolates. The fungal spore forming units which developed were further subcultured onto agar slopes or slants and incubated at 28°C for five to seven days. The isolates

Table 1. Total heterotrophic fungal count (sfu/g) of the shrimps from the markets.

Site	Head, exoskeleton and telson		Flesh	
	Before deterioration	About deterioration	Before deterioration	About deterioration
Creek Road Market	5.6 x 10^4	4.8 x 10^4	2.0 x 10^4	7.1 x 10^4
Mile One Market	6.6 x 10^4	3.1 x 10^4	6.8 x 10^4	4.6 x 10^4
Town Market	5.6 x10^4	4.8 x 10^4	3.0 x 10^4	7.1 x 10^4

Table 2. Pathogenic fungal count (sfu/g) of shrimps from the markets

Site	Head, Exoskeleton and telson		Flesh	
	Before deterioration	About deterioration	Before deterioration	About deterioration
Creek Road Market	6.4 x 10^4	4.8 x 10^4	1.7 x 10^4	5.1 x 10^4
Mile One Market	5.0 x 10^4	3.7 x 10^4	6.4 x 10^4	3.3 x 10^4
Town Market	6.4 x 10^4	4.8 x 10^4	1.7 x 10^4	7.1 x 10^4

which developed were pure cultures which were stored in the refrigerator as stock cultures for subsequent characterization tests. Identification of fungal species was done with help of manuals of Barnett and Hunter (1972) and Malloch (1997).

RESULTS

Evaluation of total heterotrophic fungi and total pathogenic fungal counts from the various markets

The results of mean count of total heterotrophic and pathogenic fungi in the various portions of the shrimps from the markets are as shown in Tables 1 and 2, respectively.

Population of heterotrophic fungi per gram of head, exoskeleton and telson/uropod (HET) of shrimps from Creek Road market, Mile One market and Town market before deterioration, ranged from 5.1 x 10^4 to 6 x 10^4 sfu/g, 4.1 x 10^4 to 6 x 10^4 g and 5.1 x 10^4 to 6 x 10^4 sfu/g, respectively while counts about deterioration ranged from 1.0 x 10^4 to 3 x 10^4 sfu/g, 4.5 x 10^4 to 9 x 10^4 sfu/g, 1.0 x 10^4 to 5 x 10^4 sfu/g.

The mean results of the total count of heterotrophic fungi per gram of flesh of shrimps from Creek Road market, Mile One market and Town market before deterioration ranged from 6.5 x 10^4 to 3 x 10^4 sfu/g, 4.1 x 10^4 to 2 x 10^4 cfu/g, and from 6.5 x 10^4 to 3 x 10^4 sfu/g while fungal counts in flesh of shrimps about deterioration ranged from 5.1 x 10^4 to 9 x 10^4 sfu/g, 3.1 x 10^4 to 6.0 x 10^6 sfu/g and from 5.1 x 10^4 to 9 x 10^{-4} sfu/g, respectively.

The mean counts of the total pathogenic fungi per gram of head, exoskeleton and telson/uropod (HET) of shrimps from Creek Road market, Mile One market and Town market before deterioration ranged from 4.8 x 10^4 cfu/g to 8.0 x 10^4 cfu/g, 3.0 x 10^4 cfu/g to 7.0 x 10^4 cfu/g and from 4.8 x 10^4 cfu/g to 8.0 x 10^4 cfu/g respectively. While counts in HET of shrimps about deterioration ranged from 3.6 x 10^4 cfu/g to 6.0 x 10^4 cfu/g, 1.4 x 10^4 cfu/g to 6.0 x 10^4 cfu/g, and from 3.6 x 10^4 to 6.0 x 10^4 cfu/g, respectively.

The mean results of the total count of pathogenic fungi per gram of flesh of shrimps from Creek Road market, Mile One market and Town market before deterioration ranged from 1.0 x 10^4 to 2.3 x 10^4 cfu/g, 4.1 x 10^4 to 8.0 x 10^4 cfu/g, and from 1.0 x 10^4 to 2.3 x 10^4 cfu/g, respectively. While counts in the flesh of the shrimps about deterioration ranged from 5.0 x 10^4 to 5.2 x 10^4 sfu/g, 1.5 x 10^4 to 5.0 x 10^4 sfu/g and from 5.2 x 10^4 to 9.0x 10^4 sfu/g, respectively.

Analysis of variance (F-distribution) showed that generally, except for the fungal count in the flesh of the shrimps before deterioration which showed significant difference at $P \geq 0.05$ (with Mile One shrimps recording the highest counts); there is no significant difference in the fungal count of shrimps between the various markets.

The frequencies of occurrence (%) of heterotrophic and pathogenic fungi in shrimps from the markets are as shown in Tables 3 and 4, respectively. The heterotrophic fungi and their percentage occurrence in the H/E/T (head, exoskeleton and telson/uropod) before deterioration and about deterioration were: *Aspergillus clavatus* (20%), *Aspergillus flavus* (20%), *Penicillium* sp. (20%), *Rhizopus* sp. (20%), *Rhodotorula* sp. (20%), and *Aspergillus flavus* (28.6%), *Mucor hiemalis* (14.3%), *Penicilllium* sp. (14.3%), *Rhizopus* sp. (14.3%), *Rhizopus stolonifer* (14.3%), yeast sp. (14.3%), respectively. While the pathogenic fungal occurrence before deterioration and about deterioration in the H/E/T were: *A. clavatus* (11.1%), *A. flavus* (22.2%), *Penicillium* sp. (33.3%), *R. oryzae* (11.1%), *Rhodotorula* sp. (11.1%), yeast sp. (11.1%), and *A. clavatus* (12.5%), *A. flavus* (12.5%), *Mucor hiemalis* (12.5%), *Penicillium* sp. (12.5%), *Rhodotorula* sp. (12.5%) and yeast sp. (37.5%),

Table 3. Occurrence (%) of heterotrophic fungi in shrimps from the markets.

Fungus	Head, exoskeleton and telson		Flesh	
	Before deterioration	About deterioration	Before deterioration	About deterioration
Aspergillus clavatus	20			
Aspergillus flavus	20	28.6	33.3	22.2
Aspergillus niger				11.1
Mucor heimalis		14.3		
Mucor plumbeus				11.1
Penicillium sp.	20	14.3	33.3	22.2
Phialophora fastigiata				11.1
Rhizopus stolonifer		14.3		11.1
Rhizopus sp.	20	14.3		
Rhodotorula sp.	20			11.1
Yeast sp.		14.3	33.3	

Table 4. Occurrence (%) of pathogenic fungi in shrimps from the markets

Fungus	Head, exoskeleton and telson		Flesh	
	Before deterioration	About deterioration	Before deterioration	About deterioration
Aspergillus clavatus	11.1	12.5		
Aspergillus flavus	22.2	12.5	28.6	
Aspergillus niger				10
Mucor heimalis		12.5		10
Mucor plumbeus				20
Penicillium sp.	33.3	12.5	14.3	10
Rhizopus oligosporus			28.6	11.1
Rhizopus oryzae	11.1			
Rhizopus stolonifer			14.3	10
Rhodotorula sp.	11.1	12.5		20
Yeast sp.	11.1	37.5	14.3	20

respectively. Heterotrophic fungi in the flesh before deterioration and about deterioration were, *A. flavus* (33.3%), *Penicillium* sp. (33.3%), yeast sp. (33.3%) and *Aspergillus flavus* (22.2%), *Aspergillus niger* (11.1%), *Mucor plumbeus* (11.1%), *Penicillium* sp. (22.2%), *Phialophora fastigiata* (11.1%), *Rhizopus stolonifer* (11.1%) and *Rhodotorula* sp. (11.1%) respectively. While the pathogenic fungal occurrence before deterioration and about deterioration in the flesh were *Aspergillius flavus* (28.6%), *Penicillium* sp. (14.3%), *Rhizopus oligosporus* (28.6%), *Rhizopus stolonifer* (14.3%), yeast sp. (14.3%), *Aspergillus niger* (10%), *Mucor hiemalis* (10%), *Mucor plumbeus* (20%), *Penicillium* sp. (10%), *Rhizopus stolonifer* (10%), *Rhodotorula* sp. (20%), and yeast sp. (20%) respectively

Among the heterotrophs isolated, only *A. flavus* and *Penicillium* sp. were isolated from the H/E/T and flesh, respectively in all three (3) markets. Among the Pathogenic fungi isolated from the H/E/T and flesh, only yeast sp., *Penicillum* sp., *R. stolonifer,* and *A. flavus* were isolated from the H/E/T in all three (3) markets, while others were isolated from either one or two of the three markets. *Phialophora fastigiata* was isolated from flesh of shrimp about deterioration from only Mile One market.

DISCUSSION

In the present study a number of fungal species recorded found to be associated with the head, exoskeleton, tail/telson (H/E/T) and flesh of shrimps before and about deterioration. The different fungal genera isolated were *Aspergillus, Mucor, Penicillium, Phialophora, Rhizopus,* and yeast spp including *Rhodotorula*. These fungal genera appear to be associated with the spoilage of fresh shrimps. These fungi have also been associated with spoilage of fruits Okigbo and Obire, 2009).

The fungal counts recorded in the shrimps are considered very high ranging from 1.7×10^4 sfu/g to 7.1×10^4

sfu/g. Generally in all the markets, the population or mean count of total heterotrophic fungi in the H/E/T were higher before deterioration Except for Mile One market, counts were lower in the flesh of the shrimps before deterioration than about deterioration. The high fungal count is an indication of gross contamination of the environment where the shrimps were harvested and of the ecological niche where the fungi has developed. The results of both heterotrophic and pathogenic fungal counts were almost exactly the same in shrimps from the Creek Road and Town markets. This is not surprising because the distance between both markets is less than 200 m; in-fact, they have virtually merged together. It is also a fact that shrimps from both market were harvested from the same environment. The isolation of *Phialophora fastigiata* from only Mile One market indicated that shrimps from this market were harvested from a different environment and hence the observed disparity in the fungal counts of shrimps from the three markets.

There was also an increase in the types of heterotrophic and pathogenic fungi in the shrimps about deterioration. These observations clearly indicate the utilization and exhaustion of nutrients in the shrimps and change in prevailing conditions such as pH which in turn resulted in the succession of fungi over time during deterioration of the shrimps

Among the pathogenic fungi isolated *Aspergillus, Penicillium* and yeast were more prevalent in H/E/T from all the markets, while *Aspergillus flavus, Mucor, Rhizopus oligosporus* and yeast sp. were prevalent in the flesh. Apart from some *Rhizopus* species frequency of isolation of the fungi were higher in the H/E/T than in the flesh of the shrimps. This shows that the presence of these fungi in the shrimps is as a result of direct contamination from the environment and shrimp handlers.

Statistical analysis showed that generally, except for the fungal count in the flesh of the shrimps before deterioration which showed significant difference at $P \geq 0.05$ (with Mile One shrimps recording the highest counts), there is no significant difference in the fungal count (heterotrophic and pathogenic) of shrimps between the various markets. This is a proof that pathogenicity of some fungi is not as a result of their numbers but in the production of mycotoxins as most of the isolates in this investigation produce mycotoxins.

The potential for disease outbreaks attributed to sea food ingestion has been recognized for over a century (Richards, 1985). It is important to state here that most of the fungi are potential pathogens. Some diseases caused by these fungi in humans are aspergillosis, liver tumor, fungal balls in the lungs, chronic productive cough and haemoptysis, bronchial asthma, infection of the ear or paranasal sinuses, fungal cells within histiocytes, necrosis and eventual abscess formation, multiple brain abscesses and Verucose dermatitis, a chronic human mycosis (Richards, 1985; Okigbo and Ajalie, 2005; Okigbo et al. 2009). Others include rhinocerebral mucor-mycosis, infection of the nasal turbates and paranasal sinuses spreading rapidly to the eyes and brain; necrosis and thrombosis and invasive mucormycosis (Singleton and Sainsbury, 2001).

Conclusion and recommendation

The presence of fungal species in the shrimps is also attributed to contamination from shrimp handlers or processors or sellers (mongers). The fungi isolated from the shrimps in this study are potential pathogens and they are capable of causing chronic illnesses in humans upon ingestion of food contaminated by them.

Owing to the health hazards associated with the fungi isolated from the brown shrimps used for this study, it is important that brown shrimps and other sea food product be properly and adequately cooked. Also unfavourable conditions should be created to prevent fungal growth, examples of such conditions are: regulation of the water holding capacity of the food (a_w), temperature and pH. Adjustments in these areas will help prevent mycotoxin contamination of the shrimp as moulds have strains which produce toxins that are capable of causing serious chronic illnesses (e.g. liver tumor caused by *A. flavus)* in humans, if consumed.

Being that some fungi contaminated the shrimps through shrimps handlers or processors, there is therefore the need for the adoption, practice, and maintenance of good personal hygiene as regards handling of the shrimp from the moment of catch till it reaches the consumer, so as to ensure good quality and long storage life of the shrimp. The maintenance of high personal and environmental hygiene as well as proper heating and cooking will improve fresh shrimp quality and prevent food-borne diseases.

REFERENCES

Dall W (1968). Food and Feeding of some Australian penaeid shrimp. Proc. World Science Conference on Bio. and Cult. Of shrimps and prawns. FAO Fisheries Reports 57(2): 251-258.

Harrigan WF, McCance ME (1990). Laboratory Methods in Food and Dairy Microbiology. 8th Edition. Academic Press, London.

ICMSF (International Commission on Microbiology Specifications for foods) 1996. Microorganisms in Food 5. Characteristics of Microbial pathogens. Blackie Academy & Professionals.

Jones RR (1973). Utilization of Louisiana estuarine seiments as a source of nutrition for the brown shrimps, Penaeus aztecus. Ph.D. Dissertation, Louisiana State University, Baton Rouge.

Malloch D (1997). Moulds Isolation, Cultivation and Identification, Department of Botany University of Toronto, Toronto USA.

Odum WE, Herald E (1972). Trophic analysis of an estuarine mangrove community. Bull. Mar. Sci. 22(3): 671-738.

Okigbo RN, Anuagasi CL, Amadi JE, Ukpabi UJ (2009). Potential inhibitory effects of some African tuberous plant extracts on *Escherichia coli, Staphylococcus aureus* and *Candida albicans*. Int. J. Integr. Biol. (India) 6 (2):91-98

Okigbo RN, Ajalie AN (2005). Inhibition of some human pathogens with tropical plants extracts *Chromolinaena odorata* and *Citrus aurantifolia* and some antibiotics. Int. J. Mol. Med. Adv. Sci. (Pakistan) 1 (1): 34 - 40.

Okigbo RN, Obire O (2009). Mycoflora and production of wine from fruits of soursop (*Annona muricata* L.) Int. J. Wine Res. (Auckland, New Zealand) 1: 1-9. Available online at http://www.dovepress.com/articles.php?article_id=2731

Richards GP(1985). Outbreak of shellfish-associated enteric virus illness in the United States: Requisite for development of viral guidelines. J. Food Prot. 48: 815 - 823.

Singleton P, Sainsbury D (2001). Dictionary of Microbiology and Molecular Biology. 3rd Edn. John Wiley and Sons Ltd. New York. pp. 895.

Williams, A. B (1984). Shrimps, lobsters and crabs of the Atlantic coast of the Eastern United States, Marine to Florida. Smithsonian Institution Press. Washington, D C. pp. 550.

Towards food security and improved nutrition in Nigeria: Taro (*Colocacia antiquorum*) grit as carbohydrate supplement in energy food drink

Onyenekenwa Cyprian Eneh

Institute for Development Studies, Enugu Campus, University of Nigeria, Nsukka, Enugu State, Nigeria.

Nigeria's population growth rates (3.2% in the late 1980s, 2.83% in the early 1990s and 3% in the mid-2000s) have been high, while subsistence farming, characterised by inadequacies, low productivity and high post-harvest losses, dictates rising food insecurity and dwindling nutrition in the country. Taro (*Colocasia antiquorum*)- a widely cultivated cocoyam, has little demand as food and is vulnerable to post-harvest rot. Towards addressing food insecurity and improving nutrition, this study investigated the potentials of taro as carbohydrate supplement in energy food drink. Chemical analyses were employed to obtain the compositions of prepared taro grit and its energy food drink prepared with sorghum malt grist. A randomly selected panel of tasters appraised the sensory qualities of the prepared taro energy food drink, while a public analyst assessed its fitness to be sold as energy food drink. Results show that the proximate percentage compositions of taro grit were 77.93% starch and 0.63% fat, among others. These were comparable with the literature values of 63-79% starch and 1.8-2.9% fat for cocoa- a popular carbobydrate supplement in common commercial energy food drinks. The malted sorghum grist had CWE value of 25.82 and HWE value of 59, which fell within the literature acceptable ranges of 16.07-26.54 (CWE) and 17.5-63.1 (HWE). The taro energy food drink was water-soluble and had similar physico-chemical characteristics and compositions to those of "Ovaltine" (a product of Cadbury Nigeria Plc), which was used as standard. The panel of tasters could not distinguish the aroma and flavour of the standard from those of the taro energy food drink, which was also adjudged suitable to be sold as energy food drink by the public analyst. It was recommended that the bench-scale production of taro energy food drink should be scaled up to pilot level with a view to commercialising it, in order to minimise post-harvest losses of taro to rot; save storage difficulties, cost and time; increase the nutritional value, presentations and demand for taro; and reduce the price to enhance the affordability of energy food drink in Nigeria.

Key words: Food security, improved nutrition, post-harvest loss, Nigeria, taro, carbohydrate supplement, energy food drink.

INTRODUCTION

The smallness of cocoyam starch grains aid digestion, and qualifies cocoyam as a composite in the manufacture of infant meals, convalescing patients requiring easily diggestible carbohydrate as a source of energy without cumbersomeness on the digestive system and stress on the metabolic process. Thus, cocoyam starch is good for peptic ulcer patients, patients with pancreatic disease, chronic liver problems, inflammatory bowel disease and gall bladder disease (Emmanuel-Ikpemel et al., 2007).

Onwuka and Eneh (1998) reported that Nigeria is one of the leading producers of root crops, including cocoyam, which is one of the under-exploited tropical plants with promising value. Its cultivars are Tannia (*Xanthosoma*) and Taro (*Colocasia*). Widely cultivated varieties in Nigeria are *Xanthosoma sagittifollium, Colocasia esculenta* and *Colocasia antiquorum.* All three varieties are rich in easily digestible starch grains. The percentage carbohydrate fractions of taro are starch (77.9), pentosans (2.6), crude fibre (1.4), reducing sugars (0.5), dextrin (0.5) and sucrose (0.1). The starch has industrial potentialities. Taro also contains a host of mineral elements, such as calcium, phosphorus, potassium, chlorine, magnesium, sulphur, iron, copper, manganese and zinc. Besides, taro has certain medicinal values and has been used as a specialty food for potentially allergic infants and for persons suffering from gastro-intestinal disorders. The study also noted that cocyam cultivation, harvest and preparation involve less labour and fewer processes than those of other tuber crops. The demand for *C. antiquorum* as food was still poor, even though it could be boiled, baked, roasted, fried, or boiled and pounded into paste (*fufu*) for food or as soup thickener. Cocoyam flour is highly digestible, and therefore, suitable for feeding invalids, for making confectionaries and baby food.

Food and Agricultural Organisation (FAO, 2010) reported that the world total output of taro was 11.3 milliom metric tons (mmt) in 2009. Nigeria was the world leading producer of taro with 4.4 mmt, followed (with a wide margin) by China and Cameroon (1.7 mmt each), Ghana (1.5 mmt) and Papua New Guinea (0.3 mmt). All the other countries put together contributed the remaining 1.7 mmt.

But, the crop yields remain under-exploited, with vulnerability to post-harvest rot in Nigeria, where population growth rates have been high at 3.2% in the late 1980s, 2.83% in the early 1990s and 3% in the mid-2000s. The character of agricultural practice, fraught with subsistence farming, low productivity and high post-harvest losses, dictates rising food insecurity. Nigeria's national average figure for household difficulty satisfying food need was 12.7% in 2005, with the Southest geo-political zone recording the highest figure of 25.4% (Eneh, 2011a).

The UNDP Human Development Index (HDI) Report 2010 ranked Nigeria 142nd out of 169 countries. Mauritius- another African country was ranked 72nd and has life expectancy of 72.1 years, while Nigeria has 48.4 years. Other African countries that ranked ahead of Nigeria were Tunisia (81st), Algeria (84th), Gabon (93rd), Egypt (101st), Namibia (105th), South Africa (110th), Morrocco (114th), Equatorial Guinea (117th), Cape Verde (118th), Congo (126th), Sao Tome and Principe (127th), Kenya (128th), Togo (129th), Ghana (130th), Cameroon (131st), and Benin Republic (134th). In terms of life expectancy, Togo (63.3 years), Benin Republic (62.3 years), Ghana (57.1 years) and Cameroon (51.7 years) were also better than Nigeria, the most populous country in sub-Sahara Africa, the region with the lowest HDI indicators. The population of the extremely poor Nigerians rose from 6.2 to 38.7 %, while non-poor decreased from 72.8 to 31% between 1980 and 2010 (NBS, 2012).

Absolute poverty, which measures not only low income, malnutrition, poor health, clothing, shelter, and lack of education, is reflected in the low living standards of Nigerians. The average daily calorie intake per capita hardly exceeds 2,000 in Nigeria (Eneh, 2011b). Food insecurity and malnutrition are, therefore, issues. National infant and maternal mortality rates are still high at 105 per 1,000 and 800 per 100,000 live births, respectively (Ugwu, 2009).

Availability and affordability of energy food drink will contribute to good nutrition and food security. Standard energy food drinks are made with malted cereal, which is rich in carbohydrate (energy source), proteins (amino acids), many vitamins and enriched with minerals. Cocoa (grit) is the widely used supplementary carbohydrate (energy) source. The rich carbohydrate and nutritional value of malt is released when the malt interacts with heat and water. Common energy food drinks contain more than 50% natural malt. Energy value is about 413 kcal/100 g. The average market price is N1.25/g for canned products and N1.15/g for sachet products (personal observations and market survey).

Sorghum, a cereal cultivated in commercial quantity in Nigeria has been malted and used for beer brewing (Onwuka and Eneh, 1998). It is suspected that sorghum malt can be applied to energy food drink production and that taro can replace cocoa as a carbohydrate supplement in the drink. This would minimise post-harvest losses of taro to rot; save storage difficulties, cost and time; increase the nutritional value, presentations and demand for taro; and lower the price to enhance affordability of energy food drinks.

So far, much of taro harvest is wasted in Nigeria and other developing countries to technological backwardness characterising the developing countries (Jhingan, 2007), which hampers food preservation and processing. Globally, about one-third of annual food production (1.3 billion tons) goes to post-harvest losses.

In West Africa, about 50% of perishable food commodities are lost after harvest. Post-harvest food loss is estimated at 25% per year in Nigeria (Aworh, 2008). It militates against food security and availability (Mrema and Rolle, 2002). If not addressed, a high-calorie and nutritious crop harvest can rot to deplorable waste, thereby limiting the supply and increasing the cost of food. As a post-harvest loss, rot wastes food, human efforts, farm inputs, livelihoods, investments and scarce resources. Enhancing food processing and preservation will minimise post-harvest losses in developing countries, where seasonal food shortages and nutritional deficiency diseases are still a major concern (World Resource Institute, 1998).

Emmanuel-Ikpemel et al. (2007) observed that physical, physiological and pathological factors are responsible for the post-harvest rot of cocoyam tubers. *Fusarium solani*, *Botryodiplodia theobromae*, *Rhizopus stolonifer*, *Aspergillus Niger* and *Sclerotium rolfsii* participate in the spoilage. The high respiratory activity of the corm also causes spoilage of cocoyam.

The need to address food insecurity and improved nutrition for the growing population of Nigerians motivated this study. Specifically, it investigated the potentials of *C. antiquorum* grit as a carbohydrate supplement in energy food drink produced with sorghum malt.

MATERIALS AND METHODS

The study was conducted between December 2005 and August 2012 in Enugu, Enugu State, Nigeria. Enugu State is one of the 5 States of the Southeast geo-political zone of Nigeria. This zone is a leading producer of cocoyam among the 6 zones in the country. Sorghum grain is produced in commercial quantity in neighbouring Kogi and Benue States. *C. antiquorum* and white sorghum variety (*Sorghum vulgare)* were bought from the Ogbete Market, Enugu, the capital city of Enugu State and the erstwhile headquarters of the defunct Eastern Region. The Vitamins Mixture Pack was also bought from a licensed veterinary shop in the city. The standard energy drink ('Ovaltine', a product of Cadbury Nigeria Plc) was bought from an accredited distributor for Cadbury Nigeria Plc in Enugu city. The chemicals (all of analytical reagent grade), apparatus and equipment were available in the laboratory of the Department of Medical Biochemistry, College of Medicine, Enugu Campus, University of Nigeria, Nsukka.

Preparation of taro grit

The procedure adopted by Onwuka and Eneh (1996) was followed.

Malting of sorghum grains and preparation of sorghum malt grist

The procedure adopted by Onwuka and Eneh (1998) was followed.

Determination of carbohydrate content of taro grit

The method adopted by Onwuka and Eneh (1996) was followed.

Determination of reducing sugars content of taro grit

The reducing sugars content of taro grit was determined by the method adopted by Onwuka and Eneh (1998).

Determination of ether extract of taro grit

The ether extract of taro grit was determined according to the method adopted by Onwuka and Eneh (1996).

Determination of fibre content of taro grit

Fibre content of taro grit was determined by the method adopted by Onwuka and Eneh (1998).

Nitrogen and protein content determination in taro grit

The nitrogen and crude protein in taro grit were determined by the method adopted by Onwuka and Eneh (1996).

Cold water extract (CWE) of sorgum malt Grist

The Onwuka and Eneh (1998) method was adopted.

Hot water extract (HWE) of sorgum malt Grist

The Onwuka and Eneh (1996) method was adopted.

Preparation of taro energy food drink

Several ratios of taro grit and sorghum malt grist were tried out to select one with sensory evaluation that approximated that of the commercial standard. To 97 g of this mixture was added 3 g of vitamins from the Vitamins Mixture Pack to get the taro energy food drink.

Water-solubility test for taro energy food drink

Fifteen grams of the taro energy food drink was stirred in 100 ml of water at room temperature (30°C) to see if it would dissolve.

Analysis of the taro energy food drink

The prepared taro energy food drink was sent to a public analyst for assessment and remarks.

Sensory evaluation of taro energy food drink

Using a commercial energy food drink ('Ovaltine' by Cadbury Nigeria Plc) as the standard, which was set at 100%, the prepared taro energy food drink was evaluated by staff and students at the University of Nigeria, Enugu Campus, whose consent had been sought and obtained. Aroma and flavour were the sensory attributes evaluated.

RESULTS AND DISCUSSION

Steeping and couching the sorghum grains resulted in sweet grains. This was in agreement with earlier findings. reproted that steeping and couching grains increased the moisture content to 25 to 40% to induce germination and to develop or activate enzyme systems important in subsequent commercial uses of the malt. Sweetness and softness indicated good seedlings modifications, characteristic of food preparations contain-ing malts and malt extracts. Kilning dried the green malt to 4-6% moisture content to stop further growth of the seedling. It also

Table 1. Proximate analysis of taro grit (%).

	Nitrogen	Crude protein	Fibre	Ether extract	Reducing sugars	Total carbohydrate
Taro grit	1.38	8.13	2.69	0.63	33.55	77.93

Table 2. Sensory evaluation of taro energy food drink.

Aroma	Rating with standard	Flavour	Rating with standard
Beverage-like	99.98%	Beverage-like	84.68%

The term, 'beverage,' used by tasters, refers to commercial energy food drinks in Nigeria made from cocoa and malt.

resulted in the development of a control-lable degree of colour. The Maillard reaction during kilning resulted in the formation of amino acid-carbonyl compounds, which underwent further transformation to yield the desired flavour, colour and aroma of the incompletely characterized compounds known as the melanoidins. In a hydrolysis process, the enzyme (dias-tase) or acid converted the carbohydrate (starch) to sweet fermentable sugars (e.g. maltose, a dissacharide), as respresented in the equation below:

$$\underset{\text{(starch)}}{2C_6H_{10}O_5 + H_2O} \xrightarrow[\text{(acid or diastase)}]{\text{Hydrolysis}} \underset{\text{(maltose, a sugar)}}{C_{12}H_{22}O_{11}}$$

The heat-dried *C. antiquorum* turned brown. After milling, the touch, odour and appearance of the taro grit agreed with earlier findings. Onwuka and Eneh (1998) reported that, on heating, a mixture of sugar and amino acids rapidly darkened, liberating carbon (IV) oxide (CO_2) gas and water. Heating treatment introduced desired colour, aroma and flavour into the carbohydrate substance by removal of water to make it friable. This process is also called caramelisation. Where dark colour is desired, special attention is given to the maintenance of the autolysis period with sufficient temperature. 'Stewing' favours darker products. The resulting colouring and flavouring materials, called melanoidins, are water-soluble colloidal materials of characteristic aroma and flavor, and weakly acidic. Having the power to act as oxidation-reduction buffers, and to enhance the stability of the product, they are representatives of the class of stabilising reductones.

The odour and appearance of the taro grit and its energy food drink persisted beyond five (5) years of storage in a tightly covered bottle at normal room temperature, while the touch was lost after one-year storage under the same conditions. Foods loose one characteristic or the other after a period of storage, hence regulatory agencies require that 'expiry' and 'best before' or 'use before' dates should be inscribed on food packages to guide consumers. Therefore, the 'best before' or 'use before' date for taro energy food drink should reflect one year gap from manufacturing date.

The cold water extract (CWE) value is a percentage measure of ready-formed soluble material in malt. The CWE of the sorghum malt grist was 25.82, which fell within the acceptable range of 16.07-26.54 (Onwuka and Eneh, 1998). The hot water extract (HWE) value is a measure of all materials brought into water solution by enzymic action in malt grist. The HWE value of the sorghum malt grist was 59, which fell within the acceptable range of 17.5-63.1 (Onwuka and Eneh, 1998). Like the the standard, the taro grit, sorghum malt powder and taro energy food drink were water-soluble. Table 1 shows the data on the proximate analyses of taro grit.

The total carbohydrate content of taro grit (77.93%) was comparable to carbohydrate contents of cocoa (63-79%) quoted in common literatures (Onwuka and Eneh, 1998, 1996), showing that taro has the potential of providing the carbohydrate supplement in energy food drink. The fat content (ether extract) of taro grit (0.63 %) was also comparable to the fat contents of cocoa (1.8-2.9%) quoted in literatures (Onwuka and Eneh, 1998, 1996), showing that taro is not only suitable for carbohydrate supplement, but also much less disposed to rancidity than cocoa. The percentage reducing sugars content (33.55), nitrogen (1.38) and crude protein (8.13) of taro grit fell within acceptable range for carbohydrate supplement in energy food drink, when compared with cocoa's 63-7% (starch) and 1.13-2.10% (fat) (Onwuka and Eneh, 1998, 1996).

Table 2 shows the sensory evaluation data of the taro energy food drink. The aroma (99.98%) and flavour (84.68%) were excellent. The report of the public analyst on the analyses of taro energy food drink is shown on Table 3. It is subdivided into physical inspection, composition and remarks. Each parameter was compared with that of the universal standard. All parameters for physical inspection of the taro energy food drink were found to be similar to those of the universal energy food drinks. While the physical appearance and colour of taro energy food

Table 3. Report on the analysis of taro energy food drink.

Parameter		**Taro energy food drink**	**Universal standard**
Physical Inspection			
1.	Appearance	Brown coloured grit	Brown to deep amber coloured grit
2.	Odour	Similar to energy food drink	Pleasant
3.	Colour	Brown	Brown to deep amber colour
4.	Texture	Crispy	Crispy
Compositions			
1.	Energy	1,733 kJ (411 kcal) /100 g	1,501 (356 kcal) - 1,792 kJ (425 kcal)/100 g
2.	Protein	9.97 %	8-13%
3.	Carbohydrate	82.18 %	74-85%
4.	Sugars	47.78 %	42-49%
5.	Fat	2.99 %	2.5-11%
6.	Fibre	2.10 %	1.3-3.7%
7.			
8.	Minerals	Sodium (0.3 %) Calcium (0.1 %) Phosphorus (0.25 %) Magnesium (0.13 %) Iron (0.15 %) Zinc (0.06 %)	Sodium (0.2-0.35%) Calcium (0.08-0.14%) Phosphorus (0.18-0.3%) Magnesium (0.1-0.22%) Iron (0.09-0.2%) Zinc (0.03-0.09%)
9.	Vitamins/100 g	A (0.8 mg) B_1 (1.8 mg) B_2 (1.7 mg) B_6 (1.5 mg) B_{12} (0.001 mg) D (0.006 mg) E (3.6 mg) Pantothenic acid (15 mg) Follic acid (0.004 mg) Niacin (22.9 mg)	A (0.5-1.3 mg) B_1 (1.2-2.2 mg) B_2 (1.1-2.3 mg) B_6 (0.9-2.1 mg) B_{12} (0.0007-0.008 mg) D (0.001-0.009 mg) E (2.9-4.2 mg) Pantothenic acid (9-18 mg) Follic acid (0.001-0.009 mg) Niacin (22-23.1 mg)

drink were brown, the universal standard is brown to dark amber. The odour of the taro energy food drink was found to be as pleasant as the odour of the universal standard. Like the universal standard energy food drink, the texture of the taro energy food drink was crispy.

The compositions of the taro energy food drink compared favourably with those of the universal energy food drinks. The energy value of the taro energy food drink was 411 kcal, against 356-425 kcal universal range of values.

The protein content was 9.97%, against 8-13% universal standard. The total carbohydrate content was 82.18%, against 74-85% universal range of values. The sugars content was 47.78%, against 42-49% universal standard. The fat content was 2.99%, against 2.5-11% universal range of values. The fibre content was 2.1%, against 1.3-3.7% universal standard.

The minerals content also compared favourably at 0.3% against 0.2-0.35% universal standard for sodium, 0.1% against 0.08-0.14% universal range of values for calcium, 0.25% against 0.18-0.3% universal standard for phosphorus, 0.13% against 0.1-0.22% universal range of values for magnesium, 0.15 % against 0.09-0.2% universal standard for iron, and 0.06% against 0.03-0.09% universal range of values for zinc. Also, the vitamin contents compared favourably with those of the universal values. The taro energy food drink had 0.8 mg of vitamin A, against 0.5-1.3 universal standard; 1.8 mg vitamin B_1, against 1.2-2.2 mg universal range of values; 1.7 mg of vitamin B_2, against 1.1-2.3 mg universal standard; 1.5 mg of B_6, against 0.9-2.1 mg universal range of values; 0.0001 mg B_{12}, against 0.0007-0.008 mg universal standard; 0.006 mg of D, against 0.001-0.009 mg universal range of values; 3.6 mg of E, against 2.9-4.2 mg universal standard; 15 mg pantothenic acid, against 9-18 mg universal range of values; 0.004 mg follic acid, against 0.001-0.009 mg universal standard; and 22.9 mg niacin, against 22-23.1 mg universal range of values. Hence, it was remarked that taro energy food drink was "suitable to be sold as energy food drink".

Malted cereals also contain water-soluble vitamins, including vitamin B, ascorbic acid, pantothenic acid, biotin, niacin, nicotineamide, riboflavin, anuerin, among other nutrients that enrich energy food drink preparations. A good quality energy food drink is high in carbohydrate content with easily digestible sugars to produce energy, and low fat content to minimize rancidity. These qualities were contributed to the taro energy food drink by the taro grit, since taro is rich in digestible carbohydrate (77.93% comparable to 63-79% carbohydrate content of cocoa) and has lower fat content (0.63%) than cocoa (1.8-2.9%).

Remarks

The product anlysed, which is similar in appearance and composition to commercial energy food drinks, is deemed suitable to be sold as energy food drink.

Conclusions and recommendations

Based on the results obtained, taro (*C. antiquorum*) variety of the cocoyam has high potentials in terms of its suitability and availability to supplement the carbohydrate content of energy food drink. This bench-scale study needs to be scaled up to pilot level, with a view to commercialising the production of taro energy food drink. This would enhance the nutritional value of taro, increase its availability and varieties, increase its demand for food, minimise its post-harvest losses to rot, save storage difficulties and time, and reduce the price to enhance affordability of energy food drinks.

REFERENCES

Aworh OC (2008). Using Food Science and Technology to Improve Nutrition and Promote National Development. In Robertson GL, Lupien JR (eds.), International Union of Food Science and Technology, Chapter 3.

Emmanuel-Ikpemel CA, Eneji CA, Essiet U (2007). Storage Stability and Sensory Evaluation of Taro Chips Fried in Palm Oil, Palm Olein Oil, Groundnut Oil, Soybean Oil and Their Blends. Pak. J. Nutr. 6(6): 570-575.

Eneh OC (2011a). Crippling poverty amidst corporate social actions: a critique of peripheral corporate community involvement in the Niger Delta region of Nigeria. Asian. J. Rural Dev. 1(1): 1-20.

Eneh OC (2011b). Nigeria's Vision 20:2020 - Issues, Challenges and Implications for Development Management. Asian. J. Rural Dev. 1(1):21-40.

FAO (Food and Agricultural Organisation) (2010). Top Taro Producers - 2009. Available at http://www.faostatunfoodandagriculturalorganisation Retrieved September 19, 2012.

Jhingan ML (2007). The Economics of Development and Planning. 39ed. Delhi: Vrinda Publications (P) Ltd.

Mrema CG, Rolle SR (2002). Status of the postharvest sector and its contribution to agricultural development and economic growth. Proceedings of the 19th JIRCAS International Symposium on value-addition to agricultural products. Ibaraki, Japan. pp. 13-20.

NBS (National Bureau of Statistics) (2012). Nigerian Poverty Index 2010 Report. Available at http://www.nigerianpovertyprofilereport2010.NBS.pdf Accessed, 5/14/2012.

Onwuka ND, Eneh OC (1996). The Cocoyam, *Xanthosoma sagittifollium*, as a Potential Raw Material Source for Beer Brewing. Plant Foods Hum. Nutr. 49(4):283-293.

Onwuka ND, Eneh OC (1998). The Potential of Cocoyam (Colocesia) in Stout-Beer Brewing. J. Sci. Technol. 4:79-86.

Ugwu DS (2009). Effects of Oil Exploration on Agriculture and Natural Resources in the Niger Delta Region of Nigeria. Sustain. Hum. Dev. Rev. 1(2):117-130.

World Resource Institute (1998). Disappearing food: How big are postharvest losses? Available at http://www.wri.org/publication/content/8386.

Synergistic fermentative nutritional quality of *Lactobacillus delbrueckii* and *Bacillus pumilus* on date fruits (*Phoenix dactylifera*)

Bankefa, Emmanuel Olufemi[1] and Oyedeji, Olumide Murtala[2]

[1]Federal University, Oye, Ekiti, Ekiti State, Nigeria.
[2]Federal University of Technology, Akure, Ondo State, Nigeria.

The work examined the effect of mixed cultural fermentation on the nutritional quality of date palm fruits (*Phoenix dactylifera*). This was done by isolating microorganism from samples subjected to natural fermentation. The isolated organisms were thereafter used as starter culture in a five day controlled fermentation. *Lactobacillus delbrueckii* and *Bacillus pumilus* were identified as the predominant microorganisms and were used as starter culture. pH, titratable acidity, proximate, amino acid and mineral composition of the fermented samples were determined. The sample fermented with mixed culture gave best nutritional value of 15.3% protein as compared to the 9.62% and 11.61% yield in *B. pumilus* and *L. delbrueckii*, respectively, on the fifth day of fermentation. Arginine was found to be on the high side among the amino acids discovered and the mineral composition of the products gave an appreciable amount of sodium, potassium and calcium.

Key words: Fermentation, Nutritional quality date palm fruits, *Lactobacillus delbrueckii*, *Bacillus pumilus*.

INTRODUCTION

Fermentation is a metabolic process that converts sugar to acids, gases and/or alcohol. Fermentation is also used more broadly to refer to the bulk growth of microorganisms on a growth medium. Sugars are the most common substrate of fermentation, and typical examples of fermentation products are ethanol, lactic acid, carbon dioxide and hydrogen gas (H_2) (El-Sohaimy and Hafez, 2010). However, more exotic compounds can be produced by fermentation, such as butyric acid and acetone. Lactic acid fermentation is commonly used in many parts of the world as a method for preservation of plant materials as well as importing desirable sensory and nutritional properties to the product (Cooke et al., 1987). Fermentation has been reported to improve the nutritional values, taste and aroma (Adegbehingbe et al., 2014).

Date fruit (*Phoenix dactylifera* L.) is a palm in the genus *Phoenix* belonging to the *Arecaceae* family, cultivated for its edible sweet fruit. Dates, is identified by its clusters of oval, dark, reddish-brown drupe with the skin of the dried fruit wrinkled and covered in a sticky waxy film. The date palm (*P. dactylifera* L.) is one of mankind's oldest cultivated plants and has been used as food for about 6000 years (Sahari et al., 2007). It is an important food crop in Middle East and is considered to be one of the most important fruit tree particularly in North African, the Middle Eastern and Asian countries. The fruit contributes to the economy and social life within these regions (Bastway et al., 2008) and it is considered as a vital component of their diet (Vayalil, 2002). Date fruits are well known as a staple nutritious food and source of wealth for many years (Khan et al., 2008). Due to its high nutritional value, great yields and its long life, the date palm has been mentioned as the "tree of life" (Augstburger et al., 2002). Date fruits are considered as major source of carbohydrate which include simple sugars like glucose and fructose (Ahmed and Ahmed, 1995; Al-Hooti et al., 1997; Myhera et al., 1999) and sucrose (Guizani et al., 2010). They are good sources of dietary fiber and some important minerals which include iron, potassium, selenium, calcium and vitamins and it also contains vitamin C, B_1, B_2, A, riboflavin and niacin but it is low in fat and protein contents (Myhera et al., 1999; Guizani et al., 2010; Sawaya et al., 1983).

Al-Shahib and Marshall (2003) reported that the fruits (dates) of the date palm contain a high percentage of carbohydrate (total sugars, 44-88%), fat (0.2-0.5%), 15 salts and minerals, protein (2.3-5.6%), vitamins and a high percentage of dietary fibre (6.4-11.5%). There are at least 15 minerals in dates. The percentage of each mineral in dried dates varies from 0.1 to 916 mg/100 g date depending on the type of mineral. Additionally, the seeds contain aluminum, cadmium, chloride lead and sulphur in various proportions. The protein in dates contains 23 types of amino acids. Dates contain at least six vitamins including a small amount of vitamin C, and vitamins B_1 thiamine, B_2 riboflavin, nicotinic acid (niacin) and vitamin. Date fruit is served mainly as desserts, these date fruits are generally chopped into tiny slices and scattered across cakes and puddings to ameliorate their flavor. Some people mix date paste with yogurt, milk, bread, or butter to enhance their flavor. Dates are also deseeded and stuffed with several sweet fillings like apricot. Besides the versatility of date preparation and consumption, these dry fruits also feature various nutritional values.

Dates are opulent in vitamins, minerals, natural fiber and are also cholesterol-free. Dates may be considered as an almost ideal food, providing a wide range of essential nutrients and potential health benefits (Al-Shahib and Marshall, 2003). In Tunisia, the date palm sap from date palm is directly consumed as a fresh drink called "Legmi" or used as an alcoholic beverage after natural fermentation. The fresh sap is purgative, sweet, clear, translucent and rapidly fermented (Ben Thabet et al., 2007). Composition analysis of fresh sap from date palm revealed that sugars are the major components (92-95% dry matter basis) with the dominance of sucrose. It contains also 2.7-5% of proteins and 2.3-2.6% of minerals (Ben Thabet et al., 2007, 2009). Palm sap is rapidly fermented by autochthones microflora composed essentially of yeasts, lactic acid bacteria and acetic acid bacteria. This is why date fruit benefits are widely known across the globe. Over the years, several approaches have been adopted in order minimize production cost, make available varieties of fermented drinks in the market and also to explore the arrays of underutilized nutritional fruits. Development of new beverage from date fruits using indigenous microorganisms may be a promising idea to the beverage industries. The aim of this research was therefore to study the quality of the date fruit upon fermentation by indigenous organisms.

MATERIALS AND METHODS

Collection of date palm fruits

Date palm fruits were purchased from a main market of Akure, Nigeria and transported to the Microbiological Laboratory of Federal University of Technology, Akure, Nigeria. The fruits were sorted and the apparently healthy fruits were washed, subjected to natural fermentation at 25±2°C for five days and samples were examined at each day of fermentation. Colonies were isolated and subjected to biochemical identification (Erdoğrul and Erbilir, 2006; Gueimonde et al., 2004). New set of apparently healthy ones were subjected to controlled fermentation using the single culture and mixed culture of the predominant cultures obtained from the fruits in a modified bioreactor at 25°C for 5 days (Adegbehingbe et al., 2014). Samples were aseptically withdrawn from the solution throughout the fermentation period and subjected to chemical, nutritional and microbiological analyses. Samples were also filtered and presented for sensory evaluation.

Determination of pH

The pH of the sample was determined at each day of treatment for five consecutive days using Jenway pH meter (Jenway 3010; Jenway Ltd., Essex, UK) which had been standardized with buffer solution of 7.

Titratable acidity

This was determined using the method of Valverde et al. (2005) by titrating 20 ml of fermented sample with 0.1 M NaOH to an end point of permanent pink colour using phenolphthalein as indicator.

Proximate composition

This was analyzed using standard procedure to determine the moisture, protein, fat, fiber, ash and carbohydrate (AOAC, 2003 and 1984).

Processing of the fermented samples

Both fermented samples were processed on the fifth day of fermentation for sensory evaluation. The flow sheet showing the process used to obtain fermented samples is given in Figure 1.

Amino acid analysis

The amino acid composition of the fermented sample was

Sieving (50 mesh sieve)/clarifying

↓

Bottling

↓

Pasteurization (75°C for 15 min)

↓

Cooling

Figure 1. The flow sheet showing the process used to obtain fermented samples.

Table 1. Occurrence of organism during the fermentation periods.

Organism	Fermentation periods (days)				
	1	2	3	4	5
B. pumilus	X	X	√	√	X
L. delbrueckii	X	√	√	√	√
B. pumilus + *L. delbrueckii*	X	√	√	√	√

determined by reverse phase HPLC by Kang-Lyung (2001).

Sensory evaluation

Sensory characteristic of the fermented processed drinks were assessed by 5 untrained panelists for their colour, texture, odour, taste and overall acceptability. The panelists recorded quality characteristics of each sample on an 8- point hedonic scale (8 = like extremely, 1=dislike extremely).

Statistical analysis

All analyses were carried out in triplicates. Data were subjected to analysis of variance (ANOVA) and Duncan's multiple range test was used for comparison of means SPSS software (version 16.0 for Windows, SPSS Inc., Chicago). Significance was accepted at $p < 0.05$.

RESULTS AND DISCUSSION

This study revealed that *B. pumilus* and *L. delbrueckii* (Table 1) are the predominant fermentative microorganisms of date fruits with the occurrence even after 24 h of fermentation. This observation is in line with the findings of Okafor (2007) and On-Ong et al. (2012) who reported similar organisms as part of organisms that are associated with the fermentation of the fruit. The absence of *B. pumilus* on the fifth day of fermentation may be an indication that the environmental condition of the fermenting medium is not conducive for its metabolic activities. This observation may be supported by the result of the total acidity (Table 3) of the medium. *B. pumilus* is an alkaline active organism so the acidity produced by the fermentation of the date fruits makes the substrate unfavorable for its reproducibility. The use of starter cultures is in line with the research of Kimaryo et al. (2000) as an appropriate approach for control and optimization of the fermentation process in order to alleviate problems of variations in organoleptic quality and microbial stability observed in African fermented foods.

The observed increase in protein content could be due to the activity of *B. pumilus* which is mostly used for alkaline protease production. The reduction in the carbohydrate content can be attributed to the utilization of the sugars present in the fruits by the utilizing microorganisms (Akande et al., 2010). The decline in the carbohydrate content during the fermentation period may be connected to the ability of fermenting microorganisms to metabolize carbohydrate as carbon source in order to synthesize cell biomass (Vadivel and Janardhanan, 2005), thereby producing alcohol or acids. Ogunbanwo et al. (2004) reported the effect of fiber in food as an aid to digestion. The insignificant increase in the crude fiber may also be an indication that the product will have easy digestion when consumed. Fermentation using the mixed culture of *L. delbrueckii* and *B.pumilus* gave the best nutritional value in all the days of fermentation (Figures 2 to 4). This finding agrees with the work of Achi and Akubor (2000) who stated that the use of mixed cultures improve the nutrition of fermented foods while working on the effect of bacteriocinogenic *Lactobacillus* spp. on the shelf life of fufu, a traditional fermented cassava product.

The observed decrease in the pH of all the samples (Table 2) and subsequent increase in the titratable acidity justifies the claim that the fermentation process by the organisms is not only alcohol based but also acidic based. In food safety, fast reduction in pH is desirable in order to inhibit the growth of microorganisms especially Gram negative and acid sensitive food borne pathogens (Orji et al., 2003). The observed increase in total titratable acidity of the fermented sample during the fermentation period could also be linked to the ability of the organisms to secrete acid while utilizing the available nutrient for their metabolic activities (Holzapfel, 2002). The results of the sensory evaluation (Table 4) from this study has shown that product from combined starter cultures of *B. pumilus* and *L. delbrueckii* had higher acceptability as compared to the spontaneously fermented product. The observed amino acids content of the product is an indication that the product can be recommended or even used as fortifier in nutrient food deficient in amino acids (Table 6). The observed high value of one of the essential amino acid arginine may also make the product recommendable for patient having low erection as arginine has been reported as amino acid used by the

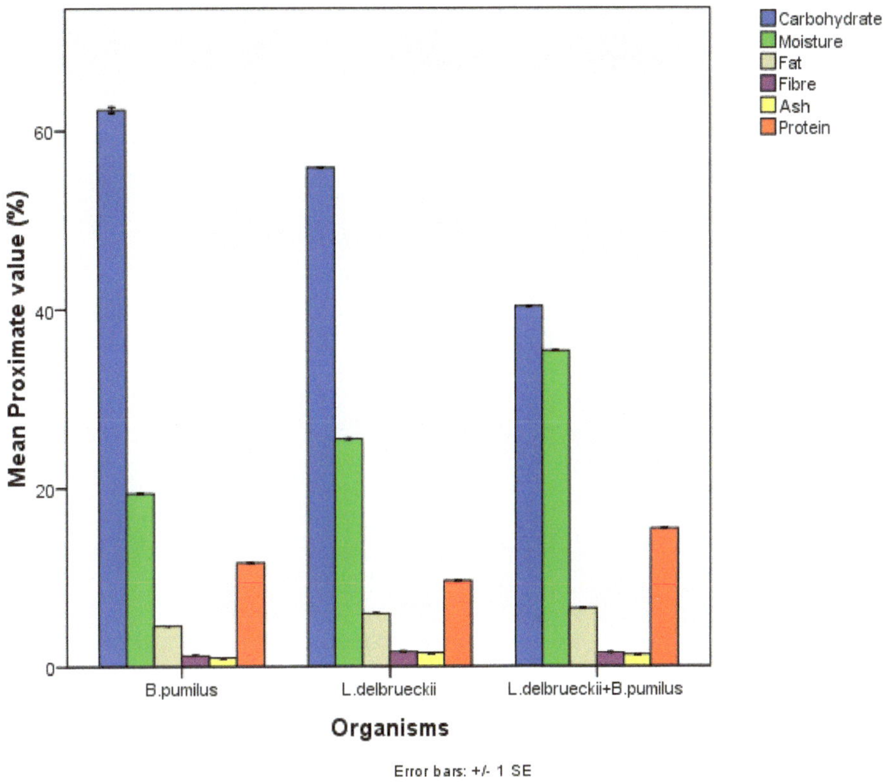

Figure 2. Proximate composition of fermented samples at the end of fermentation period (5th day).

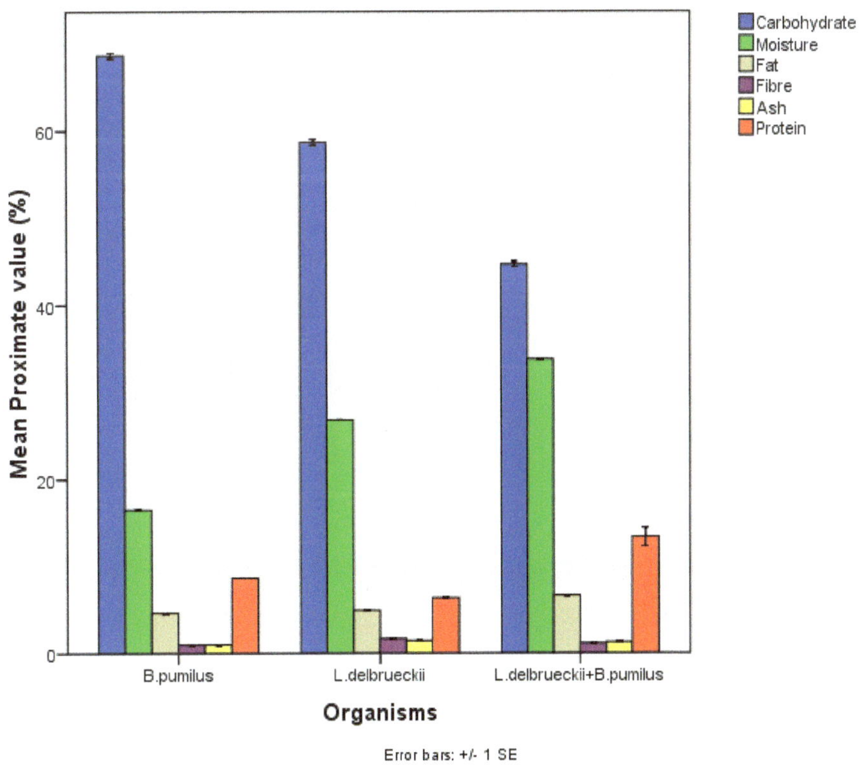

Figure 3. Proximate composition of fermented samples at the 3rd day of fermentation.

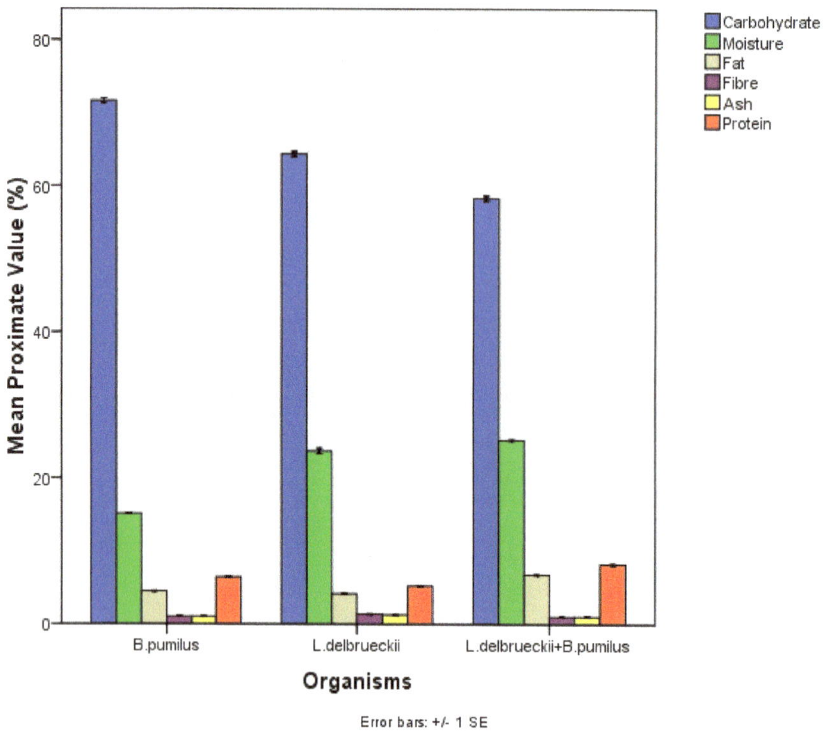

Figure 4. Proximate composition of fermented samples at the 1st day of fermentation.

Table 2. pH during fermentation period.

Organism	Fermentation periods (days)				
	1	2	3	4	5
B.pumilus	7.2±0.0^b	6.7±0.06^b	6.8±0.05^b	6.4±0.1^c	6.5±0.0^c
L.delbrueckii	7.0±0.0^b	6.7±0.06^a	6.4±0.05^a	6.0±0^b	5.7±0.1^b
B.pumilus+L. delbrueckii	6.8±0.17^a	6.5±0.14^a	6.3±0.05^a	5.4±0.0^a	4.7±0.1^a

Samples with the same superscripts down the column are not significantly different.

Table 3. Total titratable acidity during fermentation period.

Organism	Fermentation periods (days)				
	1	2	3	4	5
B.pumilus	1.1±0.0^a	1.3±0.0^a	1.33±0.1^a	1.7±0.0^a	1.7±0.0^a
L.delbrueckii	1.2±0.1^b	1.5±0.0^c	1.7±0.0^b	1.9±0.0^b	2.1±0.1^a
B.pumilus+ L.delbrueckii	1.2±0.0^{ab}	1.37±0.1^b	1.7±0.0^b	2.03±0.1^c	3.27±1.3^a

Samples with the same superscripts down the column are not significantly different.

Table 4. Sensory evaluation of the processed samples.

Fermented drinks	Appearance	Sourness	Odour	Taste	Acceptability
B. pumilus fermented drink	4.0±0.0^a	5±0^a	6.8±0.89^a	4±0.0^a	4±0.0^a
L. delbrueckii fermented drink	5.2±0.45^b	6.2±0.45^b	4.5±0^a	5.4±0.54^b	6.4±0.55^b
B. pumilus+ L. delbrueckii fermented drink	5.6±0.89^b	6.2±0.45^b	5.2±0.44^b	6.2±0.44^c	7±0.0^c

Samples with the same superscripts down the column are not significantly different.

Table 5. Mineral composition of the processed samples (μg/ml).

Fermented drinks	Mineral concentration (μg/ml)		
	Na	K	Ca
B. pumilus fermented drink	113.4±0.61^{b}	354.1 ± 0.61^{a}	118.7±0.58^{a}
L.delbrueckii fermented drink	110.3±0.21^{a}	355.0 ± 0.51^{a}	118.5±0.58^{a}
B. pumilus+ L. delbrueckii	114.8±0.5^{c}	364.0 ± 0.72^{b}	124.7±0.58^{b}

Samples with the same superscripts down the column are not significantly different.

Table 6. Amino acid composition of the processed drink.

Amino acid	Concentration (μml/ml)
Alanin	0.78
Arginine	346.45
Aspartic	126.24
Cystine	3.47
Glutamin	87.31
Methionine	7.42
Proline	286.43
Threoine	43.21
Valine	76.32
Glycine	101.32
Histidine	178.90
Isoleucine	43.29
Leucine	30.63
Lysine	30.18
Phenylalanine	26.74
Serine	13.93
Tyrosine	15.42

body to make nitric oxide (NO), which plays a role in the development of erection by enhancing the circulation of blood (Onibon et al., 2007). Fermentation of date fruit gave an appreciable mineral content (Table 5). The high potassium content observed in all the fermented products conforms with the work of Beaumont (2002) while working on nutritional and anti-nutritional composition of some Nigerian fruits. The high odour reported by the panel in the *B. pumilus* fermented product may be due to their ability to secrete ammonia a bye product during the utilization of the amino acids present in the product.

Conclusion

The result of this investigation showed that a new beverage with acceptable organoleptic properties can be produced from date palm fruits using co-cultural fermentation of *B. pumilus* and *L. delbrueckii*. This fruits can be adopted by beverage industries to introduce new product with high nutritional quality into the market.

Conflict of interests

The authors did not declare any conflict of interest.

ACKNOWLEDGEMENTS

Federal University, Oye, Ekiti, Ekiti State, Nigeria and Federal University of Technology, Akure, Ondo State, Nigeria are acknowledged for their support.

REFERENCES

Achi OK, Akubor PI (2000). Microbiological characterization of yam fermentation for elubo (yam flour) production. World J. Microbiol. Biotechnol. 16:3-7.

Adegbehingbe K, Adetuyi FC, Akinyosoye FA (2014). Effect of fermentation on nutrient and anti-nutrient contents of ground-cooked Lima bean (*Phaseolus lunatus*) seeds using *Bacillus subtilis* and *Bacillus pumilus*. Br. Microbiol. Res. J. 4(11):1285-1298.

Ahmed IA, Ahmed WK (1995). Chemical composition of date's varieties as influenced by the stage of ripening. Food Chem. 54: 305-309.

Akande KE, Abubakar MM, Adegbola TA, Bogorose SE, Doma UD (2010). Chemical evaluation of the nutritive quality of pigeon (Cajanus cajan (L.) Millsp.). Int. J. Poult. Sci. 9 (1):63-65.

Al-Hooti S, Sidhu JS, Al-Otaibi J, Al-Ameeri H, Quabazard H (1997). Date bars fortified with almonds, sesame seeds, oat flakes and skim milk powder. Plant Foods Hum. Nutr. 51: 125-135.

Al-Shahib W, Marshall RJ (2003). The fruit of the date palm: its possible use as the best food for the future?. Int. J. Food Sci. Nutr. 54 (4): 247-259.

AOAC (1984). Official methods of Analysis. Washington; DC: Association of Official Analytical Chemists.

AOAC (2003). Official methods of analysis of the association of official's analytical chemists, 17th edn. Association of official analytical chemists, Arlington, Virginia.

Augstburger F, Berger J, Censkowsky U, Heid P, Milz J, Streit C (2002). Date Palm. Naturland. Germany.

Bastway M, Hasona NA, Selemain H (2008). Protective effects of extract from dates (*Phoenix dactylifera* L.) and ascorbic acid on thioacetamide induced hepatotoxicity in rats. J. Pharm. Res. 7: 193-201.

Beaumont M (2002). Flavouring composition prepared by fermentation with Bacillus spp. Int. J. Food Microbiol. 75: 189 -196.

Ben Thabet I, Attia H, Besbes S, Deroanne C, Francis F, Drira NE, Blecker C (2007). Physicochemical and Functional Properties of Typical Tunisian Drink: Date Palm Sap (*Phoenix dactylifera* L.). Food Biophys. 2: 76-82.

Ben Thabet I, Besbes S, Attia H, Deroanne C, Francis F, Drira NE, Blecker C (2009). Physicochemical characteristics of date sap lagmi from Deglet Nour palm (*Phoenix dactylifera* L.). Int. J. Food Prop. 12: 659-670.

Cooke RD, Twiddy DR, Reilly PJA, (1987). Lactic acid fermentation as a low cost means of food preservation in tropical countries. FEMS

Microbiol. Rev 46:369-379.
El-Sohaimy SA, Hafez EE (2010). Biochemical and nutritional characterization of date palm fruits. Int. J. Appl. Sci. Res. 6(8):1060-1067.
Erdoğrul Ö, Erbilir F (2006). Isolation and characterization of *L. bulgaricus* and *L. casei* from various foods. Turk. J. Biol. 30:39-44.
Gueimonde M, Delgado S, Mayo B, Ruas-Madiedo P, Margolles A, de los Reyes-Gavilan G (2004), Viability & diversity of probiotic Lactobacillus and Bifidobacterium populations included in commercial fermented Milks. Food Res. Int. 37:839-850
Guizani N, Al-Saidi GS, Rahman MS, Bornaz S, Al-Alawi AA (2010). State diagram of dates: glass transition, freezing curves and maximal-freeze- concentration condition. J. Food Eng. 99: 92-97.
Holzapfel WH (2002). Appropriate starter culture technologies for small-scale fermentation in developing countries. Int. J. Food Microbiol. 75:197-212.
Kang-Lyung W (2001). Amino acid analysis protocol. Methods Mol. Biol. 159:141.
Khan MN, Sarwar A, Wahab F, Haleem R (2008). Physico-chemical characterization of date varieties using multivariate analysis of plums. Food Chem. 81:321-326
Kimaryo VM, Massawe GA, Olasupo NA, Holzapfel WH (2000). The use of a starter culture in the fermentation of cassava for the production of "kivunde", a traditional Tanzanian food product. Int. J. Food Microbiol. 56:106-108.
Myhera RM, Karkdas J, Taylor MS (1999). The composition of maturing Omani dates. J. Agric. Food Chem. 76: 1345-1350.
Ogunbanwo ST, Sanni AI, Onilude AA (2004). Effect of bacteriocinogenic *Lactobacillus* spp. on the shelf life of fufu, a traditional fermented cassava product. World J. Microbiol. Biotechnol. 20: 57-63.
Okafor N (2007). Modern Industrial Microbiology and Biotechnology. Science Publishers, Enfield. pp. 263-308.
Onibon VO, Abulude FO, Lawal LO (2007). Nutritional and anti-nutritional composition of some Nigerian fruits. J. Food Technol. 5:120-122.
On-Ong C, Kornwika P, Donalphora T (2012). Pineapple wine fermentation with yeast isolated from fruit as a single and mixed starter culture. Asian J. Food Agro-Ind. 5 (02): 104-111.
Orji MU, Mbaka TI, Aniche GN, Ahonkhai I (2003). The use of starter cultures to produce 'pito', a Nigerian fermented alcoholic beverage. J. Microbiol. Biotechnol. 19:733-736.
Sahari MA, Barzegar M, Radfar R (2007). Effect of varieties on the composition of dates (*Phoenix dactylifera* L.). Food Sci. Technol. Int. 13:269-275.
Sawaya W, Miski A, Khalil J, Khatchadonian H, Mashadi A (1983). Physical and chemical characterization of the major date varieties grown in Saudi Arabia; I Morphological measurements, proximate and mineral analyses. Date Palm J. 2: 1-25.
Vadivel V, Janardhanan K (2005). Nutritional and antinutritional characteristics of seven south Indian wild legumes. Plant Food Hum. Nutr. 60 (2): 69-75.
Valverde JM, Valero D, Martinez-Romero D, Guillen F, Castillo S, Serrano M (2005). Novel edible coating based on *Aloe vera* gel to maintain table grape quality and safety. J. Agric. Food Chem. 53:7807-7813.
Vayalil PK (2002). Antioxidant and antimutagenic properties of aqueous extract of date fruit (*Phoenix dactylifera* L. *Arecaceae*). J. Agric. Food Chem. 50: 610-617.

Prevention of konzo in the Democratic Republic of Congo (DRC) using the wetting method and correlation between konzo incidence and percentage of children with high urinary thiocyanate level

Jean Pierre Banea[1], J. Howard Bradbury[2] , Chretienne Mandombi[3], Damien Nahimana[1], Ian C. Denton[2], Matthew P. Foster[2], N'landa Kuwa[1] and Desire Tshala Katumbay[4]

[1]Programme National de Nutrition (PRONANUT), Kinshasa, Democratic Republic of Congo (DRC).
[2]Evolution, Ecology and Genetics Research School of Biology, Australian National University, Canberra, ACT 0200, Australia.
[3]Hospital General de Reference, Zone de Sante de Popokabaka, Democratic Republic of Congo (DRC).
[4]Centre Neuropsychopathologique, Universite de Kinshasa, Democratic Republic of Congo (DRC).

There were 68 konzo cases in three villages in Boko Health Zone, Bandundu Province, Democratic Republic of Congo (DRC), where yearly incidence had increased greatly since 2009. The nine months long intervention to prevent konzo started in the wet season. Village women were taught the wetting method, after which there were no new konzo cases and urinary thiocyanate levels were low. Using data from four different interventions in March 2010, July 2011, September 2013 and the present intervention in October-November 2012, a correlation was found between the monthly cyanide intake (measured by percentage of children with high urinary thiocyanate level) and the monthly konzo incidence. This dose response relationship between cyanide intake and konzo incidence, together with the prevention of konzo in many villages using the wetting method to reduce cyanogen intake, shows that konzo is very likely to be due to high cyanide/low sulfur amino acid intake in a diet of bitter cassava.

Key words: Dose-response relation, konzo prevalence, urinary thiocyanate, cyanide, cassava flour, wetting method.

INTRODUCTION

In the Democratic Republic of Congo (DRC) the roots of cassava are peeled, soaked (retted) in water for 3-4 days, and then dried in the sun. After drying, the roots are pounded and sieved. If the time of retting is reduced to 1-

2 days (called short soaking), because of the families need for food or fear that the roots may be stolen or pressure to sell the flour (Banea et al., 1992; Tylleskar et al., 1992), then cyanogens are incompletely removed from the flour and village people may become poisoned by cyanide and konzo may occur. Konzo is an irreversible paralysis of the legs that occurs particularly in children and young women after childbirth who live on a monotonous diet of high cyanide cassava, and is associated with high cyanide and low sulfur amino acid intake (Ministry of Health Mozambique, 1984; Cliff et al., 1985; Howlett et al., 1990). This association is supported by many studies at an aggregated level and is strengthened by the recent prevention of konzo by greatly reducing the intake of cyanide from cassava flour, by use of the wetting method (Banea et al., 2012a, 2013). Further support for the hypothesis is given in this paper by showing a month by month correlation between the percentage of school children with high urinary thiocyanate levels and incidence of konzo.

Konzo occurs amongst poor village people in DRC, Mozambique, Tanzania, Cameroon, Central African Republic and Angola. Konzo was first reported by Trolli (1938) in Popokabaka Health Zone in Bandundu Province, DRC and was controlled for the first time in Kay Kalenge village in the same Health Zone in 2010, through the use of the wetting method by village women (Banea et al., 2012a). Fourteen months after the intervention ceased, Kay Kalenge was visited again and it was found that there were no new cases of konzo, the wetting method was still being used by the women and its use had spread by word of mouth to three adjacent villages (Banea et al., 2014).

One way to get rid of cyanogens from cassava flour would be to ensure that short soaking of cassava (Banea et al., 1992; Tylleskar et al., 1992) does not occur, but this would be difficult to achieve in practice because of the reasons given above. Therefore, a better procedure would be to use the wetting method on cassava flour as an additional reliable method to make sure that cyanogens are removed, just before the flour is cooked in the traditional way in boiling water to make a stiff porridge called fufu. The wetting method consists of mixing the flour with water, spreading the wet flour in a thin layer and standing it in the sun for 2 h or in the shade for 5 h to allow hydrogen cyanide gas to escape (Bradbury, 2006; Cumbana et al., 2007; Bradbury and Denton, 2010). The advantages of using the wetting method to remove cyanogens from flour, rather than attempting to increase the time of retting are that (1) it is an additional processing method that removes cyanogens and makes the flour safe to eat, (2) it is fast (2-5 h) when compared with several extra days of retting and (3) being located inside the house or just outside in the sun, it is much more secure from thieves than retting cassava roots in the river (Banea et al., 2013). The wetting method is popular because it is simple to use, requires only a bowl, a knife and a mat and the fufu produced from it can be stored for up to three days, when compared with only one day for fufu made from untreated flour. Its taste is much superior to that of fufu prepared from untreated flour, which is bitter because of the bitter linamarin that it contains (King and Bradbury, 1995).

In this paper, we reported the prevention of konzo using the wetting method in three villages and we use the data, along with that from three other interventions, to obtain a month by month correlation between incidence of konzo and the urinary thiocyanate content of school children.

MATERIALS AND METHODS

Study area

In September 2012, the team visited the Kwango District health team and nutrition coordinator in Kenge, who want to expand interventions to control konzo, which is increasing in this district. A further visit was made to the Boko Health Zone office where two possible sites were proposed for further work. After visiting the two sites, three villages in the savanna were chosen, Tsakala Mbewa, Tsakala Kenia and Mabaka (collectively called Boko 2) and located north of three previous villages (Figure 1) in which konzo was successfully prevented in 2011-2012 (collectively called Boko 1, Banea et al., 2013). Tsakala Mbewa is on a main road, has a primary and secondary school, a health centre and a maize mill. Tsakala Kenia has a primary and secondary school. Mabaka is 8 km from the main road, has a primary school and a small health centre. Water is available about 1-4 km away.

Agriculture and processing

Bitter (high cyanide) cassava is the main crop, along with sweet (low cyanide) cassava which is often stolen. Fields are located in the forest and are now fallowed for 3-5 years, when compared with 10 years previously. There is a shortage of fields to grow cassava because now only women are engaged in agriculture, and it is not carried out like before. Cassava, corn, peanut and cowpea are major crops and there are chickens, goats and pigs. Traditional varieties of cassava are bomba, mbengi and kayayi which are resistant to plant diseases and there are new varieties mwuazi, musanzu and F-100 which give roots after six months. Cassava roots are retted (soaked) for 2 days and nights instead of the 3-4 days recommended to remove cyanogens and the soaked roots are dried for only 0.5-1 day, because of problems of food security. Cassava is processed into cossettes or chickwangue (Banea et al., 1992). Traders visit these villages mostly during the dry season (May-September), at corn harvest and caterpillar season.

Second visit to the three villages

On the second visit by the full team (Banea team, Caritas team and Boko Health Zone team) to the villages in October-November 2012, a population census was made and suspected konzo cases examined in their homes by community health workers following a standardised WHO protocol. The four diagnostic criteria for konzo are as follows: (1) a spastic visible walk, (2) sudden onset within a week in a formerly healthy person, (3) exaggerated bilateral patellar

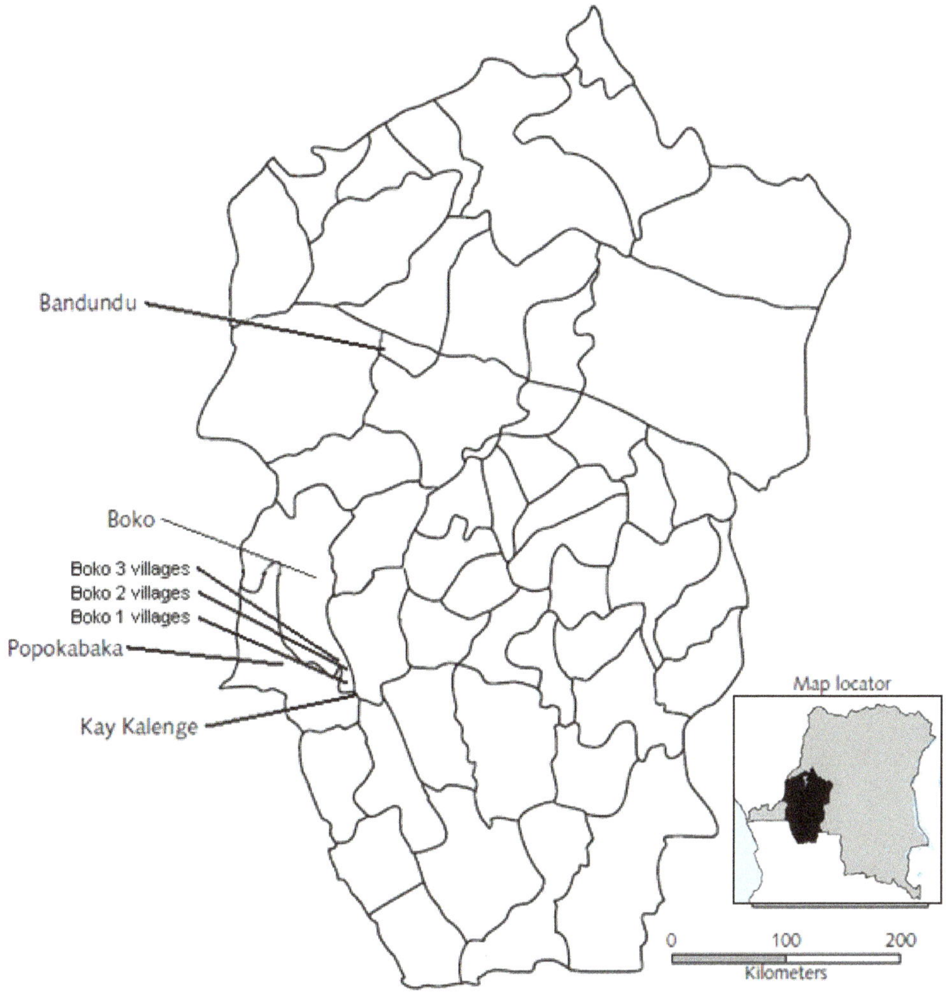

Figure 1. Map showing the health zones of Bandundu Province, DRC, including the location of the Boko 2 villages (this study), Kay Kalenge village (Banea et al., 2012a), three Boko 1 villages (Banea et al., 2013), and the six Boko 3 villages.

or achillian reflexes and (4) non-progressive evolution of the disease (WHO, 1996). Konzo cases were given multivitamins and antiinflammatory drugs. A food consumption survey was conducted among 58 households with one or more konzo cases. Cassava flour samples were collected from households and urine samples were collected from school children, about 85 from boys and 85 from girls with about 120 from 6-9 years old and 50 from 10-14 years old, with permission from their parents. All samples were analysed on site using cassava cyanide kit B 2 and urinary thiocyanate kit D1 supplied from Australia.

The wetting method was taught to 10-12 of the leading women in each village and they in turn each taught 10-15 others in the village. The training reached a total of 379 women in the three villages and a committee was formed to ensure follow up. Each woman received a plastic bowl, knife and a mat, but the delivery of the mat was delayed to a later visit. The wetting method involved adding cassava flour to a bowl and marking the level of the flour on the inside of the bowl. Water was added and the level of the wet flour initially decreased and then increased as more water was added with mixing, until it reached the mark. The wet flour was spread in a thin layer about 1 cm thick on a mat or a basket and allowed to stand for 2 h in the sun or 5 h in the shade to allow hydrogen cyanide gas to escape (Bradbury, 2006; Cumbana et al., 2007; Bradbury and Denton, 2010). The damp flour was then cooked in boiling water in the traditional way to make a thick porridge called fufu. Fufu was normally eaten with pounded, boiled cassava leaves or something else to give it flavour.

Subsequent visits to the three villages

Every month after the second visit, the three villages were visited by the Caritas team to check on use of the wetting method and in March 2013 the full team visited for the third time. Nine focus groups were conducted to check on the continued use of the wetting method during the wet season. Training sessions on use of the wetting method were held for untrained women. Difficulties raised by the women were lack of sun, houses made of straw which oozed water during rain, and a shortage of mats. During this period, people ate cassava flour, cassava leaves, mushrooms and some hunting families ate meat. Cassava flour and urine samples were collected and analysed on site. The fourth and final visit of the full

Table 1. Population, number of konzo cases and konzo prevalence (%) in Boko 2 villages in October-November 2012.

Village	Population	Number of konzo cases	Konzo prevalence (%)
Tsakala Mbewa	752	26	3.5
Tsakala Kenia	314	21	6.7
Mabaka	972	21	2.2
Total	2038	68	3.3

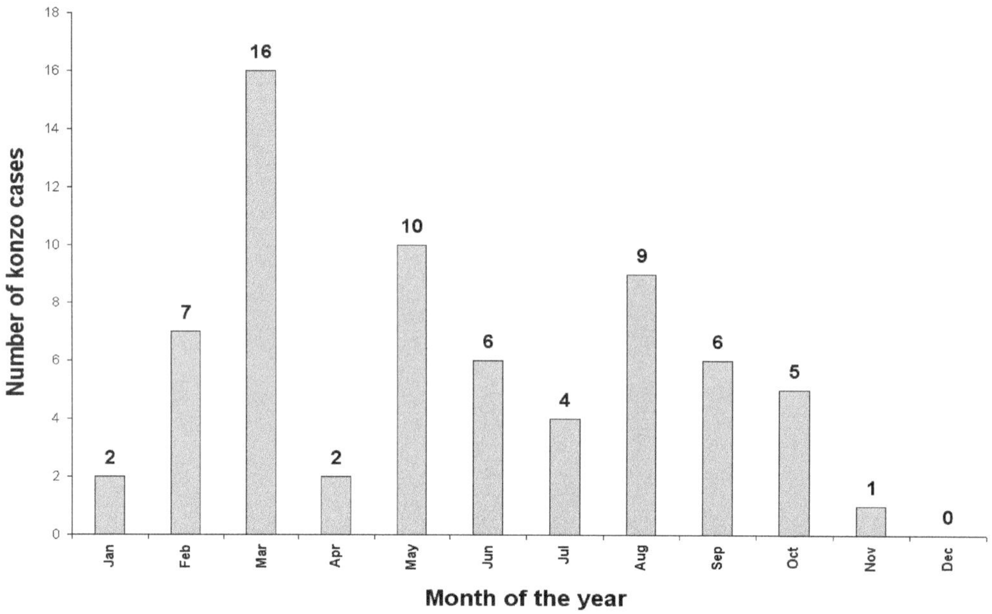

Figure 2. Monthly distribution of konzo cases in the Boko 2 villages.

team occurred in June 2013 and the same procedure was followed as in the third visit, with nine focus groups, training sessions for untrained women to use the wetting method, and collection and analysis of flour and urine samples. Bowls, knives and mats were given to women after training to encourage them to continue using the wetting method.

Urinary thiocyanate analysis

Urine samples (normally 55-60 but 45 in one case) were collected randomly from school age children in each village with consent of their parents and a record was made of their age and sex. These samples were analysed on site using the simple picrate thiocyanate kit D1 (Haque and Bradbury, 1999; http://biology.anu.edu.au/hosted_sites/CCDN/). A colour chart with 10 shades of colour from yellow to brown was used, which corresponded to 0 to 1720 µmole thiocyanate/L.

Total cyanide analysis

Thirty samples of cassava flour about to be used to prepare fufu were collected randomly from households in each village before teaching the wetting method in October-November 2012 and at the subsequent visits in March and June 2013. Analyses for total cyanide content were made on site using a simple picrate kit B2 (Egan et al., 1998; Bradbury et al., 1999; http://biology.anu.edu.au/hosted_sites/CCDN/). A colour chart was used with 10 shades of colour from yellow to brown corresponding to 0-800 mg HCN equivalents/kg cassava flour (ppm).

Ethics statement

Prior to the study, the protocol was approved by the DRC Ministry of Public Health Ethics Committee. The day before the collection of cassava flour and urine samples, each household gave their free oral consent for the sampling to be carried out.

RESULTS

In Table 1, the distribution of population, number of konzo cases and konzo prevalence (%) in the three villages are shown. There are 68 konzo cases, 42 females (62%) and 26 (38%) males, in a population of 2038 with a mean konzo prevalence of 3.3%. The monthly occurrence of konzo cases is shown in Figure 2 and the years in which they occurred in Figure 3. Onset of konzo occurred in less than one day in 91% of cases. Two people with konzo (3%) were unable to walk, 3% used two sticks,

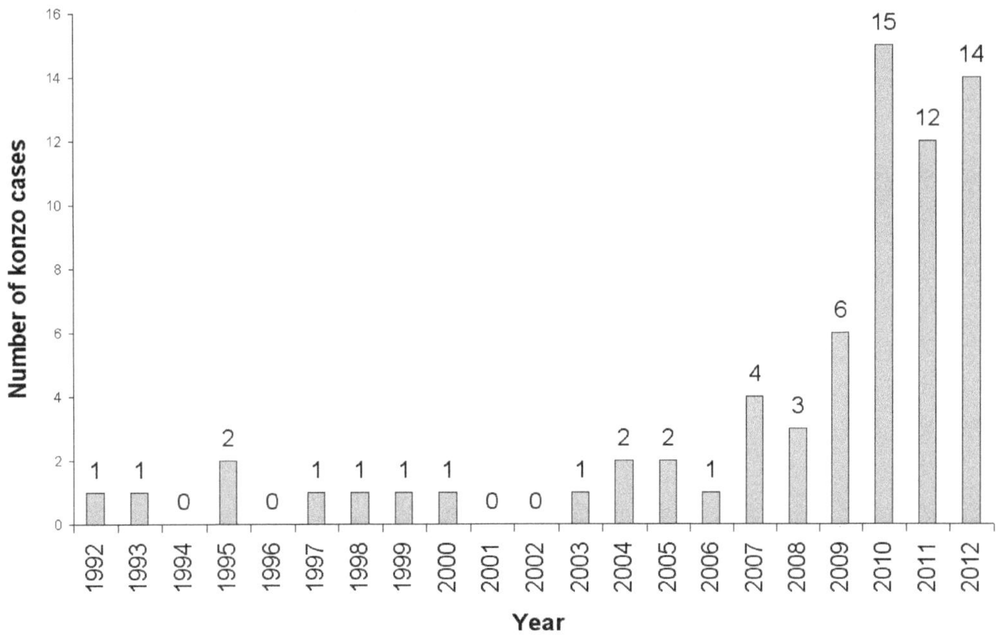

Figure 3. Annual distribution of konzo cases in the Boko 2 villages.

Table 2. Mean urinary thiocyanate level (µmole/L) of school children in Boko 2 villages before introducing the wetting method (October-November, 2012) and during the intervention[a].

Village	Mean urinary thiocyanate level (µmole/L)		
	October-November 2012	March 2013	June 2013
Tsakala Mbewa	51(60)	104(81)	100(96)
Tsakala Kenia	224(129)	161(128)	121(96)
Mabaka	240(171)	139(151)	137(110)
Mean	172(105)	135(29)	119(19)

[a]Mean values based on sample sizes of 45-60 with most sample sizes of 60. Standard deviation shown in brackets.

12% one stick and 82% limped but did not need a stick. Speech disorders were observed in 12% of patients and 16% had impaired vision. Six patients (9%) had goitre, determined by palpation. Patellar and achillian reflexes were exaggerated bilaterally in all cases.

A food consumption survey in October-November of 58 households with at least one konzo case showed that the mean number of days that foods were consumed per week were as follows: cassava flour 7, collection products (caterpillars, roots) 5, vegetables 4, cooking oil 2.8, fruits (bananas, pineapple, orange, avocado) 2.5, cereals (maize, rice, wheat flour) 1.9, sugar and sugar products 1.9, meat and fish 1.3, milk 0.5, legumes (beans, cowpeas) 0.4. The people of Tsakala Mbewa consumed more maize flour than those of the other villages because there was a maize mill in their village. Many people made fufu from a mixture of cassava flour and corn flour. In March 2013, people consumed mainly cassava flour, cassava leaves and mushrooms and hunting families (mainly in Mabaka) ate some meat. In June 2013, there was trade from Kinshasa and Kikwit through Tsakala Mbewa and these people had a better diet.

No new case of konzo occurred in the Boko 2 villages after the intervention commenced in October-November 2012. A total of 37 women gave birth over the nine month intervention and none of them contracted konzo after child birth. In March 2013, only 71% of women were using the wetting method partly due to heavy rains, but by June 2013 this had risen to 95%.

The mean urinary thiocyanate levels of school children are shown in Table 2. There were no significant differences between the mean urinary thiocyanate levels of boys and girls or between the two age groups, 5-9 and

Table 3. Percentage of school children in Boko 2 villages with urinary thiocyanate level of >350 µmole/L.

Village	Percentage of children with urinary thiocyanate level of > 350 µmole/L		
	October-November 2012[a]	March 2013	June 2013
Tsakala Mbewa	0	0	0
Tsakala Kenia	1	0	0
Mabaka	4	2	0

[a]Before introducing the wetting method.

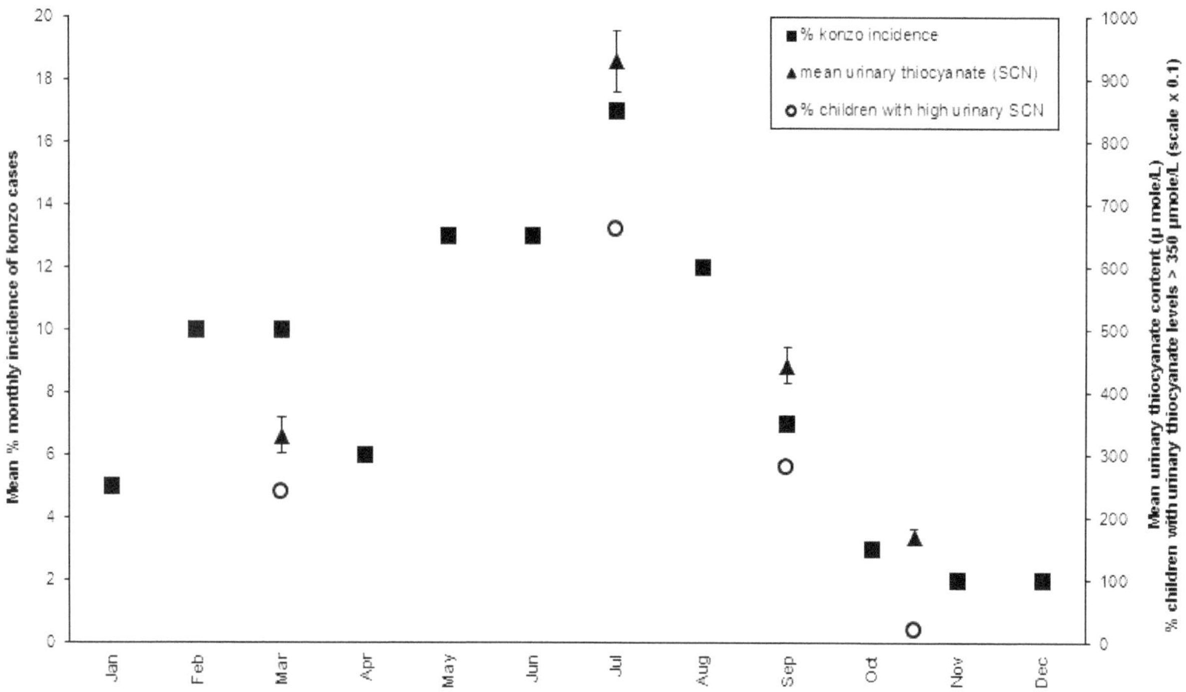

Figure 4. Graphs of the mean percentage monthly incidence of konzo cases (■), the mean urinary thiocyanate levels of school children with standard error bars (▲) and the percentage of children with urinary thiocyanate levels >350 µmole/L (○) for the four interventions made in March 2010, July 2011, September 2013 and the current intervention in October-November 2012.

10-14 years. In Table 3, the percentage of school children with a high urinary thiocyanate level of >350 µmole/L is given. The mean total cyanide content for Tsakala Mbewa, Tsakala Kenia and Mabaka villages before teaching the wetting method in October-November 2012 were 9 (19), 12 (14) and 7 (12) ppm, in March 2013 they were 9 (10), 9(8) and 8 (6) ppm and in June 2013, they were 6 (5), 6 (5) and 7 (9) ppm, respectively. Standard deviations are shown in brackets. In Figure 4, the mean percentage monthly incidence of konzo averaged over four interventions, viz. Kay Kalenge village in March 2010 (Banea et al., 2012a); three villages, Boko 1, in July 2011 (Banea et al., 2013); six villages, Boko 3, in September 2013 and this study in three villages in October-November 2012 is plotted. On the same graph, is plotted the mean urinary thiocyanate level of school children before intervention with standard error bars included and the percentage of children with mean urinary thiocyanate level >350 µmole/L. A correlation analysis showed that the mean percentage monthly incidence of konzo is significantly correlated with the percentage of children with mean urinary thiocyanate level >350 µmole/L, with P = 0.037 and correlation coefficient of 0.963. A weaker correlation was found between mean percentage monthly incidence of konzo and mean urinary thiocyanate content of school children, because the latter is a poorer measure of children in danger of contracting konzo than is the percentage of children with very high urinary thiocyanate levels (Banea et al., 2013).

DISCUSSION

As shown in Table 1, there were a total of 68 konzo cases in the three villages with a konzo prevalence rate ranging from 2.2% in Mabaka to 6.7% in Tsakala Kenia. There were 3% severely disabled patients (unable to walk), 15% moderately disabled (needing one or two sticks) and 82% mildly disabled, who walked without a stick. This data was combined with that from five previous konzo studies (Banea et al., 2012a, 2013) and the mean percentage results (standard deviations in brackets) were as follows: severely disabled 8 (5), moderately disabled 26 (14), mildly disabled 66 (13).

Figure 2 shows a maximum incidence of konzo in March (near the end of the wet season) with smaller peaks in May and August. This is the first time that konzo incidence has peaked in the wet season. Perhaps the retted cassava could not be dried sufficiently in the wet conditions and short drying times of about one day, could produce elevated residual acetone cyanohydrin content (Banea et al., 1992; Tylleskar et al., 1992). This was not observed in the mean total cyanide content of cassava flour samples, which were low before the wetting method was taught in October-November 2012 and did not significantly exceed the WHO safe level of 10 ppm (FAO/WHO, 1991). However, detailed inspection showed that 20% of the total cyanide results obtained before the wetting method was taught were ≥ 20 ppm and gave rise to much higher standard deviations from the mean (see above) than with the later results. This showed that 20% of the village women were not processing their flour correctly before the wetting method was taught.

In Figure 3 it is shown, that konzo cases occurred every year for ten successive years from 2003-2012, with large numbers since 2009, as with the Boko 1 villages (Banea et al., 2013) and Kay Kalenge (Banea et al., 2012a). The total number of konzo cases in these three locations with incidence since 1992 was 150, of which 102 (68%) occurred from 2009 to 2012. This confirmed anecdotal evidence from Kwango District Health Team (see above) that konzo was getting worse in this district. There are two reasons for this. Before 2009, the production of cassava was higher than it is now because of the present shortage of fields to grow cassava (see above), hence there was more food available and they sold more cassava and were able to buy other food which reduced their total intake of cyanogens. The second reason is that they are still obliged to sell some cassava and when they sell cassava they reduce the soaking time. They also eat the short-soaked cassava with increased cyanide content, which increases greatly their intake of poisonous cyanogens, and leads to konzo.

In Figure 4, are the results of four interventions in the same geographic region (Figure 1) at Kay Kalenge in March 2010 (Banea et al., 2012a), Boko 1 in July 2011 (Banea et al., 2013), Boko 3 in September 2013 and the present intervention in October-November 2012. In Figure 4, in October-November, the small mean urinary thiocyanate level and the small percentage of children with urinary thiocyanate level >350 μmole/L before intervention is consistent with the low recorded incidence of konzo, whereas in July, there is a high urinary thiocyanate level consistent with a large incidence of konzo (Banea et al., 2013). The monthly konzo incidence curve would be expected to change in different regions and countries, being dependent on the seasons of the year and other factors, which would explain the wide range of urinary thiocyanate levels reported in the literature from children with konzo.

There is a significant correlation between the percentage of children with urinary thiocyanate levels >350 μmole/L and the mean percentage monthly incidence of konzo. Therefore, there is a dose response relationship (Tylleskar, 1994) between cyanide intake (measured by percentage of children with urinary thiocyanate content >350 μmole/L) and the incidence of konzo. A dose-response relationship is further evidence of the correctness of the high cyanide, low S-containing amino acid association as an explanation of the occurrence of konzo (Banea et al., 2012b). Thus, the best measure of the likely incidence of konzo in a population are urinary thiocyanate measurements on school children, but because these values are dependent on the seasons of the year (Figure 4), which vary from one region and country to another, it is necessary to make comparisons in the same region. This has been done here, since the four studies are all located geographically close to one another (Figure 1).

The low mean urinary thiocyanate level before the intervention started was due to the fact that it commenced in the wet season, when cyanogen intake from cassava and konzo incidence was very low (Figure 4). For this reason the mean urinary thiocyanate level actually increased in Tsakala Mbewa from October-November 2012 to March 2013, but decreased for the other villages (Table 2), but most importantly was low in all three villages in the dry season when the level is usually high (June 2013). Furthermore, the percentage of children with urinary thiocyanate content >350 μmole/L had reduced to zero in the three villages in June 2013 (Table 3). The total time of intervention was progressively reduced from 1.5 years in the first intervention (Banea et al., 2012a) to one year in the second intervention (Banea et al., 2013) to nine months in this intervention. The progressive reduction in intervention time from eighteen to nine months still produced good results, reduced the cost of the intervention per person and allowed better use of limited funding.

Conclusion

This intervention to control konzo was commenced in the wet season when konzo incidence was low, and therefore gave much lower urinary thiocyanate and cassava cyanide values than in previous interventions (Banea et al., 2012a, 2013). The intervention was completed satisfactorily in 9 months when compared with previous interventions of 18 and 12 months and this reduced the cost per person in the village. The yearly records of konzo incidence in all three interventions showed that konzo cases had greatly increased since 2009, in agreement with anecdotal evidence that konzo was increasing in this area.

It was found that the mean monthly incidence of konzo was significantly correlated with the percentage of children with very high urinary thiocyanate levels, which is a measure of cyanide intake. This represents a dose response relation between konzo incidence and cyanide intake and shows that konzo is very likely to be due to high cyanide intake from bitter cassava combined with low sulfur amino acid intake (Tylleskar, 1994). Since detoxification of ingested cyanide occurs by formation of thiocyanate which is removed in the urine, the best measure of children in danger of contracting konzo is by urinary thiocyanate determinations, but the results should be related to the season of the year. It is good to make urinary thiocyanate measurements during the dry season when cassava consumption is high and urinary thiocyanate levels of >350 µmole/L in individual children indicate a child in danger of contracting konzo (Banea et al., 2012a, b, 2013).

Conflict of Interests

The author(s) have not declared any conflict of interests.

ACKNOWLEDGEMENTS

We wish to thank the women of Tsakala Mbewa, Tsakala Kenia and Mabaka for readily accepting and using the wetting method to remove cyanogens from their cassava flour thereby preventing konzo. We thank the health care staff of Kwango District and Boko Health Zone for their collaboration. We thank Ms Pauline Ding of the Statistical Consulting Unit at Australian National University for statistical analysis of our results. We are grateful to the Australian Agency for International Development (AusAID) for funding this intervention.

REFERENCES

Banea JP, Bradbury JH, Mandombi C, Nahimana D, Denton IC, Kuwa N, Tshala Katumbay D (2013). Control of konzo by detoxification of cassava flour in three villages in the Democratic Republic of Congo. Food Chem. Toxicol. 60:506-513.

Banea JP, Bradbury JH, Mandombi C, Nahimana D, Denton IC, Kuwa N, Tshala Katumbay D (2014). Effectiveness of wetting method for control of konzo and reduction of cyanide poisoning by removal of cyanogens from cassava flour. Food Nutr. Bull. 35(1):28-32.

Banea JP, Bradbury JH, Mandombi C, Nahimana G, Denton IC, Kuwa N (2012b). Preventive control of konzo in the Democratic Republic of Congo. Food Chem. Toxicol. 50:4234-4235.

Banea JP, Nahimana G, Mandombi C, Bradbury JH, Denton IC, Kuwa N (2012a). Control of konzo in DRC using the wetting method on cassava flour. Food Chem. Toxicol. 50:1517-1523.

Banea M, Poulter NH, Rosling H (1992). Shortcuts in cassava processing and risk of dietary cyanide exposure in Zaire. Food Nutr. Bull. 14:137-143

Bradbury JH (2006). Simple wetting method to reduce cyanogen content of cassava flour. J. Food Comp. Anal. 19: 388-393.

Bradbury JH, Denton IC (2010). Rapid wetting method to reduce cyanogen content of cassava flour. Food Chem. 121: 591-594.

Bradbury MG, Egan SV, Bradbury JH (1999). Determination of all forms of cyanogens in cassava roots and cassava products using picrate paper kits. J. Sci. Food Agric. 79:593-601.

Cliff J, Martensson J, Lundquist P, Rosling H, Sorbo B (1985). Association of high cyanide and low sulphur intake in cassava induced spastic paraparesis. Lancet 11:1211-13.

Cumbana A, Mirione E, Cliff J, Bradbury JH (2007). Reduction of cyanide content of cassava flour in Mozambique by the wetting method. Food Chem. 101:894-897.

Egan SV, Yeoh HH, Bradbury JH (1998). Simple picrate paper kit for determination of the cyanogenic potential of cassava flour. J. Sci. Food Agric. 76:39-48.

FAO/WHO (1991). Joint FAO/WHO Food Standards Programme, Codex Alimentarius Commission XII, Supplement 4, FAO, Rome, Italy.

Haque MR, Bradbury JH (1999). Simple method for determination of thiocyanate in urine. Clin. Chem. 45:1459-1464.

Howlett WP, Brubaker GR, Mlingi N, Rosling H (1990). Konzo, an epidemic upper motor neuron disease studied in Tanzania. Brain 113:223-35.

King NLR, Bradbury JH (1995). Bitterness of cassava: identification of a new apiosyl glucoside and other components that affect its bitter taste. J. Sci. Food Agric. 68:223-30.

Ministry of Health, Mozambique (1984). Mantakassa; an epidemic of spastic paraparesis associated with chronic cyanide intoxication in a cassava staple area in Mozambique. 1. Epidemiology and clinical laboratory findings in patients. Bull. WHO 62:477-84.

Trolli G (1938). Paraplegie spastique epidemique, "Konzo" des indigenes du Kwango. In: Trolli G (Ed) Resume des observations reunies, au Kwango, au sujet de deux affections d'origine indeterminee. Fonds reine Elisabeth, Brussels. pp. 1-36.

Tylleskar T (1994). The association between cassava and the paralytic disease konzo. Acta Hort 375:331-339.

Tylleskar T, Banea JP, Bikangi N, Cooke RD, Poulter NH, Rosling H (1992). Cassava cyanogens and konzo, an upper motoneuron disease found in Africa. Lancet 339:208-211.

World Health Organisation (WHO) (1996). Konzo a distinct type of upper motor neuron disease, Weekly Epidem. Record 7: 225-232.

Survey of the konzo prevalence of village people and their nutrition in Kwilu District, Bandundu Province, DRC

Jean Pierre Banea[1], J. Howard Bradbury[2] , Damien Nahimana[1], Ian C. Denton[2], N. Mashukano[1] and N'landa Kuwa[1]

[1]Programme National de Nutrition (PRONANUT), Kinshasa, Democratic Republic of Congo (DRC).
[2]EEG, Research School of Biology, Australian National University, Canberra, ACT 0200, Australia.

Konzo is a sudden spastic paraparesis that causes permanent paralysis of the legs and occurs mainly in children and young women. Konzo results from high cyanogen intake and malnutrition caused by a monotonous diet of bitter cassava. The known incidence of konzo in DRC up to 2009 is 3469 cases, but an estimate in 2002 was 100,000 cases. To help resolve this question a konzo survey was made in three health zones in Kwilu District, Bandundu Province, Democratic Republic of Congo (DRC), and the nutrition of those with konzo recorded. Thirty villages (population 22793) in Kwilu District were surveyed for konzo cases, and food consumption scores and mid upper arm circumferences obtained. There were 172 konzo cases with village konzo prevalences of 0.1-17%. The mean konzo prevalence in Masimanimba and Kingandu health zones was much less than in Payikongila health zone, probably because of the higher rate of malnutrition in Payikongila. Since 2009, konzo incidence has increased greatly in Kwilu District and also in 13 villages in nearby Kwango District, where incidence of new konzo cases has been prevented by use of the wetting method. Averaging the data over 495 konzo cases, 48% occurred from 2009 onwards and 52% occurred in the 20+ years before 2009. The very large increase of konzo incidence since 2009 is a public health problem in Bandundu Province, that could be solved by training women to use the wetting method, which removes cyanogens from cassava flour. The wetting method is more direct, effective and cheaper in preventing konzo than broad based interventions.

Key words: Konzo survey, cassava, cyanide, nutrition, sulfur amino acids, food consumption score, mid upper arm circumference.

INTRODUCTION

Cassava contains cyanogens (linamarin and a small amount of lotaustralin) that on consumption liberate cyanide in the body, which can lead to konzo, an irreversible spastic paraparesis of the legs that occurs suddenly amongst very poor village people who live on a monotonous diet of bitter cassava (Cliff et al., 1985; Howlett et al., 1990; Nzwalo and Cliff, 2011). Konzo is an upper motor neuron disease that causes permanent paralysis of the legs and affects mainly children and young women after childbirth. Many of these people suffer from malnutrition with low protein intake, in particular a shortfall of the essential sulfur amino acids methionine and cysteine/cystine that are needed for detoxification of cyanide (CN) to thiocyanate (SCN) in the body (Cliff et al., 1985). Konzo occurs in the Democratic Republic of Congo (DRC), Mozambique, Tanzania, Cameroon, Central African Republic and Angola (Bradbury et al., 2011; Allen, 2010) and because of the very large per capita consumption of cassava in Congo (Nhassico et al., 2008) konzo probably occurs there as well. The total number of konzo cases reported up to 2009 is 6788 (Nzwalo and Cliff, 2011), with 3469 cases in DRC, but this is very much less than the true number, because konzo occurs in isolated areas and during crises such as war and drought. An estimate of 100,000 konzo cases in DRC was made in 2002 (Diasolua Ngudi, 2005). In 2010 Action Against Hunger (ACF) in Kwango District, Bandundu Province, DRC, located 2218 konzo cases (Kasonga and Calo, 2011; Delhourne et al., 2012). Konzo occurs in at least four provinces, Bandundu, Kasai Oriental, Kasai Occidental and South Kivu (Chabwine et al., 2011) (Figure 1), but there has been no country wide survey of konzo in the DRC.

Konzo has been prevented in 13 villages in the DRC with a population of nearly 10000 people by use of the wetting method by village women to remove cyanogens from cassava flour (Banea et al., 2012, 2013, 2014a). The wetting method consists of adding water to cassava flour, spreading the wet flour in a thin layer on a mat for 2 h in the sun or 5 h in the shade to allow the escape of hydrogen cyanide gas produced by breakdown of linamarin by the enzyme linamarase (Bradbury, 2006; Cumbana et al., 2007; Bradbury and Denton, 2010). The damp flour is then cooked in the traditional way to make a thick porridge called fufu. The women readily accepted the wetting method and still continued to use it more than a year after the intervention ceased and the method has spread by word of mouth to other villages (Banea et al., 2014b). Furthermore, a correlation has been found between the monthly percentage incidence of konzo and the monthly cyanide intake as measured by the urinary thiocyanate content of children (Banea et al., 2014a). It is therefore likely that konzo results from high cyanide intake from bitter cassava, coupled with a low intake of sulfur amino acids.

It is important to have a good estimate of the total number of people with konzo in DRC and other countries, so that the prevention methods that we have developed can be systematically applied towards the complete prevention of this disease. In this paper we report the konzo prevalence and nutrition of konzo cases in 30 villages in three health zones in Kwilu District, Bandundu Province.

MATERIALS AND METHODS

Study area

In order to discover the most affected konzo areas in Kwilu District, discussions were held with Kwilu District health authorities in Kikwit in July 2012. Four health zones were identified as most affected Masimanimba, Kingandu, Payikongila and Djuma, but Djuma health zone was excluded for logistical reasons (Figure 1). Thirty four villages were identified as most affected and konzo cases were found in thirty villages.

Food and diseases of the people

The most important food crops are cassava, maize, peanuts, squash and cowpeas. Cassava is the staple food eaten as fufu, mainly with cassava leaves and vegetables. There is some small livestock, poultry, small scale fishing and fish farming for sale. The dry season is from mid-May to mid-August and there is another short dry season in February-March. There are two lean periods in August to November and in March, when household food security is critical and the number of meals is reduced to one meal per day. The most common diseases are malaria, acute respiratory infection, diarrhea, malnutrition and trypanosomiasis.

Health zone visits and survey methods

Discussion were conducted with the health authorities in each of the three health zones and supervisors were identified in each zone. Four supervisors were given training in Kikwit City by the konzo team of Programme National de Nutrition (PRONANUT). Each trained supervisor was responsible for collecting data on konzo cases in the health areas of his health zone. In each health zone the most affected villages were identified first. A census was taken of each affected village and suspected cases of konzo or those with gait difficulties were identified by the community health workers, who are trained to recognise common diseases including konzo and to refer the case to a health centre. Konzo cases were examined by doctors from PRONANUT following a standardised protocol (World Health Organisation, 1996) as follows: (1) a visible spastic walk or run, (2) an abrupt history of the onset of the disease in less than one week in a person in good health, with no futher progression of the disease, (3) a bilateral exaggeration of knee jerks and/or Achilles tendon (ankle clonus). The month and year were recorded when konzo occurred. Five surveyors were trained to administer the food household consumption questionnaire for each family with a case of konzo.

Food consumption score (FCS) in konzo household

The food consumption data of 155 households with konzo (some had multiple cases of konzo) were surveyed in the three health zones and this was used to calculate the World Food program food consumption score (FCS) using the equation:

$$FCS = a_{staple}\, x_{staple} + a_{pulse}\, x_{pulse} + a_{veg}\, x_{veg} + a_{fruit}\, x_{fruit} + a_{animal}\, x_{animal} + a_{sugar}\, x_{sugar} + a_{oil}\, x_{oil},$$

Where x_i = number of days for which each food group was consumed during the past seven days, a_i = weight of each food

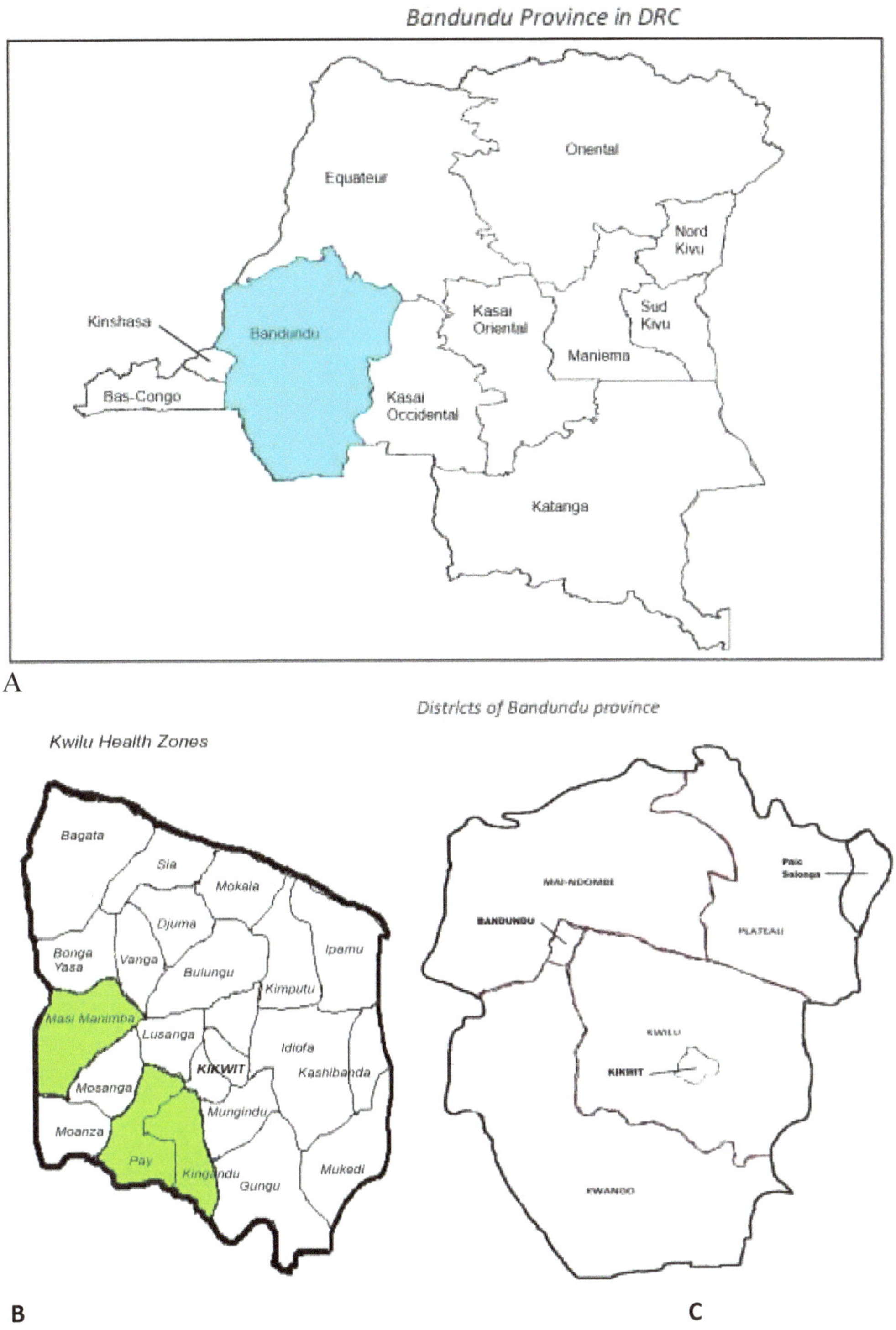

Figure 1. A. The provinces of DRC showing Bandundu, Kasai Occidental, Kasai Oriental and South Kivu where konzo is present; **B.** Health zones of Kwilu District with the survey ones marked; **C.** Districts of Bandundu Province.

group as follows: staple (cassava) 2, pulses (beans, peas, groundnuts) 3, vegetables and leaves 1, fruit 1, meat and fish 4, milk 4, sugar 0.5, oils, fats and butter 0.5. The thresholds used were: 0-23.9 for poor food consumption, 24.0-37.9 for borderline food consumption, and >=38 for acceptable food consumption (Interagency Work. Rep. WFP-FAO 2008).

Mid upper arm circumference (MUAC) and nutritional status

The mid upper arm circumference was used in order to identify high-risk malnourished (> 18 years old) konzo patients (Unicef 2012; Projet Sphere). A MUAC of < 210 mm indicated a person with malnutrition. There were 64 konzo patients measured in

Table 1. Percentage konzo prevalence in descending order for thirty villages in Kwilu District.

Village	Population	Number of konzo cases	% Konzo
Kongila Ndola[a]	106	18	17.0
Kongila Kianfu[a]	495	38	7.7
Lubamba[a]	341	9	2.6
Kinzefa[a]	420	8	1.9
Kabanga[b]	564	6	1.1
Kiwawa Kapa[c]	1453	12	0.83
Mulasi[c]	248	2	0.81
Kizefo[c]	1153	9	0.78
Kiwawa Pont[c]	1805	13	0.72
Mvudi[b]	564	4	0.71
Makanga[c]	1047	7	0.67
Mashini[b]	734	4	0.54
Kipongi[c]	1104	6	0.54
Kikonga[c]	751	4	0.53
Kimeso[c]	231	1	0.43
Yoshi[c]	467	2	0.43
Cite Masi[c]	734	3	0.41
Sakani[b]	734	3	0.41
Mbeko[c]	1030	4	0.39
Lunza[c]	805	3	0.37
Bibamba[c]	549	2	0.36
Kifunga Centre[c]	2420	6	0.25
Kikongo[c]	400	1	0.25
Kilombo M[c]	405	1	0.25
Kiwamba[c]	534	1	0.19
Kikiama[c]	564	1	0.18
Kimpembe[c]	564	1	0.18
Kiko[c]	734	1	0.14
Kinzenga Q3[c]	950	1	0.11
Mudigongo[c]	887	1	0.11
Total	22793	172	

[a] Payikongila health zone; [b]Kingandu health zone; [c] Masimanimba health zone

Masimanimba, 13 in Kingandu and 46 in Payikongila.

Ethics statement

The Ministry of Public Health Ethics Committee in DRC approved the survey protocol, which was also approved by Kwilu district health authorities and health zone health authorities. The chiefs of the villages accepted the census and medical examination of their people for konzo and distribution of a questionnaire on food consumption.

RESULTS

Thirty villages were identified in Masimanimba, Kingandu and Payikongila health zones to be surveyed, 22 in Masimanimba, four in Kingandu and four in Payikongila health zones. In a population of 22793 people there were 266 people with walking difficulties of whom 172 had konzo, giving a mean konzo prevalence of 0.75% (Table 1). The four villages in Payikongila health zone had the highest konzo prevalences ranging from 17% down to 1.9% with 42% of the total konzo cases. The main group of villages in Kingandu and Masimanimba health zones had konzo prevalences ranging from 1.1% down to 0.11% (with 9 villages recording only one konzo case) and accounting for 58% of the total number of konzo cases. Overall 53% were female and 47% male. Of the women who contracted konzo, 12% were nursing a baby at the time of the konzo attack. The age at onset was 2-4 years, 9%; 5-14 years, 53%; 15-44 years, 34%; 45-55 years, 4%. There was an abrupt onset in less than one day in 83% of cases and the remainder over 2-7 days.

Table 2. Comparison of % konzo prevalence, % malnutrition[a] and % nutrition (poor, borderline, acceptable)[b] of people in three Kwilu health zones.

Health zone	% Konzo prevalence	% Malnutrition[a]	Nutrition (%) based on food consumption score		
			Poor	borderline	acceptable
Masimanimba	0.44	20	11	28	61
Kingandu	0.66	8	18	6	76
Payikongila	5.4	50	22	38	40

[a]Based on mid upper arm circumference (MUAC) measurement of adult konzo patients (Unicef, 2012; Projet Sphere). [b] Based on food consumption score (FCS) of konzo families, see methods (Interagency Work. Rep., WHO-FAO, 2008).

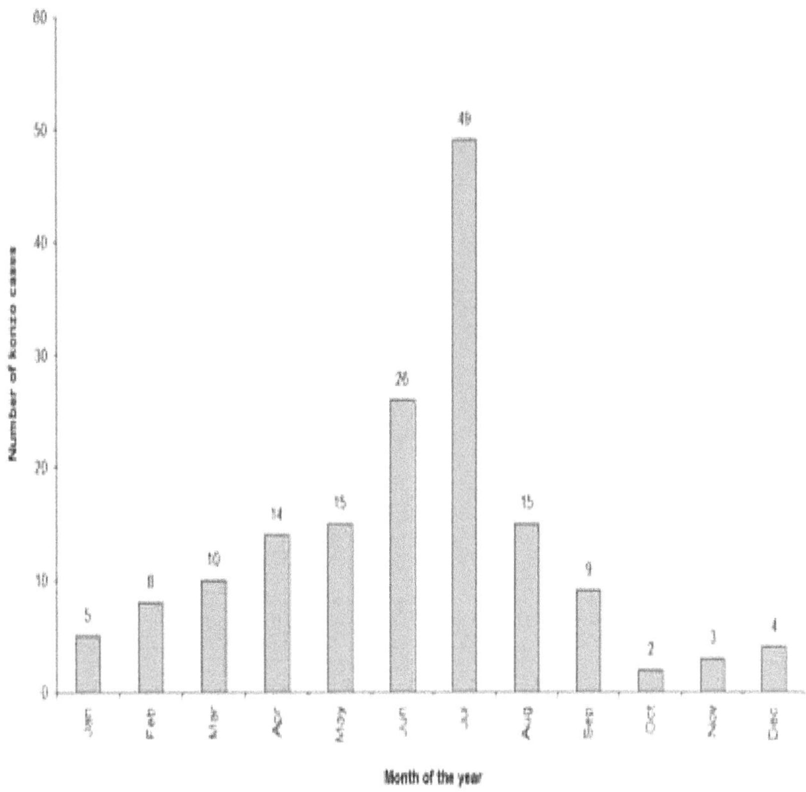

Figure 2. Monthly distribution of onset of konzo cases.

Bilateral exaggerated knee jerks were observed in 96% of cases. Speech disorders occurred in 22% of konzo cases, 17% had blurred vision and 2% goitre. It was found that 7% of people with konzo were unable to walk, 25% needed one or two sticks and 68% walked unaided but with a limp. These numbers have been combined with those from six other konzo studies in DRC and Tanzania (Banea et al., 2014a) and the results (mean +/- SD) are 8+/-5 severe, unable to walk; 26+/-13 moderate, needing one or two sticks; 66+/-12 mild, no sticks needed.

Table 2 shows the mean konzo prevalence in each health zone, % malnutrition among konzo cases measured by MUAC and the nutrition measured by a food consumption survey of the konzo families. The monthly occurrence of konzo cases is shown in Figure 2 and the years in which they occurred in Figure 3. The peeled cassava roots are soaked (retted) in a pond or a stream or at home for 2-3 days, but only for 1-2 days during the dry season. About one third of the konzo households surveyed had difficulty to ensure proper soaking, due to theft during soaking in 87% of cases or lack of water in 13% of cases. One quarter of households had consumed only one meal the day before the survey in July. Survival strategies used during periods of food shortage (August to November and in March) were to fall back on food gathering, use of less preferred foods and also to harvest crops early. Nearly all households (93%) with cases of konzo had houses made of straw and only12% had

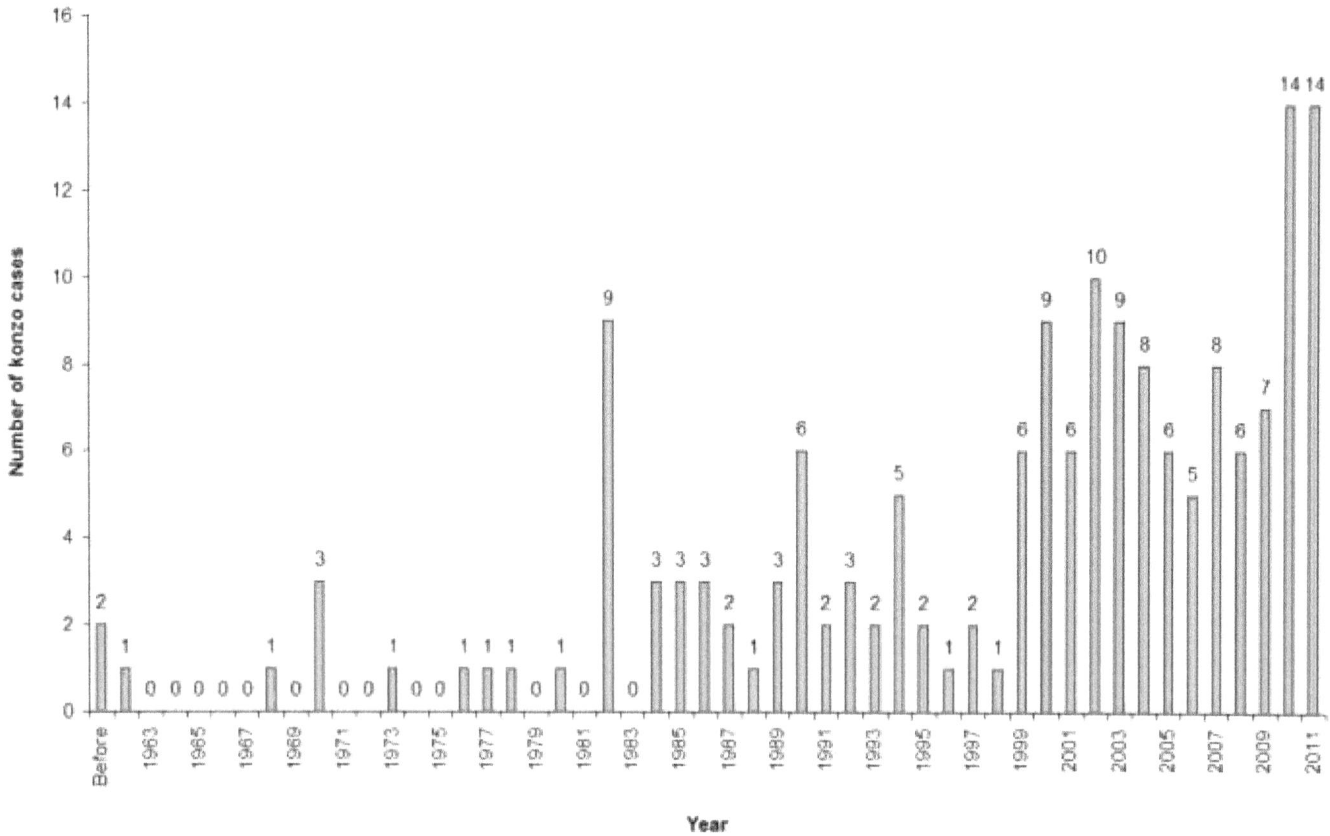

Figure 3. Annual distribution of onset of konzo cases.

access to a source of safe drinking water.

DISCUSSION

The data in Table 1 show that konzo prevalence is very high in four villages of Payikongila health zone with 42% of total konzo cases and much less in the 26 villages of Kingandu and Masimanimba health zones with 58% of total konzo cases. The high konzo prevalence in Payikongila health zone is probably due to the much poorer nutrition and higher degree of malnutrition of konzo cases in Payikongila compared with konzo patients from the other health zones (Table 2). Unfortunately, we do not have the urinary thiocyanate content of the school children, which would have allowed comparison between the cyanide intake of children from each health zone.

In Figure 2, the monthly incidence of konzo throughout the year shows a strong maximum incidence in July, the peak cassava harvesting season in the dry season, which agrees with previous work (Banea et al., 2012, 2013). The survey of konzo families showed that short soaking for 1-2 days instead of 3-4 days (Banea et al., 1992) was common especially in the dry season, which together with a peak in consumption of cassava, caused a peak in both cyanogen intake and konzo. The yearly incidence of konzo (Figure 3) shows that konzo cases have occurred every year since 1984, with much larger annual numbers since 1999, particularly in 2010 and 2011. This confirms the very concerning results found in 13 villages in Popokabaka and Boko health zones in nearby Kwango district, where konzo incidence has greatly increased since 2009 in this area of Bandundu Province (Banea et al., 2012, 2013, 2014a). By combining the data from this survey and the four interventions, it was found that 48% of the konzo incidences were in the three years from 2009 onwards and 52 % in the preceding 20+ years.

By the successful use of the wetting method by village women to remove cyanogens from cassava flour we have prevented konzo in 13 villages with 9191 people (Banea et al., 2012, 2013, 2014a). Furthermore, we have shown that there is a significant correlation between month by month percentage konzo incidence and monthly cyanide intake, as measured by the percentage of children with high urinary thiocyanate content (Banea et al., 2014a). It is therefore likely that konzo arises from high cyanide intake and malnutrition, specifically a shortfall of sulfur amino acids, also being a factor (Nzwalo and Cliff, 2011). We have prevented konzo by decreasing cyanide intake, but the results in Table 2 suggest that malnutrition may be an important factor. The importance of malnutrition was evident in three unrelated konzo epidemics in which people of the same ethnic group as those who had high konzo prevalence, but living only about 5 km away had virtually zero konzo prevalence. Those who were protected from konzo in the Mozambique konzo epidemic lived near the sea and consumed fish (Ministry of Health, 1984), as did those in Tanzania who lived near Lake Victoria (Howlett et al., 1992) and those protected in DRC lived in the forest and ate animals from the forest (Banea Mayambu, 1993). Thus, improvement of the diet particularly by inclusion of animal protein is probably a means of preventing konzo. This is part of the cross-sectoral approach used by Action Against Hunger (ACF) (Kasongo and Calo, 2011). The intervention of ACF has reduced konzo incidence by 84% between 2010 and 2011 (Delhourne et al., 2012). The reduction of cyanide intake by educating the women about the cause of konzo and how to use the wetting method to remove cyanogens, is direct, effective and less expensive than the cross-sectoral approach which involves improvement of the diet and other factors. Ideally, reduction of cyanide intake and reduction of malnutrition should be used together (Cardoso et al., 2005).

Conclusion

In three health zones of Kwilu District there were 172 cases of konzo in 30 villages. The villages of Payikongila health zone had much higher konzo prevalences and the konzo cases much poorer nutrition, than in the villages of Masimanimba and Kingandu health zones. Since 2009 there has been a large increase of yearly konzo incidence in four interventions to prevent konzo using the wetting method in nearby Kwango District and also in this survey in Kwilu District. Using the data from the four interventions and this survey it was found that 48% of konzo cases occurred from 2009 onwards and 52% of cases in 20+ years preceding 2009. The use of the wetting method to reduce cyanogen intake has prevented konzo in 13 villages costing an average $27 per person and is more effective in preventing konzo and cheaper than broad based interventions.

Conflict of interests

The authors did not declare any conflict of interest.

ACKNOWLEDGMENTS

We thank the Kwilu district health authorities and the health zone authorities for Masimanimba, Kingandu and Payikongila health zones. We also thank the village chiefs and all the people involved in this konzo and nutrition survey.

REFERENCES

Allen TJ (2010). Konzo in Angola. CCDN News No 16, 3-4.

Banea JP, Bradbury JH, Mandombi C, Nahimana D, Denton IC, Foster MP, Kuwa N, Tshala Katumbay D (2014a). Prevention of konzo in the DRC using the wetting method and correlation between konzo incidence and percentage of children with high urinary thiocyanate level. Afr. J. Food Sci. 8:297-304.

Banea JP, Bradbury JH, Mandombi C, Nahimana D, Denton IC, Kuwa N, Tshala Katumbay D (2013). Control of konzo by detoxification of cassava flour in three villages in the Democratic Republic of Congo. Food Chem. Toxicol. 60:506-513.

Banea JP, Bradbury JH, Mandombi C, Nahimana D, Denton IC, Kuwa N, Tshala Katumbay D (2014b). Effectiveness of wetting method for control of konzo and reduction of cyanide poisoning by removal of cyanogens from cassava flour. Food Nutr. Bull. 35:28-32.

Banea JP, Nahimana D, Mandombi C, Bradbury JH, Denton IC, Kuwa N (2012). Control of konzo in DRC using the wetting method on cassava flour. Food Chem. Toxicol. 50:1517-23.

Banea JP, Poulter NH, Rosling H (1992). Shortcuts in cassava processing and risk of dietary cyanide exposure in Zaire. Food Nutr. Bull. 14:137-143.

Banea Mayambu JP (1993). Cassava processing, dietary cyanide exposure and konzo in Zaire. Masters Thesis. Uppsala University, Uppsala, Sweden, p. 55.

Bradbury JH (2006). Simple wetting method to reduce cyanogen content of cassava flour. J. Food Comp. Anal. 19:388-93.

Bradbury JH, Cliff J, Denton IC (2011). Uptake of wetting method in Africa to reduce cyanide poisoning and konzo from cassava. Food Chem. Toxicol. 49:539-542.

Bradbury JH, Denton IC (2010). Rapid wetting method to reduce cyanogen content of cassava flour. Food Chem. 121:591-94.

Cardoso AP, Mirione E, Ernesto M, Massaza F, Cliff J, Haque MR, Bradbury JH (2005). Processing of cassava roots to remove cyanogens. J. Food Comp. Anal. 18:451-460.

Chabwine JN, Masheka C, Balol'ebwami Z, Maheshe B, Balegamire S, Rutega B, wa Lola M, Mutendela K, Bonnet MJ, Shangalume O, Balegasmire JM, Nemery B (2011). Appearance of konzo in South Kivu, a wartorn area in the Democratic Republic of Congo. Food Chem. Toxicol. 40:644-649.

Cliff J, Martensson J, Lundquist P, Rosling H, Sorbo B (1985). Association of high cyanide and low sulphur intake in cassava induced spastic paraparesis. Lancet 11:1211-13.

Cumbana A, Mirione E, Cliff J, Bradbury JH (2007). Reduction of cyanide content of cassava flour in Mozambique by the wetting method. Food Chem. 101:894-97.

Delhourne MM, Mayans J, Calo M, Guyot-Bender C (2012). Impact of cross-sectoral approach to addressing konzo in DRC. Field Exch. 44:50-54.

Diasolua Ngudi D (2005). Konzo and cassava toxicity: a study of associated nutritional factors in the Popokabaka District, DRC. Ph D Thesis, Universiteit Gent, Belgium.

Howlett WP, Brubaker G, Mlingi N, Rosling H (1992). A geographical cluster of konzo in Tanzania. J. Trop. Geogr. Neurol. 2:102-108.

Howlett WP, Brubaker GR, Mlingi N, Rosling H (1990) Konzo, an epidemic upper motor neuron disease studied in Tanzania. Brain 113:223–35.

Interagency Workshop Report WFP-FAO (2008). Measures of Food Consumption–Harmonizing Methodologies. April 2008, Rome. Italy.

Kasonga E, Calo M (2011). A cross-sectoral approach to addressing konzo in DRC. Field Exch. No 41, 3-5.

Ministry of Health Mozambique (1984). Mantakassa: an epidemic of spastic paraparesis associated with chronic cyanide intoxication in a cassava staple area in Mozambique. 1. Epidemiology and clinical laboratory findings in patients. Bull. WHO 62:477-484.

Nhassico D, Muquingue H, Cliff J, Cumbana A, Bradbury JH (2008). Rising African cassava production, diseases due to high cyanide intake and control measures. J. Sci. Food Agric. 88:2043-49.

Nzwalo H, Cliff J (2011). Konzo: from poverty, cassava and cyanogen intake to toxico-nutritional neurological disease. PloS Negl. Trop. Dis. 5(6): e1051.doi:10.1371/journal.pntd.0001051.

Projet Sphere : La charte humanitaire et les standards minimum de

l'intervention humanitaire, page 258.
Unicef (2012). Unicef/Ministère de la Santé publique:Protocole national de Prise en charge integrée de la malnutrition aiguë (PCIMA), Kinshasa, Avril 2012.
World Health Organisation (1996). Konzo: a distinct type of upper motor neuron disease. Wkly. Epidemiol. Rec. 71:225-232.

Effect of flooding and salinity as a result of climate change on tomato production in the coastal zone of Benin

Vincent Ezin[1] , Ibouraïma Yabi[2], Emeric G. M. Kochoni[3] and Adam Ahanchédé[1]

[1]Department of Crop Production, Faculty of Agricultural Sciences, University of Abomey-calavi, Benin.
[2]Department of Geography, University of Abomey-calavi, Benin.
[3]Institut National de la Recherche Scientifique - Eau Terre Environnement (INRS-ETE), Canada.

The production of tomato in the coastal regions is affected by flooding and salinity conditions. A survey was conducted in this region to know the impact of change in climate on tomato production. Two types of flooding conditions are prevalent: flash flooding and severe flooding where water remains stagnant for weeks. The results also show that climate change is exacerbating the existing a-biotic factors (flooding and salinity) by significantly affecting tomato development, growth and yield and yield components as reported by farmers. Flooding and salinity pose a serious threat to some producers and forced them to abandon their agricultural lands in severe cases. The repeated yield losses in some other areas caused tomato producers to move from their field close to the sea to the field far away. The producers said that flooding conditions commence at the end of June till middle of August. The producers previously grew their tomato in the areas of study in May but because of repeated flooding every year they were compelled to shift their sowing calendar. A total of 16 tomato varieties were recorded in the areas of study. Gbamingbo variety was moderately resistant to flooding conditions while Aclinkonkoui and Petomèche varieties appear to be moderately tolerant to salinity due to their average performance in terms of yield and yield components.

Key words: Coastal areas, tomato production, flooding, salinity, climate change.

INTRODUCTION

In recent years, global warming and its effect on crop production has become perceptible. As a matter of fact, agriculture is highly sensitive to environmental factors and weather extremes, such as flood, salinity, temperature and drought (de la Peña and Hughes, 2007). Human activity has already changed atmospheric characteristics and such trends are expected to continue which will pose many problems to agricultural production and farmers will be faced with a lot of challenges. Ceccarelli et al. (2010) reported that climate change that are taking place at present will have – and are already having – a negative impact on food production and food quality with the poorest farmers and the poorest countries most at risk. Acquaah (2010) stated that a-biotic environmental stresses are responsible for about 70% of yield reduction of crops in production. In recent years, Benin Republic has been devastated by flooding. Two-thirds of the country has been affected and all the growing crops in the regions were swept off by floods.

Flooding, saline soil, drought, and heat are major environmental factors crops must endure and overcome in order to produce good yields. Farmers' worries about these occurrences are for good reason. They cause massive crop-yield losses every year even more, in fact, than losses from insects and weeds (Serrano, 1999). The increase in salinity and floods negatively affect crop yields beyond the impacts of mean climate change. Climatic changes will influence the severity of environmental stress imposed on vegetable crops (de la Pena and Hughes, 2007). It seems obvious that any significant change in climate on a global scale should impact local agriculture, and therefore affect the world's food supply (IPCC, 2007). Extreme climatic conditions will also negatively impact soil fertility and increase soil erosion (Brinkman, 1990). Global warming is predicted to lead to thermal expansion of sea water resulting in a rise of sea level which may range from 0.1 to 0.5 m (4 to 20 inches) according to present estimates of the Intergovernmental Panel on Climate Change (IPCC, 2007). The increase in the level of sea and the high irrigation of water will definitely bring about high salinity in the coastal regions in Africa and particularly in Benin Republic.

Climate change will seriously affect crop production as years go on. More erratic rainfall patterns and increased salinity caused by climate change will consequently be expected to further reduce crop productivity, and developing countries in the tropics are particularly vulnerable. In these areas, increasing salinity and flooding will be major factors limiting sustaining and increasing vegetable production. Climate change effects on Benin's agriculture are evident in the coast, particularly in valley of Ouémé and Mono. A farm investigation conducted with producers to understand socio-economic factors related to innovation needs showed that all vegetable crops grown in coastal regions of Benin republic have been heavily impinged by salinity (Ezin et al., 2012).

A large majority of new varieties currently under cultivation have been bred for improved resistance to pest and diseases, as opposed to tolerance to abiotic conditions, the primary cause of crop loss (Wang et al., 2003). Habash et al. (2009) demonstrated that plant breeding might help in developing new cultivars with enhanced traits better suited to adapt to climate change conditions using both conventional and genomic technologies.

Climate change has started influencing greatly rainfall patterns where the suitability of land for different types of crops is affected. In the southern part of Benin rainfall water remains stagnant due to severe flooding.

Moderate water movement can reduce flood damage keeping them respiring and alive. Drainage within one to two days increases the chance of plant survival.

The objective of the present study was to evaluate the impact of flooding and salinity on tomato production in coastal zone of Benin.

MATERIALS AND METHODS

Study sites

The study was conducted in the coastal regions of Benin Republic (Figure 1). All the four departments along the coastal regions were surveyed due to the fact that they are tomato producing areas. Benin is bounded by Togo to the west, Nigeria to the east, and the Bight of Benin to the south. Benin's latitude ranges from 6°30′ N to 12°30′ N and its longitude from 1° E to 3°40′ E. The climate is of equatorial type with alternation of two rainy seasons (April-July and August end-November) and two dry seasons. Annual rainfall in the coastal area averages 1,360 mm. From December to January, there is harmattan: a dry and dusty West African trade wind.

Data source

The local governmental centers for agricultural development in the areas of study were identified and were approached to get information on tomato producers in the district. Then quantitative and qualitative data were collected through sampling method using questionnaires. Structure questionnaire was used in this study.

Sampling and data collection

During the exploratory phase, the production sites were visited. Group interviews were conducted with producers in different villages and cities visited to gather general information on the effects of flooding and salinity on tomato production. Additional information was obtained from local governmental centers for agricultural development on tomato production and the producers in each district surveyed.

Data were collected from 130 respondents of tomato producers. The respondents were selected randomly in each district. The in-depth survey was conducted based on interviews, observations, and structured questionnaires. Parameters such as gender, age groups, education level, social status, number of dependents, credit accessibility, agricultural technician's support obtained, input use, the number and diversity of cultivated varieties, land availability, cultivated area, cost price, effects of flooding and salinity on growth, development and yield of tomato, salinity and flooding control strategies, etc were collected.

Data analysis

Data collected were processed in Microsoft Access 2003(11). Then, the entered data were verified in order to avoid errors and inconsistencies. Descriptive statistics were essentially used in analyzing the data. Standard deviation, percentage of age, sex, education level, social economic status, input use, land availability, cultivated areas in 2011, and cost price were computed through descriptive statistics. SPSS (Statistical Package for Social Science) was used for descriptive analysis, correlation of independent variable.

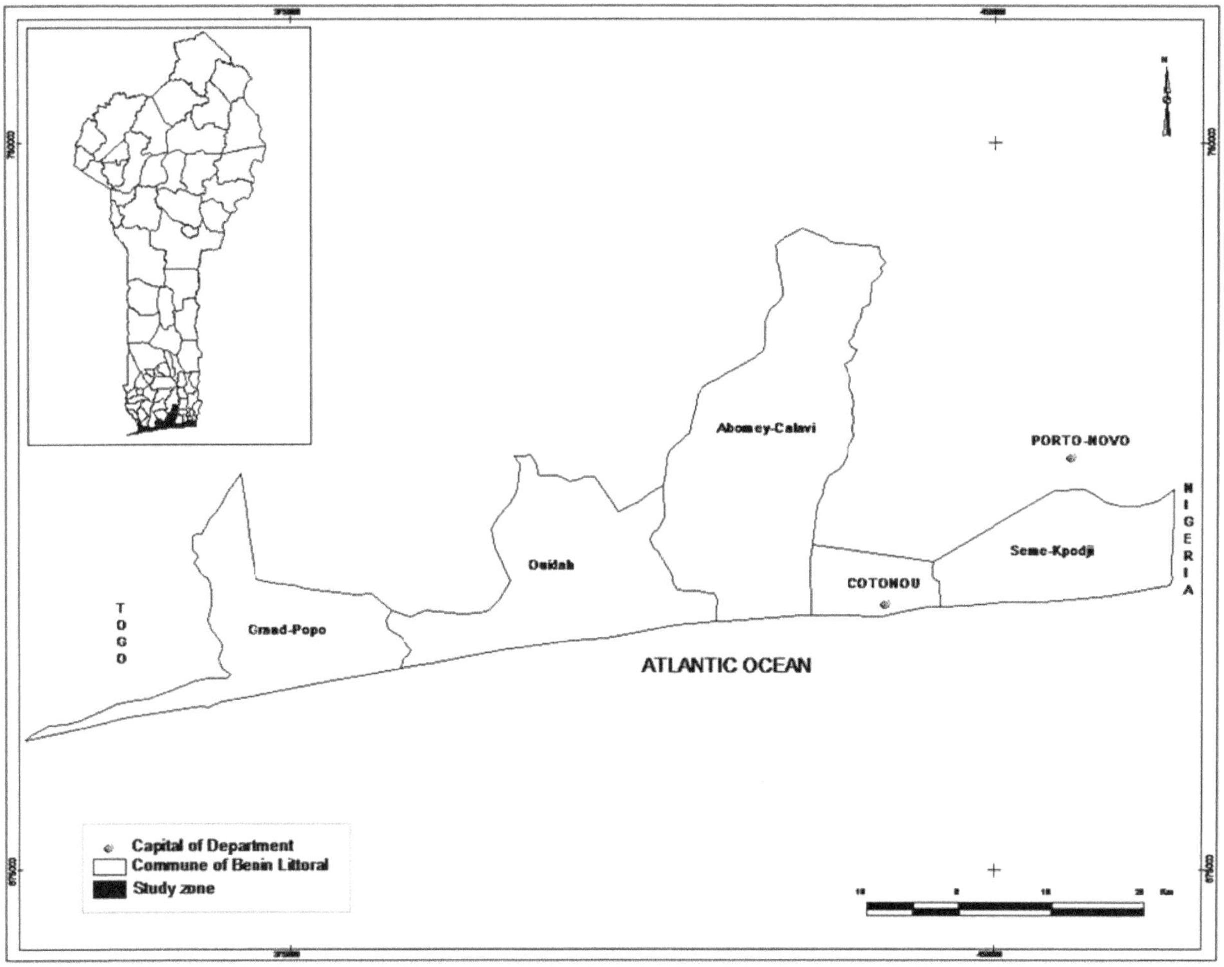

Figure 1. Benin Republic map and communes investigated.

RESULTS AND DISCUSSION

Coastal areas

The coastal regions (Figure 1) play a vital role in Benin economy. There are three main activities in the area: (1) agriculture is practiced in all the coastal areas while livestock is only practiced in Sèmè district, (2) Fishery is practiced throughout the coastal areas and has been a lucrative activity (3) the only sea port is in Cotonou city and is of huge important for Benin economy development.

Flooding and salinity have been a problem for agriculture activities. Over the last decade, many houses were lost due to flooding of sea water especially in Cotonou. Unfortunately, the protective embankment along the coastal areas that the government launched 4 years ago has failed due to lack of funding. Only 2 km out of 125 km was done. This protective embankment with big stones could help to some extent to prevent further loss of lands due to flooding of sea water.

Two types of lands characterized the areas of study: low lands and medium low lands. Most of lands encountered are low lands. Low land (75%) is higher than medium land (25%).

Characteristics of the respondents

The randomly selected respondents were tomato producers but occasionally some do the fishing. From the 130 respondents 82.5 and 17.5% were male and female, respectively. About 71.5% of producers were less than 45 years old, 24.2% from 45 to 60 years old, and 4.3% more than 60 years old.

The average cultivated lands for the production of tomato ranged from 1.3 to 0.75 ha. The production of tomato provides with the majority of farmers (59%) an annual income of less than 700000F CFA. Tomato contribution to annual income represents 1 to 65%. The average size of the household varies between 6 and 12 persons. Our results are consistent with that of Adorgloh-

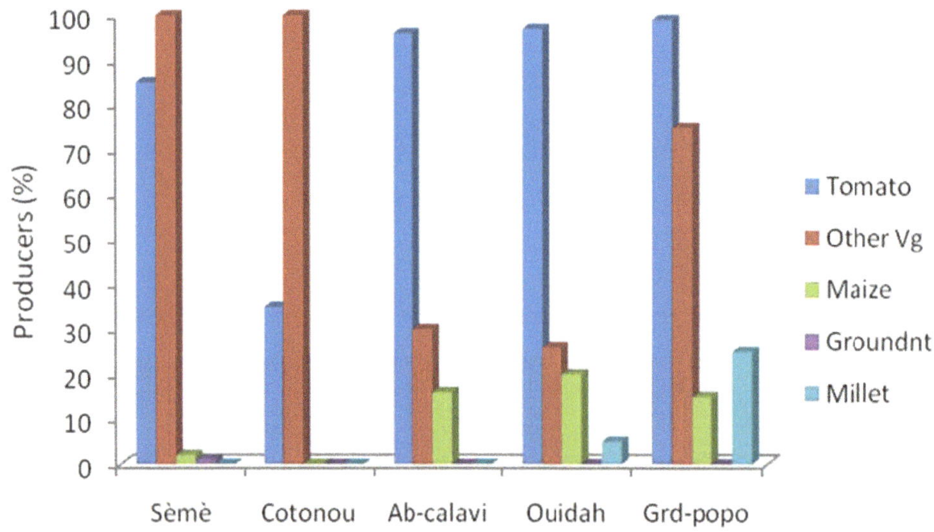

Figure 2. The main crops grown in the study sites.

Hessou (2006) who reported that tomato production plays a vital role in the economy of its producers in the South of Benin.

Crops grown in the area of study

Figure 2 shows the types of crops grown and demonstrates that tomato was grown in all the communes investigated. All the producers surveyed in Abomey-calavi, Ouidah and Grand-Popo mainly cultivated tomato while in Sèmè and Cotonou other vegetables are majorly produced when compared with tomato. Similarly, FAOSTAT (2012) reported that tomato crops are produced in urban and peri-urban areas of Benin.

Tomato varieties cultivated

Sixteen varieties were recorded in the communes investigated during our study: tounvi, gbataki, aclinkon, kekefo, pomme, adaka, gbamingbo, sonafel, ouaga, karaibo, ps royal, petomèche, mongal, tropimèche, 3fs and Brondelle. According to the respondents, ouga, ps royal, petomèche, mongal, tropimèche, 3fs, karaibo and Brondelle were imported from Ghana recently. Aclinkon, gbataki, kekefo and tropimèche were recorded in Cotonou; gbamingbo, kekefo, sonafel, adaka, tounvi in Abomey-calavi; Brondelle, tropimèche, ps royal, petomèche, 3fs, tounvi, karaibo in Ouidah, and other varieties Grand-popo, Sèmè et others.

Flood from 2010 to 2013 in Benin

The impact of climate change on abiotic stress such as flooding has been remarkable over the past five years. The flooding that has been occurring since 2010 is dramatic and devastating. In 2010, the following areas (Figure 3): Ketou, Zangnanado, Ouinhi, Adja-Ouere, Bonou, Adjohoun, Akpro-misseretè, Dangbo, So-Ava, Abomey-calavi, Sèmè-Kpodji, Cotonou, Oiudah, Grand-popo, Come, Kpomassè, Athiemé, Lalo were seriously affected. More than 130,000 hectares of crops were lost due to the heavy flooding across the country. All the tomato crops produced in those areas were damaged by the flood. Severe crops losses recorded lead to food insecurity and high food price. Some farmlands in those areas are still uncultivable. Our study is consistent with that of Khan et al. (2012) who reported that recent year of flooding in Hoar areas of Bangladesh were severe and damaged agricultural crops in a large amount. They further stated that 2010 flood in this region severely damage huge amount of agricultural production. Cumhur and Malcolm (2008) reported that environmental factors such as flooding will negatively impact agricultural crops.

Municipalities and communes of Lokossa, Bopa, Athiemé and Cotonou, were most affected by flooding in 2011 leading to the destruction of farmland and crops. Our results are in agreement with that of Thomas and Prasad (2003) who reported that the modified environ-ment resulting from global warming and climate change, will drastically affect the production and productivity of food crops.

In 2012, the southern and the northern parts of the country were severely impeded by flooding causing the destruction of crops and farmlands. In each case, many lives were lost and roads and infrastructures devastated. As years go by the flooding is a serious problem in Benin causing food price to rise even in the rain season where the price of commodities is supposed to come down. The effects of flooding have significantly increased and get worsened as a result of climate change. Atkinson et al. (2008) reported that climate change will impact negatively food production, food quality and food security. Altieri and

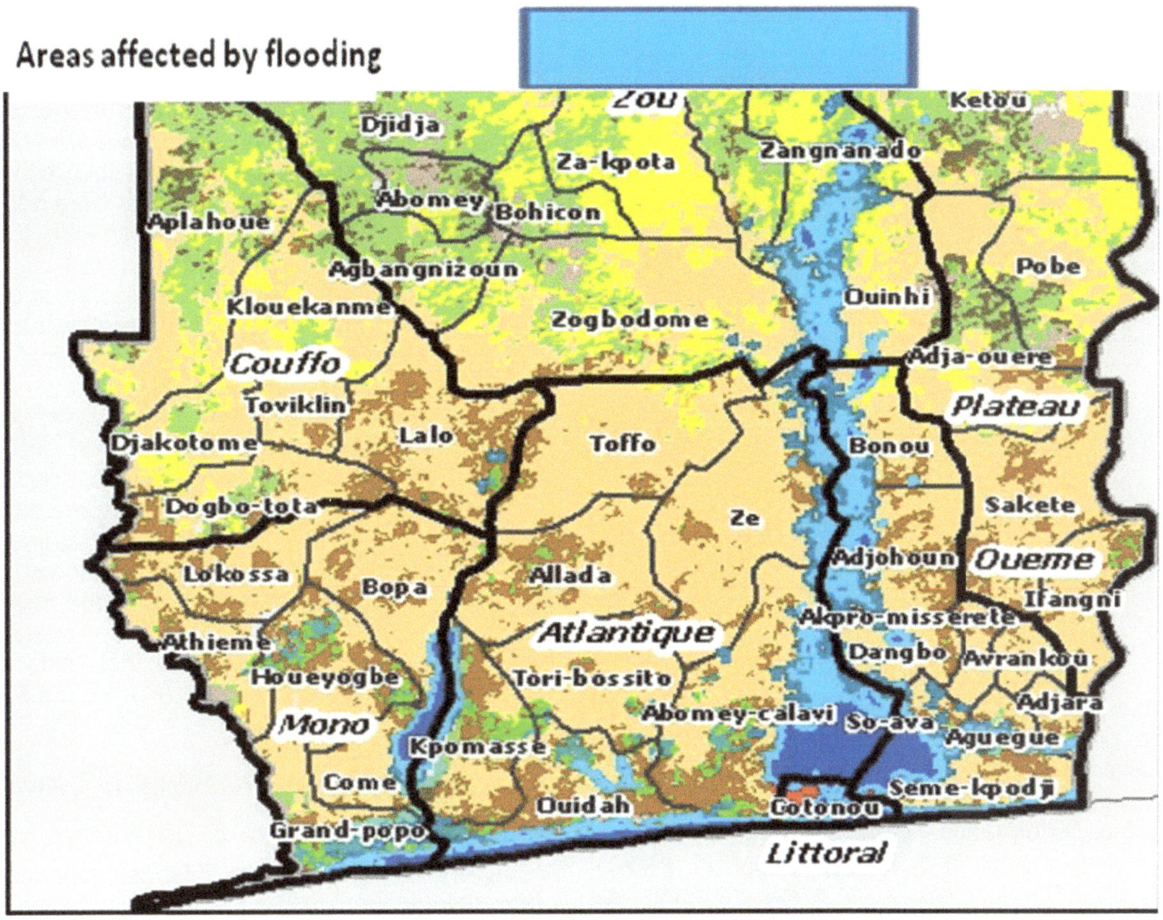

Figure 3. The villages and cities affected by flooding in 2010 and 2011.

Koohafkan (2003) also stressed that food security is potentially in danger than ever before.

Types of flooding

Most of tomato producers (87.3%) investigated said that flooding of their crops is mainly due to rainfed flood while 12.7% of respondents said it is due to ocean storm and tidal waves. They further stated that it does occur from the end of June till middle of August. The producers grew their tomato in the areas of study in May but because of repeated flooding every year they have shifted the sowing calendar. Till 2008, they still grew tomato in May but from 2011 their planting period has been August in order to avoid heavy rainfed flooding and losses of their crops. It is obvious to everybody now that climate change is occurring and the threat has become real and the question about this climate change occurrence is how to cope with it. Sidi (2012) reported that the flooding of September and October 2012 in Nigeria disrupted various sectors in the country and damaged crops like cassava, rice, yam, maize and tuber crops.

All the respondents acknowledged that due to flooding of sea water, a lot of lands were lost and the sea is progressing at alarming rate.

Flood damages tomato crops

According to tomato producers, flooding conditions caused the reduction of plant height and yield, total crop failure, yellowing of leaves, fruit rot and reduction of tomato market value and death of sensitive varieties. The same results were obtained by Kramer (1951), Kozlowski (1997), Núñnez-Elisea et al. (1999), Ashraf (2003), Walter et al. (2004) and Ezin et al. (2010) who stressed that 2 days of flooding caused reduction of plant height, wilting, leaf senescence, yield reduction and 8 days of continuous flooding lead to the death of tomato genotypes.

As shown in Figure 2, tomatoes and vegetables are the main crop grown in the areas of study. All of the tomato producers surveyed said that tomato and other vegetables were devastated by flooding. Table 1 show that only Gbamingbo is moderately tolerance to flooding amongst the 16 tomato varieties recorded. Farmers

Table 1. Evaluation of tomato varieties by farmers.

variety	**Flooding effects**	**Salinity effects**
Brondelle	Sensitive	Sensitive
Gbamingbo	Tolerant	Sensitive
Aclinkonkoui	Sensitive	Moderately tolerant
Kekefo	Sensitive	Sensitive
Touinvi	Sensitive	Sensitive
Gbataki	Sensitive	Sensitive
Sonafel	Sensitive	Moderately tolerant
Petomèche	Sensitive	Moderately tolerant
Ps royal	Sensitive	Sensitive
Mongal	Sensitive	Sensitive
tropimèche	Sensitive	Moderately tolerant
Ouaga	Sensitive	Sensitive
Adaka	Sensitive	Sensitive
3FS	Sensitive	Sensitive
Pomme	Sensitive	Sensitive
Karaibo	Sensitive	Sensitive

further said that severe flooding kill Gbamingbo variety as well. The other tomato varieties are very sensitive to flooding even 2 days of flooding are enough to damage them or render them unproductive. At 3 to 4 days of flooding, the performance of tomato varieties including Gbamingbo was poor and farmers failed to harvest tomato crops in the field, that is, there was complete crop failure.

Normally as from June the price of tomato, the most consumed vegetable, is affordable even to the poor households but unfortunately the price of tomato across the nation is still high for many reasons. On the one hand, the farmlands affected by previous flooding are yet to be available for crop production, and producers have shifted their sowing calendar due to the fact that across the country precipitation patterns have changed. On the other hand, the supply does not meet the demand due to the growing population.

In the recent survey conducted in June in the areas of study, due to the cultural practices put in place by farmers to avoid the deleterious effects of flooding on tomato, they are growing maize in the lands which were used for tomato production 4 years ago. The adjusted timing of sowing allows producers to steer clear of the flooding conditions but it is not beneficial to people who are expecting affordable tomato price. Therefore, adequate measures must be taken to avoid food insecurity now that climate change is inevitable. This study is similar to that of Sasson (2012) food security for Africa which is an urgent global challenge.

Salinity

Out of the 16 varieties recorded, only aclinkonkoui, petomèche and sonafel (Table 1) were moderately tolerant to salinity while the other varieties are sensitive. The key hindrance to the increase of crop production in the coastal areas is seasonally high content of salts in the root zone of the soil (Haque, 2006). In our study, the wet breeze from high tide and soil salinity were limiting factor to normal production and high yield as reported by farmers. For this study, soil samples were not collected and analyzed in order to determine the electrical conductivity.

Figure 4 illustrates the annual income of some tomato producers. Field 1 is the field where tomato was grown close to the sea while field 2 is where tomato was grown far away from the sea without the effect of salinity. From the graph annual incomes recorded in field 2 were significantly higher than those obtained and recorded in field 1. This indicates that salinity reduced the yield of tomato and lead to the reduction of tomato market value and consequently low income. In the last survey carried out some farmers have abandoned some fields where they produced tomato 4 years ago due to the repeat of low yield recorded. Salinity causes unfavorable environment and hydrological situation that limits normal crop production throughout the year (Haque, 2006).

Solutions used by producers to alleviate salinity effects

Farmers are still using their former methods to mitigate the deleterious effect of salinity on tomato production along the coastal regions despite the fact that climate change get more severe with time. They expressed their deception on the lack of improved varieties which could withstand the effect of climate change. Three different methods: fence with palisade (Figure 5), fence with maize/sorghum and intercropping are still employed to alleviate the effect of the tidal breeze loaded with salt which settled on tomato production in the coastal areas of the country. Most of the respondents (60%) used fence with palisade while 12 and 2% used fence with maize/sorghum and intercropping, respectively, to reduce the deleterious effect of tidal breeze which constitutes a constraint for tomato production in the regions. Only 26% of producers grew tomato in an open air without any measure of protection. Well-read people were amongst the respondents investigated and said that the rising of the sea level and waves are evident to them because the strategies of the fencing with palisade is no longer as efficient as before. They further stated that the level of the sea is higher between June and September as compared to past years. In Benin, more 10% of agricultural land in the area of study got lost due to the projection of sea and soil salinity. This is consistent with the results of Nicholls and Leatherman (1995): a 1 m sea-level rise would affect 6 million people in Egypt, with 12 to 15% of agricultural land lost, 13 million in Bangladesh, with 16% of national rice production lost, and 72 million in China and "tens of thousands" of hectares of agricultural

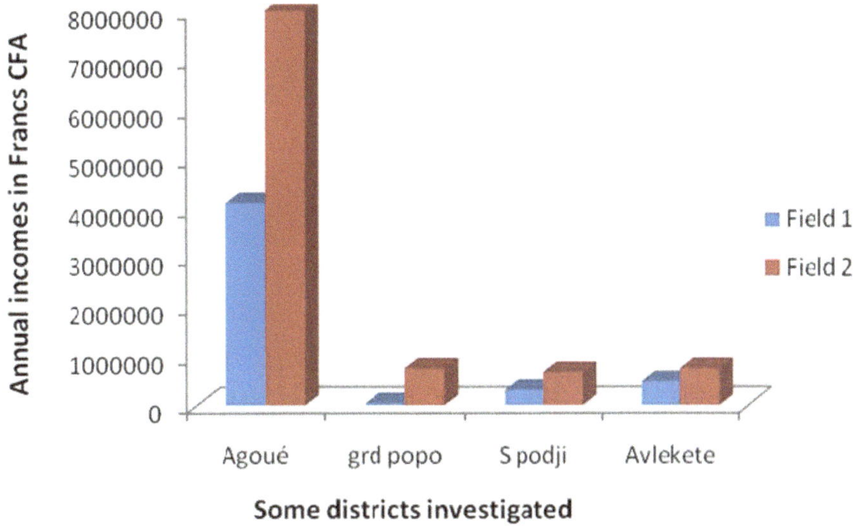

Figure 4. The Annual incomes at different locations, field 1: close to the sea and field 2: far away from the sea.

Figure 5. Method of palisade set up to abate the effect of tidal breeze

land. Tomato producers are desperate about getting other varieties which could withstand the negative impact of salinity on tomato production. Avlo district in the commune of Grand-Popo is yet to start growing vegetables and tomato in particular after they have given up their production due to high level of soil salinity.

Conclusions

The impact of climate change is real and preventive and adaptive measures must be taken to lessen its effects. Adequate cultural practices need to be adjusted and applied to avoid losses of tomato production to flooding and salinity. More tomato production is required to meet the demand of the growing population. New tomato varieties tolerant to flooding and salinity are needed to intensify the production of tomato. Tomato varieties that are resistant to both flooding and salinity will be of great advantage to the producers in the areas of study.

Conflict of interests

The authors did not declare any conflict of interests.

REFERENCES

Acquaah G (2010). Principles of plant genetics and breeding. Blackwell publishing 350 main Street, malden, MA 02148-5020, USA. p.385.

Adorgloh-Hessou RA (2006).Guide for the development of production system and marketing of quality vegetables in the urban and suburban regions of southern Benin. Report of consultation, IITA-Benin. p 82.
Altieri MA, Koohafkan P (2003). Enduring Farms: Climate Change, Smallholders and Traditional Farming Communities. Third World Network Environmental & Development Series 6. Penang, Malaysia: TWN.
Ashraf M (2003). Relationships between leaf gas exchange characteristics and growth of differently adapted populations of Blue panicgrass (*Panicum antidotale* Retz.) under salinity or waterlogging. Plant Sci. 166:69-75.
Atkinson MD, Kettlewell PS, Poulton PR, Hollins PD (2008). Grain quality in the Broadbalk wheat experiment and the winter North Atlantic oscillation. J. Agric. Sci. Cambridge 146:541-549.
Brinkman R (1990). Resilience against climate change? Soil minerals, transformations and surface properties, Eh, pH. In: Scharpenseel *et al.* (eds.). 1990. pp. 51-60.
Ceccarelli S, Grando S, Maatougui M, Michael M, Slash M, Haghparast R, Rahmanian M, Taheri A, Alyassin A, Benbelkacem A, Labdi M, Mimoun H, Nachit M (2010). Plant breeding and climate changes. J. Agric. Sci. 148:627-637.
Cumhur A, Malcolm SC (2008). The effects of global climate change on agriculture. Am. Eurasian J. Agric. Environ. Sci. 3 (5): 672-676
de la Peña R, Hughes J (2007). Improving Vegetable Productivity in a Variable and Changing Climate. SAT eJournal 4(1):1-22.
Ezin V, De la Pena R, Ahanchede A (2010). Flooding tolerance of tomato genotypes during vegetative and reproductive stages. Braz. J. Plant Physiol. 22(1):131-142.
Ezin V, Yabi I, Ahanchede A (2012). Impact of salinity on tomato production along the coastal area of Benin Republic. Afr. J. Environ. Sci.Technol. 6(4):214-223.
FAOSTAT (2012). (http://faostat.fao.org/).
Habash DZ, Kehel Z, Nachit M (2009). Genomic approaches for designing durum wheat ready for climate change with a focus on drought. J. Exp. Bot. 60: 2805-2815.
Haque AS (2006). Salinity problems and crop production in coastal regions of Bangladesh. Pak. J. Bot. 38(5):1359-1365.
IPCC (2007). Climate change 2007: Impacts, adaptation and vulnerability. Contribution of Working Group II to the Fourth Assessment Report of the Intergovernmental Panel on Climate Change. Cambridge, UK: Cambridge University Press. p. 976.
Khan MNH, Mia MY, Hossain MR (2012). Impacts of flood on crop production in *Haor* areas of two Upazillas in Kishoregonj. J. Environ. Sci. Nat. Resour. 5(1):193-198.
Kozlowski TT (1997). Response of woody plants to flooding and salinity. Tree Physiol. Monogr. 1: 1-29.
Kramer PJ (1951). Causes injury to plants resulting from flooding of the soil. Plant Physiol. 26:722-736.
Nicholls RJ, Leatherman SP (1995) Global sea-level rise. (Strzepek and Smith). pp. 92-123.
Núñez-Elisea R, Schaffer B, Fisher JB, Colls AM, Crane JH (1999). Influence of flooding on net CO_2 assimilation, growth and stem anatomy of *Annona* species. Ann. Bot. 84: 771-780.
Sasson A (2012). Food security for Africa: an urgent global challenge. Agriculture & Food Security, 1:2. http://www.agricultureandfoodsecurity.com/content/1/1/2.
Serrano R (1999). A glimpse of the mechanisms of ion homeostasis during salt stress. J. Exp. Bot. 50: 1023-1036.
Sidi MS (2012). The impact of the 2012 flooding on agriculture and food security in Nigeria using GIS. United Nations International Conference on Space-based Technologies for Disaster Management - "Risk Assessment in the Context of Global Climate Change", 7-9 November 2012, Beijing, China.
Thomas JMG, Prasad PVV (2003). Plants and the Environment /Global Warming Effects. University of Florida, Gainesville, FL, USA.
Walter S, Heuberger H, Schnitzler WS (2004). Sensibility of different vegetables of oxygen deficiency and aeration with H_2O_2 in the rhizosphere. Acta Hortic. 659:499-508.
Wang WX, Vinocur B, Altman A (2003). Plant responses to drought, salinity and extreme temperatures: towards genetic engineering for stress tolerance. Planta 218:1-14.

Hygiene practices and food contamination in managed food service facilities in Uganda

Sylvia Angubua BALUKA[1], RoseAnn MILLER[2] and John Baligwamunsi KANEENE[2]

[1]College of Veterinary Medicine, Animal Resources and Biosecurity (COVAB), Makerere University, Kampala, Uganda.
[2]Center for Comparative Epidemiology, Michigan State University, USA.

A longitudinal study was conducted to examine individual worker and institutional hygiene practices and bacterial contamination in food service facilities at Makerere University. Questionnaires regarding food service knowledge, attitudes, and practices were administered to 94 individual and institutional respondents from 16 facilities through in-person interviews. A total of 48 samples (3 per facility) were analyzed for evidence of contamination (total aerobic mesophilic bacteria, coliforms, *Escherichia coli, Salmonella*). Respondents with higher education levels had better knowledge and attitudes regarding food safety, but knowledge in specific areas were varied. The majority of individual workers used safe food handling practices, but the majority of institutions did not practice good environmental hygiene. The majority of food samples tested had APC and total coliform levels higher than acceptable, but only two tested positive for *Salmonella.* Food service worker training and managerial improvement of environmental hygiene are needed to improve food safety in these facilities.

Key words: Food safety, food service workers, coliforms, *E. coli.*

INTRODUCTION

Foodborne diseases are a challenge for both developed and developing countries (Da Cunha et al., 2012), and are a leading cause of illness and death in developing countries (Hassan et al., 2010). Despite concerted efforts for several decades, foodborne diseases remain a major global public health issue with substantial morbidity and mortality associated with the consumption of contaminated foodstuffs (Havelaar et al., 2010).

The measurement of the safety of foods has relied on evaluation of the microbiological quality of foods (Havelaar et al., 2010; Jacxsens et al., 2010). Bacterial counts in prepared food or water is a key factor in assessing the quality and safety of food, and can reveal the hygiene level adopted by food handlers in the course of preparation of such foods (Nkere et al., 2011). In a recent review, *E. coli*, *Shigella, Salmonella,* and *Campylobacter* spp. were the most commonly reported causes of gastrointestinal disease in sub-Saharan Africa (Fletcher et al., 2011), and all have been associated with foodborne disease (FDA, 2012). One study conducted in Kampala, Uganda, concluded that the microbiological safety of salads was unsatisfactory due to high bacterial counts and the presence of *Staphylococcus aureus*, which were attributed to inappropriate food handling hygiene and sanitation during preparation (Oiko, 2000, unpublished). However, in developing countries, monitoring the microbial safety of foods is not routinely practiced, due to a lack of infrastructure and effective food safety regulations and standards (Nguz, 2007).

Outbreaks of foodborne illnesses have been linked to improper food handling practices at food serving establishments (Baş et al., 2006; Çakiroğlu and Uçar, 2008; Hassan et al., 2010; Da Cunha et al., 2012). The most commonly reported food preparation practices that contribute to foodborne diseases include poor environmental hygiene, inadequate cooking, contaminated equipment, improper holding temperatures and food from unsafe sources (Guzewich and Ross, 1999; Da Cunha et al., 2012).

Food handlers, the people who are employed directly in the production and preparation of foodstuffs, are integral to reducing food safety risks (De Sousa, 2008; Chapman, 2009). Lack of personal hygiene among food handlers is one of the most commonly reported practices that contribute to foodborne illness (Taulo et al., 2009). The majority of foodborne outbreaks associated with food workers have involved transmission of the pathogens to food by the food workers' hands (Guzewich and Ross, 1999; Çakiroğlu and Uçar, 2008), and ensuring personal hygiene, particularly hand washing, has been cited the most effective tool in preventing the spread of foodborne infections (NHS Plus, 2008).

Developing and educating a workforce knowledgeable in food safety and hygiene is necessary to improve food safety in food service establishments. Studies of the knowledge, attitudes and practices of food service workers in developing countries have been conducted, and several trends have been reported in this workforce. Factors that were associated with better food safety knowledge include the level of education of the worker (Zeru and Kumie, 2007; Kibret and Abera, 2012; Onyeheho and Hedberg, 2013), and training in food safety (Garin et al., 2002; Baş et al., 2006; Kibret and Abera, 2012; Olumankaiye and Bakare, 2013; Onyeheho and Hedberg, 2013). Workers in food service establishments in hospitals and schools were also found to have better food safety knowledge and practices than workers in other restaurants, street vendors, and other small food service establishments (Baş et al., 2006; Onyeheho and Hedberg, 2013). Training programs are effective, and improved environmental and worker hygiene practices in a study of fast food and street food vendors in Nigeria (Olumankaiye and Bakare, 2013). However, despite knowledge and awareness of safe food handling methods, several studies have found that food handlers often do not use safe food handling practices, based on observation and microbial food testing (Baş et al., 2006; Zeru and Kumie, 2007; Kibret and Abera, 2012).

Although there is a growing body of research on food safety and food worker hygiene and practices in developing countries, the majority of these studies focus on street food vendors rather than larger institutional food service establishments. Given the reported differences in food safety knowledge and practices between food service facilities in schools and restaurants, this study was conducted to test two research hypotheses: 1) food contamination levels will be significantly lower in facilities where kitchen staff or food handlers practice good hygiene than those that practice poor hygiene; and 2) food contamination levels will be lower in university food service establishments due to better worker hygiene and food safety practices. The objectives of the study were to: describe the knowledge, attitudes and practices of food handlers in food service facilities at a university campus (Makerere University, Kampala, Uganda) and restaurants in its neighborhood; measure levels of food contamination at these facilities through microbiological analyses; and determine the effect of individual hygiene practices on food contamination levels.

MATERIALS AND METHODS

A longitudinal study involving two groups of food service workers and facilities was conducted between September, 2012 and July, 2013. The kitchen staff at Makerere University, working in student halls of residence, comprised Group A (facilities A1-A9), and the neighborhood restaurants that operated in improvised makeshift structures comprised Group B (facilities B1-B7). A proportional random sample was selected for the study: 75 subjects were selected from 95 workers in nine different kitchens at Makerere University and 25 subjects were selected from 35 workers in seven restaurants outside the University. Food samples, consisting of all the items served at the time of sampling, were collected from each facility on three different points during the study period.

Data collection

Face-to-face interviews were conducted to collect information on workers' hygiene practices and attitudes using questions adopted from Baş et al., 2006 and Giritlioglu et al., 2011. Information on food safety practices was collected through a semi-structured questionnaires through in-person interviews by trained study personnel at two levels: the institution level (from the 16 facilities), and the individual level (from food service workers). The institution and individual-level information on food safety practices was collected by questionnaires administered to kitchen supervisors, and individual food service workers. All responses were validated by interviewers' observations of the facilities and respondents, and responses were corrected in situations where observations did not agree with the responses (e.g., there was no soap at hand washing basins but the respondent indicated that soap was provided). Individual respondents were assigned food safety attitude and hygiene practice scores:

$$Attitude_{ij} = \frac{\sum_{j=1}^{16} Response_{ij}}{\sum_{j=1}^{16} MaxResponse_j} X\,100 \qquad Practice_{ik} = \frac{\sum_{k=1}^{12} Response_{ik}}{\sum_{k}^{12} MaxResponse_k} X\,100$$

Where, *Attitude*$_{ij}$ = attitude score for respondent *i*; *Response*$_{ij}$ = response of respondent *i* to attitude question *j*; *MaxResponse*$_j$ = the maximum value of possible responses to attitude question *j* (2 for yes/no responses); $Practice_{ik}$ = hygiene practice score for respondent i, $Response_{ik}$ = response of respondent *i* to hygiene practice question *k*; and $MaxResponse_k$ = the maximum value of possible responses to practice question *k* (6 for Likert scale questions regarding the frequency of personal hygiene practices; 2 for yes/no responses). The attitude and practice scores could range in value from 0 to 100.

Food sample collection

Food samples were collected from the study kitchens at serving point. Approximately 250-500 g of food were collected and sealed in sterile stomacher bags, placed in cool boxes, and transported to the food safety laboratory at the College of Veterinary Medicine, Animal Resources and Biosecurity (COVAB), Makerere University, within 30 minutes after collection. Samples were processed within 6 hours of collection.

Microbiological analyses

Food contamination was measured by total aerobic mesophilic bacteria plate counts (APC), enumeration of *Escherichia coli (E. coli)* and total coliforms, and presence of *Salmonella.* A 25 g sample was collected from each food sample, and prepared using sterile surgical blades and adding 225 ml of sterile buffered peptone water. Each sample was divided into two, and each sub-sample was placed in sterile stomacher bags and homogenized using a pulsifier. After homogenization, each sub-sample was divided into two, and serial 10-fold dilutions were made, up to 10^{-7} dilution (Harrigan, 1998). Selected dilutions of the food samples were mixed by vortexing, and inoculations were made within 25 min of processing, using methods adapted from Downes and Ito (2001).

For total aerobic mesophilic bacteria (APC) counts, 0.1 ml of the processed food samples of specified dilutions were inoculated onto sterile Plate Count Agar (HiMedia Laboratories, India), using the surface spread method, and incubated for 24 h at 37°C (Refai, 1979; Harrigan, 1998). After incubation, plates containing 25-250 colonies were selected for counting. Counts obtained were characterized by the reciprocal of the dilution factors used, and additionally by 10^1. The bacteria population was expressed as a number of colony forming units per gram (CFU/g).

In processing for total coliforms and *E. coli*, 0.1 ml of the processed food samples of specified dilutions were inoculated onto Chromocult coliform agar (Merck, Germany), a selective indicator media for the enumeration of *E. coli* and other coliforms. After incubation for 24 h at 37°C, dark blue colonies were classified as *E. coli*, while pink colonies were classified as other coliforms (Merck, 1996). Gram staining was carried out on suspected *E. coli* colonies, and all cultures with Gram negative short rods were biochemically confirmed as *E. coli* using the IMVIC tests (Downes and Ito, 2001). Indole production was regarded as positive for *E. coli* if there was an appearance of a distinct red color in the upper layer; the Voges-Proskauer (VP) test was considered positive if the eosin pink color developed; the methyl red test was considered positive if a distinct red color developed; and the citrate utilization test was considered positive if Simmon's citrate media changed from green to blue.

To enhance the recovery of *Salmonella*, samples underwent pre-enrichment for 24 h at 37°C (Refai, 1979). Enrichment in selective media [Tetrathionate Brilliant green broth (Merck, Germany)] was conducted in triplicate for each sample, for 24 h at 37°C (Downes and Ito, 2001). After enrichment, enriched samples were streaked on plates of Xylose Lysine Desoxycholate (XLD) (Oxoid, UK) selective medium, incubated at 37°C for 24 – 48 h, and examined for *Salmonella* colonies which exhibited the typical appearance "i.e. pink with large black centers which made them appear black" (FDA, 2011). Suspect colonies were subcultured to get a pure culture, and were subjected to confirmatory tests (Harrigan, 1998). Confirmation of *Salmonella* was based on results of Gram staining, urease, indole, citrate, and TSI tests (Oxoid, UK).

Statistical analyses

The primary outcomes of interest were the safety status of food in each facility, and the use of unsafe food handling practices at the individual and institutional level. Each employee and institution was classified as having safe food handling practices if they had food hygiene practices and attitude scores of 75% or more. For each facility, foods were classified as safe if the mean APC of the three samples was less than 100,000 CFU/g, and the mean counts for coliforms and *E. coli* were less than 100 CFU/g, using cutpoints for microbiological food safety (Gilbert et al., 2000). There were only two food samples where *Salmonella* was detected, so statistical analyses regarding *Salmonella* were not conducted.

Descriptive statistics were generated to characterize the study respondents, responses to questionnaires, attitude and practice scores, and laboratory testing results. Associations between food service worker characteristics (age, gender, level of education and working experience measured by years in employment) and their attitudes on hygiene were assessed using the chi-square test for associations. Associations between safe food handling status and categorical risk factors were evaluated through odds ratios (OR), with 95% confidence intervals and p-values. Associations between mean APC, coliforms and *E. coli* with different types of food were assessed using the non-parametric Kruskall-Wallis X^2 of the Wilcoxon Rank-Sum test, and associations between unacceptable levels of bacteria with different types of food were assessed using Fisher's Exact 2-tailed test, and the strength of association reported through odds ratios with 95% confidence intervals.

Multivariable logistic regression was used to identify risk factors associated with the safety of foods. Risk factors significantly associated with food safety from the univariable analyses ($p \leq 0.05$) were considered for inclusion in multivariable analysis. Binary logistic regression was used to screen these risk factors – those that were statistically significant ($p \leq 0.05$) were used in the multivariable analysis. Stepwise logistic regression was used to develop the final model for food safety, with adjustment for confounders (Kleinbaum and Klein, 2010).

RESULTS

A total of 94 interviews were conducted, including 54 individual respondents (41 from Group A, 13 from Group B) and 40 institutional respondents (Table 1). The majority of individual respondents were females from 26 to 60 years in age, received secondary level educations, and had worked with food service facilities for more than twelve years. The majority of the individual respondents were food servers and cooks, and only 40.7% of respondents had attended on-the-job training since they joined employment. The 40 institutional respondents (33 from Group A, 7 from Group B) were males from 26 to 60 years in age, received secondary and tertiary educations, and had over 5 years of experience in food service. Chiefs and cooks made up the majority of institutional respondents, and only 35% had attended any on-the-job training.

Safe food handling attitudes and practices

When asked about attitudes regarding food safety, the

Table 1. Characteristics of study respondents.

Factor	Level	Individual (n = 54)		Institutional (n = 40)	
		N	%	N	%
Facility type	A	41	75.9	33	82.5
	B	13	24.1	7	17.5
Gender	Male	19	35.2	24	60.0
	Female	35	64.8	16	40.0
Age	18 – 25	6	11.1	2	5.0
	26 – 35	11	20.4	13	32.5
	36 – 45	17	31.5	8	20.0
	46 – 60	19	35.2	17	42.5
	> 60	1	1.9	0	0
Highest Level of Education	None	1	1.9	0	0
	Primary	19	35.2	2	5.0
	Secondary	32	59.3	19	47.5
	Tertiary	1	1.9	14	35.0
	Other	1	1.9	5	12.5
Food Service Industry Experience	1 – 5 years	12	22.2	8	20.0
	6 – 12 years	5	9.3	10	25.0
	13 – 20 years	17	31.5	5	12.5
	21 – 30 years	18	33.3	11	27.5
	> 30 years	2	3.7	6	15.0
On-the-job training		22	40.7	14	35.0
Role	Cook	24	44.4	9	22.5
	Server	25	46.3	8	20.0
	Caterer	1	1.9	-	-
	Kitchen Supervisor	4	7.4	-	-
	Chief	-	-	12	30.0
	Bursar	-	-	7	17.5
	Warden	-	-	4	10.0

majority of individual respondents indicated they understood the importance of safe food handling and their personal responsibility for food safety, but respondent attitudes about proper food storage and holding temperatures and food safety were mixed (Table 2). The majority of institutional respondents reported employees washed hands properly and frequently, appeared in good health, kept fingernails short, unpolished and clean, and wore little jewelry (Table 3).

Microbiological testing of foods

A total of 48 food samples (27 from the University, 21 from restaurants outside the University) were collected (Table 4). The most commonly served foods were posho, rice, beans, and beef, and there was a larger variety of foods served in Group B facilities.

The mean APC for all food samples was 937,165 (Figure 1). There were no statistically significant differences in the mean APC for Group A (925,626 CFU/g), versus Group B samples (952,000 CFU/g). Using the APC cutpoint of 100,000 CFU/g, 66.7% of Group A facilities and 71.4% of Group B facilities served meals in violation of food safety standards. Of facilities with violations, all three samples from one facility in Group A (A3) and two facilities in Group B (B3, B4) had APCs higher than 100,000 CFU/g; and three in Group A (A5, A6, A7) and two in Group B (B1, B7) had two samples with unacceptably high APCs.

Total coliforms were found in samples from all facilities, and total coliform plate counts were significantly higher (Kruskall-Wallis $X^2 = 3.88$, 1 d.f., $p = 0.0489$) in samples from Group B (7,965.2 CFU/g) than Group A (5,271.1 CFU/g) (Figure 2). Six Group A facilities and one Group B facility had at least one sample with no total coliforms. The highest mean total coliform CFU/g was 40,490 CFU/g, from one Group A facility (A4). Using a cutpoint of 100 CFU/g, there were no statistically significant differences in the numbers of samples with unacceptable

Table 2. Individual respondent hygiene practices and attitudes (n = 54).

Parameter	Response	#	%
Employee practices			
Meat can be chopped with vegetables	Yes	14	25.9
Raw and cooked foods should be kept separate	Yes	43	79.6
	Yes	7	13.0
Defrosted foods may be frozen only once	No	33	61.0
	Not sure	14	25.9
PPE use reduces food contamination risk	Yes	50	92.6
Knowing fridge temperature reduces food contamination risk	Yes	24	44.4
	Yes	24	44.2
Checking fridge thermometer settings once a day is important	No	26	48.4
	Not sure	4	7.4
Improper heating of food causes foodborne diseases	Yes	52	96.3
Improper food storage may be hazardous to health	Yes	52	96.3
Employee attitudes			
Safe food handling is important	Yes	50	92.6
Learning more about food safety is important	Yes	51	94.4
Food preparation without hygiene rules causes foodborne diseases	Yes	50	92.6
Employee personal hygiene			
How I handle food relates to food safety	Yes	45	83.3
Do not come to work if I have flu or diarrhea	Yes	29	53.7
Do not wear jewelry during food preparation	Yes	29	53.7
Food service staff with cuts on fingers or hands shouldn't touch cooked or unwrapped foods	Yes	30	55.6

Table 3. Institutional respondent hygiene practices and attitudes (n = 40).

Parameter	# Yes	%
Employee practices		
Employees use effective hair restraints	10	25.0
Hands are washed properly and frequently	34	85.0
Employees cover wounds completely	15	37.5
Food preparation activities only in designated zones	15	37.5
Employees wear clean and proper uniforms	10	25.0
Fingernails are short, unpolished, and clean	27	67.5
Jewelry is limited to a plain ring	34	57.5
Employees use disposable tissues	4	10.0
Facilities		
Employees appear in good health	32	80.0
Sinks are unobstructed	22	55.0
Sinks are stocked with soap	10	25.0
Hand washing reminder signs are posted	6	15.0
Employee toilets are operational and clean	35	87.5

coliform levels between the two groups. Only three Group A facilities and one Group B facility had acceptable mean total coliform levels.

Isolation of *E. coli* was not common: it was detected in 6 samples from five facilities from Group A, and 7 samples from three facilities from Group B (Figure 3). The highest mean *E. coli* CFU/g (7,530 CFU/g) was found in the same Group A facility with the highest mean

Table 4. Types of foods in samples collected from study facilities

Food	Group A (n=9)		Group B (n=7)	
	# Facilities serving	# Samples	# Facilities serving	# Samples
Posho	9	26	7	15
Rice	9	25	6	13
Beans	9	13	5	8
Beef	8	8	7	10
Matooke	0	0	7	17
Cabbage	2	2	5	8
Sweet Potato	1	1	6	9
Irish Potato	1	1	3	7
Chicken	4	4	2	2
Fried Chicken	4	4	2	2
Soup	4	4	1	1
Other Foods*	0	0	7	25

*Fried rice (5 samples from 4 sites); groundnut stew (4 samples from 4 sites); greens (4 samples from 3 sites); cassava, millet (3 samples from 2 sites); pumpkin (2 samples from 2 sites); and avocado, fish, goat, spaghetti (1 sample from 1 site).

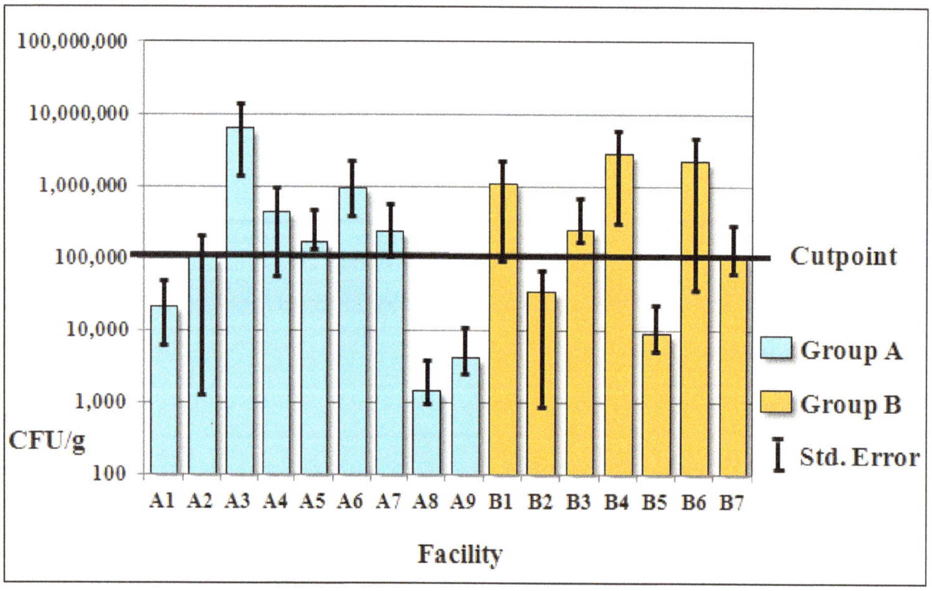

Figure 1. Mean Aerobic Plate Counts for food samples from university residence halls (n=9) and local restaurants (n=7), with cutpoints for microbiological food safety.

total coliforms (A4). Again, using the 100 CFU/g cutpoint for food safety, only one Group A and two Group B facilities had unacceptable levels of *E. coli*. There were no statistically significant differences in *E. coli* levels between the two sites.

Two food samples from two Group A facilities (A2, A9), collected on the second sampling visits, were positive for *Salmonella*. Both halls served the same food items (boiled rice, posho and beans) on that date. Facility A2 had APC (104,433 CFU/g) and total coliforms (107 CFU/g) slightly above the food safety cutpoints, while facility A9 had an acceptable APC (4,200 CFU/g) but unacceptably high total coliforms (1,007 CFU/g). Both facilities had no *E. coli*.

When examining specific foods, there were several foods that were associated with increasing or decreasing the risk of unacceptable levels of bacteria (Table 5). Food samples from both groups that contained beef were at

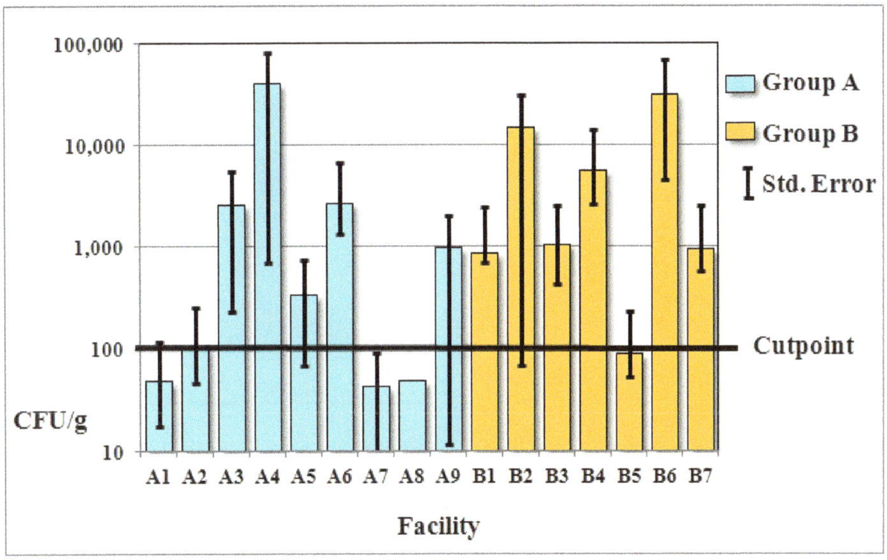

Figure 2. Mean total coliform counts for food samples from university residence halls (n=9) and local restaurants (n=7), with cutpoints for microbiological food safety.

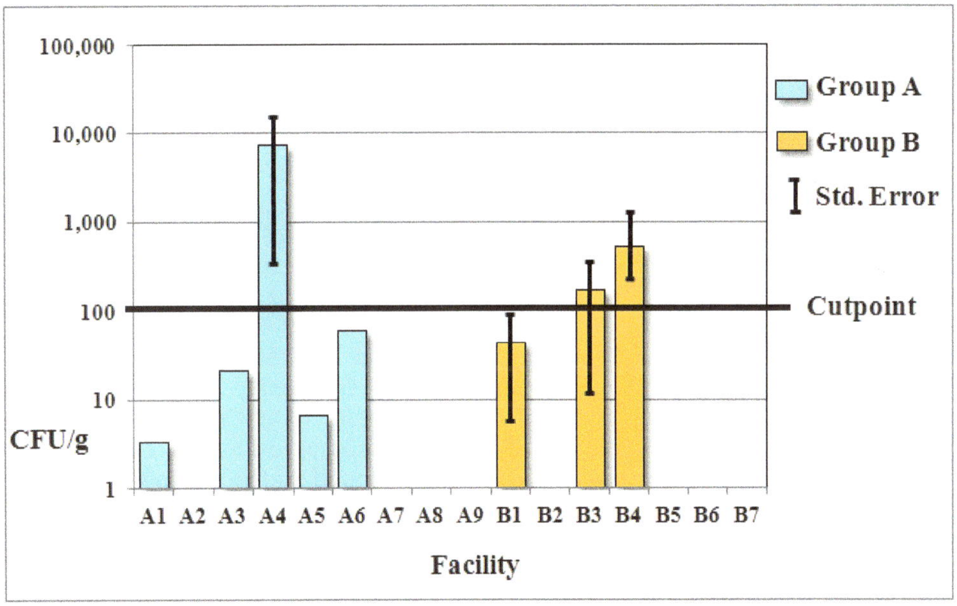

Figure 3. Mean *E. coli* counts for food samples from university residence halls (n=9) and local restaurants (n=7), with cutpoints for microbiological food safety.

higher risk for unacceptable coliform levels, and mean coliform counts for samples with beef (13,117 CFU/g) were significantly higher (Kruskall-Wallis X^2 – 4.5, 1 d.f., p = 0.0345) than samples that did not contain beef (2,449 CFU/g). Samples from Group B facilities that contained beans were at higher risk for unacceptable levels of *E. coli*, and had significantly higher mean *E. coli* (217.5 CFU/g) than meals that did not contain beans (29.2 CFU/g) (Kruskall-Wallis X^2 = 4.8, 1.d.f., p = 0.0278). Meals that contained sweet potato were at lower risk for unacceptable APC levels for both groups combined, and this association was statistically significant when considering samples from Group B alone (89,056 CFU/g with sweet potato versus 1,605,968 CFU/g without; Kruskall-Wallis X^2 = 5.9, 1.d.f., p = 0.0156). Samples from Group B that contained matooke had statistically lower mean levels of *E. coli* than samples without matooke (37.1 CFU/g and 405.0 CFU/g, respectively;

Table 5. Associations (p < 0.1) between food type and unsafe levels of aerobic mesophilic bacteria, coliforms, and *E.coli*

Group	Food	# Tested	Bacteria	% Unsafe	Odds Ratio	95% C.I.
Both Groups	Posho	41	*E. coli*	12.2	0.24	0.04 – 1.46
	Beef	18	Coliforms	37.5	3.91	0.91 – 16.82
	Sweet potato	10	APC	20.0	0.10	0.02 – 0.63
	Chicken	12	Coliforms	41.7	0.29	0.07 – 1.17
Group B	Beans*	8	*E. coli*	50.0	12.00	1.02 – 14.34
	Beef	10	APC	80.0	7.00	0.97 – 50.57
	Matooke*	17	*E. coli*	11.8	0.04	0.003 – 0.66
	Sweet potato*	9	APC	22.2	0.06	0.01 – 0.51

* Association significant at $p \leq 0.05$

Kruskall-Wallis $X^2 = 5.3$, 1.d.f., p = 0.0218), and were significantly less likely to have unacceptable levels of *E. coli.*

Institutional and individual-level food safety attitudes and practices

A minority of institutional respondents were classified as using safe food hygiene practices (26.9% from Group A; 42.9% from Group B). There were no significant differences in age and work experience, but respondents with secondary and higher level educations were more likely to use safe food hygiene practices (O.R. = 4.75; 95% C.I. = 2.57 – 8.79). For specific practices, the use of effective hair restraints (O.R. = 56.0; 95% C.I. = 6.78 – 462.64) and posting hand washing reminder signs (O.R. = 9.33; 95% C.I. = 1.38 – 63.20) were positively associated with safe hygiene practices.

Over 90% of the 54 individual respondents (95.1% in Group A; 84.6% in Group B) were classified as using safe food handling practices. There were no significant differences in age, education, work experience, position in kitchen, or facility between safe and unsafe food handling respondents. For specific hygiene practices and attitudes, there were significant differences between safe and unsafe handling groups for wearing jewelry during food preparation (OR = 0.84, 95% C.I. = 0.71 – 1.0), and believing that food preparation without hygiene rules causes foodborne diseases (OR = 24.0, 95% C.I. = 2.14 – 269.12).

The association between individual worker risk factors and food classified as safe through microbiological analyses were tested, and odds ratios computed for nine variables were considered for inclusion in multivariable analyses (Table 6). After multivariable model building was completed, the final model for safe food (based on microbiological testing) contained only one variable: not wearing jewelry was associated with decreasing risk for unsafe foods (adjusted OR = 0.05, 95% CI: 0.00 – 1.05). Overall, 48.6% of the variation in safety of food served in Makerere University halls of residence and neighborhood restaurants was correctly explained by the model (Hosmer and Lemeshow goodness-of-fit $X^2 = 4.45$, 5 d.f., p = 0.49).

DISCUSSION

Food service employees

The majority of individual respondents in this study were female. Although this study found no associations between gender and safe food handling, female workers had better hygiene scores and practices than males (Çakiroğlu and Uçar, 2008; Kibret and Abera, 2012), and female clients in a study of food service outlets in Nigeria were more likely to patronize outlets with better hygiene levels (Olumakaiye and Bakare, 2013). Most respondents were from 18 to 45 years of age, and had worked in food service facilities for more than five years. Other studies have found that older workers had better hygiene scores than younger workers (Çakiroğlu and Uçar, 2008; Olumakaiye and Bakare, 2013), and workers with more than seven years of food service work experience had better hygiene scores than younger and newer workers(Çakiroğlu and Uçar, 2008).

There were differences in level of education between individual food service workers and institutional respondents. The majority (59.3%) of individual workers possessed secondary level educations, which is higher than reported in studies of food handlers (Baş et al., 2006) The majority of institutional respondents had attained higher education levels than the individual respondents. Higher levels of education have been associated with better food safety knowledge awareness, and better sanitary conditions in other studies (Zeru and Kumie, 2007; Olumakaiye and Bakare, 2013; Onyeneho and Hedberg, 2013), and a study of the environmental hygiene of food service outlets were significantly

Table 6. Association between risk factors and food safety determined by microbiological testing, from 41 workers in 9 kitchens from Makerere University, and 13 workers in 7 kitchens from local restaurants.

Variable	Level	Odds Ratio	95% C.I.
Education Level	Primary and Below	4.75*	2.57 – 8.79
	Secondary and above	(baseline)	
Use effective hair restraints	No	56.0**	6.78 – 462.64
	Yes	(baseline)	
Clean, short, unpolished fingernails	No	6.0	0.67 – 53.68
	Yes	(baseline)	
Wear jewelry during food preparation	No	0.84*	0.71 – 1.00
	Yes	(baseline)	
Cuts and abrasions covered	No	0.86	0.18 – 4.04
	Yes	(baseline)	
Hand washing reminders	No	9.33	1.38 – 63.20
	Yes	(baseline)	
Personal protective equipment reduces food contamination risk	No	1.08	1.00 – 1.18
	Yes	(baseline)	
Prepare food when sick	No	3.82	0.37 – 39.28
	Yes	(baseline)	
Prepare food disregarding hygiene rules	No	24.0**	2.14 – 269.11
	Yes	(baseline)	

*p < 0.05; **p < 0.01

associated with the age, education level and income of operators (Olumankaiye and Bakare, 2013). The requirements for employee education and training by food service facilities has been shown to have positive influences on food hygiene. Studies have reported that food service workers from hospitals and schools had better food safety knowledge, attitudes and practice scores than workers from smaller food service establishments, fast-food outlets, and vending stalls or street food vendors (Baş et al., 2006; Onyeneho and Hedberg, 2013).

Specific training for food service workers was not common in this study. Only 40.7% of individual respondents and 35% of institutional respondents had attended any on-the-job training since they joined employment, which is lower but comparable to other studies (47.8%, Baş et al., 2006). In several studies, food service workers that received training had better hygiene scores and safe food handling practices than those that did not receive training (Baş et al., 2006; Kibret and Abera, 2012; Onyeneho and Hedberg, 2013; Ababio and Lovatt, 2014). This indicates that on-the-job and short term training sessions for food service workers would be beneficial to facilities in this study. A study of 277 food handlers in Ethiopia found the most common sources of information about food safety were mass media (50%) and health centers (42.7%) (Zeru and Kumie, 2007), which also suggests that using less traditional venues for education, such as radio or newspaper campaigns, can also provide useful information to food service workers and the population in general.

Food service employee knowledge, attitudes, and practices

The majority of individual respondents indicated they understood the importance of safe food handling and their personal responsibility for food safety, but knowledge and attitudes about some aspects of food safety were mixed. The majority of individual respondents acknowledge the importance of separating raw and cooked food, using personal protective equipment (PPE) to reduce food contamination, and the hazards of improper food heating and food storage, which has been seen in other studies of food service workers (Baş et al., 2006). However, less than 45% of respondents indicated that knowing refrigerator temperatures was important, and that checking refrigerator thermometer settings was important, which was seen in 35% of head chefs and managers of restaurants in Nigeria (Onyeneho and Hedberg, 2013).

Over 80% of individual respondents reported that food handling affected food safety, but only 54% of respondents understood that working when sick or wearing jewelry during food preparation were practices to be avoided for food safety. Approximately 44% of respondents did not acknowledge that food service workers with cuts on their fingers or hands should avoid handling cooked or unwrapped foods, which is similar to findings from other studies in Turkey and Nigeria (Baş et al., 2006; Onyeneho and Hedberg, 2013). One study reported that 47% of food service chefs and managers had a lack of awareness that sick persons can spread foodborne illness (Onyeneho and Hedberg, 2013), and surveys of food vendors found that only 42% of workers in Nigeria (Olowogbon et al., 2012) and 23% in Ethiopia (Zeru and Kumie, 2007) had at least one medical examination per year. The health status of these workers could have serious implications for food safety.

Environmental hygiene is important for food safety, and necessary to support safe food handling and hygiene by employees. It was the duty of the employer to provide and enforce use of facilities and tools for the safe handling of food, including PPE. The majority of institutional respondents reported employees washed hands properly and frequently, appeared in good health, kept fingernails short, unpolished and clean, and wore little jewelry, but this study revealed inadequate provision of uniforms, hygiene and food handling equipment. Most facilities had sinks with no soap, hand washing reminder signs were not posted, and kitchen staff did not wear proper or full uniforms (including hair restraints or a cap). In sites with poor access to clean running water, hand washing water (Taulo et al., 2009) and dish washing water (Nkere et al., 2011) can harbor fecal bacteria and serve as a source of bacterial contaminants for both food and workers. Other studies have found that foods that have been properly prepared can become contaminated by serving utensils washed in contaminated water, or handled by unwashed hands (Taulo et al., 2009; Nkere et al., 2011).

From the results of the multivariable analysis, food service employees who wore jewelry (36% of employees) during food preparation were more likely to be associated with the serving of unsafe food. It is possible that, when jewelry becomes contaminated, the lack of soap at sinks and absence of hand washing reminder signs allows jewelry to stay contaminated for longer periods of time, and increase the opportunities for cross- contamination during food handling. It is also likely that the insistence on wearing large amounts of jewelry during food preparation is reflective of a lack of awareness of potential hazards: workers may not realize that this can be a source of contamination, or may not be aware of the hazards that avoiding hand washing to avoid damaging jewelry can pose.

Although the majority of individual respondents indicated they understood the importance of safe food handling and their personal responsibility for food safety, they performed poorly in important food safety and hygiene practices, which has been widely reported in food hygiene practice studies in Ghana (Ababio and Lovatt, 2014), Nigeria (Ondyeneho and Hedberg, 2013), and other countries (Fulham and Mullan, 2011). In one study of the "intention-behavior gap" in hygienic food handling (Fulham and Mullan, 2011), researchers found that subjective norms (food handling practices by co-workers, "peer pressure") and perceived behavioral control (the worker's perception of the ease or difficulty of performing the behavior) predicted their intentions to follow good food handling practices and these intentions predicted behavior. However, the study found that behavioral prepotency (old habits, past behaviors) was the best predictor of intention and behavior, regardless of worker knowledge or attitude (Fulham and Mullan, 2011). Hygienic food handling has immediate negative perceived behavioral control (takes more time, costs more, inconvenient to use, uncomfortable to wear), and the benefits of hygienic food handling are not immediate and personal, but are more long-term and general, which contributes to why workers often fail to use best practices even when they are aware of their importance. Some possible solutions to overcome the "intention-behavior" gap are to change the working atmosphere of food service workers (subjective norms) or to change the perceptions of the ease or difficulty of using safe food handling practices (perceived behavioral controls). Individual respondents in this study indicated their willingness to wear proper uniforms and use hair restraints if the employer provided them.

Safety of food served in residence halls and neighborhood restaurants

The majority of food samples tested in this study had APC and total coliform CFU/g counts higher than acceptable, which indicates that there are ample areas for improvement in safe food handling. In the case of the samples that tested positive for *Salmonella*, both came from two halls that served the same food items on that sampling date, and were collected by different individuals. This suggests a common source of *Salmonella* for these two food service facilities, but identifying the common source is difficult, given that the two halls are far apart and unlikely to share personnel or facilities.

The findings that there were significantly higher levels of coliforms in samples from neighborhood restaurants (Group B) than in samples from University food service facilities (Group A) were expected, given that the residence halls have better facilities than the restaurants, of which some operate in improvised structures. However, the lack of significant differences in APC or *E. coli* CFU/g in food samples from Makerere University food service facilities (A1-A9) than in food samples from neighborhood restaurants (B1-B7) was unexpected. This may be explained by laxity in supervision in the halls as compared to the restaurants, where the owners seem to supervise their staff more keenly. However, given the generally low levels of good hygiene practices and problems in environmental hygiene, there are likely other factors beyond the scope of this study which influenced the levels of bacteria found in food samples from food service facilities in this study.

Conclusions

In general, this study has found that there is a critical need for improving food safety at restaurants and dining halls at Makerere University. The APC levels in samples of all facilities were higher than desired. The finding that the majority of respondents did not follow good hygiene practices (e.g., use of hair restraints), indicates laxity or lack of supervision, and a need to overcome the problems of the "intention-behavior gap" in hygienic food handling. Employees in food service facilities are aware of proper food handling hygiene practices and have positive attitudes towards food safety, but inadequate facilitation (e.g., lack of soap at sinks) prevents them from observing good hygiene practices when handling food.

Based on results from this study, there are specific areas for improvement in both university dining halls and local restaurants. For individual kitchen workers, providing training on personal hygiene and proper food handling techniques will be helpful, particularly in raising awareness of some specific practices (for example working while sick, handling food when there are cuts or wounds on the hand).

At the institutional level, food service facility managers should improve environmental hygiene, including steps such as ensuring that sinks or hand washing basins in kitchens and toilets have running water, and are stocked with soap at all times, and hand washing reminder signs should be posted. To improve the workforce itself, food service facility managers should establish and maintain minimum qualifications for employees above primary education, as better-educated staff are more likely to adhere to good hygienic practices. Managers can support safe food handling by their employees by providing and enforcing wearing of uniforms, including clothing, hair restraints, aprons and gum boots, by all food service workers.

There is a need for governmental support to improve food safety management systems, and education and awareness programs (Onyeneho and Hedberg, 2013; Ababio and Lovatt, 2014). Regular inspection of food service facilities is critical: facilities subject to regular inspection had better sanitary conditions than uninspected facilities (Zeru and Kumie, 2007). Finally, government and leaders in the food service industry should strive to institute a thorough assessment of the food processing chain to identify and address areas that are responsible for food contamination.

Conflict of interests

The authors did not declare any conflict of interest.

REFERENCES

Ababio FW, Lovatt P (2014). A review on food safety and food hygiene studies in Ghana. Food Control 47:92-97.

Baş M, Ersun AS, Kivanc G (2006). The evaluation of food hygiene knowledge, attitudes and practices of food handlers in food businesses in Turkey. Food Control 17:317-322.

Çakiroğlu FP, Uçar A (2008). Employees' perception of hygiene in the catering industry in Ankara, Turkey. Food Control 19:9-15.

Chapman BJ (2009). Development and evaluation of a tool to enhance positive food safety practices amongst food handlers: Food Safety Infosheets. Ph.D. dissertation. University of Guelph, Guelph, Ontario, Canada.

Da Cunha DT, Stedefeldt E, De Rosso VV (2012). Perceived risk of foodborne disease by school food handlers and principals: The influence of frequent training. J. Food Saf. 32:219-225.

De Sousa CP (2008). The Impact of Food Manufacturing Practices on Foodborne Diseases. Braz. Arch. Biol. Technol. 51:815-825.

Downes FP, Ito K (2001). Compendium of Methods for Microbiological Examination of Foods. Fourth Edition. American Public Health Association, Washington D.C., USA.

Fletcher SM, Stark D, Ellis J (2011). Prevalence of gastrointestinal pathogens in sub-Saharan Africa: systematic review and meta-analysis. J. Public Health Afr. 2:127-137.

Fulham E, Mullan B (2011). Hygienic Food Handling Behaviors: Attempting to bridge the intention-behavior gap using aspects from temporal self-regulated theory. J. Food Prot. 74:925-932.

Gilbert RJ, De Louvois J, Donovan T, Little C, Nye K, Ribeiro CD, Richards J, Roberts D, Bolton FJ (2000). Guidelines for the microbiological quality of some ready-to-eat foods sampled at the point of sale. Commun. Dis. Public Health 3:163-167.

Giritlioglu I, Batman O, Tetik N (2011). The knowledge and practices of food safety and hygiene of cookery students in Turkey. Food Control 22:838-842.

Guzewich J, Ross MP (1999). Evaluation of risks related to microbiological contamination of ready-to-eat foods by food preparation workers and the effectiveness of interventions to minimize those risks. FDA White Paper, Center for Food Safety and Applied Nutrition.

Harrigan WF (1998). Laboratory Methods in Food Microbiology, 3rd edition. Academic Press, London.

Hassan AN, Farooqui A, Khan A, Yahya KA, Kazmi SU (2010). Microbial contamination of raw meat and its environment in retail shops in Karachi, Pakistan. J. Infect. Dev. Ctries. 4:382-388.

Havelaar AH, Brul S, De Jonge A, De Jonge R, Zwietering MH, Terkuile BH (2010). Future challenges to microbial food safety. Int. J. Food Microbiol. 139:S79 – S94.

Jacxsens L, Uyttendaele M, Devlieghere F, Rovira J, Gomez SO, Luning PA (2010). Food safety performance indicators to benchmark food safety output of food safety management systems. Int. J. Food Microbiol. 141:S180-S178.

Kibret M, Abera B (2012). The sanitary conditions of food service establishments and food safety knowledge and practices of food handlers in Bahir Dar town. Ethiop. J. Health Sci. 22:27-35.

Kleinbaum DG, Klein M (2010). Logistic Regression, a self-learning text, 3rd ed. Springer, New York.

Merck Microbiology (1996). Microbiology Manual. 10th edition. Merck KGaA, Darmstadt, Germany.

Nguz K (2007). Assessing food safety system in sub-Saharan African countries: An overview of key issues. Food Control 18:131-134.

NHS Plus (2008). Royal College of Physicians, Faculty of Occupational Medicine. Infected food handlers: occupational aspects of management. A national guideline. RCP, London.

Nkere CK, Ibe NI, Iroegbu CU (2011). Bacteriological quality of foods and water sold by vendors and in Restaurants in Nsukka, Enugu State, Nigeria: A Comparative Study of three Microbiological Methods. J. Health Popul. Nutr. 29:560-566.

Oiko S (2000). Microbiological quality of selected foods in fast food outlets in Kampala City, Uganda. B.S. research project in Food Science and Technology. Makerere University, Kampala, Uganda.

Olowogbon ST, Adnrelete YA, Uhunmwangho A. (2012). Attitudes and practices of local food vendors regarding food hygiene and handling.

African Newsletter 22:43-45.
Olumakaiye MF, Bakare KO (2013). Training of food providers for improved environmental conditions of food service outlets in urban area Nigeria. Food Nutr. Sci. 4:99-105.
Onyeneho SN, Hedberg CW (2013). An assessment of food safety needs of restaurants in Owerri, Imo State, Nigeria. Int. J. Environ. Res. Public Health. 10:3296-3309.
Refai MK (1979). Manuals of Food Quality Control: 4. Microbiological Analysis. Food and Agricultural Organization of the United Nations, Rome.
Taulo S, Wetlesen A, Abrahamsen RK, Narvhus JA, Mkakosya R (2009). Quantification and variability of *Escherichia coli* and *Staphylococcus aureus* cross-contamination during serving and consumption of cooked thick porridge in Lungwena rural households, Malawi. Food Control 20:1158-1166.
U.S. Food and Drug Administration (FDA) (2011). Bacteriological Analytical Manual (BAM). http://www.fda.gov/Food/FoodScienceResearch/LaboratoryMethods/ucm2006949.htm
U.S. Food and Drug Administration (FDA) (2012). *Foodborne* Pathogenic Microorganisms and Natural Toxins Handbook, The bad bug book, 2nd edition. http://www.fda.gov/downloads/Food/FoodborneIllnessContaminants/UCM297627.pdf.
Zeru K, Kumie A (2007). Sanitary conditions of food establishments in Mekelle town, Tigray, north Ethiopia. Ethiop. J. Health Dev. 21:3-11.

Storage stability of tomato paste as influenced by oil-citric acid and packaging materials

Alakali Joseph[1], J. K. Agomuo[2] , I. C. Alaka[3] and J. Faasema[1]

[1]Department of Food Science and Technology, University of Agriculture, Makurdi, P.M.B 2373, Makurdi, Benue State, Nigeria.
[2]Department of Food Science and Technology, Federal University Dutsinma, P.M.B 5001, Dutsinma, Katsina State, Nigeria.
[3]Department of Food Science and Technology, Ebonyi State University, P.M.B 501, Ebonyi, Ebonyi State, Nigeria.

Tomato pulp was concentrated by boiling at 90°C for 1 h. A mixture of 0.1% citric acid and 25 ml of vegetable oil were added to 100 g of the paste and packaged in aluminum foil (AF), low density polyethylene (LDP), plastic containers (PC) and stored under ambient temperature (30±0.1°C). Chemical analyses were carried out to determine pH, total solids, total acidity and refractive index as well as viscosity. Microbial analyses were also carried out after 8 weeks of storage. Results obtained showed significant ($p<0.05$) decreased in the pH value and increase in the titrable acidity (TTA) with storage time in all the packaging materials and samples treated with oil and citric acid (WOC) and those with no oil and citric acid (NOC). However, the decrease in pH and increase in TTA was more rapid in NOC than WOC. The total solid, vitamin A and C decreased significantly ($p<0.05$) with storage time following the trend as pH. Yeast and total viable count increased significantly ($p<0.05$) as the storage time increased. Tomato paste stored using AF had lower total plate count, ranging from 1.25×10^5 to 4.51×10^5 as compared to LDP (1.21×10^5-4.41×10^5) and PC (1.21×10^5-6.02×10^5). Generally, tomato paste in AF retained higher quality after 8 weeks of storage as compared to samples stored in LDP and PC. Also, NOC samples were more prone to spoilage than WOC. These findings can be applied for better preservation of tomato paste in rural communities where there is no electricity for cold storage.

Key words: Packaging materials, tomato paste, citric acid, vegetable oil, storage period.

INTRODUCTION

Tomato (*Lycopersicon esculentum*) is one of the most widely consumed fresh vegetable in the world (Thybo et al., 2006). Fresh-market tomatoes are a popular and versatile fruit vegetable, making significant contributions to human nutrition throughout the world (Simonne et al., 2006). Tomato is rich in water soluble vitamin such as carotenoids, vitamins B and C; it could provide 12.2% recommended daily allowance of vitamin C (Smith and Hull, 2004). It has 20-25 mg ascorbic acid per 100 g (Ahmed and Shivhare, 2001). Tomato is also a rich source of natural lycopene. Tomato contains diseases fighting phytochemical (Robert et al., 2002) and antioxidant agent (Ahmed and Shivhare, 2001). The fruit consist of 98.1 g water/100 g, 0.7 protein/100 g, 3.1 g carbohydrate/100 g, 1.3 g dietary fibre/100 g and 0.3 g fat/100 g (Ahmed Shivhare, 2001).

FAO reports indicate that tomato production is on the increase globally (FAO, 2007), leading to rapid development of tomato processing industries with a series of inter-linked activities such as production of salad, soup, juice, puree, paste and powder and extraction of oil from the pulp. The demand for tomato pastes is increasing rapidly both in domestic and in international market (Tyssandie et al., 2004).

Tomato production in Nigeria at present is estimated at about one million tonnes (FAO, 2007). Due to its highly perishable nature, huge wastage and losses occur during the harvesting period. Therefore, the prevention of this losses and wastage is of major interest especially in Nigeria where there is imbalance in supply and demand at the harvesting and offseason (Akanbi et al., 2006). The imbalance is largely due to inability of the large population of Nigerians, especially the rural dwellers, to have access and use readily available and appropriate packaging materials for storage of tomato pastes, rather depending mainly on fresh tomato.

Packaging of the pastes is very important because of protection of the product from contamination by macro or micro-organism, reduction of oxidation (Matsuoka et al., 2002). Appropriate packaging of tomato pastes locally will serve purposes which are essential to extend the shelf life of the product and create a brand that is appealing to the consumers as well as add value to the product (Matsuoka et al., 2002).

Several works have been reported in literature on the use of plastic containers and low density polyethylene in the packaging of tomato paste (Famurewa et al., 2013; Akanbi and Oludemi, 2004); information is lacking on the use of Aluminum foil which are readily available and cheap. This work seeks to investigate the use of Aluminum foil in tomato paste packaging and storage.

Citric acid and oil has being reported to have inhibitory effect on the growth of microorganism especially in canned tomato (USDA, 2009). There is however scanty information on the use of citric acid and oil to extend the shelf life of packaged tomatoes. Thus, oil and citric acid were used in this study to investigate the effect of the pre-treatments on shelf life of packaged tomato pastes.

The objective of this study was to evaluate the effect of three different storage materials on the shelf stability of tomato paste pre-treated with citric acid and vegetable oil.

MATERIALS AND METHODS

Source of materials

The tomatoes used in this study were ripe and free from any form of mechanical injury, visible disease or rot. The fresh tomatoes were purchase from a farm on the bank of River Benue at Wurukum Makurdi, Benue State, Nigeria. Aluminum foil, low density polyethylene and plastic container were purchased from Modern Market Makurdi.

Sample preparation

Tomato paste was processed by the traditional methods described by (Iwe, 2002) with slight modifications. The ripe and wholesome tomatoes were sorted out, selected and washed thoroughly to remove dirt and contaminants. The fresh tomatoes were blanched in hot water at about 90°C for 3-5 s to inactivate enzymes and easy peeling of the skin. The blanched tomatoes were sprayed with cold water and the skin was peeled off by pulling it back from the blossom end. The seed cavity was removed neatly with a knife and pulped in a blender. The pulp was heated in an open pot at 90°C for 1 h to concentrate the paste.

The paste was divided into three lots of 100 g each and filled into plastic containers, low density polyethylene and aluminium foil respectively, to a head space of about 10 mm in each case. Each lot was divided into two sublots. The headspace of one of the sublots was overlaid with vegetable oil and citric acid, while the sub lots without oil and citric acid act as control. The plastic container, low density polyethylene and aluminium foil were covered with air tight-lid and stored at room temperature 30 ± 1°C. Samples were checked at two weeks intervals for chemical and microbiological changes.

Determination physicochemical analysis

Total acidity

The total acidity was determined according to AOAC (2005) official method by direct titration of 2 g of the puree with 0.1 M sodium hydroxide using phenolphthalein as indicator. The total acidity (as percentage citric acid) was calculated as follows:

Citric Acid=(mole of NaOH x volume of base x molar mass of acid x 100)/(Milli equivalent of acid x volume of sample)

pH

Twenty grammes of samples were dissolved in water and made up to 100 ml. The pH of the pastes was estimated using the electrode of a portable pH meter. The pH meter was standardized using buffer solution prior to sample analysis.

Vitamin A

This was determined according to the method of James (1995)

Ascorbic acid content

5 ml of standard solution of ascorbic acid was pipetted into 100 ml conical flask. 10 ml of oxalic acid was added and the solution titrated against the dye (V1 ml) until a pink colour persisted for 15 s.

The dye consumed is equivalent to the amount of ascorbic acid. Also, 0.5g of the sample was extracted in 4% oxalic acid and made up to 100 ml. The solution was filtered. 10 ml of oxalic acid was added to 5ml of the filtrate above. The solution was then titrated against the dye solution (2,6-dichlorophenol indophenol). The volume of dye used was recorded as (V2 ml) (Ibitoye, 2005).

$$\text{Ascorbic acid (mg/100 g)} = \frac{0.5\text{ mg} \times V_2 \times 100\text{ ml}}{V_1 \times 5\text{ml} \times W} \times 100$$

Where W = sample weight.

Table 1. Physicochemical parameters and vitamin content of tomatoes paste stored in aluminum foil (AF) without and with oil/citric acid for varying period of time.

Storage time	PH		Total solid		Viscosity (centipoise)		Total Acidity		Refractive index		Yeast count (cfu/g)		TVC (10^5) (cfu/g)		Vitamin C		Vitamin A	
	NOC	WOC	NOC	WOC	NOC	WOC	NOC	WOC	NOC	WOC	NOC	WOC	NOC	WOC	WOC	NOC	NOC	WOC
0 week	4.51^a	4.22^a	0.90^a	0.90^a	91.25^a	80.10^a	0.006^d	0.071^c	1.35^a	1.33^a	0.62^d	0.39^c	1.81^d	1.21^e	2.31^a	2.31^a	1.61^a	1.32^a
2 weeks	4.48^a	4.21^a	0.87^b	0.90^a	90.10^a	80.20^a	0.05^c	0.06^d	1.35^a	1.33^a	0.61^d	0.31^d	1.51^e	1.22^d	2.22^b	2.12^b	1.51^b	1.31^b
4 weeks	4.41^a	4.20^a	0.85^c	0.87^b	90.01^a	80.10^a	0.05^c	0.07^c	1.35^a	1.35^a	0.65^c	0.39^c	2.92^c	1.82^c	1.96^c	2.11^c	1.49^c	1.31^b
6 weeks	4.35^a	4.11^a	0.81^d	0.85^c	85.15^b	75.10^b	0.01^b	0.17^b	1.36^a	1.34^a	1.21^b	0.73^b	3.41^b	2.52^b	1.86^d	2.06^e	1.32^d	1.30^c
8 weeks	4.05^b	3.99^b	0.78^e	0.81^d	80.10^c	70.20^c	0.12^a	0.18^a	1.35^a	1.34^a	2.22^a	1.81^a	4.52^a	3.31^a	1.86^d	2.07^d	1.30^e	1.31^b
P-Value	0.210		0.001		0.787		0.004		0.224		0.001		0.001		0.004		0.002	

Mean in the same column with different superscripts differ significantly (P<0.05). NOC, Without oil and citric acid; WOC, with oil and citric acid.

Viscosity

The viscosity of the tomato was measured using A Brook field viscometer (Lv – 8 viscometer) with spindle No. 4 at a speed of 6 rpm.

Total solid

This was determined by the use of a refractometer (IHP series Brownel London, Belligham and Stanly Limited). Ten grammes of the paste were mixed with 20g of water and filtered through muslin. The total solid was then determined by a drop of the filtered paste on the splint with the corresponding reading on the refractometer as percentage total solid AOAC (2005).

Refractive index

Ten grammes of the puree was mixed with 20 ml of the water and strained directly through muslin. The refractive index of the drop of the filtrate at 30°C was obtained by taking the corresponding reading on the refractometer (Okanlawon et al., 2002).

Total soluble solids

The total dissolved solids of the various samples were analyzed by weighing 10 g of the sample into 20ml of small Whatman filter paper. The refractive index of a drop of the filtrate was taking at 20°C (AOAC, 2005).

Microbiological analysis

Preparation of media

Acidified potato dextrose agar for mould count was prepared according to the method of Ogbulie et al. (2001).

Microbial culturing and examination

Methods of standard pour plate technique were used as according to Ogbulie et al. (2001). After autodialing and cooling of the agar media to a temperature of 45°C about 20-25 ml of agar was poured into each dried plate.

1 ml portion of each sample serial dilution was added to the plate and the plate tilted from side to side until the liquid was evenly distributed across the surface of the agar. The culture plates were incubated at 37°C for 24 h for bacteria and mould count respectively prior to counting of colonies.

Statistical analysis

Determination were done in triplicate and all data were subjected to analysis of variance (ANOVA) and the mean separated using Duncan's multiple range test (DMRT) using version 16.0.

RESULTS AND DISCUSSION

Tables 1, 2 and 3 respectively showed that the pH values of samples treated with oil and citric acid (WOC) were generally lower than that with no oil and citric acid (NOC), indicating that WOC samples were more acidic. Addition of citric acid may have increased acidity of WOC samples. Also, the pH of samples stored in aluminum foil (AF) was higher than that stored in low density LDP and PC. The differences could be due to variation in rate of permeability of oxygen in the packaging materials. Aluminium foil has higher resistance to oxygen permeability (Smith and Hull, 2004) and therefore experienced less oxidation, evidenced by low pH values compared with LDP and PC.

There was general decrease in the pH value as the storage time increased for both WOC and NOC samples. pH values decreased from 4.51- 4.05; 4.51-3.95 and 4.51-3.98 for NOC samples stored in AF, LDP and PC respectively within the storage time of 8 weeks. Similarly, WOC samples showed more slow decrease in pH values from 4.22-3.99; 4.44-3.55 and 4.46-3.20 for AF, LDP and PC respectively. Values were not significantly different ($p<0.05$) from 0-6 weeks but differed significantly ($p>0.05$) at the 8th week across storage period.

Table 2. Physicochemical parameters and vitamin content of tomatoes paste stored in Low density polyethylene (LDP) without and with oil/citric acid for varying period of time.

Storage time	pH		Total solid		Viscosity (centipoise)		Total Acidity		Refractive index		Yeast count (cfu/g)		TVC (10^5) (cfu/g)		Vitamin C		Vitamin A	
	NOC	WOC	NOC	WOC	NOC	WOC	NOC	WOC	NOC	WOC	NOC	WOC	NOC	WOC	WOC	NOC	NOC	WOC
0 week	4.51^a	4.44^a	0.90^a	0.90^a	91.15^a	87.00^a	0.006^d	0.07^c	1.34^a	1.34^a	0.85^c	0.14^e	1.81^e	1.21^e	3.21^a	2.85^a	1.37^a	1.31^a
2 weeks	4.35^a	4.31^{ab}	0.88^b	0.90^a	91.15^a	80.10^b	0.05^a	0.06^d	1.36^a	1.35^a	0.73^e	0.56^c	2.52^d	1.72^d	3.21^a	2.83^b	1.30^b	1.31^a
4 weeks	4.22^b	4.20^b	0.87^c	0.88^b	85.20^b	80.20^b	0.01^c	0.02^e	1.36^a	1.35^a	0.82^d	0.43^d	4.11^c	2.81^c	2.96^b	1.76^d	1.30^b	1.31^b
6 weeks	4.21^b	4.00^b	0.83^d	0.86^c	80.20^c	70.25^d	0.01^c	0.17^b	1.37^a	1.37^a	2.01^b	1.61^b	4.51^b	3.01^b	2.72^c	1.77^c	1.28^d	1.24^d
8 weeks	3.95^c	3.55^c	0.70^e	0.76^d	70.25^d	75.15^c	0.02^b	0.80^a	1.37^a	1.35^a	4.15^a	2.62^a	4.91^a	4.41^a	2.72^c	1.77^c	1.29^c	1.25^c
P-Value	0.210		0.001		0.787		0.004		0.224		0.001		0.001		0.004		0.002	

Mean in the same column with different superscripts differ significantly (P<0.05). NOC, without oil and citric acid; WOC, with oil and citric acid.

Table 3. Physicochemical parameters and vitamin content of tomatoes paste stored in Paste container (PC) without and with oil/citric acid for varying period of time.

Storage time	pH		Total solid		Viscosity (centipoise)		Total Acidity		Refractive index		Yeast count (cfu/g)		TVC (10^5) (cfu/g)		Vitamin C		Vitamin A	
	NOC	WOC	NOC	WOC	NOC	WOC	NOC	WOC	NOC	WOC	NOC	WOC	NOC	WOC	WOC	MOC	NOC	WOC
0 week	4.51^a	4.46^a	0.90^a	0.93^a	91.15^a	80.06^a	0.006^d	0.009^e	1.33^a	1.34^a	0.63^e	0.39^d	1.81^e	1.21^e	2.31^b	3.03^a	1.51^a	1.31^a
2 weeks	4.46^a	4.24^b	0.89^b	0.91^b	91.10^a	70.10^b	0.006^d	0.09^c	1.35^a	1.34^a	1.71^d	0.83^c	3.31^d	1.82^d	2.32^a	2.83^b	1.50^a	1.31^a
4 weeks	4.31^{ab}	4.19^b	0.86^c	0.90^c	91.02^a	65.15^c	0.07^c	0.08^d	1.36^a	1.34^a	1.41^c	0.73^d	4.21^c	3.71^c	2.22^d	2.79^c	1.50^a	1.31^a
6 weeks	4.21^b	3.99^b	0.85^d	0.88^d	90.02^b	60.10^d	0.16^b	0.18^b	1.36^a	1.34^a	2.81^b	2.22^b	5.71^b	4.21^b	2.22^d	2.66^d	1.49^b	1.30^b
8 weeks	3.98^c	3.20^c	0.79^e	0.84^e	88.10^c	60.05^d	0.17^a	0.19^a	1.37^a	1.34^a	4.21^a	3.72^a	6.81^a	6.02^a	2.23^c	2.52^e	1.49^b	1.30^b
P-Value	0.210		0.001		0.787		0.004		0.224		0.001		0.001		0.004		0.002	

Mean in the same column with different superscripts differ significantly (P<0.05). NOC, without oil and citric acid; WOC, with oil and citric acid.

The decrease in the pH values with storage time may be due to increase in microbial activities with storage time (Giordano et al., 2000). According to Campos et al. (2006), pH is a key element in tomato quality and for tomato paste, pH below 4.5 is appropriate for tomato paste, but above 4.5 is undesirable trait, because it will not halts the proliferation of microorganisms in the final product.

Total solid (TS) content is a measure of the solid particles after concentration. The results of the total solid as presented in Tables 1, 2 and 3 for samples stored in AF, LDP and PC respectively revealed that WOC samples had higher TS values compared with NOC samples. Similarly, samples stored in PC were higher in TS than those stored in AF while those in LDP had lower in values. Generally, there was significant (p<0.05) decrease in TS as the storage period increased. From Table 1, for samples packaged in AF, the TS value of NOC decreased significantly (p<0.05) from 0.90 - 0.78 during the period of storage (0-8 weeks). On the other hand, the values for WOC decreased more steadily from0.90-0.81. Similar trend were also observed for samples stored in LDP and PC (Tables 2 and 3). Smith and Hull (2004) reported that AF, LDP and PC permit the diffusion of gases, vapour and volatile flavour of stored products. Therefore, the observed trend could be due to differences in the rate of diffusion of gasses through the packaging materials.

According to earlier researchers (Saliba-Colombani et al., 2001), there is a connection between sugar content and total solid in tomato fruit. Soluble solid content measurements may give a fair estimate of the sugar level in tomato fruit, thus, the reduction in TS over storage time may be a function of the degradation of sugar by microorganism which was more noticeable in NOC than WOC samples. This gives indication of the inhibitory effect of citric acid and oil on microbial activities.

The viscosity of tomato products depends on fibre, fat, protein, and total solids (Tamburini et al., 1999). The viscosities of NOC samples stored in AF were higher and decreased more steadily than WOC samples (Tables 1, 2 and 3). Viscosity was observed also to decrease significantly ($p<0.05$) with storage time especially from the 2nd week of storage. Similar trend was also observed in samples stored in LDP and PC packaging materials, attributable to microbial activity on the stored tomato paste

The total titrable acidity (TTA) increased significantly ($p<0.05$) with the storage period for both NOC and WOC samples stored in AF, LDP and PC packaging materials. In Tables 1, 2 and 3, the TTA of WOC samples were generally higher than NOC, owing to the presence of citric acid and oil used in the treatment procedure. The TTA of NOC samples stored in PC were higher (0.006-0.17) than LDP (0.006-0.05) and AF (0.006-0.12). On the other hand, TTA of WOC sample was higher in LDP (0.07-0.80) compared to the other packaging materials.This result is in line with the work of Okanlawon et al. (2002). The authors observed that, decrease in acidity is probably due to the effect of organisms responsible for the spoilage, some of which can release basic substances into the samples.

The refractive index of WOC samples were generally lower than NOC, although values were not significantly different ($p<0.05$) across the packaging materials and storage period (Tables 1, 2 and 3). Refractive index is a measure of the soluble solid (Famurewa et al., 2013), hence the presence of citric acid and oil in WOC samples reduced the breakdown of solids leading to less available soluble solids compared to NOC samples.

Tables 1, 2 and 3 show that the yeast counts in WOC samples were generally lower than NOC in the three packaging materials namely AF, LDP and PC. The yeast count of WOC samples ranged from 0.39-1.81; 0.14-2.62 and 0.39-3.72 for AF, LDP and PC respectively, while that of NOC was higher ranging from 0.63-4.21; 0.85-4.15 and 0.62-2.22. Samples were observed to increase significantly ($p>0.05$) in yeast count with storage time in all the packaging materials. This trend could be attributed to the effect of the citric acid and oil which inhibited the multiplication of the microorganism in WOC samples. Samples stored in LDP were higher in yeast count than AF and PC for NOC and WOC. This higher population of the microorganisms in LDP may be due to higher permeability of O_2, CO_2 and other gases (Smith and Hull, 2004). The growth of yeast and mould increased with storage time in accordance with the report of (Mozumber et al., 2012)

Tables 1, 2 and 3 also show that the total viable count (TVC) followed the same trend as the yeast and mould count. Total viable count was higher in NOC samples than WOC across all samples stored in AF, LDP and PC.This is also attributable to the inhibitory action of oil and citric acid.

Values obtained for samples stored in AF, were lower than those stored in LDP and PC, indicating the relative resistance of AF to microbial growth. The total viable count values generally increased significantly ($p<0.05$) as the storage time increased in according with (Mozumber et al., 2012) who reported that microbial growth increases as the storage time prolongs.

The results presented in Tables 1, 2 and 3 show that there was significant difference ($p>0.05$) in the vitamin C content among the pre-treatments and the storage period.

The vitamin C values of samples treated with oil and citric acid (WOC) were generally higher than samples with no citric acid and oil (NOC). Higher retention of vitamin C in WOC may be as a result of inaction of the endogenous enzymes such as ascorbic acid oxidase, cytochrome oxidase and peroxidase as reported in similar by (Okorie et al., 2004).

The vitamin C of NOC samples was lower than WOC. Vitamin C was higher in LDP compared to the other packaging material. Generally, vitamin C decreased significantly ($p<0.05$) with storage period owing to the reactive nature of vitamin C, to storage environment (Davey et al., 2000). The result of this work agrees with earlier work by Leonardi et al. (2001) who reported that storage condition has significant impact on the loss of ascorbic acid.

The data obtained for vitamin A showed higher values in NOC than WOC samples. This may be due to the concentration of citric acid used in the pre-treatment procedure of the samples. Vitamin A samples stored in PC degraded more slowly than those stored in LP and AF; this may be as a result of heat generated internally by the PC. Vitamin A values were observed to decrease more gradually, relative to the other parameters as the storage time increased. This may be due to degradation of vitamin A by light and heat in the storage environment.

Conflict of interest

The authors did not declare any conflict of interest.

REFERENCES

Ahmed J, Shivhare US (2001). Effect of Pre-treatment on Drying. Characteristics and Colour of Dehydrated Green Chilies. J. Food Sci. Technol. 38(5): 504-506.

Akanbi CT, Oludemi FO (2006). Effects of Processing and Packaging on the Lycopene Content of Tomato Products, International Journal of Food Properties, 7:39-152.

Akanbi CT, Oludemi FO (2004). Effect of Processing and packaging on the lycopene content of tomato products. Int. J. Food Properties,

7(1):139-152.
Akanbi CT, Adeyemi RS, Ojo A (2006). Drying Characteristics and Sorption Isotherm of Tomato Slices. J.Food Eng. 73,157-163.
AOAC, (2005).Official Method of Analysis.Association of Analytical Chemistry.Washington D.C. 91(1)112-122.
Campos CAB, Fernandes PD, Gheyi HR, Blanco FF, Goncalves CB, Campos SAF (2006). Yield and fruit quality of industrial tomato under saline irrigation. J. Sci. Agric. 2:63-69.
Famurewa JAV, Ibidapo PO, Olaifa Y (2013). Storage Stability of Tomato Paste Packaged in Plastic Bottle and Polythene Stored in Ambient Temperature. Int.J. Applied Sci.ce and Technol. 3(6):34-42.
FAO (2007). Technical Manual on Small Scale Processing of Fruit and Vegetables. FAO Regional Office for Latin America and the Caribbean, Santiago, Chile (CTA Practical).
Ibitoye AA (2005). Basic methods in Plant analysis. Concept IT and educational consults. Akure, Nigeria. Pp 6, 28, 41.
Iwe MO (2002). Handbook of Sensory Methods and Analysis, Rejoint Com.Services Ltd pp.76-77.
James CS (1995). Analytical Chemistry of Foods. Blackie Academic and Proffessional, London. Pp. 41-46.
Leonardi C, Ambrosino P, Esposito F, Fogliano V (2001). Antioxidant activity and carotenoid and tomatine contents in different typologies of fresh'consumption tomatoes. J. Agric. and Food Chem. 48:4723-4727.
Matsuoka K, Kakemura T, Kshima H, Seki T, Tsujino (2002). Development of high barrierplastic bottles, Worldpack 2002, Improving the quality of life through packaging innovation, Eastlancing, MI, USA, 23–28 June, 1, pp.393–399.
Mozumber NHMR, Rahman MA, Kamal MS, Mustafa AKM, Rahman MS (2012). Effects of pre-drying treatment on quality of cabinet dried tomato powder. J.Environ. and Natural Res 5:253-265.
Ogbulie JN, Uwaezuoke JC, Ogieche SI (2001). Introductory Microbiology Practical(2nd ed). Concave Publishers, Owerri Nigeria pp 63-64.
Okanlawon SO, Ibrahim MH, Oyebani AO (2002). Effect of pre drying treatment on the storage of dried tomato. Tropical Science, 42:40-41.
Okorie SU, Nwanekezi EC, Okoro CC (2004). The Quality properties of Tomatoes as influenced byprocessing with a chemical preservative and storage. Nigerian Food Journal, Vol. 22, 2004 pp 195-197.
Robert PK, Sargent SA, Fox AJ (2002). Effect of storage temperature on ripening and postharvest quality of grape and mini-pear tomatoes. Proc Fla State Horticulturel Society, 115:80-84.
Saliba-Colombani V, Causse M, Langlois D, Philouze J, Buret M (2001). Genetic analysis of organoleptic quality in fresh market tomato. 1. Mapping QTLs physical and chemical traits. Theoretica. Application. Genet. 102:259-272.
Simonne AH, Behe BK, Marshall MM (2006). Consumers prefer low-priced and high-lycopene-content fresh-market tomatoes. Horticultural Technology, 16 674-681.
Smith JS, Hull YH (2004). Food processing, principles and applications,Blackwell publishing, London
Tamburini RL, Sandel AA, De Sio F, Leoni C (1999). Effect of storage conditions on lycopene content in tomato puree obtained with different processing techniques. Industria-conserve 74(4):341-357
Thybo AK, Christiansen J, Kaack K, Petersen MA (2006). Effect of cultivars, wound healing and storage on sensory quality and chemical components in pre-peeled potatoes. LWT. J Food Sci. and Technol. 39:166-176.
Tyssandie V, Feillet-Coudray C, Caris-Veyrat C, Guilland J, Coudray C,Bureau S, Reich M, Amiot-Carlin M, Bouteloup-Demange C, Boirie Y, Borel P (2004). Effect of tomato product consumption on the plasma status of antioxidant microconstituents and on the plasma total antioxidant capacity in healthy subjects. Journal of American College Nutrition, 23:148-156
USDA (2009). "Complete Guide to Home Canning," Agriculture Information Bulletin No. 539.

Permissions

All chapters in this book were first published in AJFS, by Academic Journals; hereby published with permission under the Creative Commons Attribution License or equivalent. Every chapter published in this book has been scrutinized by our experts. Their significance has been extensively debated. The topics covered herein carry significant findings which will fuel the growth of the discipline. They may even be implemented as practical applications or may be referred to as a beginning point for another development.

The contributors of this book come from diverse backgrounds, making this book a truly international effort. This book will bring forth new frontiers with its revolutionizing research information and detailed analysis of the nascent developments around the world.

We would like to thank all the contributing authors for lending their expertise to make the book truly unique. They have played a crucial role in the development of this book. Without their invaluable contributions this book wouldn't have been possible. They have made vital efforts to compile up to date information on the varied aspects of this subject to make this book a valuable addition to the collection of many professionals and students.

This book was conceptualized with the vision of imparting up-to-date information and advanced data in this field. To ensure the same, a matchless editorial board was set up. Every individual on the board went through rigorous rounds of assessment to prove their worth. After which they invested a large part of their time researching and compiling the most relevant data for our readers.

The editorial board has been involved in producing this book since its inception. They have spent rigorous hours researching and exploring the diverse topics which have resulted in the successful publishing of this book. They have passed on their knowledge of decades through this book. To expedite this challenging task, the publisher supported the team at every step. A small team of assistant editors was also appointed to further simplify the editing procedure and attain best results for the readers.

Apart from the editorial board, the designing team has also invested a significant amount of their time in understanding the subject and creating the most relevant covers. They scrutinized every image to scout for the most suitable representation of the subject and create an appropriate cover for the book.

The publishing team has been an ardent support to the editorial, designing and production team. Their endless efforts to recruit the best for this project, has resulted in the accomplishment of this book. They are a veteran in the field of academics and their pool of knowledge is as vast as their experience in printing. Their expertise and guidance has proved useful at every step. Their uncompromising quality standards have made this book an exceptional effort. Their encouragement from time to time has been an inspiration for everyone.

The publisher and the editorial board hope that this book will prove to be a valuable piece of knowledge for researchers, students, practitioners and scholars across the globe.

List of Contributors

Karima M. F. Ahmed
Milk Hygiene and Control, Department of Food Hygiene and control, Faculty of Veterinary Medicine, Cairo University, Egypt

Ragaa S. Hafez
Milk Hygiene and Control, Department of Food Hygiene and control, Faculty of Veterinary Medicine, Cairo University, Egypt

S. D.Morgan
Milk Hygiene and Control, Department of Food Hygiene and control, Faculty of Veterinary Medicine, Cairo University, Egypt

Abeer A. Awad
Milk Hygiene and Control, Department of Food Hygiene and control, Faculty of Veterinary Medicine, Cairo University, Egypt

Banderas-Tarabay José Antonio
Doctorado en Ciencias Biológicas, Universidad Autónoma de Tlaxcala. Tlaxcala, México
Laboratorio de Química de Alimentos, Facultad de Nutrición y Facultad de Ciencias de la Salud, Universidad Autónoma de Tlaxcala, Tlaxcala, México

Cervantes-Rodríguez Margarita
Laboratorio de Química de Alimentos, Facultad de Nutrición y Facultad de Ciencias de la Salud, Universidad Autónoma de Tlaxcala, Tlaxcala, México

Grada-Sánchez Miriam
Laboratorio de Química de Alimentos, Facultad de Nutrición y Facultad de Ciencias de la Salud, Universidad Autónoma de Tlaxcala, Tlaxcala, México

Espíndola-Lozano Marlen
Laboratorio de Química de Alimentos, Facultad de Nutrición y Facultad de Ciencias de la Salud, Universidad Autónoma de Tlaxcala, Tlaxcala, México
Maestría en Ciencias Biológicas, Universidad Autónoma de Tlaxcala. Tlaxcala, México

Cuevas-Romero Estela
Centro Tlaxcala de Biología de la Conducta, Universidad Autónoma de Tlaxcala. Tlaxcala, México

Navarro-Ocaña Arturo
Facultad de Química, Universidad Nacional Autónoma de México, CU, México DF

Méndez-Iturbide Daniel
Laboratorio de Química de Alimentos, Facultad de Nutrición y Facultad de Ciencias de la Salud, Universidad Autónoma de Tlaxcala, Tlaxcala, México

A. A. Doutoum
Laboratory of Hygiene and Food Industry of Animal Origin (HIDAOA). EISMV, P.O Box 5077, Dakar, Senegal

A. Tidjani
Faculty of Human Health, University of N'Djamena, P. O. Box: 1117, N'Djamena, Chad

I. Abdelaziz Arada
Laboratory of Microbiology Immunology Infectious Pathology (MIPI). EISMV, P.O. Box 5077, Dakar, Senegal

S. M. T. Tidjani
Laboratory of Microbiology Immunology Infectious Pathology (MIPI). EISMV, P.O. Box 5077, Dakar, Senegal

M. Balde
Laboratory of Hygiene and Food Industry of Animal Origin (HIDAOA). EISMV, P.O Box 5077, Dakar, Senegal

K. S. B. Sylla
Laboratory of Hygiene and Food Industry of Animal Origin (HIDAOA). EISMV, P.O Box 5077, Dakar, Senegal

Mg Seydi
Laboratory of Hygiene and Food Industry of Animal Origin (HIDAOA). EISMV, P.O Box 5077, Dakar, Senegal

B. S. Toguebaye
Department of Animal Biology, Faculty of Science and Technology, University Cheikh Anta Diop (UCAD), Senegal

Yéo Mohamed Anderson
Laboratoire de Biochimie Sciences des Aliments, UFR Biosciences, Université Félix Houphouet Boigny de Cocody, Abidjan, Côte d'Ivoire

O. Chatigre
Laboratoire de Biochimie Sciences des Aliments, UFR Biosciences, Université Félix Houphouet Boigny de Cocody, Abidjan, Côte d'Ivoire

E. Elleingand
Laboratoire de Genie chimie et Agroalimentaire, Institut Nationale Polytechnique Félix Houphouet Boigny de Yamoussoukro, Côte d'Ivoire

Koni Pascal
Société Ivoirienne de Technologie Tropicale (I2T), Côte d'Ivoire

Koffi Ernest
Laboratoire de Biochimie Sciences des Aliments, UFR Biosciences, Université Félix Houphouet Boigny de Cocody, Abidjan, Côte d'Ivoire

I. B Okey
Cross River University of Science and Technology, Obubra Campus, Cross River State, Nigeria

R. I Keremah
Department of Fisheries/ Livestock Production Technology, Niger Delta University, Amasssoma, P. M.B. 071, Yenaogoa, Bayelsa State, Nigeria

G. U Ikpi
Cross River University of Science and Technology, Obubra Campus, Cross River State, Nigeria

L. A Ebeniro
Federal College of Agriculture, Ishiagu Ebonyi State, Nigeria

Amira Ben Aoun
Institut des Régions Arides, Laboratoire d'aridoculture et cultures oasiennes, 4119 Médenine, Tunisia

Belgacem Lechiheb
Institut des Régions Arides, Laboratoire d'aridoculture et cultures oasiennes, 4119 Médenine, Tunisia

Leila Benyahya
Institut des Régions Arides, Laboratoire d'aridoculture et cultures oasiennes, 4119 Médenine, Tunisia

Ali Ferchichi
Institut des Régions Arides, Laboratoire d'aridoculture et cultures oasiennes, 4119 Médenine, Tunisia

Saiqa Bashir
Department of Food Technology, The University of Poonch Rawalakot, Pakistan

Muhammad Siddique Awan
Department of Food Technology, The University of Poonch Rawalakot, Pakistan

Shahzad Akbar Khan
Faculty of Veterinary and Animal Sciences, The University of Poonch Rawalakot, Pakistan

Habib Ahmed Rathore
Department of Food Technology, The University of Poonch Rawalakot, Pakistan

Muhammad Akhtar Qureshi
Faculty of Veterinary and Animal Sciences, The University of Poonch Rawalakot, Pakistan

Zulfiqar Hussain Kathu
Department of Animal Husbandry, Government of AJK, Pakistan

Mustafa Alhaji Isa
Department of Microbiology, University of Maiduguri, P. O. Box 1069, Maiduguri, Borno State, Nigeria

Ibrahim Alkali Allamin
Department of Microbiology, University of Maiduguri, P. O. Box 1069, Maiduguri, Borno State, Nigeria

Haruna Yahaya Ismail
Department of Microbiology, University of Maiduguri, P. O. Box 1069, Maiduguri, Borno State, Nigeria

Abubakar Shettima
Department of Microbiology, University of Maiduguri, P. O. Box 1069, Maiduguri, Borno State, Nigeria

Sonchieu Jean
Department of Social Economy and Family Management, Higher Technical Teachers' Training College (HTTTC), University of Bamenda, Cameroon

Ngassoum Martin Benoit
Department of Applied Chemistry, ENSAI, University of Ngaoundere, P. O. Box: 455, Ngaoundere, Cameroon

Tchatchueng Jean Bosco
Department of Applied Chemistry, ENSAI, University of Ngaoundere, P. O. Box: 455, Ngaoundere, Cameroon

Ashutosh Kumar Srivastava
Pesticides Toxicology Laboratory, Indian Institute of Toxicology Research, (CSIR), P B No 80, M G Marg, 226001, Lucknow, India

Laxman Prasad Srivastava
Pesticides Toxicology Laboratory, Indian Institute of Toxicology Research, (CSIR), P B No 80, M G Marg, 226001, Lucknow, India

Mohammed A. Satter
Institute of Food Science and Technology (IFST), Bangladesh Council of Scientific and Industrial Research (BCSIR), Dr. Qudrat- E-. Khuda Road, Dhanmondi, Dhaka -1205 Bangladesh

Syeda Absha Jabin
Institute of Food Science and Technology (IFST), Bangladesh Council of Scientific and Industrial Research (BCSIR), Dr. Qudrat- E-. Khuda Road, Dhanmondi, Dhaka -1205 Bangladesh

Nusrat Abedin
Institute of Food Science and Technology (IFST), Bangladesh Council of Scientific and Industrial Research (BCSIR), Dr. Qudrat- E-. Khuda Road, Dhanmondi, Dhaka -1205 Bangladesh

Taslima Arzu
Institute of Food Science and Technology (IFST), Bangladesh Council of Scientific and Industrial Research (BCSIR), Dr. Qudrat- E-. Khuda Road, Dhanmondi, Dhaka -1205 Bangladesh

Kanika Mitra
Institute of Food Science and Technology (IFST), Bangladesh Council of Scientific and Industrial Research (BCSIR), Dr. Qudrat- E-. Khuda Road, Dhanmondi, Dhaka -1205 Bangladesh

A. M. Abdullah
BCSIR Laboratories, Dhaka, Bangladesh Council of Scientific and Industrial Research (BCSIR), Dr. Qudrat- E-. Khuda Road, Dhanmondi, Dhaka -1205 Bangladesh

D. K. Pau
Islamic University, Kushtia, Bangladesh

R. Mohanlall
Department of Biotechnology and Food Technology, Faculty of Applied Science, Durban University of Technology, P. O. Box 1334, Durban 4001, South Africa

B. Odhav
Department of Biotechnology and Food Technology, Faculty of Applied Science, Durban University of Technology, P. O. Box 1334, Durban 4001, South Africa

V. Mohanlall
Department of Biotechnology and Food Technology, Faculty of Applied Science, Durban University of Technology, P. O. Box 1334, Durban 4001, South Africa

Alexandre Carvalho de MOURA
University of the Federal Border South, Campus Royalty Pr. Agricultural Engineering doctoral program, Brazil

Fabiana Gisele da Silva PINTO
State University of West Paraná - UNIOESTE, Cascavel Pr Campus Dr in Microbiology, Brazil

Eliana Almeida Mira de BONA
Technique in the Itaipu Technological Park FoundationUNIOESTE, Campus Pastor Rattlesnake Master in Conservation And Management of Natural Resources, Brazil

Luciana Pagliosa Carvalho GUEDES
State University of West Paraná - UNIOESTE, Cascavel Pr Campus Dr in Statistics, Brazil

Izabel Aparecida SOARES
Federal University of South Border Royalty Campus Brazil

O. P. Bamidele
Department of Food Science and Technology, Obafemi Awolowo University, Ile-Ife, Nigeria

C. T. Akanbi
Department of Food Science and Technology, Obafemi Awolowo University, Ile-Ife, Nigeria

Fijelu Frank
School of Food Science and Technology, Jiangnan University, Wuxi, Jiangsu 214122, China

Yanshun Xu
School of Food Science and Technology, Jiangnan University, Wuxi, Jiangsu 214122, China

Qixing jiang
School of Food Science and Technology, Jiangnan University, Wuxi, Jiangsu 214122, China

Wenshui Xia
School of Food Science and Technology, Jiangnan University, Wuxi, Jiangsu 214122, China

Boujemaa El Marnissi
Laboratoire Agroalimentaire et Sécurité Sanitaire des Aliments (LASSA), Université Sidi Mohamed Ben Abdallah, Ecole Supérieure de Technologie Fès, Route d'Imouzzer B.P : 2427 - Fez, Morocco

Laila Bennani
Laboratoire Régional de Diagnostic Epidémiologique et d'Hygiène du Milieu, Direction Régionale de la Santé, Hôpital EL Ghassani-Fez, Morocco

Nezha Cohen
Division de Microbiologie et Hygiène des Aliments et de l'Environnement Institut Pasteur du Maroc, 1 Place Louis Pasteur, Casablanca 20360, Morocco

Abdelhakim El Ouali Lalami
Laboratoire Régional de Diagnostic Epidémiologique et d'Hygiène du Milieu, Direction Régionale de la Santé, Hôpital EL Ghassani-Fez, Morocco

Rajae Belkhou
Laboratoire Agroalimentaire et Sécurité Sanitaire des Aliments (LASSA), Université Sidi Mohamed Ben Abdallah, Ecole Supérieure de Technologie Fès, Route d'Imouzzer B.P : 2427 - Fez, Morocco

O. A Ashaye
Institute of Agricultural Research and Training P.M.B 5029, Moor-Plantation Ibadan, Oyo State, Nigeria

O. T. Olanipekun
Institute of Agricultural Research and Training P.M.B 5029, Moor-Plantation Ibadan, Oyo State, Nigeria

E. A.Ige
Institute of Agricultural Research and Training P.M.B 5029, Moor-Plantation Ibadan, Oyo State, Nigeria

E. O. Farinde
Institute of Agricultural Research and Training P.M.B 5029, Moor-Plantation Ibadan, Oyo State, Nigeria

S. O. Akinkunmi
Federal College of Agriculture, Ibadan, Oyo State, Nigeria

Ngando-Ebongue Georges Frank
IRAD, Centre for Oil Palm Research of La Dibamba, P.O. Box 243, Douala, Cameroon

Mpondo-Mpondo Emmanuel Albert
Department of Biochemistry, Faculty of Science, University of Douala, P.O. Box 24 157, Douala, Cameroon

Ewane Marcelle Astride
Department of Biochemistry, Faculty of Science, University of Douala, P.O. Box 24 157, Douala, Cameroon

Hozan Jalil Hamasalim
Animal Production, Faculty of Agricultural Science, University of Sulaimani, Iraq

Hemin Nuradden Mohammed
Animal Production, Faculty of Agricultural Science, University of Sulaimani, Iraq

Yin Zhang
Key Laboratory of Meat Processing of Sichuan, Chengdu University, Chengdu, 610106, China

Xinhui Wang
Key Laboratory of Meat Processing of Sichuan, Chengdu University, Chengdu, 610106, China

Wei Wang
Key Laboratory of Meat Processing of Sichuan, Chengdu University, Chengdu, 610106, China

Jiaming Zhang
Key Laboratory of Meat Processing of Sichuan, Chengdu University, Chengdu, 610106, China

O. O. Olaniyi
Department of Microbiology, Federal University of Technology, Akure, Nigeria

Olusegun T Afolabi
Department of Community Health, Obafemi Awolowo University, Ile-Ife, Osun State, Nigeria

Olufemi O Aluko
Department of Community Health, Obafemi Awolowo University, Ile-Ife, Osun State, Nigeria

O Olaniran
Department of Microbiology and Parasitology, Obafemi Awolowo University, Ile-Ife, Osun State, Nigeria

Olusola Ajao
Department of Community Health, Obafemi Awolowo University, Ile-Ife, Osun State, Nigeria

Bolade K Ojumu
Department of Community Health, Obafemi Awolowo University, Ile-Ife, Osun State, Nigeria

Oluwatoyosi Olawande
Institute of Human Virology, P. O. Box 9396, Garki, Abuja, Nigeria

CO Adebayo
Department of Science Technology, Federal Polytechnic, Ado-Ekiti, Ekiti State, Nigeria

BI Aderiye
Department of Microbiology, Ekiti State University, Ado-Ekiti, Ekiti State, Nigeria

OB Akpor
Department of Biological Sciences, Landmark University, Omu-Aran, Kwara State, Nigeria

Kiprotich Felix
Biochemistry and Molecular Biology Department, Egerton University, P.O. Box 563-20115, Egerton, Kenya

Mwendia M. Charles
Biochemistry and Molecular Biology Department, Egerton University, P.O. Box 563-20115, Egerton, Kenya

Cheruiyot K. Erick
Department of Crops, Horticulture and Soil, Egerton University, P.O. Box 536-20115, Egerton, Kenya

Wachira N. Francis
Association for Strengthening Agricultural Research in Eastern and Central Africa (ASARECA) P.O. Box 765, Entebbe, Uganda

L. K. Makelele
Department of Biotechnology, Faculty of Sciences, University of Kisangani, PO. Box 2012, Kisangani, Democratic Republic of Congo

Z. A. Kazadi
Department of Biotechnology, Faculty of Sciences, University of Kisangani, PO. Box 2012, Kisangani, Democratic Republic of Congo

R. W. Oleko
Department of Biotechnology, Faculty of Sciences, University of Kisangani, PO. Box 2012, Kisangani, Democratic Republic of Congo

R. Foma
Department of Food Chemistry, Agronomy Institute Faculty, P.O. Box 1232, Yangambi, Democratic Republic of Congo

Rosette Kabwang a Mpalang
Laboratory of Expertise, Hygiene and Technology of Foods of Animal Origin, Faculty of Veterinary Medicine, University of Lubumbashi, P.O. Box 1825, Lubumbashi, Democratic Republic of Congo

Koto - te - Nyiwa Ngbolua
Department of Biology, Faculty of Sciences, University of Kinshasa, P.O. Box 190, Kinshasa, Democratic Republic of Congo

Bongo Ngiala Gédeon
Department of Biology, Faculty of Sciences, University of Kinshasa, P.O. Box 190, Kinshasa, Democratic Republic of Congo
Sokoine University of Agriculture, Faculty of Veterinary Medicine, P.O. Box 3000, Chuo Kikuu, Morogoro, United Republic of Tanzania

B. Djeri
Laboratoire de Microbiologie et de Contrôle de qualité des Denrées Alimentaires, Ecole Supérieure des Techniques Biologiques et Alimentaires (ESTBA), Université de Lomé BP 1515, Lomé Togo
Unité de Recherche en Génie enzymatique et Alimentaire de l'Ecole Polytechnique d'Abomey Calavi URGEA/ EPAC Université d'Abomey Calavi, Benin

P. F Tchobo
Unité de Recherche en Génie enzymatique et Alimentaire de l'Ecole Polytechnique d'Abomey Calavi URGEA/ EPAC Université d'Abomey Calavi, Benin

Y. Adjrah
Laboratoire de Microbiologie et de Contrôle de qualité des Denrées Alimentaires, Ecole Supérieure des Techniques Biologiques et Alimentaires (ESTBA), Université de Lomé BP 1515, Lomé Togo

D. S. Karou
Laboratoire de Microbiologie et de Contrôle de qualité des Denrées Alimentaires, Ecole Supérieure des Techniques Biologiques et Alimentaires (ESTBA), Université de Lomé BP 1515, Lomé Togo

Y. Ameyapoh
Laboratoire de Microbiologie et de Contrôle de qualité des Denrées Alimentaires, Ecole Supérieure des Techniques Biologiques et Alimentaires (ESTBA), Université de Lomé BP 1515, Lomé Togo

M. M Soumanou
Unité de Recherche en Génie enzymatique et Alimentaire de l□Ecole Polytechnique d□Abomey Calavi URGEA/ EPAC Université d□Abomey Calavi, Benin

C. Souza
Laboratoire de Microbiologie et de Contrôle de qualité des Denrées Alimentaires, Ecole Supérieure des Techniques Biologiques et Alimentaires (ESTBA), Université de Lomé BP 1515, Lomé Togo

Samuel A. Donkoh
Department of Agricultural and Resource Economics, University for Development Studies, Tamale, Ghana

Hamdiyah Alhassan
Department of Agricultural and Resource Economics, University for Development Studies, Tamale, Ghana

Paul K. Nkegbe
Department of Economics and Entrepreneurship Development, University for Development Studies, Ghana

Debela Hunde Feyssa
Jimma University, College of Agriculture and Veterinary Medicine, Jimma, Ethiopia

T. J. Njoka
Nairobi University, College of Agriculture and Veterinary Sciences, Faculty of Agriculture, Départementof LARMAT, Nairobi, Kenya

Zemede Asfaw
Addis Ababa University, Collège of Natural Science, Department of Biology and Biodiversity Management, Addis Abeba, Ethiopia

M. M. Nyangito
Nairobi University, College of Agriculture and Veterinary Sciences, Faculty of Agriculture, Départementof LARMAT, Nairobi, Kenya

O Obire
Department of Applied and Environmental Biology, University of Science and Technology, P.M.B. 5080, Port Harcourt, Nigeria

R. N Okigbo
Department of Botany, Nnamdi Azikiwe University, Awka, Nigeria

T. R. Minimah
Department of Applied and Environmental Biology, University of Science and Technology, P.M.B. 5080, Port Harcourt, Nigeria

Onyenekenwa Cyprian Eneh
Institute for Development Studies, Enugu Campus, University of Nigeria, Nsukka, Enugu State, Nigeria

Emmanuel Olufemi Bankefa
Federal University, Oye, Ekiti, Ekiti State, Nigeria

Olumide Murtala Oyedeji
Federal University of Technology, Akure, Ondo State, Nigeria

Jean Pierre Banea
Programme National de Nutrition (PRONANUT), Kinshasa, Democratic Republic of Congo (DRC)

J. Howard Bradbury
Evolution, Ecology and Genetics Research School of Biology, Australian National University, Canberra, ACT 0200, Australia

Chretienne Mandombi
Hospital General de Reference, Zone de Sante de Popokabaka, Democratic Republic of Congo (DRC)

Damien Nahimana
Programme National de Nutrition (PRONANUT), Kinshasa, Democratic Republic of Congo (DRC)

Ian C.Denton
Evolution, Ecology and Genetics Research School of Biology, Australian National University, Canberra, ACT 0200, Australia
Matthew P. Foster
Evolution, Ecology and Genetics Research School of Biology, Australian National University, Canberra, ACT 0200, Australia

N landa Kuwa
Programme National de Nutrition (PRONANUT), Kinshasa, Democratic Republic of Congo (DRC)

Desire Tshala Katumbay
Centre Neuropsychopathologique, Universite de Kinshasa, Democratic Republic of Congo (DRC)

Jean Pierre Banea
Programme National de Nutrition (PRONANUT), Kinshasa, Democratic Republic of Congo (DRC)

J. Howard Bradbury
EEG, Research School of Biology, Australian National University, Canberra, ACT 0200, Australia

Damien Nahimana
Programme National de Nutrition (PRONANUT), Kinshasa, Democratic Republic of Congo (DRC)

Ian C. Denton
EEG, Research School of Biology, Australian National University, Canberra, ACT 0200, Australia

N. Mashukano
Programme National de Nutrition (PRONANUT), Kinshasa, Democratic Republic of Congo (DRC)

N'landa Kuwa
Programme National de Nutrition (PRONANUT), Kinshasa, Democratic Republic of Congo (DRC)

Vincent Ezin
Department of Crop Production, Faculty of Agricultural Sciences, University of Abomey-calavi, Benin

Ibouraïma Yabi
Department of Geography, University of Abomey-calavi, Benin

Emeric G. M. Kochoni
Institut National de la Recherche Scientifique - Eau Terre Environnement (INRS-ETE), Canada

Adam Ahanchédé
Department of Crop Production, Faculty of Agricultural Sciences, University of Abomey-calavi, Benin

Sylvia Angubua BALUKA
College of Veterinary Medicine, Animal Resources and Biosecurity (COVAB), Makerere University, Kampala, Uganda

RoseAnn MILLER
College of Veterinary Medicine, Animal Resources and Biosecurity (COVAB), Makerere University, Kampala, Uganda

John Baligwamunsi KANEENE
Center for Comparative Epidemiology, Michigan State University, USA

Alakali Joseph
Department of Food Science and Technology, University of Agriculture, Makurdi, P.M.B 2373, Makurdi, Benue State, Nigeria

J. K. Agomuo
Department of Food Science and Technology, Federal University Dutsinma, P.M.B 5001, Dutsinma, Katsina State, Nigeria

I. C. Alaka
Department of Food Science and Technology, Ebonyi State University, P.M.B 501, Ebonyi, Ebonyi State, Nigeria

J. Faasema
Department of Food Science and Technology, University of Agriculture, Makurdi, P.M.B 2373, Makurdi, Benue State, Nigeria